Micromechanics of defects in solids

Monographs and textbooks on mechanics of solids and fluids
editor-in-chief: G.Æ. Oravas

Mechanics of elastic and inelastic solids
editor: S. Nemat-Nasser

Books published under this series are:

1. G.M.L. Gladwell
   Contact problems in the classical theory of elasticity
2. G. Wempner
   Mechanics of solids with applications to thin bodies
3. T. Mura
   Micromechanics of defects in solids

Micromechanics of defects in solids

Toshio Mura

Department of Civil Engineering and
Materials Research Center,
Northwestern University, Evanston, Ill., U.S.A.

1982

MARTINUS NIJHOFF PUBLISHERS

THE HAGUE/BOSTON/LONDON

Distributors:

for the United States and Canada

Kluwer Boston, Inc.
190 Old Derby Street
Hingham, MA 02043
USA

for all other countries

Kluwer Academic Publishers Group
Distribution Center
P.O. Box 322
3300 AH Dordrecht
The Netherlands

Library of Congress Cataloging in Publication Data

Mura, Toshio, 1925-
Micromechanics of defects in solids.

(Monographs and textbooks on mechanics of solids and fluids)
Includes bibliographical references and indexes.
1. Micromechanics. 2. Solids. 3. Crystals--Defects. 4. Strains and stresses. I. Title. II. Series.
QC176.8.M5M87 620.1'05 82-2152
AACR2

ISBN 978-94-011-8548-6 ISBN 978-94-011-9306-1 (eBook)
DOI 10.1007/978-94-011-9306-1

Preface

This book stems from a course on Micromechanics that I started about fifteen years ago at Northwestern University. At that time, micromechanics was a rather unfamiliar subject. Although I repeated the course every year, I was never convinced that my notes have quite developed into a final manuscript because new topics emerged constantly requiring revisions, and additions. I finally came to realize that if this is continued, then I will never complete the book to my total satisfaction. Meanwhile, T. Mori and I had coauthored a book in Japanese, entitled *Micromechanics*, published by Baifu-kan, Tokyo, in 1975. It received an extremely favorable response from students and researchers in Japan. This encouraged me to go ahead and publish my course notes in their latest version, as this book, which contains further development of the subject and is more comprehensive than the one published in Japanese.

Micromechanics encompasses mechanics related to microstructures of materials. The method employed is a continuum theory of elasticity yet its applications cover a broad area relating to the mechanical behavior of materials: plasticity, fracture and fatigue, constitutive equations, composite materials, polycrystals, etc. These subjects are treated in this book by means of a powerful and unified method which is called the 'eigenstrain method.' In particular, problems relating to inclusions and dislocations are most effectively analyzed by this method, and therefore, special emphasis is placed on these topics. When this book is used as a text for a graduate course, Sections 3, 11, and 22 should be emphasized.

Eigenstrain is a generic name given by the author to such nonelastic strains as thermal expansion, phase transformation, and misfit strains. J.D. Eshelby, who is a pioneer in this area, refers to eigenstrains as stress-free transformation strains in his celebrated papers (1957, 1959).

The term eigenstrain should not be confused with the term 'eigenvalue' which occurs in mathematical physics, and relates to an entirely different concept.

No particular background is required of readers of this book because necessary mathematics and physics are explained in the text and Appendix. Although I have tried to be fair in citing literature, I have to apologize if some papers do not receive proper credit or are not cited.

The sections and subsections marked with an asterisk (*), can be skipped in the first reading, since the subjects discussed there are peripheral to the main theme.

I wish to express my thanks to all the people who have helped me during the course of the preparation of the manuscript: my previous graduate students, Zissis A. Moschovidis, Minoru Taya, Carl R. Vilmann, and Ronald B. Castles, as well as my friends R. Furuhashi, N. Kinoshita, T. Morita, M. Inokuti, and T. Mori. Mori receives my special thanks for having advised me on the subject matter, for discussing with me whole chapters, and for helping me to write Chapter 7. The manuscript reached the final form in his hands. I also wish to thank S. Nemat-Nasser who has read through the manuscript and has given valuable comments.

I give my thanks to Vera Fisher for her skillful typing and her great patience with me, to the secretaries whom I involved in various aspects of the work over the years: Erika Ivansons, Miriam Littell, Masa Sumikura, and Carolyn Andrews, and to my family for their patience and understanding.

Finally, I acknowledge the National Science Foundation and the U.S. Army Research Office for their support of my research in the area of micromechanics.

November 13, 1980 T.M.

Contents

1

General theory of eigenstrains

The definition of the eigenstrains is given first. Then the associated general solutions for elastic fields for given eigenstrains are expressed by Fourier integrals and Green's functions. Some details of calculations for Green's functions are described for static and dynamic cases.

As fundamental formulae for the subsequent chapters, general expressions of elastic fields are given for inclusions, dislocations, and disclinations. The stress discontinuity on boundaries of inclusions and the incompatibility of eigenstrains are discussed as general theories.

Throughout this work, a fixed rectangular Cartesian coordinate system with coordinate axes x_i, $i = 1, 2, 3$, is used.

1. Definition of eigenstrains

'Eigenstrain' is a generic name given by the author to such nonelastic strains as thermal expansion, phase transformation, initial strains, plastic, misfit strains. 'Eigenstress' is a generic name given to self-equilibrated internal stresses caused by one or several of these eigenstrains in bodies which are free from any other external force and surface constraint. The eigenstress fields are created by the incompatibility of the eigenstrains.

This new English terminology was adapted from the German 'Eigenspannungen und Eigenspannungsquellen,' which is the title of H. Reissner's paper (1931) on residual stresses. Eshelby (1957) referred to eigenstrains as stress-free transformation strains in his celebrated paper which has stimulated the present author to work on inclusion and dislocation problems. The term 'elastic polarization' was used by Kröner (1958) for the eigenstrains in a slightly different context—when the nonhomogeneity of polycrystal deformation is under consideration.

Engineers have used the term 'residual stresses' for the self-equilibrated internal stresses when they remain in materials after fabrication or plastic deformation. Eigenstresses are called thermal stresses when thermal expansion is a cause of the corresponding elastic fields. For example, when a part Ω of material (Fig. 1.1) has its temperature raised by T, thermal stress σ_{ij} is induced in the material D by the constraint from the part which surrounds Ω. The thermal expansion αT, where α is the linear thermal expansion coefficient, constitutes the thermal expansion strain,

$$\epsilon_{ij}^* = \delta_{ij}\alpha T, \tag{1.1}$$

where δ_{ij} is the Kronecker delta (see Appendix 1). The thermal expansion strain is the strain caused when Ω can be expanded freely with the removal of the constraint from the surrounding part.

The actual strain is then the sum of the thermal and elastic strains. The elastic strain is related to the thermal stress by Hooke's law. The thermal expansion strain (1.1) is a typical example of eigenstrains. In the elastic theory of eigenstrains and eigenstresses, however, it is not necessary to attribute ϵ_{ij}^* to any specific source. The source could be phase transformation, precipitation, plastic deformation or a fictitious source necessary for the equivalent inclusion method (to be discussed in Section 22).

When an eigenstrain ϵ_{ij}^* is prescribed in a finite subdomain Ω in a homogeneous material D (see Fig. 1.1), and it being zero in the matrix D-Ω, then Ω is called an inclusion. The elastic moduli of the material are assumed to be homogeneous when inclusions are under consideration.

If a subdomain Ω in a material D has elastic moduli different from those of the matrix, then Ω is called an inhomogeneity. Applied stresses will be disturbed by the existence of the inhomogeneity. This disturbed stress field will be simulated by an eigenstress field by considering a fictitious eigenstrain ϵ_{ij}^* in Ω in an *homogeneous* material.

When Ω in Fig. 1.1 is a plane embedded in a material D and ϵ_{ij}^* is given on Ω as a plastic strain caused by a finite slip **b**, the boundary of Ω

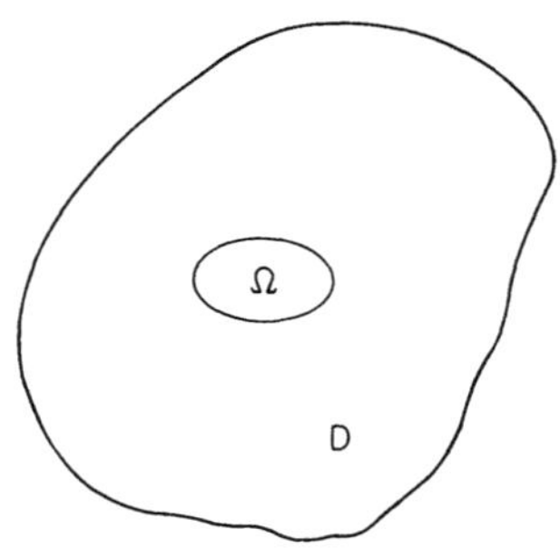

Fig. 1.1. Inclusion Ω

is called a dislocation loop. If ϵ_{ij}^* is created by a rigid rotation of plane Ω by $\boldsymbol{\omega}$, the boundary of Ω is called a disclination loop.

2. Fundamental equations of elasticity

In this section the field equations for elasticity theory will be reviewed with particular reference to solving eigenstrain problems. These problems consist of finding displacement u_i, strain ϵ_{ij}, and stress σ_{ij} at an arbitrary point $\mathbf{x}(x_1, x_2, x_3)$ when a free body D is subjected to a given distribution of eigenstrain ϵ_{ij}^*. A free body is one which is free from any external surface of body force.

Hooke's law

For infinitesimal deformations considered in this book, the total strain ϵ_{ij} is regarded as the sum of elastic strain e_{ij} and eigenstrain ϵ_{ij}^*,

$$\epsilon_{ij} = e_{ij} + \epsilon_{ij}^*. \tag{2.1}$$

The total strain must be compatible,

$$\epsilon_{ij} = (1/2)(u_{i,j} + u_{j,i}), \tag{2.2}$$

where $u_{i,j} = \partial u_i/\partial x_j$.

The elastic strain is related to stress σ_{ij} by Hooke's law;

$$\sigma_{ij} = C_{ijkl}e_{kl} = C_{ijkl}(\epsilon_{kl} - \epsilon_{kl}^*) \tag{2.3}$$

or

$$\sigma_{ij} = C_{ijkl}(u_{k,l} - \epsilon_{kl}^*), \tag{2.4}$$

where C_{ijkl} are the elastic moduli (constants) (see Appendix 2), and the summation convention for the repeated indices is employed (see Appendix 1). Since C_{ijkl} is symmetric ($C_{ijlk} = C_{ijkl}$), we have $C_{ijkl}u_{l,k} = C_{ijkl}u_{k,l}$. In the domain where $\epsilon_{ij}^* = 0$, (2.4) becomes

$$\sigma_{ij} = C_{ijkl}\epsilon_{kl} = C_{ijkl}u_{k,l}. \tag{2.5}$$

The inverse expression of (2.3) is

$$\epsilon_{ij} - \epsilon_{ij}^* = C_{ijkl}^{-1}\sigma_{kl}, \tag{2.6}$$

where C_{ijkl}^{-1} is the elastic compliance.

For isotropic materials, (2.3) and (2.6) can be written as

$$\begin{aligned} \sigma_{ij} &= 2\mu(\epsilon_{ij} - \epsilon_{ij}^*) + \lambda\delta_{ij}(\epsilon_{kk} - \epsilon_{kk}^*), \\ \epsilon_{ij} - \epsilon_{ij}^* &= \{\sigma_{ij} - \delta_{ij}\sigma_{kk}\nu/(1+\nu)\}/2\mu, \end{aligned} \tag{2.7}$$

where λ and μ are the Lamé constants, and ν is Poisson's ratio. Young's modulus E, the shear modulus μ, and the bulk modulus K are connected by $2\mu = E/(1+\nu)$, $K = E/3(1-2\nu)$ and $\lambda = 2\mu\nu/(1-2\nu)$. The alternative expressions for (2.7) are

$$\begin{aligned}
\sigma_x &= \frac{E}{1+\nu}\left\{(\epsilon_x - \epsilon_x^*) + \frac{\nu}{1-2\nu}(\epsilon_{kk} - \epsilon_{kk}^*)\right\},\\
\sigma_y &= \frac{E}{1+\nu}\left\{(\epsilon_y - \epsilon_y^*) + \frac{\nu}{1-2\nu}(\epsilon_{kk} - \epsilon_{kk}^*)\right\},\\
\sigma_z &= \frac{E}{1+\nu}\left\{(\epsilon_z - \epsilon_z^*) + \frac{\nu}{1-2\nu}(\epsilon_{kk} - \epsilon_{kk}^*)\right\},\\
\sigma_{xy} &= \frac{E}{1+\nu}(\epsilon_{xy} - \epsilon_{xy}^*), \qquad (2.8)\\
\sigma_{yz} &= \frac{E}{1+\nu}(\epsilon_{yz} - \epsilon_{yz}^*),\\
\sigma_{zx} &= \frac{E}{1+\nu}(\epsilon_{zx} - \epsilon_{zx}^*),
\end{aligned}$$

and

$$\begin{aligned}
\epsilon_x - \epsilon_x^* &= \{\sigma_x - \nu(\sigma_y + \sigma_z)\}/E,\\
\epsilon_y - \epsilon_y^* &= \{\sigma_y - \nu(\sigma_z + \sigma_x)\}/E,\\
\epsilon_z - \epsilon_z^* &= \{\sigma_z - \nu(\sigma_x + \sigma_y)\}/E,\\
\epsilon_{xy} - \epsilon_{xy}^* &= \frac{1+\nu}{E}\sigma_{xy}, \qquad (2.9)\\
\epsilon_{yz} - \epsilon_{yz}^* &= \frac{1+\nu}{E}\sigma_{yz},\\
\epsilon_{zx} - \epsilon_{zx}^* &= \frac{1+\nu}{E}\sigma_{zx},
\end{aligned}$$

where $\epsilon_{kk} = \epsilon_x + \epsilon_y + \epsilon_z$ and $\epsilon_{kk}^* = \epsilon_x^* + \epsilon_y^* + \epsilon_z^*$. It is convenient to use (2.8) for the plane strain case where $\epsilon_z = 0$. Expression (2.9) is recommended for the plane stress case where $\sigma_z = \sigma_{zx} = \sigma_{zy} = 0$. It should be noted that solutions for the plane stress can be obtained directly from those for the plane strain by replacing $E/(1-\nu^2)$ by E and $\nu/(1-\nu)$ by ν.

Equilibrium conditions

When eigenstresses are calculated, material domain D is assumed to be free from any external force and any surface constraint. If these conditions for the free body are not satisfied, the stress field can be constructed from the superposition of the eigenstress for the free body and the solution of a proper boundary value problem.

The equations for equilibrium are

$$\sigma_{ij,j} = 0 \quad (i = 1, 2, 3). \tag{2.10}$$

The boundary conditions for free external surface forces are

$$\sigma_{ij}n_j = 0, \tag{2.11}$$

where n_i is the exterior unit normal vector on the boundary of D.

By substituting (2.4) into (2.10) and (2.11), we have

$$C_{ijkl}u_{k,lj} = C_{ijkl}\epsilon^*_{kl,j} \tag{2.12}$$

and

$$C_{ijkl}u_{k,l}n_j = C_{ijkl}\epsilon^*_{kl}n_j. \tag{2.13}$$

It can be seen that the contribution of ϵ^*_{ij} to the equations of equilibrium is similar to that of a body force since the equations of equilibrium under body force X_i with zero ϵ^*_{ij} are $C_{ijkl}u_{k,lj} = -X_i$. Similarly, $C_{ijkl}\epsilon^*_{kl}n_j$ behaves like a surface force on the boundary. Thus, it can be said that the elastic displacement field caused by ϵ^*_{ij} in a free body is equivalent to that caused by body force $-C_{ijkl}\epsilon^*_{kl,j}$ and surface force $C_{ijkl}\epsilon^*_{kl}n_j$.

In subsequent chapters, D in most cases is considered as an infinitely extended body (infinite body), and condition (2.11) is replaced by the condition $\sigma_{ij}(\mathbf{x}) \to 0$ for $\mathbf{x} \to \infty$.

Compatibility conditions

The strain tensor ϵ_{ij} has six components, while the displacement vector u_i has three components. The tensor and the vector are related to each other through the relation (2.2), which can be called the condition for the compatibility of strain ϵ_{ij}. Generally, however, the equations of compatibility are referred to the relations which are derived from (2.2) by eliminating u_i,

$$\epsilon_{pki}\epsilon_{qlj}\epsilon_{ij,kl} = 0, \tag{2.14}$$

where ϵ_{pki} is the permutation tensor (see Appendix 1). Relation (2.14) will be discussed in Section 10.

The displacement differential equations of elasticity theory are given by (2.12). In some cases, however, it is more convenient to consider (2.10), (2.3), and (2.14). Boundary conditions and various side conditions, such as singularity conditions, continuity conditions, etc., arise in problems from time to time, we can say at this point that the fundamental equations to be solved are equations (2.12).

Eigenstresses are caused by constraint from the surrounding elastic medium which prohibits the geometrically incompatible deformation of ϵ^*_{ij}. The incompatibility of ϵ^*_{ij} was discussed by Reissner (1931), and Neményi (1931). Dislocations due to incompatibility were studied by Weingarten (1901), Cesáro (1906), Volterra (1907) and Moriguti (1947) from the viewpoint of elasticity theory in connection with the multiple values of displacements and rotations. Another viewpoint of dislocations, from plasticity theory, was developed by Kondo (1955), Bilby (1960), and Kröner (1958).

In the following sections we investigate the methods of finding the associated elastic fields (displacements, strains, stresses) and the related problems for given distributions of ϵ^*_{ij}. Particular emphasis will be placed on the case when a uniform ϵ^*_{ij} is given in an ellipsoidal domain Ω in an infinitely extended medium D. The results are useful for the study of the mechanical properties of solids which may contain precipitates, inclusions, voids, cracks, etc. The most fundamental contribution to this study was made by Eshelby (1951, 1956, 1957, 1959 and 1961).

3. General expressions of elastic fields for given eigenstrain distributions

The case where a given material is infinitely extended is of particular interest for the mathematical simplicity of the solution as well as for its practical importance. When the solution is applied to inclusion problems, it can be assumed with sufficient accuracy that the materials are infinitely extended since the size of inclusions is relatively small compared to the size of the macroscopic material samples.

The fundamental equations to be solved for given ϵ^*_{ij}, (2.12), are

$$C_{ijkl}u_{k,lj} = C_{ijkl}\epsilon^*_{kl,j}. \tag{3.1}$$

Periodic solutions

Suppose $\epsilon^*_{ij}(\mathbf{x})$ is given in the form of a single wave of amplitude $\bar{\epsilon}^*_{ij}(\boldsymbol{\xi})$, where $\boldsymbol{\xi}$ is the wave vector corresponding to the given period of the wave,

$$\epsilon_{ij}^*(\mathbf{x}) = \bar{\epsilon}_{ij}^*(\boldsymbol{\xi}) \exp(i\boldsymbol{\xi} \cdot \mathbf{x}), \tag{3.2}$$

where $i = \sqrt{-1}$ and $\boldsymbol{\xi} \cdot \mathbf{x} = \xi_k x_k$.

The solution of (3.1) corresponding to this wave may also be expressed in the form of a single wave of the same period, that is,

$$u_i(\mathbf{x}) = \bar{u}_i(\boldsymbol{\xi}) \exp(i\boldsymbol{\xi} \cdot \mathbf{x}). \tag{3.3}$$

Substituting (3.2) and (3.3) into (3.1), we have

$$C_{ijkl}\bar{u}_k\xi_l\xi_j = -iC_{ijkl}\bar{\epsilon}_{kl}^*\xi_j \tag{3.4}$$

where in the derivation $(i\boldsymbol{\xi} \cdot \mathbf{x})_{,l} = i\xi_l$ is used. Expression (3.4) stands for three equations ($i = 1, 2, 3$) to determine the three unknowns $\bar{u}_i$ for the given $\bar{\epsilon}_{ij}^*$.

Using the notation

$$\begin{aligned} K_{ik}(\boldsymbol{\xi}) &= C_{ijkl}\xi_j\xi_l, \\ X_i &= -iC_{ijkl}\bar{\epsilon}_{kl}^*\xi_j, \end{aligned} \tag{3.5}$$

we can write (3.4) as

$$\begin{aligned} K_{11}\bar{u}_1 + K_{12}\bar{u}_2 + K_{13}\bar{u}_3 &= X_1, \\ K_{21}\bar{u}_1 + K_{22}\bar{u}_2 + K_{23}\bar{u}_3 &= X_2, \\ K_{31}\bar{u}_1 + K_{32}\bar{u}_2 + K_{33}\bar{u}_3 &= X_3. \end{aligned} \tag{3.6}$$

Then, $\bar{u}_i$ is obtained as

$$\bar{u}_i(\boldsymbol{\xi}) = X_j N_{ij}(\boldsymbol{\xi})/D(\boldsymbol{\xi}), \tag{3.7}$$

where N_{ij} are cofactors of the matrix

$$\mathbf{K}(\boldsymbol{\xi}) = \begin{pmatrix} K_{11} & K_{12} & K_{13} \\ K_{21} & K_{22} & K_{23} \\ K_{31} & K_{32} & K_{33} \end{pmatrix} \tag{3.8}$$

and $D(\boldsymbol{\xi})$ is the determinant of $\mathbf{K}(\boldsymbol{\xi})$. Note that $K_{ki} = C_{kjil}\xi_j\xi_l = C_{klij}\xi_l\xi_j = C_{ijkl}\xi_l\xi_j = K_{ik}$ due to the symmetry of the elastic constants $C_{ijkl} = C_{klij}$, and that $N_{ij} = N_{ji}$. The explicit expressions for $D(\boldsymbol{\xi})$ and $N_{ij}(\boldsymbol{\xi})$ are,

$$\begin{aligned} D(\boldsymbol{\xi}) &= \epsilon_{mnl} K_{m1} K_{n2} K_{l3}, \\ N_{ij}(\boldsymbol{\xi}) &= \tfrac{1}{2} \epsilon_{ikl} \epsilon_{jmn} K_{km} K_{ln} \\ &= K_{im} K_{mj} - K_{mm} K_{ij} \\ &\quad + (\epsilon_{mn1} K_{m2} K_{n3} + \epsilon_{mn2} K_{m3} K_{n1} + \epsilon_{mn3} K_{m1} K_{n2}) \delta_{ij}, \end{aligned} \tag{3.9}$$

where ϵ_{ijk} is the permutation tensor.

Substituting (3.7) into (3.3), we have

$$u_i(\mathbf{x}) = -iC_{jlmn} \bar{\epsilon}^*_{mn}(\boldsymbol{\xi}) \xi_l N_{ij}(\boldsymbol{\xi}) D^{-1}(\boldsymbol{\xi}) \exp(i\boldsymbol{\xi} \cdot \mathbf{x}). \tag{3.10}$$

The corresponding strain and stress are obtained from (2.2) and (2.4) as

$$\epsilon_{ij}(\mathbf{x}) = (1/2) C_{klmn} \bar{\epsilon}^*_{mn}(\boldsymbol{\xi}) \xi_l \{\xi_j N_{ik}(\boldsymbol{\xi}) + \xi_i N_{jk}(\boldsymbol{\xi})\} D^{-1}(\boldsymbol{\xi}) \exp(i\boldsymbol{\xi} \cdot \mathbf{x}) \tag{3.11}$$

and

$$\sigma_{ij}(\mathbf{x}) = C_{ijkl} \{C_{pqmn} \bar{\epsilon}^*_{mn}(\boldsymbol{\xi}) \xi_q \xi_l N_{kp}(\boldsymbol{\xi}) D^{-1}(\boldsymbol{\xi}) \exp(i\boldsymbol{\xi} \cdot \mathbf{x}) - \epsilon^*_{kl}(\mathbf{x})\}, \tag{3.12}$$

where $D^{-1} = 1/D$. The above result was used by Mura (1964) for periodic distributions of dislocations and by Khachaturyan (1967) for a coherent inclusion of a new phase.

Method of Fourier series and Fourier integrals

The linear theory of elasticity allows for the superposition of solutions. If $\epsilon^*_{ij}(\mathbf{x})$ is given in the Fourier series form,

$$\epsilon^*_{ij}(\mathbf{x}) = \sum \bar{\epsilon}^*_{ij}(\boldsymbol{\xi}) \exp(i\boldsymbol{\xi} \cdot \mathbf{x}), \tag{3.13}$$

the corresponding displacement, strain and stress are then obtained as superpositions of the solutions for single waves of the form (3.2), namely,

$$\begin{aligned} u_i(\mathbf{x}) &= -i \sum C_{jlmn} \bar{\epsilon}^*_{mn}(\boldsymbol{\xi}) \xi_l N_{ij}(\boldsymbol{\xi}) D^{-1}(\boldsymbol{\xi}) \exp(i\boldsymbol{\xi} \cdot \mathbf{x}), \\ \epsilon_{ij}(\mathbf{x}) &= (1/2) \sum C_{klmn} \bar{\epsilon}^*_{mn}(\boldsymbol{\xi}) \xi_l \{\xi_j N_{ik}(\boldsymbol{\xi}) + \xi_i N_{jk}(\boldsymbol{\xi})\} D^{-1}(\boldsymbol{\xi}) \exp(i\boldsymbol{\xi} \cdot \mathbf{x}), \\ \sigma_{ij}(\mathbf{x}) &= C_{ijkl} \left\{ \sum C_{pqmn} \bar{\epsilon}^*_{mn}(\boldsymbol{\xi}) \xi_q \xi_l N_{kp}(\boldsymbol{\xi}) D^{-1}(\boldsymbol{\xi}) \exp(i\boldsymbol{\xi} \cdot \mathbf{x}) - \epsilon^*_{kl}(\mathbf{x}) \right\}, \end{aligned} \tag{3.14}$$

where the summations in (3.13) and (3.14) are taken with respect to $\boldsymbol{\xi}$.

Similarly, if ϵ^*_{ij} is given by the Fourier integral form (see Appendix 3),

$$\epsilon^*_{ij}(\mathbf{x}) = \int_{-\infty}^{\infty} \bar{\epsilon}^*_{ij}(\boldsymbol{\xi}) \exp(i\boldsymbol{\xi} \cdot \mathbf{x}) d\boldsymbol{\xi}, \tag{3.15}$$

where

$$\bar{\epsilon}^*_{ij}(\boldsymbol{\xi}) = (2\pi)^{-3} \int_{-\infty}^{\infty} \epsilon^*_{ij}(\mathbf{x}) \exp(-i\boldsymbol{\xi} \cdot \mathbf{x}) d\mathbf{x}, \tag{3.16}$$

we have

$$u_i(\mathbf{x}) = -i \int_{-\infty}^{\infty} C_{jlmn} \bar{\epsilon}^*_{mn}(\boldsymbol{\xi}) \xi_l N_{ij}(\boldsymbol{\xi}) D^{-1}(\boldsymbol{\xi}) \exp(i\boldsymbol{\xi} \cdot \mathbf{x}) d\boldsymbol{\xi},$$

$$\epsilon_{ij}(\mathbf{x}) = (1/2) \int_{-\infty}^{\infty} C_{klmn} \bar{\epsilon}^*_{mn}(\boldsymbol{\xi}) \xi_l \{\xi_j N_{ik}(\boldsymbol{\xi}) + \xi_i N_{jk}(\boldsymbol{\xi})\} D^{-1}(\boldsymbol{\xi}) \exp(i\boldsymbol{\xi} \cdot \mathbf{x}) d\boldsymbol{\xi}, \tag{3.17}$$

$$\sigma_{ij}(\mathbf{x}) = C_{ijkl} \left\{ \int_{-\infty}^{\infty} C_{pqmn} \bar{\epsilon}^*_{mn}(\boldsymbol{\xi}) \xi_q \xi_l N_{kp}(\boldsymbol{\xi}) D^{-1}(\boldsymbol{\xi}) \exp(i\boldsymbol{\xi} \cdot \mathbf{x}) d\boldsymbol{\xi} - \epsilon^*_{kl}(\mathbf{x}) \right\},$$

where

$$\int_{-\infty}^{\infty} d\boldsymbol{\xi} = \int_{-\infty}^{\infty} d\xi_1 \int_{-\infty}^{\infty} d\xi_2 \int_{-\infty}^{\infty} d\xi_3, \quad \int_{-\infty}^{\infty} d\mathbf{x} = \int_{-\infty}^{\infty} dx_1 \int_{-\infty}^{\infty} dx_2 \int_{-\infty}^{\infty} dx_3. \tag{3.18}$$

When (3.16) is substituted into (3.17), we have

$$\begin{aligned} u_i(\mathbf{x}) &= -i(2\pi)^{-3} \int_{-\infty}^{\infty} \int_{-\infty}^{\infty} C_{jlmn} \epsilon^*_{mn}(\mathbf{x}') \xi_l N_{ij}(\boldsymbol{\xi}) D^{-1}(\boldsymbol{\xi}) \exp\{i\boldsymbol{\xi} \cdot (\mathbf{x} - \mathbf{x}') d\boldsymbol{\xi} d\mathbf{x}' \\ &= -(2\pi)^{-3} \frac{\partial}{\partial x_l} \int_{-\infty}^{\infty} \int_{-\infty}^{\infty} C_{jlmn} \epsilon^*_{mn}(\mathbf{x}') N_{ij}(\boldsymbol{\xi}) D^{-1}(\boldsymbol{\xi}) \\ &\quad \times \exp\{i\boldsymbol{\xi} \cdot (\mathbf{x} - \mathbf{x}')\} d\boldsymbol{\xi} d\mathbf{x}' \end{aligned} \tag{3.19}$$

$$\begin{aligned} \epsilon_{ij}(\mathbf{x}) = (2\pi)^{-3} \int_{-\infty}^{\infty} \int_{-\infty}^{\infty} &(1/2) C_{klmn} \epsilon^*_{mn}(\mathbf{x}') \xi_l \{\xi_j N_{ik}(\boldsymbol{\xi}) + \xi_i N_{jk}(\boldsymbol{\xi})\} D^{-1}(\boldsymbol{\xi}) \\ &\times \exp\{i\boldsymbol{\xi} \cdot (\mathbf{x} - \mathbf{x}')\} d\boldsymbol{\xi} d\mathbf{x}', \end{aligned} \tag{3.20}$$

$$\begin{aligned} \sigma_{ij}(\mathbf{x}) = C_{ijkl} \Big\{ (2\pi)^{-3} \int_{-\infty}^{\infty} \int_{-\infty}^{\infty} &C_{pqmn} \epsilon^*_{mn}(\mathbf{x}') \xi_q \xi_l N_{kp}(\boldsymbol{\xi}) D^{-1}(\boldsymbol{\xi}) \\ &\times \exp\{i\boldsymbol{\xi} \cdot (\mathbf{x} - \mathbf{x}')\} d\boldsymbol{\xi} d\mathbf{x}' - \epsilon^*_{kl}(\mathbf{x}) \Big\}. \end{aligned} \tag{3.21}$$

Method of Green's functions

When Green's functions $G_{ij}(\mathbf{x}-\mathbf{x}')$ are defined as

$$G_{ij}(\mathbf{x}-\mathbf{x}') = (2\pi)^{-3}\int_{-\infty}^{\infty} N_{ij}(\boldsymbol{\xi})D^{-1}(\boldsymbol{\xi})\exp\{i\boldsymbol{\xi}\cdot(\mathbf{x}-\mathbf{x}')\}d\boldsymbol{\xi}, \tag{3.22}$$

(3.19) can be written as

$$u_i(\mathbf{x}) = -\int_{-\infty}^{\infty} C_{jlmn}\epsilon^*_{mn}(\mathbf{x}')G_{ij,l}(\mathbf{x}-\mathbf{x}')d\mathbf{x}', \tag{3.23}$$

where $G_{ij,l}(\mathbf{x}-\mathbf{x}') = \partial/\partial x_l G_{ij}(\mathbf{x}-\mathbf{x}') = -\partial/\partial x'_l G_{ij}(\mathbf{x}-\mathbf{x}')$.

The corresponding expressions for the strain and stress become

$$\epsilon_{ij}(\mathbf{x}) = -(1/2)\int_{-\infty}^{\infty} C_{klmn}\epsilon^*_{mn}(\mathbf{x}')\{G_{ik,lj}(\mathbf{x}-\mathbf{x}') + G_{jk,li}(\mathbf{x}-\mathbf{x}')\}d\mathbf{x}' \tag{3.24}$$

and

$$\sigma_{ij}(\mathbf{x}) = -C_{ijkl}\left\{\int_{-\infty}^{\infty} C_{pqmn}\epsilon^*_{mn}(\mathbf{x}')G_{kp,ql}(\mathbf{x}-\mathbf{x}')d\mathbf{x}' + \epsilon^*_{kl}(\mathbf{x})\right\}. \tag{3.25}$$

Mura (1963) rewrote (3.25) in the form

$$\sigma_{ij}(\mathbf{x}) = C_{ijkl}\int_{-\infty}^{\infty} \epsilon_{sth}\epsilon_{lnh}C_{pqmn}G_{kp,qt}(\mathbf{x}-\mathbf{x}')\epsilon^*_{sm}(\mathbf{x}')d\mathbf{x}', \tag{3.26}$$

which will be useful for the dislocation theory given in later sections. It is easy to prove that (3.26) is equivalent to (3.25). Since $\epsilon_{sth}\epsilon_{lnh} = \delta_{sl}\delta_{tn} - \delta_{sn}\delta_{tl}$, (3.26) becomes

$$\sigma_{ij}(\mathbf{x}) = C_{ijkl}\int_{-\infty}^{\infty} C_{pqmn}(G_{kp,qn}\epsilon^*_{ml} - G_{kp,ql}\epsilon^*_{mn})d\mathbf{x}'. \tag{3.27}$$

In Section 5 it is shown that

$$C_{mnpq}G_{pk,qn}(\mathbf{x}-\mathbf{x}') = -\delta_{mk}\delta(\mathbf{x}-\mathbf{x}'), \tag{3.28}$$

where $\delta(\mathbf{x}-\mathbf{x}')$ is Dirac's delta function having the property

$$\int_{-\infty}^{\infty} \epsilon^*_{ml}(\mathbf{x}')\delta(\mathbf{x}-\mathbf{x}')d\mathbf{x}' = \epsilon^*_{ml}(\mathbf{x}); \tag{3.29}$$

therefore, (3.25) follows from (3.27).

It is seen from (3.28) that Green's function $G_{pk}(\mathbf{x}-\mathbf{x}')$ is the displacement component in the x_p-direction at point $\mathbf{x}$ when a unit body force in the x_k-direction is applied at point $\mathbf{x}'$ in the infinitely extended material. By this definition of Green's function we can directly derive

(3.23) from (3.1). As was mentioned in Section 2, the displacement u_i in (3.1) can be considered as a displacement caused by the body force–$C_{ilmn}\epsilon^*_{mn,l}$ applied in the x_i-direction. Since $G_{ij}(\mathbf{x}-\mathbf{x}')$ is the solution for a unit body force applied in the x_j-direction, the solution for the present problem is the product of G_{ij} and the body force–$C_{jlmn}\epsilon^*_{mn,l}$, namely,

$$u_i(\mathbf{x}) = -\int_{-\infty}^{\infty} G_{ij}(\mathbf{x}-\mathbf{x}')C_{jlmn}\epsilon^*_{mn,l}(\mathbf{x}')d\mathbf{x}'. \tag{3.30}$$

Integrating by parts and assuming that the boundary terms vanish, we have

$$u_i(\mathbf{x}) = \int_{-\infty}^{\infty} C_{jlmn}\epsilon^*_{mn}(\mathbf{x}')\frac{\partial}{\partial x'_l}G_{ij}(\mathbf{x}-\mathbf{x}')d\mathbf{x}'. \tag{3.31}$$

For an infinite body it holds that $(\partial/\partial x'_l)G_{ij}(\mathbf{x}-\mathbf{x}') = -(\partial/\partial x_l)G_{ij}(\mathbf{x}-\mathbf{x}')$; (3.23) is thereby obtained.

Expression (3.31) or (3.23) is preferable to expression (3.30). When ϵ^*_{mn} is constant in Ω and is zero in D-Ω, it can be seen that the integrand in (3.30) vanishes except on the boundary of Ω.

As will be seen in (5.9), $G_{ij}(\mathbf{x}-\mathbf{x}')$ has a singularity at $\mathbf{x}=\mathbf{x}'$ with the order of $|\mathbf{x}-\mathbf{x}'|^{-1}$. Thus, the integrals in (3.24) and (3.25) do not exist in the sense of Riemann integrals. This difficulty can be avoided by writing (3.25) in the form

$$\sigma_{ij}(\mathbf{x}) = -C_{ijkl}\left\{\frac{\partial}{\partial x_l}\int_{-\infty}^{\infty} C_{pqmn}\epsilon^*_{mn}(\mathbf{x}')G_{kp,q}(\mathbf{x}-\mathbf{x}')d\mathbf{x}' + \epsilon^*_{kl}(\mathbf{x})\right\}. \tag{3.32}$$

Expressions (3.25) or (3.17) is permissible in the context of generalized functions (Lighthill 1964).

Expressions (3.23), (3.25) and their equivalents were developed by Fredholm (1900). In connection with the solution for dislocations, many papers have discussed these expressions more extensively: Weingarten (1901), Volterra (1907), Somigliana (1914), Burgers (1939a,b), Leibfried (1953), Eshelby (1961), Kröner (1958), Steketee (1958), deWit (1960) and Indenbom (1966) among others.

As will be seen in Section 5, Green's functions have been obtained explicitly only for isotropic and transversely isotropic materials. Therefore, for practical calculations the usage of Green's functions as seen in (3.23) ~ (3.25) is limited, and the use of Fourier integral expressions (3.17) is much more convenient. For this reason the integrands appearing in (3.17) are written down in detail.

Isotropic materials

$$\begin{aligned} D(\boldsymbol{\xi}) &= \mu^2(\lambda+2\mu)\xi^6 \\ N_{ij}(\boldsymbol{\xi}) &= \mu\xi^2\{(\lambda+2\mu)\delta_{ij}\xi^2-(\lambda+\mu)\xi_i\xi_j\}, \end{aligned} \tag{3.33}$$

where $\xi^2=\xi_k\xi_k$,

$$\begin{aligned} C_{jlmn}\xi_l N_{ij}(\boldsymbol{\xi})D^{-1}(\boldsymbol{\xi}) = (\lambda+2\mu)^{-1}\xi^{-4}\{\lambda\delta_{mn}\xi_i\xi^2+(\lambda+2\mu)\delta_{im}\xi_n\xi^2 \\ +(\lambda+2\mu)\delta_{in}\xi_m\xi^2-2(\lambda+\mu)\xi_i\xi_m\xi_n\}, \end{aligned} \tag{3.34}$$

$$\begin{aligned} C_{ijkl}C_{pqmn}\xi_q\xi_l N_{kp}(\boldsymbol{\xi})D^{-1}(\boldsymbol{\xi}) = (\lambda+2\mu)^{-1}\xi^{-4}\{\lambda^2\delta_{ij}\delta_{mn}\xi^4+2\lambda\mu\delta_{mn}\xi_i\xi_j\xi^2 \\ +2\lambda\mu\delta_{ij}\xi_m\xi_n\xi^2+\mu(\lambda+2\mu) \\ \times(\delta_{im}\xi_j\xi_n+\delta_{jm}\xi_i\xi_n+\delta_{in}\xi_j\xi_m+\delta_{jn}\xi_i\xi_m)\xi^2 \\ -4\mu(\lambda+\mu)\xi_i\xi_j\xi_m\xi_n\}. \end{aligned}$$

Cubic crystals

$$\begin{aligned} D(\boldsymbol{\xi}) = \mu^2(\lambda+2\mu+\mu')\xi^6+\mu\mu'(2\lambda+2\mu+\mu')\xi^2(\xi_1^2\xi_2^2+\xi_2^2\xi_3^2+\xi_3^2\xi_1^2) \\ +\mu'^2(3\lambda+3\mu+\mu')\xi_1^2\xi_2^2\xi_3^2, \end{aligned} \tag{3.35}$$

$$\begin{aligned} N_{11}(\boldsymbol{\xi}) &= \mu^2\xi^4+\beta\xi^2(\xi_2^2+\xi_3^2)+\gamma\xi_2^2\xi_3^2, \\ N_{12}(\boldsymbol{\xi}) &= -(\lambda+\mu)\xi_1\xi_2(\mu\xi^2+\mu'\xi_3^2), \end{aligned} \tag{3.36}$$

and the other components are obtained by the cyclic permutation of 1, 2, 3, where

$$\begin{aligned} \xi^2 &= \xi_i\xi_i, \\ \beta &= \mu(\lambda+\mu+\mu'), \\ \gamma &= \mu'(2\lambda+2\mu+\mu'), \\ \lambda &= C_{12}, \\ \mu &= C_{44}, \\ \mu' &= C_{11}-C_{12}-2C_{44}. \end{aligned} \tag{3.37}$$

Hexagonal crystals (transversely isotropic)

$$\begin{aligned} D(\boldsymbol{\xi}) &= (\alpha'\eta^2+\gamma\xi_3^2)\{\alpha\gamma\eta^4+(\alpha\beta+\gamma^2-\gamma'^2)\eta^2\xi_3^2+\beta\gamma\xi_3^4\} \\ &= (\alpha'\eta^2+\gamma\xi_3^2)\{(\gamma\eta^2+\beta\xi_3^2)(\alpha\eta^2+\gamma\xi_3^2)-\gamma'^2\eta^2\xi_3^2\}, \\ N_{11}(\boldsymbol{\xi}) &= (\alpha'\xi_1^2+\alpha\xi_2^2+\gamma\xi_3^2)(\gamma\eta^2+\beta\xi_3^2)-\gamma'^2\xi_2^2\xi_3^2, \\ N_{12}(\boldsymbol{\xi}) &= \gamma'^2\xi_1\xi_2\xi_3^2-(\alpha-\alpha')\xi_1\xi_2(\gamma\eta^2+\beta\xi_3^2), \end{aligned} \tag{3.38}$$

$$\begin{aligned}
N_{13}(\boldsymbol{\xi}) &= (\alpha-\alpha')\gamma'\xi_1\xi_2^2\xi_3 - \gamma'\xi_1\xi_3(\alpha'\xi_1^2+\alpha\xi_2^2+\gamma\xi_3^2),\\
N_{22}(\boldsymbol{\xi}) &= (\alpha\xi_1^2+\alpha'\xi_2^2+\gamma\xi_3^2)(\gamma\eta^2+\beta\xi_3^2) - \gamma'^2\xi_1^2\xi_3^2,\\
N_{23}(\boldsymbol{\xi}) &= (\alpha-\alpha')\gamma'\xi_1^2\xi_2\xi_3 - \gamma'\xi_2\xi_3(\alpha\xi_1^2+\alpha'\xi_2^2+\gamma\xi_3^2),\\
N_{33}(\boldsymbol{\xi}) &= (\alpha\xi_1^2+\alpha'\xi_2^2+\gamma\xi_3^2)(\alpha'\xi_1^2+\alpha\xi_2^2+\gamma\xi_3^2) - (\alpha-\alpha')^2\xi_1^2\xi_2^2,
\end{aligned} \tag{3.39}$$

where

$$\begin{aligned}
&\alpha = C_{11} = C_{22}, && \alpha' = C_{66} = (1/2)(C_{11}-C_{12}),\\
&\beta = C_{33}, && \gamma' - \gamma = C_{13} = C_{23},\\
&\gamma = C_{44} = C_{55}, && \eta^2 = \xi_1^2 + \xi_2^2.
\end{aligned} \tag{3.40}$$

For isotropic materials, $\alpha = \beta = \lambda + 2\mu$, $\gamma' = \lambda + \mu$, and $\gamma = \alpha' = \mu$.

4. Exercises of general formulae

General formulae are given in Section 3 for the elastic fields associated with prescribed eigenstrains. This section provides exercises in the usage of these formulae: the best way to understand general statements is to work out specific examples.

A straight screw dislocation

Let us consider the case where Ω in Fig. 1.1 is the half plane ($x_2 = 0$, $x_1 < 0$) (see Fig. 4.1), and ϵ_{23}^* is prescribed on Ω. Other components of ϵ_{ij}^* are zero. The boundary of Ω (the x_3 axis) is called a straight positive screw dislocation when $\epsilon_{23}^* = \epsilon_{32}^*$ is given by

$$\epsilon_{23}^*(\mathbf{x}) = (b/2)\delta(x_2)H(-x_1), \tag{4.1}$$

where $H(-x_1)$ is the Heaviside step function (see Appendix 3),

$$H(-x_1) = \begin{cases} 1 & x_1 < 0 \\ 0 & x_1 > 0, \end{cases} \tag{4.2}$$

and $\delta(x_2)$ is Dirac's delta function. The plastic strain (eigenstrain) (4.1) is caused by a relative slip $\mathbf{b}$ on the half plane ($x_2 = 0, x_1 < 0$) in the x_3-direction (Fig. 4.1).

The Fourier transform of (4.1) is, from (3.16),

$$\bar{\epsilon}_{23}^*(\boldsymbol{\xi}) = (2\pi)^{-3}\iiint_{-\infty}^{\infty} (b/2)\delta(x_2)H(-x_1) \times \exp\{-i(\xi_1x_1+\xi_2x_2+\xi_3x_3)\}dx_1dx_2dx_3. \tag{4.3}$$

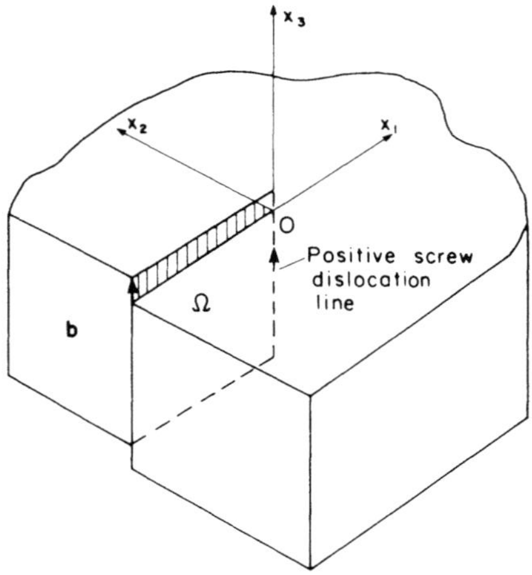

Fig. 4.1. A screw dislocation.

By using

$$\int_{-\infty}^{\infty} H(-x_1)\exp(-i\xi_1 x_1)dx_1 = -1/i\xi_1,$$
$$\int_{-\infty}^{\infty} \exp(-i\xi_3 x_3)dx_3 = 2\pi\delta(\xi_3), \tag{4.4}$$

we have

$$\bar{\epsilon}^*_{23}(\boldsymbol{\xi}) = -\frac{(b/2)}{(2\pi)^2}\frac{\delta(\xi_3)}{i\xi_1}. \tag{4.5}$$

Formulae (3.17) are used as follows: assuming, for simplicity, that the material is isotropic, (3.34) is used in (3.17); then, we have

$$u_3(\mathbf{x}) = \int_{-\infty}^{\infty} \frac{\xi_2 N_{33} + \xi_3 N_{23}}{(\lambda + 2\mu)\xi^6}\frac{b}{(2\pi)^2}\frac{\delta(\xi_3)}{\xi_1}\exp(i\boldsymbol{\xi}\cdot\mathbf{x})d\boldsymbol{\xi}$$
$$= \frac{b}{(2\pi)^2}\iint_{-\infty}^{\infty} \frac{\xi_2}{\xi_1(\xi_1^2+\xi_2^2)}\exp\{i(\xi_1 x_1 + \xi_2 x_2)\}d\xi_1 d\xi_2. \tag{4.6}$$

The following integrals are useful for further calculations; these are obtained from formulae 444, 633, 415, 632, and 638·1 in the Table of

Fourier Integrals by Campbell and Foster (1948):

$$\iint_{-\infty}^{\infty} \frac{\exp\{i(\xi_1 x_1 + \xi_2 x_2)\}}{\xi_1^2 + \xi_2^2}\, d\xi_1 d\xi_2 = -\pi \log(x_1^2 + x_2^2),$$

$$\iint_{-\infty}^{\infty} \frac{\xi_2 \exp\{i(\xi_1 x_1 + \xi_2 x_2)\}}{\xi_1(\xi_1^2 + \xi_2^2)}\, d\xi_1 d\xi_2 = 2\pi \tan^{-1}(x_2/x_1),$$

$$\iint_{-\infty}^{\infty} \frac{\xi_1\xi_2 \exp\{i(\xi_1 x_1 + \xi_2 x_2)\}}{(\xi_1^2 + \xi_2^2)^2}\, d\xi_1 d\xi_2 = \frac{-\pi x_1 x_2}{x_1^2 + x_2^2}, \tag{4.7}$$

$$\iint_{-\infty}^{\infty} \frac{\xi_2^2 \exp\{i(\xi_1 x_1 + \xi_2 x_2)\}}{(\xi_1^2 + \xi_2^2)^2}\, d\xi_1 d\xi_2 = -(\pi/2)\log(x_1^2 + x_2^2) - \frac{\pi x_2^2}{x_1^2 + x_2^2}.$$

Then, (4.6) becomes

$$u_3(\mathbf{x}) = (b/2\pi)\tan^{-1}(x_2/x_1) \tag{4.8}$$

which has been obtained by Burgers (1939). Other components of u_i are zero. Equation (4.8) can also be obtained from (3.23), where G_{ij} for isotropic materials is given in Section 5.

When (4.1) is substituted into (3.23), one arrives at

$$\begin{aligned} u_3(\mathbf{x}) &= -\iiint_{-\infty}^{\infty} C_{3232} b\delta(x_2')H(-x_1')\{G_{33,2}(\mathbf{x}-\mathbf{x}') + G_{32,3}(\mathbf{x}-\mathbf{x}')\}dx_1'dx_2'dx_3' \\ &= \frac{b}{4\pi}\int_{-\infty}^{\infty} dx_1' \int_{-\infty}^{\infty} dx_3' \left(\frac{u}{\lambda+2\mu}\frac{x_2}{\bar{x}^3} + 3\frac{\lambda+\mu}{\lambda+2\mu}\frac{x_2(x_3')^2}{\bar{x}^5}\right), \end{aligned} \tag{4.9}$$

where

$$\bar{x} = [(x_1 - x_1')^2 + x_2^2 + (x_3')^2]^{1/2}.$$

Now, with the aid of integrals

$$\int_{-\infty}^{\infty} \frac{dx_3'}{\bar{x}^3} = \frac{2}{(x_1 - x_1')^2 + x_2^2},$$

$$\int_{-\infty}^{\infty} \frac{(x_3')^2 dx_3'}{\bar{x}^5} = \frac{2/3}{(x_1 - x_1')^2 + x_2^2}, \tag{4.10}$$

(4.9) is reduced to (4.8). The stress field can be obtained from the displacement field by the use of (2.4).

A straight edge dislocation

Let Ω be the half plane $(x_2 = 0, x_1 < 0)$, and ϵ^*_{21} be prescribed by

$$\epsilon^*_{21}(\mathbf{x}) = (b/2)\delta(x_2)H(-x_1). \tag{4.11}$$

The x_3-axis is then a straight positive edge dislocation (see Fig. 4.2). The plastic strain ϵ^*_{21} is caused by the relative slip $\mathbf{b}$ on the half plane $(x_2 = 0, x_1 < 0)$ in the x_1-direction.

Substitution of (4.11) into (3.16) leads to

$$\bar{\epsilon}^*_{21} = -(2\pi)^{-2}(b/2)\delta(\xi_3)/i\xi_1. \tag{4.12}$$

For isotropic materials, we have, from (3.17) and (3.34),

$$\begin{aligned} u_1(\mathbf{x}) &= (2\pi)^{-2}b \iint_{-\infty}^{\infty} \left\{\frac{\xi_2}{\xi_1(\xi_1^2+\xi_2^2)} - \frac{1}{1-\nu}\frac{\xi_1\xi_2}{(\xi_1^2+\xi_2^2)^2}\right\} \\ &\quad \times \exp\{i(\xi_1x_1+\xi_2x_2)\}d\xi_1 d\xi_2 \\ &= (b/2\pi)\tan^{-1}(x_2/x_1) + (b/4\pi)\frac{1}{1-\nu}\frac{x_1x_2}{x_1^2+x_2^2}. \end{aligned} \tag{4.13}$$

This result was obtained by Koehler (1941), correcting Taylor's result (1934). Similarly, other components become

$$\begin{aligned} u_2(\mathbf{x}) &= (2\pi)^{-2}b \iint_{-\infty}^{\infty} \left\{\frac{1}{\xi_1^2+\xi_2^2} - \frac{1}{1-\nu}\frac{\xi_2^2}{(\xi_1^2+\xi_2^2)^2}\right\} \exp\{i(\xi_1x_1+\xi_2x_2)\}d\xi_1 d\xi_2 \\ &= \frac{b(2\nu-1)}{8\pi(1-\nu)}\log(x_1^2+x_2^2) + \frac{b}{4\pi(1-\nu)}\frac{x_2^2}{x_1^2+x_2^2}, \end{aligned} \tag{4.14}$$

$$u_3(\mathbf{x}) = 0.$$

Note that $u_2(\mathbf{x})$ in (4.14) differs by a constant from that in most text books on dislocations (e.g. Read 1953, p. 116), because differences in constant values of displacement components are admissible in elasticity theory.

The same result can be obtained by the use of Green's functions for isotropic materials. From (3.23) and (4.11) we have

$$u_i(\mathbf{x}) = -\int_{-\infty}^{0} dx_1' \int_{-\infty}^{\infty} C_{jl21} b G_{ij,l}(\mathbf{x}-\mathbf{x}')dx_3'. \tag{4.15}$$

The above expression becomes, from (5.10),

$$u_i(\mathbf{x}) = \frac{b}{8\pi(1-\nu)}\int_{-\infty}^{0} dx_1' \int_{-\infty}^{\infty}\left\{(1-2\nu)\frac{\delta_{2i}\bar{x}_1+\delta_{1i}\bar{x}_2}{\bar{x}^3} + 3\frac{\bar{x}_2\bar{x}_1\bar{x}_i}{\bar{x}^5}\right\}dx_3', \tag{4.16}$$

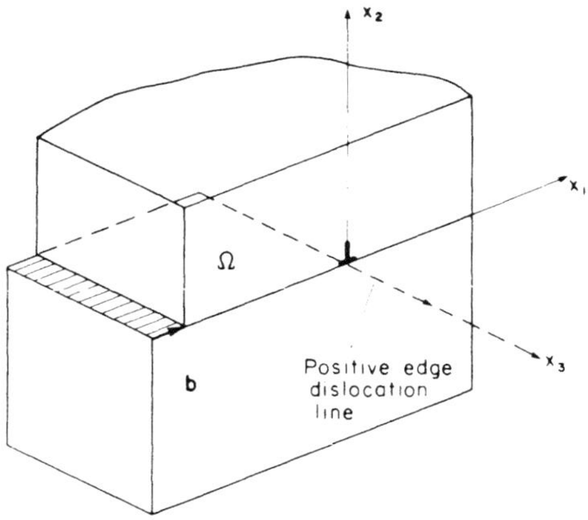

Fig. 4.2. An edge dislocation.

where $\bar{x}_1 = x_1 - x_1'$, $\bar{x}_2 = x_2$, $\bar{x}_3 = -x_3'$. By using (4.10) the integrations in (4.16) can be done easily, leading to (4.13) and (4.14).

Periodic distribution of cuboidal precipitates

Consider cuboidal regions of inelastic strain (eigenstrain) due to solute segregation to form cuboidal precipitates. Let the edge length of each be 2a (with edges along x_1, x_2, x_3), and let these regions be spaced 2L apart in all three directions. Within each region the eigenstrain will be assumed to have a constant value, ϵ, and be zero outside this region,

$$\epsilon_{ij}^*(\mathbf{x}) = \begin{cases} \delta_{ij}\epsilon & \text{in cuboidal regions} \\ 0 & \text{outside} \end{cases} \tag{4.17}$$

According to Appendix 3, the Fourier series expression for (4.17) is

$$\epsilon_{ij}^*(\mathbf{x}) = \delta_{ij}\epsilon \sum_{p,q,r=-\infty}^{\infty} a_{pqr} \exp\left\{i\frac{\pi}{L}(px_1 + qx_2 + rx_3)\right\}, \tag{4.18}$$

where p, q, r are integers, and

$$a_{pqr} = (\pi)^{-3}(pqr)^{-1} \sin\left(\frac{p\pi a}{L}\right) \sin\left(\frac{q\pi a}{L}\right) \sin\left(\frac{r\pi a}{L}\right). \tag{4.19}$$

Expressions in (3.14) are used for the elastic field, where

$$\begin{aligned} &\bar{\epsilon}_{mn}^*(p, q, r) = \delta_{mn}\epsilon a_{pqr} \\ &\xi_1 = \pi p/L, \quad \xi_2 = \pi q/L, \quad \xi_3 = \pi r/L, \end{aligned} \tag{4.20}$$

and the summations are taken with respect to p, q, and r. Namely,

$$u_i(\mathbf{x}) = -i \sum_{p,q,r=-\infty}^{\infty} C_{jlmn}\delta_{mn}\epsilon a_{pqr}\xi_l N_{ij}(\boldsymbol{\xi})D^{-1}(\boldsymbol{\xi}) \exp\left\{i \frac{\pi}{L}(px_1 + qx_2 + rx_3)\right\}. \tag{4.21}$$

For isotropic materials we have

$$u_1(\mathbf{x}) = 8 \frac{3\lambda + 2\mu}{\lambda + 2\mu} \frac{L\epsilon}{\pi^4} \sum_{p,q,r=0}^{\infty} \frac{\sin\left(\frac{p\pi a}{L}\right) \sin\left(\frac{q\pi a}{L}\right) \sin\left(\frac{r\pi a}{L}\right)}{(p^2 + q^2 + r^2)qr}$$

$$\times \sin\left(\frac{p\pi x_1}{L}\right) \cos\left(\frac{q\pi x_2}{L}\right) \cos\left(\frac{r\pi x_3}{L}\right). \tag{4.22}$$

The other components of the displacement can be obtained by the cyclic permutation of p, q, r and x_1, x_2, x_3. The summations in (4.18) and (4.22) exclude $p = q = r = 0$ in order to have the convergency of series. Since the term corresponding to $p = q = r = 0$ in (4.18) represents a uniform eigenstrain in the whole space, the term in the displacement field (4.22) corresponding to the uniform eigenstrain can be ignored. Sass, Mura and Cohen (1967) calculated the diffraction contrast of the cuboidal inclusions by using (4.22) and two-beam dynamical equations. Their electron microscope observation on Ni–Ti alloy provides good support of the theory.

5. Static Green's functions

In Section 3 Green's functions $G_{ij}(\mathbf{x} - \mathbf{x}')$ have been formally defined by (3.22),

$$G_{ij}(\mathbf{x} - \mathbf{x}') = (2\pi)^{-3} \int_{-\infty}^{\infty} N_{ij}(\boldsymbol{\xi})D^{-1}(\boldsymbol{\xi}) \exp\{i\boldsymbol{\xi} \cdot (\mathbf{x} - \mathbf{x}')\} d\boldsymbol{\xi}. \tag{5.1}$$

The elastic field quantities for a given $\epsilon^*_{ij}(\mathbf{x})$ are determined by (3.23) ~ (3.25) if $G_{ij}(\mathbf{x} - \mathbf{x}')$ are known.

It can easily be shown that G_{ij} satisfy

$$C_{ijkl}G_{km,lj}(\mathbf{x} - \mathbf{x}') + \delta_{im}\delta(\mathbf{x} - \mathbf{x}') = 0, \tag{5.2}$$

where $\delta(\mathbf{x} - \mathbf{x}')$ is the three-dimensional delta function which is everywhere zero except at point $\mathbf{x} = \mathbf{x}'$, and it gives

$$\int f(\mathbf{x}')\delta(\mathbf{x} - \mathbf{x}')d\mathbf{x}' = f(\mathbf{x}) \tag{5.3}$$

for any continuous function $f(\mathbf{x})$ where the integration domain contains point $\mathbf{x}$. The proof of (5.2) is as follows:

From (5.1), we have

$$C_{ijkl}G_{km,lj}(\mathbf{x}-\mathbf{x}') = -(2\pi)^{-3}\int_{-\infty}^{\infty} C_{ijkl}N_{km}D^{-1}\xi_l\xi_j \exp\{i\boldsymbol{\xi}\cdot(\mathbf{x}-\mathbf{x}')\}d\xi$$

$$= -(2\pi)^{-3}\int_{-\infty}^{\infty} K_{ik}N_{km}D^{-1}\exp\{i\boldsymbol{\xi}\cdot(\mathbf{x}-\mathbf{x}')\}d\boldsymbol{\xi}, \tag{5.4}$$

where $K_{ik} = C_{ijkl}\xi_j\xi_l$. Since N_{km} is the cofactor of $\mathbf{K}$, we have

$$K_{ik}N_{km}D^{-1} = \delta_{im}. \tag{5.5}$$

On the other hand, Dirac's delta function can be defined (see Appendix 3) as

$$\delta(\mathbf{x}-\mathbf{x}') = \delta(x_1-x_1')\delta(x_2-x_2')\delta(x_3-x_3')$$

$$= (2\pi)^{-3}\int_{-\infty}^{\infty} \exp\{i\boldsymbol{\xi}\cdot(\mathbf{x}-\mathbf{x}')\}d\boldsymbol{\xi}. \tag{5.6}$$

Thus, (5.2) is proved. Equation (5.2) is the equation of equilibrium with respect to a displacement $G_{km}(\mathbf{x}-\mathbf{x}')$ when the body force $X_i(\mathbf{x}) = \delta_{im}\delta(\mathbf{x}-\mathbf{x}')$ is applied. In other words, $G_{ij}(\mathbf{x}-\mathbf{x}')$ is the x_i-component of displacement at point $\mathbf{x}$ when a unit body force in the x_j-direction is applied at point $\mathbf{x}'$. The region occupied by the material is assumed to be infinitely extended. Since $G_{ij}(\mathbf{x}-\mathbf{x}') = G_{ji}(\mathbf{x}-\mathbf{x}')$, the directions x_i and x_j in the above statement are interchangeable. It should be noted that $N_{ij}(\boldsymbol{\xi})$ and $D(\boldsymbol{\xi})$ are homogeneous polynomials with degrees four and six, respectively. Thus we have, $G_{ij}(\mathbf{x}-\mathbf{x}') = G_{ij}(\mathbf{x}'-\mathbf{x})$.

The integration in (5.1) requires explicit expressions for $\mathbf{K}$, N_{ij}, and D mentioned in Section 3.

Isotropic materials

Substituting (3.33) into (5.1) it follows that,

$$G_{ij}(\mathbf{x}) = (2\pi)^{-3}\int_{-\infty}^{\infty}\frac{(\lambda+2\mu)\delta_{ij}\xi^2-(\lambda+\mu)\xi_i\xi_j}{\mu(\lambda+2\mu)\xi^4}\exp(i\boldsymbol{\xi}\cdot\mathbf{x})d\boldsymbol{\xi}, \tag{5.7}$$

where $\mathbf{x}'$ is taken as zero without loss of generality. After the triple integrations are carried out, we obtain

$$G_{ij}(\mathbf{x}) = \frac{1}{8\pi\mu x}\left\{2\delta_{ij} - \frac{\lambda+\mu}{\lambda+2\mu}(\delta_{ij}-x_ix_j/x^2)\right\}$$

$$= \frac{1}{16\pi\mu(1-\nu)x}\{(3-4\nu)\delta_{ij}+x_ix_j/x^2\}, \tag{5.8}$$

where $x = (x_i x_i)^{1/2}$, or

$$G_{ij}(\mathbf{x}-\mathbf{x}') = \frac{1}{4\pi\mu}\frac{\delta_{ij}}{|\mathbf{x}-\mathbf{x}'|} - \frac{1}{16\pi\mu(1-\nu)}\frac{\partial^2}{\partial x_i \partial x_j}|\mathbf{x}-\mathbf{x}'|, \tag{5.9}$$

where λ, μ are the Lamé constants (μ = shear modulus), ν is Poisson's ratio, δ_{ij} is the Kronecker delta, and $|\mathbf{x}-\mathbf{x}'|^2 = (x_i - x_i')(x_i - x_i')$. This result was found by Lord Kelvin (1848).

The following expression is frequently used for isotropic materials:

$$C_{jlmn}G_{ij,l}(\bar{\mathbf{x}}) = \frac{-1}{8\pi(1-\nu)}\left\{(1-2\nu)\frac{\delta_{mi}\bar{x}_n + \delta_{ni}\bar{x}_m - \delta_{mn}\bar{x}_i}{\bar{x}^3} + 3\frac{\bar{x}_m\bar{x}_n\bar{x}_i}{\bar{x}^5}\right\}, \tag{5.10}$$

where $\bar{\mathbf{x}} = \mathbf{x}-\mathbf{x}'$, $|\mathbf{x}-\mathbf{x}'| = \bar{x}$.

The details of the integration of (5.7) are as follows. The volume element $d\boldsymbol{\xi}$ is defined as

$$d\boldsymbol{\xi} = \xi^2 d\xi dS, \tag{5.11}$$

where $\xi = (\xi_i\xi_i)^{1/2}$ and dS is the surface element on the unit sphere S^2 in the $\boldsymbol{\xi}$-space, centered at the origin of the coordinates ξ_i (see Fig. 5.1). By denoting

$$\boldsymbol{\xi} = \xi\bar{\boldsymbol{\xi}}, \quad \mathbf{x} = x\bar{\mathbf{x}} \tag{5.12}$$

(note that $\bar{x}$ has been used in (5.10) with a different meaning), we can write (5.7) as

$$G_{ij}(\mathbf{x}) = (2\pi)^{-3}\int_0^\infty d\xi \int_{S^2}\frac{(\lambda+2\mu)\delta_{ij} - (\lambda+\mu)\bar{\xi}_i\bar{\xi}_j}{\mu(\lambda+2\mu)}\exp(i\xi x\bar{\boldsymbol{\xi}}\cdot\bar{\mathbf{x}})dS(\bar{\boldsymbol{\xi}}). \tag{5.13}$$

When $\boldsymbol{\xi}$ in (5.7) is replaced by $-\boldsymbol{\xi}$ as a new variable, we have

$$G_{ij}(\mathbf{x}) = (2\pi)^{-3}\int_{-\infty}^{\infty}\frac{(\lambda+2\mu)\delta_{ij}\xi^2 - (\lambda+\mu)\xi_i\xi_j}{\mu(\lambda+2\mu)\xi^4}\exp(-i\boldsymbol{\xi}\cdot\mathbf{x})d\boldsymbol{\xi}. \tag{5.14}$$

Then, similarly,

$$\begin{aligned} G_{ij}(\mathbf{x}) &= (2\pi)^{-3}\int_0^\infty d\xi \int_{S_2}\frac{(\lambda+2\mu)\delta_{ij} - (\lambda+\mu)\bar{\xi}_i\bar{\xi}_j}{\mu(\lambda+2\mu)}\exp(-i\xi x\bar{\boldsymbol{\xi}}\cdot\bar{\mathbf{x}})dS(\bar{\boldsymbol{\xi}}) \\ &= (2\pi)^{-3}\int_{-\infty}^0 d\xi \int_{S^2}\frac{(\lambda+2\mu)\delta_{ij} - (\lambda+\mu)\bar{\xi}_i\bar{\xi}_j}{\mu(\lambda+2\mu)}\exp(i\xi x\bar{\boldsymbol{\xi}}\cdot\bar{\mathbf{x}})dS(\bar{\boldsymbol{\xi}}). \end{aligned} \tag{5.15}$$

The last result has been obtained by the transformation of $\xi \to -\xi$. By adding (5.13) and (5.15), we have

$$G_{ij}(\mathbf{x}) = \frac{(2\pi)^{-3}}{2}\int_{-\infty}^{\infty} d\xi \int_{S^2} \frac{(\lambda+2\mu)\delta_{ij} - (\lambda+\mu)\bar{\xi}_i\bar{\xi}_j}{\mu(\lambda+2\mu)} \exp(i\xi x\bar{\boldsymbol{\xi}}\cdot\bar{\mathbf{x}})dS(\bar{\boldsymbol{\xi}}). \tag{5.16}$$

The integration with respect to ξ leads to Dirac's delta function. Thus,

$$G_{ij}(\mathbf{x}) = \frac{(2\pi)^{-2}}{2}\int_{S^2} \delta(x\bar{\boldsymbol{\xi}}\cdot\bar{\mathbf{x}}) \frac{(\lambda+2\mu)\delta_{ij} - (\lambda+\mu)\bar{\xi}_i\bar{\xi}_j}{\mu(\lambda+2\mu)} dS(\bar{\boldsymbol{\xi}}). \tag{5.17}$$

Denoting the angle between $\bar{\boldsymbol{\xi}}$ and $\bar{\mathbf{x}}$ by θ (see Fig. 5.1), we have

$$\begin{aligned} &\bar{\boldsymbol{\xi}}\cdot\bar{\mathbf{x}} = \cos\theta, \quad d(\bar{\boldsymbol{\xi}}\cdot\bar{\mathbf{x}}) = -\sin\theta d\theta, \\ &dS(\bar{\boldsymbol{\xi}}) = \sin\theta d\theta d\phi = -d(\bar{\boldsymbol{\xi}}\cdot\bar{\mathbf{x}})d\phi, \end{aligned} \tag{5.18}$$

where ϕ is defined on the plane perpendicular to $\mathbf{x}$ as shown in Fig. 5.1. The starting line for measuring ϕ can be arbitrary and (5.17) becomes

$$\begin{aligned} G_{ij}(\mathbf{x}) &= \frac{(2\pi)^{-2}}{2}\int_{-1}^{1} d(\bar{\boldsymbol{\xi}}\cdot\bar{\mathbf{x}})\delta(x\bar{\boldsymbol{\xi}}\cdot\bar{\mathbf{x}}) \int_0^{2\pi} d\phi \frac{(\lambda+2\mu)\delta_{ij} - (\lambda+\mu)\bar{\xi}_i\bar{\xi}_j}{\mu(\lambda+2\mu)} \\ &= \frac{(2\pi)^{-2}}{2x}\oint_{S^1} \frac{(\lambda+2\mu)\delta_{ij} - (\lambda+\mu)\bar{\xi}_i\bar{\xi}_j}{\mu(\lambda+2\mu)} d\phi, \end{aligned} \tag{5.19}$$

where S^1 is the unit circle on S^2 intersected by the plane perpendicular to $\mathbf{x}$.

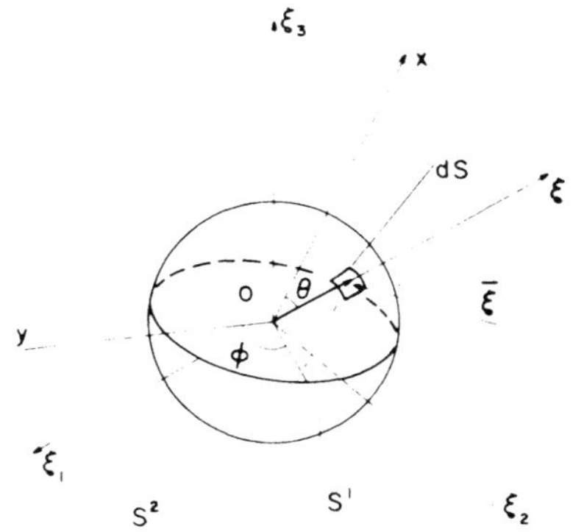

Fig. 5.1. The unit sphere S^2 in the ξ-space. Green's function at point **x** *is expressed by a line integral along S^1 which lies on the plane perpendicular to* **x**.

$\bar{\boldsymbol{\xi}}$ located on S^1 can be written as

$$\bar{\xi}_i = a_{1i}\cos\phi + a_{2i}\sin\phi, \tag{5.20}$$

where a_{1i} is the unit vector along 0y, and a_{2i} is the unit vector lying on the plane of S^1 and normal to 0y. Then,

$$\oint_{S^1} \bar{\xi}_i\bar{\xi}_j d\phi = \pi(a_{1i}a_{1j} + a_{2i}a_{2j}) = \pi(\delta_{ij} - \bar{x}_i\bar{x}_j). \tag{5.21}$$

The last result is obtained by the orthogonality of the three unit vectors a_{1i}, a_{2i}, $\bar{x}_i$,

$$a_{1i}a_{1j} + a_{2i}a_{2j} + \bar{x}_i\bar{x}_j = \delta_{ij} \tag{5.22}$$

which is the scalar product of the unit vectors along the x_i and x_j axes.

Green's functions for the two-dimensional plane strain problems can be obtained by considering the elastic field due to a distributed line force along the x_3-axis,

$$G_{ij}(x_1 - x_1', x_2 - x_2') = \int_{-\infty}^{\infty} G_{ij}(\mathbf{x} - \mathbf{x}')dx_3', \tag{5.23}$$

where $G_{ij}(\mathbf{x}-\mathbf{x}')$ is defined by (5.1). For isotropic media,

$$G_{ij}(x_1 - x_1', x_2 - x_2') = \{\bar{x}_i\bar{x}_j/\bar{R}^2 - (3-4\nu)\delta_{ij}\log\bar{R}\}/8\pi(1-\nu)\mu \tag{5.24}$$

where $\bar{R}^2 = (x_1 - x_1')^2 + (x_2 - x_2')^2$.

Green's functions for the plane stress can be derived from those for the plane strain by replacing E by $E(1+2\nu)/(1+\nu)^2$ and ν by $\nu/(1+\nu)$, since in both cases these replacements give the same form of Hooke's law.

**Anisotropic materials*

As seen in (3.9), $N_{ij}(\boldsymbol{\xi})$ and $D(\boldsymbol{\xi})$ are homogeneous polynomials with the degrees four and six, respectively. We can therefore rewrite (5.1) in the same way as we derived (5.17) and (5.19),

$$G_{ij}(\mathbf{x}) = \frac{1}{8\pi^2}\int_{S^2} \delta(x\bar{\boldsymbol{\xi}}\cdot\bar{\mathbf{x}})N_{ij}(\bar{\boldsymbol{\xi}})D^{-1}(\bar{\boldsymbol{\xi}})dS(\bar{\boldsymbol{\xi}}) \tag{5.25}$$

and

$$G_{ij}(\mathbf{x}) = \frac{1}{8\pi^2 x}\oint_{S^1} N_{ij}(\bar{\boldsymbol{\xi}})D^{-1}(\bar{\boldsymbol{\xi}})d\phi, \tag{5.26}$$

where $x = |\mathbf{x}|$ and the unit sphere S^2 and circle S^1 are shown in Fig. 5.1. The line integral expression (5.26) has been investigated by Fredholm (1900), Lifshitz and Rozenzweig (1947), and Synge (1957).

As will be seen in later sections, it is not always necessary to have explicit expressions of Green's functions. The form (5.25) is sometimes more convenient than explicit expressions or numerical representations, when it is applied to problems of inclusions and dislocations. However, much effort has been devoted to deriving explicit expressions, approximate forms, or numerical values of Green's functions. Kröner (1953) has factorized $D(\boldsymbol{\xi})$ for hexagonal crystals into three polynomials of degree two and obtained an explicit expression of Green's function for hexagonal crystals. He has also expanded the Green's function for general anisotropic materials in a series of surface harmonic functions. Further analytical investigations of Green's functions for anisotropic materials have been done by Willis (1965), Indenbom and Orlov (1968), Mura and Kinoshita (1971), among others. For cubic crystals no analytical expressions of Green's functions are available except for the two-dimensional case; see, for instance, Eshelby et al. (1953) and Foreman (1955). However, series expressions are given for approximated solutions of Green's functions, as seen in the works of Mann et al. (1961), Lie and Koehler (1968), and Bross (1968).

**Transversely isotropic materials*

The presentation in this subsection follows to a large extent the work of Willis (1965). For transversely isotropic materials, Willis has performed the integration in (5.26) by the residue calculation. Expression $D(\bar{\boldsymbol{\xi}})$ in (3.38) is written in the form

$$D(\bar{\boldsymbol{\xi}}) = \beta\gamma^2 \prod_{i=1}^{3} \{A_i(\bar{\xi}_1^2 + \bar{\xi}_2^2) + \bar{\xi}_3^2\}, \tag{5.27}$$

where

$$A_1 = \alpha'/\gamma, \quad A_2 = \frac{1}{2}\left[\frac{\gamma^2 + \alpha\beta - \gamma'^2}{\beta\gamma} + \left\{\left(\frac{\gamma^2 + \alpha\beta - \gamma'^2}{\beta\gamma}\right)^2 - 4\frac{\alpha}{\beta}\right\}^{1/2}\right],$$
$$A_3 = \frac{1}{2}\left[\frac{\gamma^2 + \alpha\beta - \gamma'^2}{\beta\gamma} - \left\{\left(\frac{\gamma^2 + \alpha\beta - \gamma'^2}{\beta\gamma}\right)^2 - 4\frac{\alpha}{\beta}\right\}^{1/2}\right]. \tag{5.28}$$

For transversely isotropic materials the direction of the x_3-axis is chosen as an axis of symmetry, but the other two directions, x_1 and x_2, can be arbitrary. For a given **x** we choose the coordinate system as shown in Fig. 5.2. The x_1, x_3 axes and vector **x** lie on the same plane, the x_2-axis is normal to the plane; that is, the x_1-axis is chosen on the plane containing the x_3-axis and vector x, and the x_2-axis is normal to the plane. A new coordinate system $\bar{\eta}_1$, $\bar{\eta}_2$, and $\bar{\eta}_3$ is taken as shown in Fig. 5.2. The $\bar{\eta}_1$- and $\bar{\eta}_2$-axes lie on the plane containing S^1, and $\bar{\eta}_1$, $\bar{\xi}_1$, $\bar{\eta}_3$ and $\bar{\xi}_3$ are on the same plane. The angle between the $\bar{\eta}_1$- and x_1-axes is denoted by θ. Then for a point $\bar{\boldsymbol{\xi}}$ on S^1,

$$\bar{\xi}_1 = \bar{\eta}_1 \cos\theta, \quad \bar{\xi}_2 = \bar{\eta}_2, \quad \bar{\xi}_3 = -\bar{\eta}_1 \sin\theta. \tag{5.29}$$

Further transformation is introduced. We set $\zeta = \bar{\eta}_1 + i\bar{\eta}_2$, where

$$\begin{aligned} &\zeta = \exp(i\phi) && \bar{\eta}_1 = (1/2)(\zeta + 1/\zeta), \\ &\bar{\eta}_2 = (1/2i)(\zeta - 1/\zeta), && d\phi = d\zeta/i\zeta. \end{aligned} \tag{5.30}$$

Then the factor in (5.27) is written as

$$A_i(\bar{\xi}_1^2 + \bar{\xi}_2^2) + \bar{\xi}_3^2 = \sin^2\theta(1 - A_i)(\zeta^4 + 2B_i\zeta^2 + 1)/4\zeta^2$$
$$\text{with } B_i = (A_i \cos^2\theta + \sin^2\theta + A_i)/\sin^2\theta(1 - A_i). \tag{5.31}$$

Now Cauchy's theory of residues is applied to (5.26). If ζ is a root of the equation

$$\zeta^4 + 2B_i\zeta^2 + 1 = 0, \tag{5.32}$$

then so are $-\zeta$, $1/\zeta$, $-1/\zeta$. Hence only two roots, α_i and $-\alpha_i$, exist inside

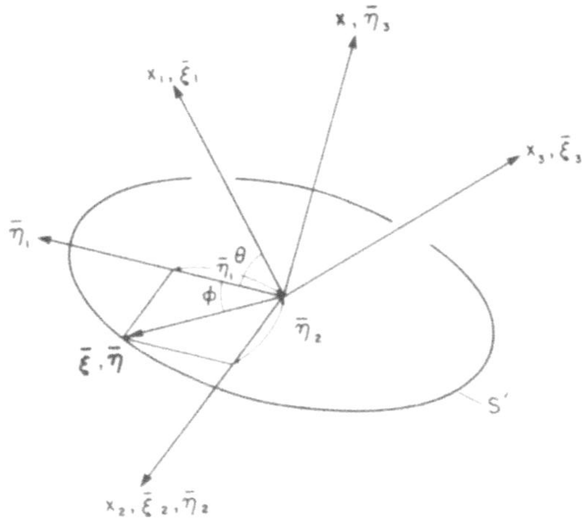

Fig. 5.2. Coordinate systems and a unit circle S^1.

the unit circle, S^1. Let us consider the case where all A_i are real so that α_1^2, α_2^2, α_3^2, are also real.

When the α_i $(i = 1, 2, 3)$ are all distinct and non-zero, the residue calculation yields

$$G_{ij}(\mathbf{x}) = \frac{C}{\pi x}\left\{\frac{h_{ij}(\alpha_1^2)}{(\alpha_1^2 - \alpha_2^2)(\alpha_1^2 - \alpha_3^2)} + \frac{h_{ij}(\alpha_2^2)}{(\alpha_2^2 - \alpha_3^2)(\alpha_2^2 - \alpha_1^2)} + \frac{h_{ij}(\alpha_3^2)}{(\alpha_3^2 - \alpha_1^2)(\alpha_3^2 - \alpha_2^2)}\right\}, \tag{5.33}$$

where

$$C = \left\{\beta\gamma^2 \sin^6\theta \prod_{i=1}^{3}(1 - A_i)\right\}^{-1},$$

$$h_{ij}(\alpha_1^2) = \frac{N_{ij}\left((\alpha_2^1 - 1)\cos\theta, \dfrac{\alpha_1^2 - 1}{i}, -(\alpha_1^2 + 1)\sin\theta\right)}{(\alpha_1^2 - \beta_1^2)(\alpha_1^2 - \beta_2^2)(\alpha_1^2 - \beta_3^2)}, \tag{5.34}$$

$$\beta_1^2 = 1/\alpha_1^2, \quad \beta_2^2 = 1/\alpha_2^2, \quad \beta_3^2 = 1/\alpha_3^2.$$

Expressions $N_{ij}(\xi_1, \xi_2, \xi_3)$ have been given by (3.39).

We now discuss a degenerate case, which occurs when $\alpha' = \gamma$, so that $A_1 = 1$. In this case, $A_1(\bar{\xi}_1^2 + \bar{\xi}_2^2) + \bar{\xi}_3^2 = 1$ everywhere on the contour of integration and may therefore be omitted from the integrands. Similar residue calculation leads to

$$G_{ij}(\mathbf{x}) = \frac{\{\beta\gamma^2 \sin^4\theta(1 - A_2)(1 - A_3)\}^{-1}}{4\pi x}\left\{N_{ij}(\boldsymbol{\xi}_1) + \frac{N_{ij}(\boldsymbol{\xi}_2)\alpha_2^2\alpha_3^2}{(\alpha_2^4 - 1)(\alpha_2^2\alpha_3^2 - 1)(\alpha_2^2 - \alpha_3^2)} + \frac{N_{ij}(\boldsymbol{\xi}_3)\alpha_3^2\alpha_2^2}{(\alpha_3^4 - 1)(\alpha_3^2\alpha_2^2 - 1)(\alpha_3^2 - \alpha_2^2)}\right\}, \tag{5.35}$$

where

$$\begin{aligned} \boldsymbol{\xi}_1 &= (\cos\theta, i, -\sin\theta), \\ \boldsymbol{\xi}_2 &= \sqrt{2}((1 - B_2)^{1/2}\cos\theta, (1 + B_2)^{1/2}, -(1 - B_2)^{1/2}\sin\theta), \\ \boldsymbol{\xi}_3 &= \sqrt{2}((1 - B_3)^{1/2}\cos\theta, (1 + B_3)^{1/2}, -(1 - B_3)^{1/2}\sin\theta). \end{aligned} \tag{5.36}$$

Alternative analytical expressions of Green's function for transversely isotropic materials have been obtained by Lifshitz and Rosenzweig (1947), Elliot (1948), Kröner (1953), Lejček (1969) and Pan and

Chou (1976). The result of Pan and Chou is written as follows:

$$G_{11}(\mathbf{x}) = \sum_{i=1}^{2} \{(A'_i - B'_i)(\nu_i/R_i - \nu_i x_1^2/R_i^3) + 2\nu_i B'_i(1/R_i^* - x_1^2/R_i R_i^{*2})\} + D(1/R_3^* - x_2^2/R_3 R_3^{*2}),$$

$$G_{12}(\mathbf{x}) = G_{21}(\mathbf{x}) = \sum_{i=1}^{2} \{-(A'_i - B'_i)\nu_i x_1 x_2/R_i^3 - 2\nu_i B'_i x_1 x_2/R_i R_i^{*2}\} + D x_1 x_2/R_3 R_3^{*2},$$

$$G_{3j}(\mathbf{x}) = \sum_{i=1}^{2} \{-(A'_i - B'_i)\nu_i^2 k_i x_j z_i/R_i^3 - (C_{11}A'_i - C_{44}\nu_i^2 B'_i) \times 2x_j/(C_{13} + C_{44})R_i R_i^*\}, \ (j = 1, 2)$$

$$G_{13}(\mathbf{x}) = \sum_{i=1}^{2} \{\nu_i A_i x_1/R_i R_i^* - \nu_i(A_i + B_i)x_1 z_i/R_i^3\}, \tag{5.37}$$

$$G_{23}(\mathbf{x}) = \sum_{i=1}^{2} \{\nu_i A_i x_2/R_i R_i^* - \nu_i(A_i + B_i)x_2 z_i/R_i^3\},$$

$$G_{33}(\mathbf{x}) = \sum_{i=1}^{2} \{-(C_{11}B_i + C_{44}\nu_i^2 A_i)/(C_{13} + C_{44})R_i - (A_i + B_i)\nu_i^2(C_{44}\rho^2 + C_{11}x_3^2)/(C_{13} + C_{44})R_i^3\},$$

$$G_{22}(\mathbf{x}) = \sum_{i=1}^{2} \{(A'_i - B'_i)(\nu_i/R_i - \nu_i x_2^2/R_i^3) + 2\nu_i B'_i(1/R_i^* - x_2^2/R_i R_i^{*2})\} + D(1/R_3^* - x_1^2/R_3 R_3^{*2}),$$

where

$$\begin{aligned}
\nu_1 &= (\tilde{C}_{13} - C_{13})^{1/2}(\tilde{C}_{13} + C_{13} + 2C_{44})^{1/2}/(4C_{33}C_{44})^{1/2} \\
&\quad + (\tilde{C}_{13} + C_{13})^{1/2}(\tilde{C}_{13} - C_{13} - 2C_{44})^{1/2}/(4C_{33}C_{44})^{1/2}, \\
\nu_2 &= (\tilde{C}_{13} - C_{13})^{1/2}(\tilde{C}_{13} + C_{13} + 2C_{44})^{1/2}/(4C_{33}C_{44})^{1/2} \\
&\quad - (\tilde{C}_{13} + C_{13})^{1/2}(\tilde{C}_{13} - C_{13} - 2C_{44})^{1/2}/(4C_{33}C_{44})^{1/2}, \\
\nu_3 &= (C_{66}/C_{44})^{1/2}, \quad k_i = (C_{11}/\nu_i^2 - C_{44})/(C_{13} + C_{44}), \\
\tilde{C}_{13} &= (C_{11}C_{33})^{1/2}, \quad z_1 = \nu_1 x_3, \quad z_2 = \nu_2 x_3, \quad z_3 = \nu_3 x_3, \\
R_i &= (x_1^2 + x_2^2 + z_i^2)^{1/2}, \quad R_i^* = R_i + z_i, \quad D = 1/4\pi C_{44}\nu_3.
\end{aligned} \tag{5.38}$$

For $\tilde{C}_{13} - C_{13} - 2C_{44} \neq 0$,

$$\begin{aligned}
A'_1 = B'_1 &= -(C_{44} - C_{33}\nu_1^2)/8\pi C_{33}C_{44}(\nu_1^2 - \nu_2^2)\nu_1^2, \\
A'_2 = B'_2 &= (C_{44} - C_{33}\nu_2^2)/8\pi C_{33}C_{44}(\nu_1^2 - \nu_2^2)\nu_2^2,
\end{aligned} \tag{5.39}$$

$$\nu_1 A_1 = -\nu_2 A_2 = (C_{13} + C_{44})/4\pi C_{33} C_{44}(\nu_2^2 - \nu_1^2),$$

$$B_i = -A_i.$$

For $\tilde{C}_{13} - C_{13} - 2C_{44} = 0,$

$$\begin{aligned} A_1' &= A_2' = 1/16\pi C_{11}, \\ B_1' &= B_2' = 1/16\pi C_{44}\nu_1^2, \\ A_1 &= A_2 = 0, \\ B_1 &= B_2 = -(C_{13} + C_{44})/16\pi C_{11} C_{44}. \end{aligned} \tag{5.40}$$

Pan and T.W. Chou (1979) have extended their solution to a two-phase material (see also Sveklo 1969).

**Kröner's formula*

Kröner's formula (1953) is a series expression of Green's functions for general anisotropic materials. Since $N_{ij}(\boldsymbol{\xi})/D(\boldsymbol{\xi})$ in (5.25) and (5.26) is a continuous function on S^2, it can be expanded in a series of surface harmonic functions which uniformly converge on S^2, as follows (Hobson, 1931):

$$N_{ij}(\boldsymbol{\xi})/D(\boldsymbol{\xi}) = \sum_{n=0}^{\infty} U_n(\boldsymbol{\xi}), \tag{5.41}$$

where

$$U_n(\boldsymbol{\xi}) = \frac{2n+1}{4\pi}\int_{S^2} P_n(\boldsymbol{\xi}\cdot\boldsymbol{\xi}')N_{ij}(\boldsymbol{\xi}')D^{-1}(\boldsymbol{\xi}')dS(\boldsymbol{\xi}'), \quad n = 0, 1, 2, \ldots, \tag{5.42}$$

and P_n is the Legendre polynomial. From the orthogonality of the Legendre polynomials, we can easily derive

$$U_n(\boldsymbol{\xi}) = \frac{2n+1}{4\pi}\int_{S^2} P_n(\boldsymbol{\xi}\cdot\boldsymbol{\xi}')U_n(\boldsymbol{\xi}')dS(\boldsymbol{\xi}'). \tag{5.43}$$

When (5.43) is integrated over S^1, we have

$$\int_{S^1} U_n(\boldsymbol{\xi})ds(\boldsymbol{\xi}) = \frac{2n+1}{4\pi}\int_{S^2} U_n(\boldsymbol{\xi}')dS(\boldsymbol{\xi}')\int_{S^1} P_n(\boldsymbol{\xi}\cdot\boldsymbol{\xi}')ds(\boldsymbol{\xi}), \tag{5.44}$$

where $ds(\boldsymbol{\xi}) = d\phi$ is a line element of S^1.

The addition theorem of the Legendre polynomials is

$$P_n(\boldsymbol{\xi}\cdot\boldsymbol{\xi}') = P_n(\boldsymbol{\xi}\cdot\bar{\mathbf{x}})P_n(\boldsymbol{\xi}'\cdot\bar{\mathbf{x}}) + 2\sum_{m=1}^{n}\frac{(n-m)!}{(n+m)!}P_n^m(\boldsymbol{\xi}\cdot\bar{\mathbf{x}})P_n^m(\boldsymbol{\xi}'\cdot\bar{\mathbf{x}})\cos m(\phi-\phi'), \quad (5.45)$$

where $P_n^m(z)$ is the associated Legendre function of degree n and of order m. When (5.45) is integrated over S^1 with respect to $\boldsymbol{\xi}$ for a fixed value of $\boldsymbol{\xi}'$, we have, since $\boldsymbol{\xi}\cdot\bar{\mathbf{x}}=0$ and $\int_{S^1}\cos m(\phi-\phi')d\phi=0$,

$$\int_{S^1} P_n(\boldsymbol{\xi}\cdot\boldsymbol{\xi}')ds(\boldsymbol{\xi}) = 2\pi P_n(0)P_n(\boldsymbol{\xi}'\cdot\bar{\mathbf{x}}). \quad (5.46)$$

Substituting (5.46) into (5.44), and using (5.43) we have

$$\int_{S^1} U_n(\boldsymbol{\xi})ds(\boldsymbol{\xi}) = 2\pi P_n(0)U_n(\bar{\mathbf{x}}). \quad (5.47)$$

Therefore, (5.26) can be written as

$$G_{ij}(\mathbf{x}) = \frac{1}{4\pi x}\sum_{n=0}^{\infty} P_n(0)U_n(\bar{\mathbf{x}}) \quad (5.48)$$

which is a modified expression of Kröner's formula (1953) by Mura and Kinoshita (1971). The integer n in (5.48) takes only even numbers, since $P_n(0)=0$ for odd numbers of n. $U_n(\bar{\mathbf{x}})$ is calculated from

$$U_n(\bar{\mathbf{x}}) = \frac{2n+1}{4\pi}\int_{S^2} P_n(\bar{\mathbf{x}}\cdot\bar{\boldsymbol{\xi}})N_{ij}(\bar{\boldsymbol{\xi}})D^{-1}(\bar{\boldsymbol{\xi}})dS(\bar{\boldsymbol{\xi}}). \quad (5.49)$$

The uniform convergency of (5.48) has been proved by Mura and Kinoshita (1953). They also showed that

$$G_{ij}(\mathbf{x}) = \frac{1}{16\pi^2}\Delta\int_{S^2} N_{ij}(\bar{\boldsymbol{\xi}})D^{-1}(\bar{\boldsymbol{\xi}})|\mathbf{x}\cdot\bar{\boldsymbol{\xi}}|dS(\bar{\boldsymbol{\xi}}), \quad (5.50)$$

where Δ is the Laplacian. The above expression is continuously differentiable.

*Derivatives of Green's functions

As seen in Eqs. (3.23) ~ (3.25), derivatives of Green's function appear in these formulae rather than Green's function itself. The plane of the unit circle S^1 in (5.26), however, is orthogonal to the direction of $\mathbf{x}$ and therefore depends on $\mathbf{x}$. Therefore, the derivatives of G_{ij} need special consideration. The following argument follows the work done by Barnett (1972) and Willis (1975). Differentiating (5.25) with respect to x_k

leads to

$$G_{ij,k}(\mathbf{x}) = \frac{1}{8\pi^2}\int_{S^2} \delta'(\bar{\boldsymbol{\xi}} \cdot \mathbf{x})\bar{\xi}_k N_{ij}(\bar{\boldsymbol{\xi}})D^{-1}(\bar{\boldsymbol{\xi}})dS(\bar{\boldsymbol{\xi}}), \tag{5.51}$$

where δ' is the derivative of δ with respect to the argument. Since

$$\int_{S^2} dS(\bar{\boldsymbol{\xi}}) = \oint_{S^1} d\phi \int_{-1}^{1} d(\bar{\boldsymbol{\xi}} \cdot \bar{\mathbf{x}}) \tag{5.52}$$

and $\delta(x\bar{\boldsymbol{\xi}} \cdot \bar{\mathbf{x}})\bar{\xi}_k N_{ij}(\bar{\boldsymbol{\xi}})D^{-1} = 0$ at $\bar{\boldsymbol{\xi}} \cdot \bar{\mathbf{x}} = 1$ and -1 $(\theta = 0, \pi)$, integration by parts of (5.51) yields

$$G_{ij,k}(\mathbf{x}) = \frac{-1}{8\pi^2}\int_{S^2} \delta(\bar{\boldsymbol{\xi}} \cdot \mathbf{x})\partial[\bar{\xi}_k N_{ij}(\bar{\boldsymbol{\xi}})D^{-1}(\bar{\boldsymbol{\xi}})]/\partial(\bar{\boldsymbol{\xi}} \cdot \mathbf{x})dS(\bar{\boldsymbol{\xi}}) \tag{5.53}$$

and therefore

$$G_{ij,k}(\mathbf{x}) = \frac{-1}{8\pi^2 x}\oint_{S^1} \partial[\bar{\xi}_k N_{ij}(\bar{\boldsymbol{\xi}})D^{-1}(\bar{\boldsymbol{\xi}})]/\partial(\bar{\boldsymbol{\xi}} \cdot \mathbf{x})d\phi. \tag{5.54}$$

On S^1 we have (see Fig. 5.1)

$$\begin{aligned}
&\partial(\bar{\boldsymbol{\xi}} \cdot \mathbf{x}) = x\partial(\bar{\boldsymbol{\xi}} \cdot \bar{\mathbf{x}}) = -x \sin\theta d\theta = -x d\theta,\\
&\partial\bar{\xi}_k/\partial\theta = -\bar{x}_k,\\
&\partial K_{lm}/\partial\theta = -C_{lpmq}(\bar{x}_p\bar{\xi}_q + \bar{\xi}_p\bar{x}_q),
\end{aligned} \tag{5.55}$$

where K_{lm} has been defined in (3.5). On the other hand, differentiating $K_{lm}N_{mj}D^{-1} = \delta_{lj}$ with respect to θ we obtain

$$\frac{\partial}{\partial\theta}(K_{lm})N_{mj}D^{-1} + K_{lm}\frac{\partial}{\partial\theta}(N_{mj}D^{-1}) = 0 \tag{5.56}$$

and

$$\frac{\partial}{\partial\theta}(N_{ij}D^{-1}) = -N_{li}D^{-1}\frac{\partial}{\partial\theta}(K_{lm})N_{mj}D^{-1} = N_{il}N_{jm}D^{-2}C_{lpmq}(\bar{x}_p\bar{\xi}_q + \bar{\xi}_p\bar{x}_q) \tag{5.57}$$

since $N_{li}D^{-1}K_{lm} = \delta_{im}$. Thus, (5.54) is written as

$$\begin{aligned}
G_{ij,k}(\mathbf{x}) = \frac{1}{8\pi^2 x^2}\oint_{S^1} &[-\bar{x}_k N_{ij}(\bar{\boldsymbol{\xi}})D^{-1}(\bar{\boldsymbol{\xi}})\\
&+ \bar{\xi}_k C_{lpmq}(\bar{x}_p\bar{\xi}_q + \bar{\xi}_p\bar{x}_q)N_{li}N_{mj}D^{-2}]d\phi.
\end{aligned} \tag{5.58}$$

Higher derivatives can be obtained in a similar manner.

Differentiating (5.25) with respect to x_k and x_l, we have

$$G_{ij,kl}(\mathbf{x}) = \frac{1}{8\pi^2}\oint_{S^1} d\phi \int_{-1}^{1} \delta''(\boldsymbol{\xi}\cdot\mathbf{x})\xi_k\xi_l N_{ij}(\boldsymbol{\xi})D^{-1}(\boldsymbol{\xi})d(\boldsymbol{\xi}\cdot\bar{\mathbf{x}}). \tag{5.59}$$

Integrating twice by parts, we arrive at

$$G_{ij,kl}(\mathbf{x}) = \frac{1}{8\pi^2 x^3}\oint_{S^1} d\phi\,\partial^2(\bar{\xi}_k\bar{\xi}_l N_{ij}D^{-1})/\partial\theta^2 \tag{5.60}$$

which is finally written as

$$\begin{aligned}G_{ij,kl}(\mathbf{x}) = \frac{1}{8\pi^2 x^3}\oint_{S^1} &[2\bar{x}_k\bar{x}_l N_{ij}(\bar{\boldsymbol{\xi}})D^{-1}(\bar{\boldsymbol{\xi}}) - 2\{(\bar{x}_k\bar{\xi}_l + \bar{\xi}_k\bar{x}_l)(\bar{x}_p\bar{\xi}_q + \bar{\xi}_p\bar{x}_q)\\ &+ \bar{\xi}_k\bar{\xi}_l\bar{x}_p\bar{x}_q\}N_{ih}(\bar{\boldsymbol{\xi}})N_{jm}(\bar{\boldsymbol{\xi}})C_{hpmq}D^{-2}(\bar{\boldsymbol{\xi}})\\ &+ \bar{\xi}_k\bar{\xi}_l C_{hpmq}(\bar{x}_p\bar{\xi}_q + \bar{\xi}_p\bar{x}_q)C_{satb}(\bar{x}_a\bar{\xi}_b + \bar{\xi}_a\bar{x}_b)\{N_{jm}(\bar{\boldsymbol{\xi}})N_{is}(\bar{\boldsymbol{\xi}})N_{ht}(\bar{\boldsymbol{\xi}})\\ &+ N_{ih}(\bar{\boldsymbol{\xi}})N_{js}(\bar{\boldsymbol{\xi}})N_{mt}(\bar{\boldsymbol{\xi}})\}D^{-3}(\bar{\boldsymbol{\xi}})]d\phi.\end{aligned} \tag{5.61}$$

**Two-dimensional Green's function*

Green's function for two-dimensional problems has been given by (5.24) for isotropic materials. For anisotropic materials, the two dimension Green's function has been developed by Eshelby, Read and Shockley (1953), Stroh (1962), and Bullough and Bilby (1954). It is particularly convenient to use Green's function for generalized plane problems (mixture of plane strain and antiplane strain).

Green's function $G_{ij}(\mathbf{x}-\mathbf{x}')$ for generalized plane problems can be obtained as the displacement $u_i(\mathbf{x})$ when a unit line force F_j acting at $\mathbf{x}'$ parallel to the x_3-axis is applied. The equations of equilibrium for u_i are

$$C_{i\beta k\alpha}u_{k,\alpha\beta} = 0, \quad i, k = 1, 2, 3;\ \alpha, \beta = 1, 2, \tag{5.62}$$

for all points except $\mathbf{x}'$. Hooke's law is

$$\sigma_{ij} = C_{ijkl}e_{kl}, \tag{5.63}$$

and the strain and displacement relation is

$$e_{ij} = (1/2)(u_{i,j} + u_{j,i}). \tag{5.64}$$

The solution is given by an arbitrary function of a linear combination of the variables x_1 and x_2 and can be written as

$$u_k = A_k f(z), \tag{5.65}$$

where

$$z = x_1 + px_2. \tag{5.66}$$

Substituting (5.65) into (5.62), we have

$$\phi_{ik}A_k = 0 \tag{5.67}$$

with

$$\phi_{ik} = C_{i1k1} + p(C_{i1k2} + C_{i2k1}) + p^2C_{i2k2}. \tag{5.68}$$

Equation (5.67) has a nontrivial solution for the vector A_k if and only if the determinant of the coefficients equals zero,

$$|\phi_{ik}| = 0. \tag{5.69}$$

This determinant is a six-order polynomial in p. Its complex roots occur in conjugate pairs. For a given root $p_{(l)}$, the corresponding values of A_k are determined from (5.67). However, the displacements must be real; this condition requires that the imaginary parts of corresponding pairs of solution cancel out, and we need take only three roots $p_{(1)}$, $p_{(2)}$ and $p_{(3)}$, no two of which are complex conjugates. The displacement equation can be written more explicitly as

$$u_k = \mathrm{Re}\left\{\sum_{l=1}^{3} A_{(l)k}f_{(l)}[z_{(l)}]\right\}, \tag{5.70}$$

where $z_{(l)} = x_1 + p_{(l)}x_2$, and Re stands for the real part of the complex function in the braces. The stress components are

$$\sigma_{ij} = \mathrm{Re}\left\{\sum_{l=1}^{3} [C_{ijk1} + p_{(l)}C_{ijk2}]A_{(l)k}f'_{(l)}[z_{(l)}]\right\}. \tag{5.71}$$

If $f_{(l)}[z_{(l)}]$ is not a single-valued function, the singular line represents a Volterra dislocation. On the other hand, if $f_{(l)}[z_{(l)}]$ is a single-valued function but the resultant force around the singular line is not zero the singular line then represents a line force per unit length, F_i. The conditions necessary for obtaining Green's function are

$$\int_c \sigma_{ij}n_j ds = F_i \tag{5.72}$$

and

$$\int_c du_k ds = 0, \tag{5.73}$$

where c is a closed curve around point $\mathbf{x}'$, and n_i is a unit vector normal to c. Condition (5.72) can be written as

$$-\int_c d\psi_i = F_i, \tag{5.74}$$

where ψ_i is defined by

$$\sigma_{i1} = -\partial\psi_i/\partial x_2, \quad \sigma_{i2} = \partial\psi_i/\partial x_1. \tag{5.75}$$

From (5.71), (5.68) and (5.75), we have

$$\psi_i = \mathrm{Re}\left\{\sum_{l=1}^{3} [C_{i2k1} + p_{(l)}C_{i2k2}]A_{(l)k}f_{(l)}[z_{(l)}]\right\}. \tag{5.76}$$

Eshelby et al. (1953) have shown that

$$f_{(l)}[z_{(l)}] = (1/\pm 2\pi i)D_{(l)} \log[z_{(l)} - z'_{(l)}] \tag{5.77}$$

with

$$z'_{(l)} = x'_1 + p_{(l)}x'_2, \tag{5.78}$$

where the sign of $2\pi i$ is taken to be the same as the sign of the imaginary part of $p_{(l)}$. Then, conditions (5.74) and (5.73) become

$$\mathrm{Re}\left\{\sum_{l=1}^{3} [C_{i2k1} + P_{(l)}C_{i2k2}]A_{(l)k}D_{(l)}\right\} = F_i \tag{5.79}$$

and

$$\mathrm{Re}\left\{\sum_{l=1}^{3} A_{(l)k}D_{(l)}\right\} = 0. \tag{5.80}$$

Equations (5.79) and (5.80) represent a system of six simultaneous linear equations for six unknowns $D_{(l)}$.

The properties of the coefficients $A_{(l)k}$ were studied in detail by Stroh (1962). $D_{(l)}$ can always be expressed as a linear combination of F_i,

$$D_{(l)} = d_{(l)m}F_m. \tag{5.81}$$

Now let us define a complex variable $X = x_1 + ix_2$ (with its conjugate $\bar{X} = x_i - ix_2$). It follows that

$$z_{(l)} = (1/2)[X + \bar{X} - ip_{(l)}X + ip_{(l)}\bar{X}]. \tag{5.82}$$

The displacement then is

$$u_k(x) = \mathrm{Re}\left\{\sum_{l=1}^{3} A_{(l)k} \frac{d_{(l)m}F_m}{\pm 2\pi i} \log \frac{1}{2} [(X + \bar{X} - ip_{(l)}X + ip_{(l)}\bar{X}) - (X' + \bar{X}' - ip_{(l)}X' + ip_{(l)}\bar{X}')]\right\}. \tag{5.83}$$

Finally, Green's function is obtained from (5.83) as

$$G_{km}(\mathbf{x}-\mathbf{x}') = \mathrm{Re}\left\{\sum_{l=1}^{3} A_{(l)k} \frac{d_{(l)m}}{\pm 2\pi i} \log \frac{1}{2}\,[(X+\bar{X}-ip_{(l)}X+ip_{(l)}\bar{X}) - (X'+\bar{X}'-ip_{(l)}X'+ip_{(l)}\bar{X}')]\right\}. \tag{5.84}$$

6. Inclusions and inhomogeneities

Inclusions

An inclusion is defined as a sub-domain Ω in domain D, where eigenstrain $\epsilon^*_{ij}(\mathbf{x})$ is given in Ω and is zero in D-Ω. The elastic moduli in Ω and D-Ω are assumed to be the same. The remaining domain D-Ω is called the matrix. From the result in Section 3, the elastic field due to the inclusion can be written as

$$\begin{aligned}
u_i(\mathbf{x}) &= -\int_\Omega C_{jlmn}\epsilon^*_{mn}(\mathbf{x}')G_{ij,l}(\mathbf{x}-\mathbf{x}')d\mathbf{x}',\\
\epsilon_{ij}(\mathbf{x}) &= -(1/2)\int_\Omega C_{klmn}\epsilon^*_{mn}(\mathbf{x}')\{G_{ik,lj}(\mathbf{x}-\mathbf{x}')+G_{jk,li}(\mathbf{x}-\mathbf{x}')\}d\mathbf{x}',\\
\sigma_{ij}(\mathbf{x}) &= -C_{ijkl}\left\{\int_\Omega C_{pqmn}\epsilon^*_{mn}(\mathbf{x}')G_{kp,ql}(\mathbf{x}-\mathbf{x}')d\mathbf{x}'+\epsilon^*_{kl}(\mathbf{x})\right\},
\end{aligned} \tag{6.1}$$

or

$$\begin{aligned}
u_i(\mathbf{x}) &= -i(2\pi)^{-3}\int_{-\infty}^{\infty}\int_\Omega C_{jlmn}\epsilon^*_{mn}(\mathbf{x}')\xi_l N_{ij}(\boldsymbol{\xi})D^{-1}(\boldsymbol{\xi})\\
&\quad\times\exp\{i\boldsymbol{\xi}\cdot(\mathbf{x}-\mathbf{x}')\}d\boldsymbol{\xi}d\mathbf{x}',\\
\epsilon_{ij}(\mathbf{x}) &= (2\pi)^{-3}\int_{-\infty}^{\infty}\int_\Omega (1/2)C_{klmn}\epsilon^*_{mn}(\mathbf{x}')\xi_l\{\xi_j N_{ik}(\boldsymbol{\xi})+\xi_i N_{jk}(\boldsymbol{\xi})\}\\
&\quad\times D^{-1}(\boldsymbol{\xi})\exp\{i\boldsymbol{\xi}\cdot(\mathbf{x}-\mathbf{x}')\}d\boldsymbol{\xi}d\mathbf{x}',\\
\sigma_{ij}(\mathbf{x}) &= C_{ijkl}\left\{(2\pi)^{-3}\int_{-\infty}^{\infty}\int_\Omega C_{pqmn}\epsilon^*_{mn}(\mathbf{x}')\xi_q\xi_l N_{kp}(\boldsymbol{\xi})D^{-1}(\boldsymbol{\xi})\right.\\
&\quad\left.\times\exp\{i\boldsymbol{\xi}\cdot(\mathbf{x}-\mathbf{x}')\}d\boldsymbol{\xi}d\mathbf{x}'-\epsilon^*_{kl}(\mathbf{x})\right\},
\end{aligned} \tag{6.2}$$

where $\epsilon^*_{kl}(\mathbf{x})=0$ for $\mathbf{x}\in D-\Omega$.

Since $\epsilon^*_{kl}(\mathbf{x})$ is discontinuous on the boundary of domain Ω, some quantities may also be discontinuous on that boundary. However, the displacement and the interfacial traction across the boundary must be continuous; that is,

$$[u_i] \equiv u_i(\text{out}) - u_i(\text{in}) = 0,$$
$$[\sigma_{ij}]n_j \equiv \{\sigma_{ij}(\text{out}) - \sigma_{ij}(\text{in})\}n_j = 0, \tag{6.3}$$

where n_i is the outward unit normal on the boundary of Ω. As discussed by Hill (1961) and by Walpole (1967), the displacement gradient (distortion) $u_{i,j}$ is continuous inside Ω and D-Ω but discontinuous at the interface between the two domains. The jump across the interface can be written as

$$[u_{i,j}] \equiv u_{i,j}(\text{out}) - u_{i,j}(\text{in}) = \lambda_i n_j, \tag{6.4}$$

λ_i is the proportionality constant (the magnitude of the jump) to be determined. The above equation is designed so that the condition $[u_{i,j}]dx_j = 0$ for $d\mathbf{x}$ along the boundary of Ω can be satisfied. This condition stems from the continuity of displacement along the boundary ($n_j dx_j = 0$ for $d\mathbf{x}$ along the boundary).

Since

$$\sigma_{ij} = C_{ijkl}(u_{k,l} - \epsilon^*_{kl}), \tag{6.5}$$

the second equation in (6.3) is written as

$$C_{ijkl}\{[u_{k,l}] - [\epsilon^*_{kl}]\}n_j = 0, \tag{6.6}$$

where $[\epsilon^*_{kl}] = \epsilon^*_{kl}(\text{out}) - \epsilon^*_{kl}(\text{in})$, $\epsilon^*_{kl}(\text{out}) = 0$, and $\epsilon^*_{kl}(\text{in}) = \epsilon^*_{kl}$. Substitution of (6.4) into (6.6) gives

$$C_{ijkl}\lambda_k n_l n_j = -C_{ijkl}\epsilon^*_{kl}n_j. \tag{6.7}$$

Equation (6.7) is a system of equations to determine $\boldsymbol{\lambda}$ for given $\mathbf{n}$ and ϵ^*_{ij}. Since the system has the same form as (3.4), we obtain

$$\lambda_i = -C_{jkmn}\epsilon^*_{mn}n_k N_{ij}(\mathbf{n})/D(\mathbf{n}), \tag{6.8}$$

where $N_{ij}(\mathbf{n})$ and $D(\mathbf{n})$ are the cofactor and determinant of the matrix $\mathbf{K}(\mathbf{n})$, namely,

$$D(\mathbf{n}) = \epsilon_{mnl}K_{m1}K_{n2}K_{l3},$$
$$N_{ij}(\mathbf{n}) = \tfrac{1}{2}\epsilon_{ikl}\epsilon_{jmn}K_{km}K_{ln}, \tag{6.9}$$

where $K_{ik} = C_{ijkl}n_j n_l$. Substituting (6.8) into (6.4), we arrive at

$$[u_{i,j}] = -C_{lkmn}\epsilon^*_{mn}n_k n_j N_{il}(\mathbf{n})/D(\mathbf{n}) \tag{6.10}$$

and, therefore,

$$[\sigma_{ij}] \equiv \sigma_{ij}(\text{out}) - \sigma_{ij}(\text{in}) = C_{ijkl}\{[u_{k,l}] - [\epsilon^*_{kl}]\}$$

$$= C_{ijkl}\{-C_{pqmn}\epsilon^*_{mn}n_q n_l N_{kp}(\mathbf{n})/D(\mathbf{n}) + \epsilon^*_{kl}\}. \tag{6.11}$$

Equation (6.11) is useful for evaluating the stress just outside the inclusion when the stress just inside the inclusion is known. As shown in later chapters, eigenstresses inside ellipsoidal inclusions are usually easy to obtain. This interesting observation, relation (6.11), was first made by Goodier (1937) in his paper on thermal stresses.

Stress concentration factors of ellipsoidal inhomogeneities are easily obtained from (6.11), as seen in Section 31. The integrations in (6.1) and (6.2) are performed in Chapters 2 and 3 for ellipsoidal inclusions.

When the eigenstrain ϵ^*_{mn} is uniform in Ω and Ω has an arbitrary shape, it is sometimes convenient to rewrite (6.1), by integration by parts. Then, the first equation in (6.1) becomes

$$u_i(\mathbf{x}) = \int_{|\Omega|} C_{jlmn}\epsilon^*_{mn}G_{ij}(\mathbf{x}-\mathbf{x}')n_l dS, \tag{6.12}$$

where $|\Omega|$ is the boundary of Ω and dS is its surface element.

Inhomogeneities

When a subdomain of a material has elastic moduli different from those of the remainder (matrix), the subdivision is called an inhomogeneity. A uniform applied stress at infinity is not uniform in the neighbourhood of the inhomogeneity. This stress disturbance due to the inhomogeneity can be simulated by an eigenstress field caused by an inclusion when the eigenstrain is chosen properly, as seen in Section 22.

For plane strain or plane stress inhomogeneity problems (composite problems), the complex potential method of Muskhelishvili (1953) is more effective than the equivalent inclusion method (see Sherman (1959), Hardiman (1954), Sendeckyj (1970) and Theocaris and Ioakimidis (1977), among others.) Numerical examples of the application of the complex potential method are given for multiple circular inclusion problems by Kim and Ukadgaonker (1971) and Yu and Sendeckyj (1974). Sendeckyj (1970, 1971) also give the exact solution of the antiplane longitudinal shear deformation of composites. In antiplane problems, the nonzero displacement and stress components are given in terms of one analytic function (see Milne-Thomson, 1962).

Effect of isotropic elastic moduli on stress

Suppose that a composite body consisting of two isotropic elastic phases is loaded by prescribed surface tractions $\lambda\mathbf{F}$, where $\mathbf{F}$ is a fixed vector and λ is the load parameter. According to the linear theory, and when proportional loading is involved, the induced stress is proportional to the magnitude of the applied tractions. It then follows from dimensional analysis (e.g., Langhaar 1951) that in three dimensional cases, the stress at any point in the composite is a function of three parameters involving the four elastic moduli. It is obvious that σ_{ij}/λ, λ/μ_1, μ_2/μ_1, ν_1, and ν_2 are dimensionless and that σ_{ij}/λ can be expressed as

$$\sigma_{ij}/\lambda = f(\lambda/\mu_1, \mu_2/\mu_1, \nu_1, \nu_2) \tag{6.13}$$

where μ_1 and ν_1 are the shear modulus and Poisson's ratio of the matrix and μ_2, ν_2 are the corresponding constants of the inhomogeneity. Since σ_{ij} is a linear function of λ, (6.13) must have the form

$$\sigma_{ij}/\lambda = f(\mu_2/\mu_1, \nu_1, \nu_2). \tag{6.14}$$

Therefore, σ_{ij}/λ is a function of three parameters μ_2/μ_1, ν_1 and ν_2, or three combinations of them.

An important reduction in the dependence of stress on the elastic moduli takes place when the geometry and loading of the composite body is such that it is in a state of either plane strain or plane stress. If

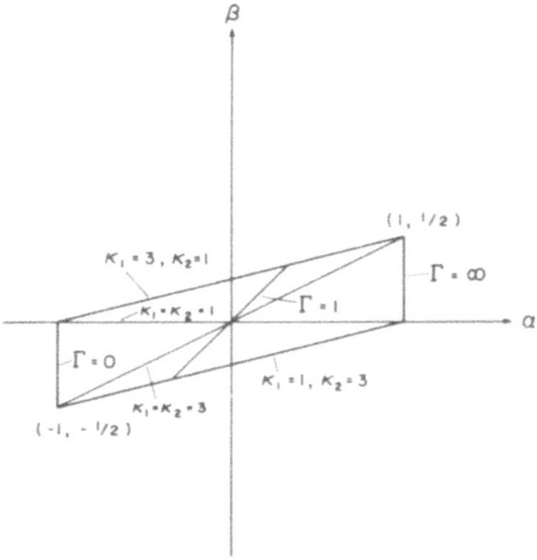

Fig. 6.1. The Dundurs constants diagram.

the composite body is multiply connected, and if the vector sums of tractions on holes vanish for every individual hole, Dundurs (1967) has shown that, under these conditions, the stress depends on only two combinations of elastic moduli. In a sense this result is the counterpart of the theorem by Michell (1899) establishing the circumstances under which stress is independent of elastic constants in a homogeneous material. Choice of these two combinations (parameters) of elastic moduli is not unique. Dundurs (1969, 1970) has proposed the following two parameters (Dundurs' constants):

$$\alpha = \frac{(\kappa_1 + 1)\Gamma - (\kappa_2 + 1)}{(\kappa_1 + 1)\Gamma + \kappa_2 + 1},$$
$$\beta = \frac{(\kappa_1 - 1)\Gamma - (\kappa_2 - 1)}{(\kappa_1 + 1)\Gamma + \kappa_2 + 1}, \tag{6.15}$$

where $\kappa = 3 - 4\nu$ for plane strain and $\kappa = (3 - \nu)/(1 + \nu)$ for plane stress, ν being Poisson's ratio. The subscripts 1 and 2 are used on the elastic moduli to distinguish between the two phases, and $\Gamma = \mu_2/\mu_1$, where μ is the shear modulus. The α, β-plane provides a convenient means for classifying composite materials regarding their physical behavior and for exhibiting such results as stress concentration factors that depend on the elastic moduli. Because of the physical limits $\Gamma \geq 0$ and $1 \leq \kappa \leq 3$, the admissible values of α and β are restricted to a bounded region or, more specifically, a parallelogram in the α, β-plane, as shown in Fig. 6.1. Identical materials are represented by $\alpha = \beta = 0$, and the case of equal shear moduli, or $\Gamma = 1$, corresponds to the straight line $\alpha = \beta$. Representation of stress concentration factors for a composite in terms of α and β is found in the paper by Gerstner and Dundurs (1969).

7. Dislocations

The mathematical theory of dislocations in elastic continua was first systematically studied by Volterra (1907) and Weingarten (1901). Taylor (1934), Orowan (1934) and Polanyi (1934) independently used dislocations to explain plastic deformation of single crystals. They considered dislocations as imperfections in crystals and explained why observed yield stresses of crystals are much lower than the theoretical values calculated from atomic theory on the basis of perfect-lattice state. The existence of dislocations was first confirmed directly by Hirsch and his coworkers (1960) by transmission electron microscopy (indirect observations by etch pits, etc., has been done earlier—see Hirth and Lothe 1968). The direct observation of dislocations has been extensively made (see Amelinckx 1964), and the dislocation theory has been developed by many scientists to explain not only the mechanical, but also the optical and electromagnetic properties of crystals (see Nabarro 1967).

As already explained by Figs. 4.1 and 4.2 in Section 4, the dislocation (line) is defined as a part of the boundary of a slip plane, which is embedded in a material. The part of the slip boundary which is exposed on the surface of the material is not called the dislocation line. To define the direction of a dislocation line in a more precise manner, consider a slip plane S inside a material as shown in Fig. 7.1. It is assumed that, for instance, the upper plane (denoted by S^+) slips by $\mathbf{b}$ relative to the lower plane (denoted by S^-). In order to specify this configuration of the slip, we define the direction $\boldsymbol{\nu}$ of the dislocation line L in the following manner: Go around a linking circuit c (the Burgers circuit) in the direction of rotation of a right-handed screw advancing along the direction of the dislocation. The surface S^+ that is displaced by $\mathbf{b}$ (the Burgers vector) relative to the other surface S^- is designated as the surface on which the end point of the Burgers circuit is located; the circuit does not cross the boundary between S^+ and S^-, as shown in Fig. 7.1.

For a crystal, the Burgers vector of a dislocation is usually a lattice vector. Such a dislocation is called a perfect dislocation. A dislocation which has the Burgers vector not equal to a lattice vector is called an imperfect or a partial dislocation. The slip deformation in crystal plasticity is caused by the movement and generation of dislocations, and is generally irreversible due to the atomic potential barrier. Due to the singularity along the dislocation line, an elastic field is introduced in the material. This dislocation state of the material is simulated by an eigenstress field caused by eigenstrain ϵ^*_{ij}, as discussed in Section 4.

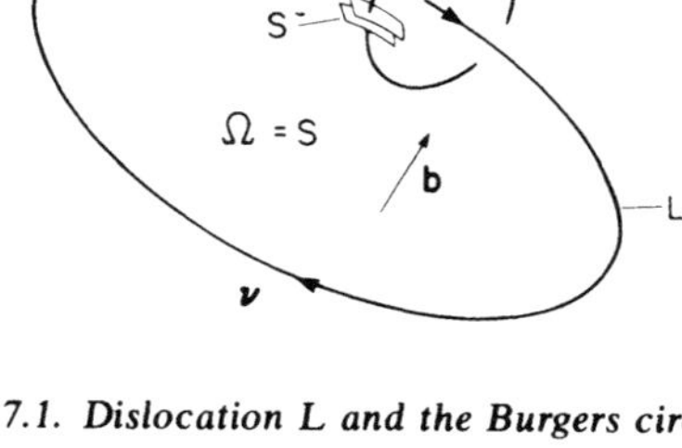

Fig. 7.1. Dislocation L and the Burgers circuit c.

Volterra and Mura formulas

According to Kröner (1958), the displacement gradient $u_{i,j}$ (total distortion) is assumed to consist of elastic distortion β_{ji} and plastic distortion β^*_{ji},

$$u_{i,j} = \beta_{ji} + \beta^*_{ji}. \tag{7.1}$$

The (total) strain ϵ_{ij}, elastic strain ϵ_{ij}, and eigenstrain ϵ^*_{ij} defined in (2.1) ~ (2.4), are

$$\begin{aligned} \epsilon_{ij} &= (1/2)(u_{i,j} + u_{j,i}), \\ e_{ij} &= (1/2)(\beta_{ij} + \beta_{ji}), \\ \epsilon^*_{ij} &= (1/2)(\beta^*_{ij} + \beta^*_{ji}). \end{aligned} \tag{7.2}$$

Since β^*_{ji} is caused by the slip b_i of plane S^+ whose normal vector is n_j, we can write

$$\beta^*_{ji}(\mathbf{x}) = -b_i n_j \delta(\mathbf{S} - \mathbf{x}) \tag{7.3}$$

where $\delta(\mathbf{S}-\mathbf{x})$ is the one-dimensional Dirac delta function in the normal direction of S, being unbounded when $\mathbf{x}$ is on $\mathbf{S}$ and zero otherwise. As in (4.1) and (4.11) we write

$$\epsilon^*_{ij}(\mathbf{x}) = -(1/2)(b_i n_j + b_j n_i)\delta(\mathbf{S} - \mathbf{x}). \tag{7.4}$$

When this expression is substituted into (6.1), and since

$$\int_\Omega \delta(\mathbf{S} - \mathbf{x}')d\mathbf{x}' = \int_S dS, \tag{7.5}$$

then we have the Volterra formula (1907),

$$u_i(\mathbf{x}) = \int_S C_{jlmn} b_m G_{ij,l}(\mathbf{x} - \mathbf{x}') n_n dS(\mathbf{x}'). \tag{7.6}$$

Differentiating (7.6) we have

$$u_{i,j}(\mathbf{x}) = \int_S C_{klmn} b_m G_{ik,lj}(\mathbf{x} - \mathbf{x}') n_n dS(\mathbf{x}'). \tag{7.7}$$

The elastic distortion is obtained from (7.1) as

$$\beta_{ji}(\mathbf{x}) = \int_S C_{klmn} b_m G_{ik,lj}(\mathbf{x} - \mathbf{x}') n_n dS(\mathbf{x}') + b_i n_j \delta(\mathbf{S} - \mathbf{x}). \tag{7.8}$$

Integrating (7.8), the Mura formula (1963) is obtained as

$$\beta_{ji}(\mathbf{x}) = \int_L \epsilon_{jnh} C_{pqmn} G_{ip,q}(\mathbf{x}-\mathbf{x}') b_m \nu_h dl(\mathbf{x}'), \tag{7.9}$$

where $\boldsymbol{\nu}$ is the direction of the dislocation line L and dl is the dislocation line element. The equivalence of (7.8) and (7.9) is easily proved by the use of Stokes' theorem;

$$\int_L f\nu_h dl = \int_S \epsilon_{klh} f_{,l} n_k dS. \tag{7.10}$$

Applying (7.10) to (7.9) we have

$$\beta_{ji}(\mathbf{x}) = -\int_S \epsilon_{klh}\epsilon_{jnh} C_{pqmn} G_{ip,ql}(\mathbf{x}-\mathbf{x}') b_m n_k dS(\mathbf{x}'). \tag{7.11}$$

The minus sign comes from $(\partial/\partial x'_l)G_{ip,q} = -G_{ip,ql}$.

Since $\epsilon_{klh}\epsilon_{jnh} = \delta_{kj}\delta_{ln} - \delta_{kn}\delta_{lj}$,

$$\beta_{ji}(\mathbf{x}) = -\int_S (C_{pqmn} G_{ip,qn} b_m n_j - C_{pqmn} G_{ip,qj} b_m n_n) dS. \tag{7.12}$$

From (5.2), $C_{pqmn}G_{ip,qn}(\mathbf{x}-\mathbf{x}') = -\delta_{mi}\delta(\mathbf{x}-\mathbf{x}')$; therefore the first term in (7.12) is written as

$$\int_S \delta(\mathbf{x}-\mathbf{x}') b_i n_j dS(\mathbf{x}') = b_i n_j \delta(\mathbf{S}-\mathbf{x}). \tag{7.13}$$

The two Dirac's delta functions in (7.13) are different. $\delta(\mathbf{x}-\mathbf{x}')$ is the three dimensional delta function defined by (5.6), and $\delta(\mathbf{S}-\mathbf{x})$ is the one-dimensional delta function in the direction normal to S. Equation $\int_S \delta(\mathbf{x}-\mathbf{x}')dS(\mathbf{x}') = \delta(\mathbf{S}-\mathbf{x})$ can be regarded as the defining equation for $\delta(S-\mathbf{x})$. Now it can be seen that (7.12) equals (7.8). Systematic derivations of (7.9) are presented by Willis (1967) and Teodosiu (1970).

The stress components are

$$\sigma_{ij} = C_{ijkl}\beta_{lk}, \tag{7.14}$$

or, from (7.9),

$$\sigma_{ij}(\mathbf{x}) = C_{ijkl}\int_L \epsilon_{lnh} C_{pqmn} G_{kp,q}(\mathbf{x}-\mathbf{x}') b_m \nu_h dl(\mathbf{x}'). \tag{7.15}$$

The fact that the elastic distortion and stress fields can be expressed by line integrals along the dislocation line proves that these fields depend only on the boundary of S. On the other hand, the displacement field cannot be expressed by such a line integral, which means that the displacement field cannot be determined uniquely by only L. It can, however, be determined by knowing S, as shown in (7.6). If a history of the origin of dislocation (i.e., how it has been created) is given, the displacement field can also be expressed by a line integral along the dislocation line (see Section 38). Giving S and $\mathbf{b}$ for a dislocation is equivalent to giving a history of creation of the dislocation line.

The Fourier integral expression for (7.9) is

$$\beta_{ji}(\mathbf{x}) = \frac{i}{(2\pi)^3} \int_L \nu_h dl(\mathbf{x}') \int_{-\infty}^{\infty} \epsilon_{jnh} C_{pqmn} \xi_q N_{ip}(\boldsymbol{\xi}) D^{-1}(\boldsymbol{\xi}) b_m \times \exp\{i\boldsymbol{\xi} \cdot (\mathbf{x} - \mathbf{x}')\} d\boldsymbol{\xi} \tag{7.16}$$

which will be useful for further calculations (see Section 38).

It should be mentioned that $u_{i,j}(\mathbf{x}) = \beta_{ji}(\mathbf{x})$ when point $\mathbf{x}$ is not on S, and therefore (7.9) or (7.16) can be regarded as expressions for $u_{i,j}(\mathbf{x})$ at such a point.

**The Indenbom and Orlov formula*

The line integral expression for $\beta_{ji}(\mathbf{x})$ or $\sigma_{ij}(\mathbf{x})$ as shown in (7.9) or (7.15) is not unique. An alternate expression was obtained by Indenbom and Orlov (1968),

$$\beta_{ji}(\mathbf{x}) = \frac{1}{2} \int_L \nu_p \nu_q \frac{\partial^2}{\partial x_p \partial x_q} \beta_{ji}(\boldsymbol{\nu}; \mathbf{t}) dl(\mathbf{x}'), \tag{7.17}$$

where

$$\beta_{ji}(\boldsymbol{\nu}; \mathbf{t}) = \frac{i}{(2\pi)^3} \int_{-\infty}^{\infty} t_h ds \int_{-\infty}^{\infty} \epsilon_{jnh} C_{klmn} \xi_l N_{ik}(\boldsymbol{\xi}) D^{-1}(\boldsymbol{\xi}) b_m \times \exp\{i\boldsymbol{\xi} \cdot (\boldsymbol{\nu} - \mathbf{t}s)\} d\boldsymbol{\xi}, \tag{7.18}$$

and $\boldsymbol{\nu}$ is the unit tangent to L at $\mathbf{x}'$. The unit vector along $\mathbf{x}'\mathbf{x}$ is denoted by $\mathbf{t}$.

It can be seen from (7.16) that $\beta_{ji}(\boldsymbol{\nu}; \mathbf{t})$ is the elastic distortion at the end point of vector $\boldsymbol{\nu}$ caused by a straight, infinitely long dislocation connecting $\mathbf{x}$ and $\mathbf{x}'$ points (Fig. 7.2).

The interesting point of expression (7.17) is that the elastic stress field of curved dislocation line L can be expressed in terms of the stress field of a straight infinite dislocation. However, the appearance of the second derivatives in the integrand makes (7.17) less attractive unless it is modified as done by Asaro and Barnett (1976). This is discussed in Section 36. The equivalence of (7.17) and (7.9) has been proved by Asaro and Barnett (1976).

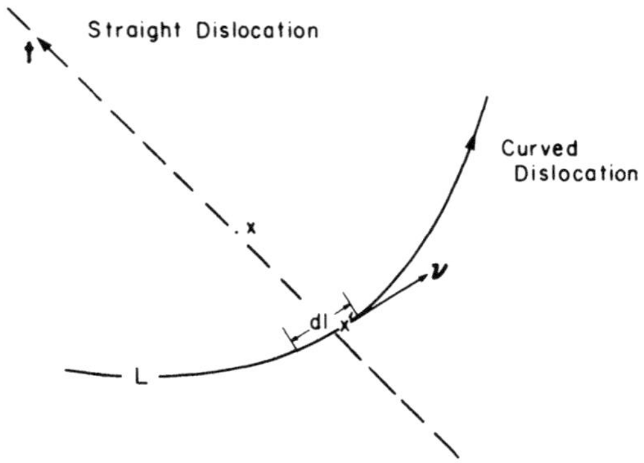

Fig. 7.2. Explanation of notations appearing the the Indenbom and Orlov formula.

*Disclinations

Volterra (1907) considered two types of dislocations: one is translational and the other one is rotational. The translational dislocation is the dislocation shown by Fig. 7.1. The rotational dislocation is called disclination and is shown in Fig. 7.3. The name "disclination" was coined by Frank (1958) in his study of cholesteric liquid crystals. Liquid crystal disclinations have been observed as twisting discontinuities, allowing discrete jumps of one half-pitch of the helicoidal texture (e.g., Orsay Liquid Crystal Group 1969, Kléman and Friedel 1969, Robinson, Ward and Beevers 1958, Cladis, Kléman and Pieranski 1971, Williams 1975). Träuble and Essman (1968) have made direct observations of disclinations in the lattice of flux lines in type II superconductors and have shown that disclinations cause strong elastic distortions and lattice bendings. Nabarro (1967, 1969) and Harris (1969) have extended the concept of the disclination to surface defects and suggested possible applications to biological structures such as some plasma membranes,

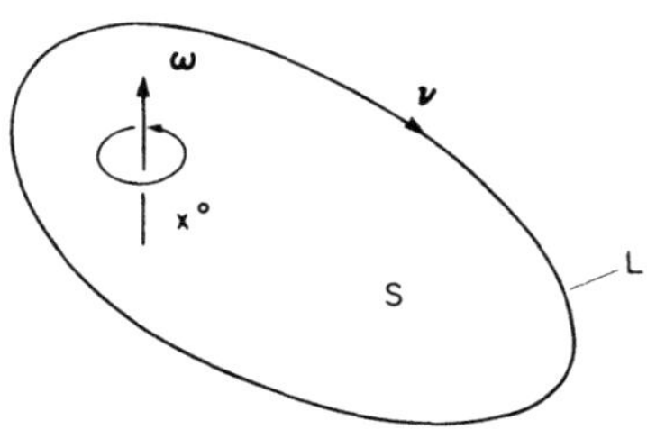

Fig. 7.3. Disclination L.

virus capsids and insect muscle (see also Harris and Scriven 1970, Harris and Thomas 1975, Chou and Ichikawa 1979 and Ichikawa and Chou 1979). Disclinations are also created by multiple twinning and have been observed in FCC crystals (Smith and Bechtoldt 1968). Li and Gilman (1970) have shown that disclinations are caused in polymers by chain kinking and twisting of molecules and that they play an important role in plastic deformation, internal friction, and glass transition.

The disclination (line) L in Fig. 7.3 is created by twisting surface S^+ against surface S^- by rotation angle $\boldsymbol{\omega}$ at point $\mathbf{x}^0$. Surfaces S^+ and S^- are defined by the direction $\boldsymbol{\nu}$ of the disclination line and the advancing rule of the right-handed screw, as employed for the definition of dislocations (see Fig. 7.1). Surface S^+ will be denoted by S. When vector $\boldsymbol{\omega}$ is normal to S, the disclination is of a twist type, and it is of a wedge type when $\boldsymbol{\omega}$ is on S.

The elastic field caused by a disclination is simulated by the eigenstress field caused by a suitable eigenstrain ϵ_{ij}^* distributed on S. Since the jump in the displacement components across S is

$$[u_i(\mathbf{x})] = \epsilon_{ipq}\omega_p(x_q - x_q^0), \tag{7.19}$$

the plastic distortion on S is defined by

$$\beta_{ji}^*(\mathbf{x}) = -\epsilon_{ipq}\omega_p(x_q - x_q^0)n_j\delta(\mathbf{S} - \mathbf{x}) \tag{7.20}$$

which is similar to the definition (7.3). The corresponding eigenstrain is defined by (7.2). When (7.20) is substituted into (6.1), where $\Omega = S$, we have

$$u_i(\mathbf{x}) = \int_S C_{jlmn}\epsilon_{mpq}\omega_p(x_q' - x_q^0)G_{ij,l}(\mathbf{x} - \mathbf{x}')n_n dS(\mathbf{x}') \tag{7.21}$$

and

$$u_{i,j}(\mathbf{x}) = \int_S C_{klmn}\epsilon_{mpq}\omega_p(x_q' - x_q^0)G_{ik,lj}(\mathbf{x} - \mathbf{x}')n_n dS(\mathbf{x}'). \tag{7.22}$$

$u_{i,j} - \beta_{ji}^* = \beta_{ji}$ cannot be fully expressed by a line integral along L as done in (7.9); however, the symmetric part, $\beta_{(ji)} \equiv (1/2)(\beta_{ji} + \beta_{ij})$, can be expressed by the line integral,

$$\begin{aligned}\beta_{(ji)} = &\int_L (\epsilon_{jnh}G_{ik,l}(\mathbf{x} - \mathbf{x}'))^s C_{klmn}\epsilon_{mpq}\omega_p(x_q' - x_q^0)\nu_n dl(\mathbf{x}') \\ &- \int_L (\epsilon_{jnh}G_{iklq}^*(\mathbf{x} - \mathbf{x}'))^s C_{klmn}\epsilon_{mpq}\omega_h\nu_p dl(\mathbf{x}'),\end{aligned} \tag{7.23}$$

where

$$G^*_{iklq}(\mathbf{x}) = -(2\pi)^{-3}\int_{-\infty}^{\infty} N_{ik}(\boldsymbol{\xi})D^{-1}(\boldsymbol{\xi})\xi_l\xi_q\xi^{-2}\exp(i\boldsymbol{\xi}\cdot\mathbf{x})d\boldsymbol{\xi}, \tag{7.24}$$

and $[\ ]^s$ is the symmetric part of $[\ ]$ with respect to i, j.

Mura (1972) found that

$$\begin{aligned} u_{i,ja}(\mathbf{x}) &= \int_L \epsilon_{jnh}C_{klmn}G_{ik,la}(\mathbf{x}-\mathbf{x}')\epsilon_{mpq}\omega_p(x'_q - x^0_q)\nu_h dl(\mathbf{x}') \\ &\quad - \int_L \epsilon_{jnh}C_{klmn}G_{ik,l}(\mathbf{x}-\mathbf{x}')\epsilon_{mab}\omega_h\nu_b dl(\mathbf{x}') \\ &\quad + \beta^*_{ji,a}(\mathbf{x}) + \epsilon_{jih}\phi^*_{ah}(\mathbf{x}), \end{aligned} \tag{7.25}$$

where

$$\phi^*_{ah}(\mathbf{x}) = -\omega_h n_a \delta(\mathbf{S}-\mathbf{x}). \tag{7.26}$$

ϕ^*_{ah} is called the plastic rotation. Contrary to the dislocation case, $u_{i,j}$ cannot be expressed by the line integral along L. One more derivative of $u_{i,j}$ is required. Quantities expressed by line integrals along L can be called state quantities since in a dislocated state of crystals the only measurable (visible) line imperfections are lines L of dislocations or disclinations. It can be said from (7.23) and (7.25) that $\beta_{(ji)}$ and $u_{i,ja} - \beta^*_{ji,a} - \epsilon_{jih}\phi^*_{ah}$ are state quantities. In the dislocation case, $u_{i,j} - \beta^*_{ji}$ is a state quantity. A derivation similar to (7.25) is presented by Kossecka and deWit (1977).

Formula (7.25) is useful for the derivation of the dislocation density tensor, α_{si}. Since $u_{i,ja} = u_{i,aj}$, it holds that

$$\epsilon_{sja}u_{i,ja} = 0. \tag{7.27}$$

When (7.25) is substituted into (7.27), we have

$$\begin{aligned} &\int_L C_{klma}G_{ik,la}\epsilon_{mpq}\omega_p(x'_q - x^0_q)dx'_s - \int_L C_{klms}G_{ik,la}\epsilon_{mpq}\omega_p(x'_q - x^0_q)dx'_a \\ &- \int_L C_{klma}G_{ik,l}\epsilon_{mab}\omega_s dx'_b + \int_L C_{klms}G_{ik,l}\epsilon_{mab}\omega_a dx'_b \\ &+ \epsilon_{sja}\beta^*_{ji,a} + \epsilon_{sja}\epsilon_{jih}\phi^*_{ah} = 0, \end{aligned} \tag{7.28}$$

where $dx'_s = \nu_s dl$. The first term becomes $-\int_L \delta_{mi}\delta(\mathbf{x}-\mathbf{x}')\epsilon_{mpq}\omega_p(x'_q - x^0_q)dx'_s$, where the property of Green's function is used. The second term becomes $-\int_L C_{klms}G_{ik,l}\epsilon_{mpq}\omega_p\delta_{qa}dx'_a$ by integrating by parts, and it cancels out with the fourth term. The third term is zero since $C_{klma}\epsilon_{mab} = 0$. If we define

$$\int_L \delta(\mathbf{x}-\mathbf{x}')dl(\mathbf{x}') = \delta(\mathbf{L}-\mathbf{x}) \tag{7.29}$$

as in (7.13), equation (7.28) can be written as

$$-\epsilon_{ipq}\omega_p(x_q - x_q^0)\nu_s\delta(\mathbf{L}-\mathbf{x}) + \epsilon_{sja}\beta^*_{ji,a} + \epsilon_{sja}\epsilon_{jih}\phi^*_{ah} = 0 \tag{7.30}$$

or

$$\alpha_{si} = -\epsilon_{saj}\beta^*_{ji,a} - \epsilon_{saj}\epsilon_{jih}\phi^*_{ah}, \tag{7.31}$$

where

$$\alpha_{si} = \epsilon_{ipq}\omega_p(x_q - x_q^0)\nu_s\delta(\mathbf{L}-\mathbf{x}). \tag{7.32}$$

Equation (7.31) states that the disclination line can be expressed in terms of the plastic distortion and rotation. Suppose that a crystal is distorted by many disclination lines. The smearing out of the density of disclination lines may be expressed in the same form as (7.31) with continuous functions of tensors, α_{si}, β^*_{ji}, and ϕ^*_{ah}. α_{si} is called the dislocation density tensor. The dislocation theory can be included in the theory of disclinations because $-\epsilon_{ipq}\omega_p x_q^0$ in (7.19) is a constant and can be replaced by the Burgers vector b_i. The dislocation theory is a special case where $\phi^*_{ah} = 0$. Then α_{si} becomes Nye's (1953) dislocation density tensor, and we have

$$\alpha_{si} = -\epsilon_{saj}\beta^*_{ji,a}. \tag{7.33}$$

The above equation was first obtained by Kröner (1955, 1956). In Section 37, (7.33) is derived by a different method.

The divergence of (7.31) leads to

$$\alpha_{si,s} = -\epsilon_{ijh}\theta_{jh}, \tag{7.34}$$

where

$$\theta_{jh} = -\epsilon_{saj}\phi^*_{ah,s}. \tag{7.35}$$

Anthony, et al. (1968) first found (7.34), and they defined θ_{ij} as the disclination density tensor. In dislocation theory the relation $\alpha_{hi,h} = 0$ is well established. It has the form of a continuity equation and has been interpreted to mean that dislocations cannot end inside the medium. deWit (1969, 1971) pointed out that (7.34) has the form of a continuity equation with a source or sink and that dislocations can start or end on disclinations (see also Das et al. 1973 and Marukawa 1974). The recent development of generalized continua (e.g., Kröner 1968) has stimulated the study of moments, transport law of a motor (Schaefer 1968), the curvature tensor of Cosserat continua, and the disclination density tensor (e.g., deWit 1972, 1973, Anthony 1970, Kröner and Anthony 1975 and Günther 1972). Elastic stress fields and strain energies of disclinations have been investigated by Huang and Mura (1970, 1972), T.W. Chou (1971), Kuo et al. (1972, 1973), Liu and Li (1971) and Minagawa (1977).

8. Dynamic solutions

The eigenstrain problems discussed in Section 3 are easily extended to dynamic problems.

Consider an elastic homogeneous material which is infinitely extended. The problem is to find the elastic field when prescribed eigenstrains ϵ_{ij}^* are given as functions of space and time. ϵ_{ij}^* is given by the Fourier integrals as

$$\epsilon_{ij}^*(\mathbf{x}, t) = \iint_{-\infty}^{\infty} \bar{\epsilon}_{ij}^*(\boldsymbol{\xi}, \omega) \exp\{i(\boldsymbol{\xi} \cdot \mathbf{x} - \omega t)\} d\boldsymbol{\xi} d\omega$$

$$\bar{\epsilon}_{ij}^*(\boldsymbol{\xi}, \omega) = (2\pi)^{-4} \iint_{-\infty}^{\infty} \epsilon_{ij}^*(\mathbf{x}, t) \exp\{-i(\boldsymbol{\xi} \cdot \mathbf{x} - \omega t)\} d\mathbf{x} dt. \tag{8.1}$$

The equations of motion are

$$\sigma_{ij,j} = \rho \ddot{u}_i, \tag{8.2}$$

where ρ is the density of the material. The stress-strain relations are

$$\sigma_{ij} = C_{ijkl}(u_{k,l} - \epsilon_{kl}^*). \tag{8.3}$$

We assume the solution of (8.2) to be

$$u_i(\mathbf{x}, t) = \iint_{-\infty}^{\infty} \bar{u}_i(\boldsymbol{\xi}, \omega) \exp\{i(\boldsymbol{\xi} \cdot \mathbf{x} - \omega t)\} d\boldsymbol{\xi} d\omega. \tag{8.4}$$

Substitution of these equations into (8.2) leads to

$$C_{ijkl}\bar{u}_k \xi_l \xi_j - \rho \bar{u}_i \omega^2 = -iC_{ijkl}\bar{\epsilon}_{kl}^* \xi_j. \tag{8.5}$$

By using the notation

$$\begin{aligned} K_{ik} &= C_{ijkl}\xi_j\xi_l, \\ X_i &= -iC_{ijkl}\bar{\epsilon}^*_{kl}\xi_j, \end{aligned} \tag{8.6}$$

$\bar{u}_i$ is obtained from (8.5) as

$$\bar{u}_i(\boldsymbol{\xi}, \omega) = X_j N_{ji}(\boldsymbol{\xi}, \omega)/D(\boldsymbol{\xi}, \omega), \tag{8.7}$$

where N_{ij} and D are the cofactor and the determinant of the matrix

$$\mathbf{K}(\boldsymbol{\xi}, \omega) = \begin{pmatrix} K_{11} - \rho\omega^2 & K_{12} & K_{13} \\ K_{21} & K_{22} - \rho\omega^2 & K_{23} \\ K_{31} & K_{32} & K_{33} - \rho\omega^2 \end{pmatrix} \tag{8.8}$$

Now the solution can be written as

$$u_i(\mathbf{x}, t) = -\iint_{-\infty}^{\infty} iC_{jlmn}\bar{\epsilon}^*_{mn}(\boldsymbol{\xi}, \omega)\xi_l N_{ij}(\boldsymbol{\xi}, \omega)D^{-1}(\boldsymbol{\xi}, \omega) \times \exp\{i(\boldsymbol{\xi}\cdot\mathbf{x} - \omega t)\}d\boldsymbol{\xi}d\omega. \tag{8.9}$$

If we define

$$G_{ij}(\mathbf{x}, t) = (2\pi)^{-4}\iint_{-\infty}^{\infty} N_{ij}(\boldsymbol{\xi}, \omega)D^{-1}(\boldsymbol{\xi}, \omega)\exp\{i(\boldsymbol{\xi}\cdot\mathbf{x} - \omega t)\}d\boldsymbol{\xi}d\omega, \tag{8.10}$$

(8.9) becomes

$$u_i(\mathbf{x}, t) = -\iint_{-\infty}^{\infty} C_{jlmn}\epsilon^*_{mn}(\mathbf{x}', t')G_{ij,l}(\mathbf{x} - \mathbf{x}', t - t')d\mathbf{x}'dt', \tag{8.11}$$

where the second equation in (8.1) has been used. $G_{ij}(\mathbf{x}, t)$ defined by (8.10) is called the dynamic Green's function. Willis (1965) used (8.11) for a body which spontaneously tends to undergo an eigenstrain ϵ^*_{mn}.

Uniformly moving edge dislocation

The plastic distortion of a uniformly moving edge dislocation with velocity v_1 in the x_1 direction is described in a form similar to (4.11),

$$\beta^*_{21}(\mathbf{x}, t) = b_1\delta(x_2)H(v_1 t - x_1), \tag{8.12}$$

where H is the Heaviside step function and $\epsilon^*_{12} = \epsilon^*_{21} = (1/2)\beta^*_{21}$. The Fourier transform of (8.12) becomes, from (8.1),

$$\bar{\beta}^*_{21} = -b_1\delta(\xi_3)\delta(\xi_1 v_1 + \omega)/(2\pi)^2 i\xi_1 \tag{8.13}$$

and $\bar{\epsilon}^*_{21} = \bar{\epsilon}^*_{12} = (1/2)\bar{\beta}^*_{21}$. Assume that the material is isotropic. Then, (8.9) becomes, by use of (8.13),

$$u_i(\mathbf{x}, t) = \iint_{-\infty}^{\infty} \frac{b\mu}{(2\pi)^2\xi_1} \frac{\xi_1 N_{i2}(\xi_1, \xi_2, 0; -\xi_1 v_1) + \xi_2 N_{i1}(\xi_1, \xi_2, 0; -\xi_1 v_1)}{D(\xi_1, \xi_2, 0; -\xi_1 v_1)} \times \exp[i\{\xi_1(x_1 - v_1 t) + \xi_2 x_2\}] d\xi_1 d\xi_2. \tag{8.14}$$

Therefore,

$$\begin{aligned} u_1 &= \frac{b_1}{(2\pi)^2}\int_{-\infty}^{\infty}\int_{-\infty}^{\infty} \frac{\xi_2}{\xi_1}\left[\frac{2c_1^2}{v_1^2}\{\xi_1^2(1 - v_1^2/c_2^2) + \xi_2^2\}^{-1} + \left(1 - \frac{2c_1^2}{v_1^2}\right)\right. \\ &\quad \left.\times\left\{\xi_1^2\left(1 - \frac{v_1^2}{c_1^2}\right) + \xi_2^2\right\}^{-1} \exp[i\{\xi_1(x_1 - v_1 t) + \xi_2 x_2\}]\right] d\xi_1 d\xi_2, \\ u_2 &= \frac{b_1}{(2\pi)^2}\int_{-\infty}^{\infty}\int_{-\infty}^{\infty} \left[\frac{2c_1^2}{v_1^2}\left(\frac{v_1^2}{c_2^2} - 1\right)\left\{\xi_1^2\left(1 - \frac{v_1^2}{c_2^2}\right) + \xi_2^2\right\}^{-1}\right. \\ &\quad \left. + \left(\frac{2c_1^2}{v_1^2} - 1\right)\left\{\xi_1^2\left(1 - \frac{v_1^2}{c_1^2}\right) + \xi_2^2\right\}^{-1} \exp[i\{\xi_1(x_1 - v_1 t) + \xi_2 x_2\}]\right] d\xi_1 d\xi_2, \\ u_3 &= 0, \end{aligned} \tag{8.15}$$

where $c_1 = (\mu/\rho)^{1/2}$ and $c_2 = \{(\lambda + 2\mu)/\rho\}^{1/2}$ are the shear and the dilatational wave velocities. The integrations in (8.15) are easily performed by use of (4.7). Then,

$$\begin{aligned} u_1 &= \frac{b_1 c_1^2}{\pi v_1^2}\left\{\tan^{-1}\frac{x_2(1 - v_1^2/c_2^2)^{1/2}}{x_1 - v_1 t} + \left(\frac{v_1^2}{2c_1^2} - 1\right)\tan^{-1}\frac{x_2(1 - v_1^2/c_1^2)^{1/2}}{x_1 - v_1 t}\right\}, \\ u_2 &= \frac{b_1 c_1^2}{\pi v_1^2}\left[-\frac{1 - v_1^2/2c_1^2}{(1 - v_1^2/c_1^2)^{1/2}}\log\left\{\frac{(x_1 - v_1 t)^2}{1 - v_1^2/c_1^2} + x_2^2\right\}^{1/2}\right. \\ &\quad \left. + (1 - v_1^2/c_2^2)^{1/2}\log\left\{\frac{(x_1 - v_1 t)^2}{1 - v_1^2/c_2^2} + x_2^2\right\}^{1/2}\right], \\ u_3 &= 0. \end{aligned} \tag{8.16}$$

The above result was first obtained by Eshelby (1949).

Uniformly moving screw dislocation

The plastic distortion of a uniformly moving screw dislocation with velocity v_1 in the x_1-direction is

$$\beta^*_{23}(\mathbf{x}, t) = b_3\delta(x_2)H(v_1 t - x_1) \tag{8.17}$$

where $\epsilon^*_{23} = \epsilon^*_{32} = (1/2)\beta^*_{23}$. In the same way the reader can easily obtain

$$u_1 = 0, \quad u_2 = 0,$$
$$u_3 = \frac{b_3}{2\pi} \tan^{-1} \frac{x_2(1 - v_1^2/c_1^2)^{1/2}}{x_1 - v_1 t} \tag{8.18}$$

which is the well-known solution found by Frank (1949).

*9. Dynamic Green's functions

The dynamic Green's function defined by (8.10) is also defined as the solution of the equation of motion

$$C_{ijkl}G_{km,lj}(\mathbf{x}, t) + \delta_{im}\delta(\mathbf{x})\delta(t) = \rho \ddot{G}_{im}(\mathbf{x}, t). \tag{9.1}$$

$G_{ij}(\mathbf{x}, t)$ is the elastic displacement component in the x_i-direction at point $\mathbf{x}$ and time t, produced by a unit impulsive force applied in the x_j-direction at point $\mathbf{x} = 0$ and time $t = 0$. G_{ij} is a symmetric tensor, and therefore the x_i- and x_j-directions in the above explanation can be interchanged. The solution of (9.1) is

$$G_{ij}(\mathbf{x}, t) = (2\pi)^{-4} \iint_{-\infty}^{\infty} N_{ij}(\boldsymbol{\xi}, \omega) D^{-1}(\boldsymbol{\xi}, \omega) \exp\{i(\boldsymbol{\xi} \cdot \mathbf{x} - \omega t)\} d\boldsymbol{\xi} d\omega, \tag{9.2}$$

where N_{ij} and D are the cofactor and the determinant of the matrix as defined by (8.8).

It is easy to show that (9.2) is the solution of (9.1). Write

$$G_{ij}(\mathbf{x}, t) = (2\pi)^{-4} \iint_{-\infty}^{\infty} \bar{G}_{ij}(\boldsymbol{\xi}, \omega) \exp\{i(\boldsymbol{\xi} \cdot \mathbf{x} - \omega t)\} d\boldsymbol{\xi} d\omega \tag{9.3}$$

and substitute it into (9.1). Then,

$$(K_{ik} - \rho\omega^2\delta_{ik})\bar{G}_{km} = \delta_{im}, \tag{9.4}$$

where

$$\delta(\mathbf{x})\delta(t) = (2\pi)^{-4} \iint_{-\infty}^{\infty} \exp\{i(\boldsymbol{\xi} \cdot \mathbf{x} - \omega t)\} d\boldsymbol{\xi} d\omega. \tag{9.5}$$

is used.

The solution of (9.4) is

$$\bar{G}_{ij}(\boldsymbol{\xi}, \omega) = N_{ij}(\boldsymbol{\xi}, \omega)/D(\boldsymbol{\xi}, \omega). \tag{9.6}$$

The explicit expressions for D and N_{ij} are

$$D(\boldsymbol{\xi}, \omega) = -(\rho\omega^2)^3 + (\rho\omega^2)^2 A - \rho\omega^2 B + C,$$
$$N_{ij}(\boldsymbol{\xi}, \omega) = K_{im}K_{mj} + (\rho\omega^2 - A)K_{ij} + \{(\rho\omega^2)^2 - \rho\omega^2 A + B\}\delta_{ij}, \tag{9.7}$$

where

$$A = K_{mm},$$
$$B = \epsilon_{mn1}K_{m2}K_{n3} + \epsilon_{mn2}K_{m3}K_{n1} + \epsilon_{mn3}K_{m1}K_{n2}, \tag{9.8}$$
$$C = \epsilon_{mnl}K_{m1}K_{n2}K_{l3}.$$

The roots of $D(\boldsymbol{\xi}, \omega) = 0$ with respect to $\rho\omega^2$ are denoted by $\rho\omega_1^2$, $\rho\omega_2^2$ and $\rho\omega_3^2$. Then we can write

$$D(\boldsymbol{\xi}, \omega) = (\rho\omega_1^2 - \rho\omega^2)(\rho\omega_2^2 - \rho\omega^2)(\rho\omega_3^2 - \rho\omega^2) \tag{9.9}$$

and therefore,

$$A = \rho\omega_1^2 + \rho\omega_2^2 + \rho\omega_3^2,$$
$$B = \rho^2\omega_1^2\omega_2^2 + \rho^2\omega_2^2\omega_3^2 + \rho^2\omega_3^2\omega_1^2, \tag{9.10}$$
$$C = \rho^3\omega_1^2\omega_2^2\omega_3^2.$$

It should be noted that $D(\boldsymbol{\xi}, \omega)$ becomes $D(\boldsymbol{\xi})$ when $\omega = 0$ and $D(\boldsymbol{\xi})$ is positive due to (9.10). $\bar{G}_{ij}$ can then be written as

$$\bar{G}_{ij}(\boldsymbol{\xi}, \omega) = \sum_{\nu=1}^{3} \frac{\overset{\nu}{\phi}_i\overset{\nu}{\phi}_j}{\rho\omega_\nu^2 - \rho\omega^2}, \tag{9.11}$$

where

$$\overset{1}{\phi}_i\overset{1}{\phi}_j = \frac{(K_{im} - \rho\omega_2^2\delta_{im})(K_{mj} - \rho\omega_3^2\delta_{mj})}{(\rho\omega_1^2 - \rho\omega_2^2)(\rho\omega_1^2 - \rho\omega_3^2)} \tag{9.12}$$

and $\overset{2}{\phi}_i\overset{2}{\phi}_j$, $\overset{3}{\phi}_i\overset{3}{\phi}_j$ are obtained by cyclic permutation of $(1, 2, 3)$. $\overset{\nu}{\phi}_i$ are eigenvectors of (8.8), and $\rho\omega_\nu$ are the corresponding eigenvalues,

$$(K_{ij} - \rho\omega_\nu^2\delta_{ij})\overset{\nu}{\phi}_j = 0,$$
$$\sum_{\nu=1}^{3} \overset{\nu}{\phi}_i\overset{\nu}{\phi}_j = \delta_{ij}. \tag{9.13}$$

It should be noted that $\overset{\nu}{\phi}_i\overset{\nu}{\phi}_j$ are homogeneous functions of $\boldsymbol{\xi}$ of degree 0, since $\rho\omega_\nu^2$ are homogeneous functions of $\boldsymbol{\xi}$ of degree 2, that is,

$$\begin{aligned}\rho\omega_\nu^2 &= \xi^2\rho\bar{\omega}_\nu^2,\\ \boldsymbol{\xi} &= \xi\bar{\boldsymbol{\xi}}.\end{aligned} \tag{9.14}$$

The integral (9.3) can be written as

$$G_{ij}(\mathbf{x},t)=\frac{1}{(2\pi)^4\rho}\int_{-\infty}^{\infty}\sum_{\nu=1}^{3}\frac{\exp(-i\omega t)}{\omega_\nu^2-\omega^2}\,d\omega\int_0^\infty\xi^2 d\xi$$

$$\times\int_{S^2}\overset{\nu}{\phi}_i(\bar{\boldsymbol{\xi}})\overset{\nu}{\phi}_j(\bar{\boldsymbol{\xi}})\exp(i\xi\bar{\boldsymbol{\xi}}\cdot\mathbf{x})dS(\bar{\boldsymbol{\xi}}) \tag{9.15}$$

where $\boldsymbol{\xi}=\xi\bar{\boldsymbol{\xi}}$ and S^2 is the unit sphere $|\bar{\boldsymbol{\xi}}|=1$. To integrate with respect to ω, one uses the Cauchy integral theorem. Since matrix (K_{ij}) is positive definite, the poles ω_ν are located on the real axes. For $t>0$, Γ is the semi-infinite half circle in the lower half-plane, and for $T<0$, Γ is the semi-infinite circle in the upper half-plane. With this Γ, the initial condition $G_{ij}=0$ for $t<0$ is satisfied. Thus, we have

$$G_{ij}(\mathbf{x},t)=\frac{H(t)}{(2\pi)^3\rho}\sum_{\nu=1}^{3}\int_0^\infty\xi^2 d\xi\,\frac{\sin(\omega_\nu t)}{\omega_\nu}\int_{S^2}\overset{\nu}{\phi}_i(\bar{\boldsymbol{\xi}})\overset{\nu}{\phi}_j(\bar{\boldsymbol{\xi}})\exp(i\xi\bar{\boldsymbol{\xi}}\cdot\mathbf{x})dS(\bar{\boldsymbol{\xi}}), \tag{9.16}$$

where $H(t)$ is the Heaviside step function. We introduce an auxiliary function $F_{ij}(\mathbf{x},t)$ defined as

$$G_{ij}(\mathbf{x},t)=-\frac{\partial^2}{\partial x_k\partial x_k}F_{ij}(\mathbf{x},t). \tag{9.17}$$

Then,

$$F_{ij}(\mathbf{x},t)=\frac{H(t)}{(2\pi)^3\rho}\sum_{\nu=1}^{3}\int_0^\infty d\xi\int_{S^2}\overset{\nu}{\phi}_i(\bar{\boldsymbol{\xi}})\overset{\nu}{\phi}_j(\bar{\boldsymbol{\xi}})\,\frac{\sin(\omega_\nu t)}{\omega_\nu}\exp(i\xi\bar{\boldsymbol{\xi}}\cdot\mathbf{x})dS(\bar{\boldsymbol{\xi}}). \tag{9.18}$$

Since the factor of $\exp(i\xi\bar{\boldsymbol{\xi}}\cdot\mathbf{x})$ in the integrand of (9.18) is an even function of $\bar{\boldsymbol{\xi}}$, it can be written as

$$F_{ij}(\mathbf{x},t)=\frac{H(t)}{(2\pi)^3\rho}\sum_{\nu=1}^{3}\int_0^\infty d\xi\,\frac{\sin(\bar{\omega}_\nu t\xi)}{\bar{\omega}_\nu\xi}\cos(\xi\bar{\boldsymbol{\xi}}\cdot\mathbf{x})\int_{S^2}\overset{\nu}{\phi}_i(\bar{\boldsymbol{\xi}})\overset{\nu}{\phi}_j(\bar{\boldsymbol{\xi}})dS(\bar{\boldsymbol{\xi}}), \tag{9.19}$$

where $\omega_\nu=\xi\bar{\omega}_\nu$. By using

$$\int_0^\infty\frac{\sin mx\cos nx}{2}\,dx=\frac{\pi}{4}\{\operatorname{sgn}(m+n)+\operatorname{sgn}(m-n)\}\quad\text{for } m,n>0, \tag{9.20}$$

we have

$$F_{ij}(\mathbf{x}, t) = \frac{H(t)}{(2\pi)^3\rho} \sum_{\nu=1}^{3} \int_{S^2} \frac{\pi}{4} \{\operatorname{sgn}(\bar{\omega}_\nu t + \bar{\boldsymbol{\xi}} \cdot \mathbf{x}) + \operatorname{sgn}(\bar{\omega}_\nu t - \bar{\boldsymbol{\xi}} \cdot \mathbf{x})\}$$
$$\times \frac{\overset{\nu}{\phi}_i(\bar{\boldsymbol{\xi}})\overset{\nu}{\phi}_j(\bar{\boldsymbol{\xi}})}{\bar{\omega}_\nu} dS(\bar{\boldsymbol{\xi}}). \tag{9.21}$$

Since

$$\frac{d}{du}(\operatorname{sgn} u) = 2\delta(u) \tag{9.22}$$

(9.17) becomes

$$G_{ij}(\mathbf{x}, t) = \frac{-H(t)}{8\pi^2\rho} \frac{\partial}{\partial t} \sum_{\nu=1}^{3} \int_{S^2} \delta(\bar{\omega}_\nu t - \bar{\boldsymbol{\xi}} \cdot \mathbf{x}) \overset{\nu}{\phi}_i(\bar{\boldsymbol{\xi}})\overset{\nu}{\phi}_j(\bar{\boldsymbol{\xi}})\bar{\omega}_\nu^{-2}(\bar{\boldsymbol{\xi}}) dS(\bar{\boldsymbol{\xi}}), \tag{9.23}$$

where

$$\int_{S^2} \delta'(\bar{\omega}_\nu t + \bar{\boldsymbol{\xi}} \cdot \mathbf{x}) \overset{\nu}{\phi}_i(\bar{\boldsymbol{\xi}})\overset{\nu}{\phi}_j(\bar{\boldsymbol{\xi}})\bar{\omega}_\nu^{-1} dS(\bar{\boldsymbol{\xi}}) = \int_{S^2} \delta'(\bar{\omega}_\nu t - \bar{\boldsymbol{\xi}} \cdot \mathbf{x})$$
$$\times \overset{\nu}{\phi}_i(\bar{\boldsymbol{\xi}})\overset{\nu}{\phi}_j(\bar{\boldsymbol{\xi}})\bar{\omega}_\nu^{-1} dS(\bar{\boldsymbol{\xi}}). \tag{9.24}$$

is used.

**Isotropic materials*

In this case,

$$\begin{aligned} K_{ij} &= (\lambda + \mu)\xi_i\xi_j + \mu\delta_{ij}\xi^2, \\ D(\boldsymbol{\xi}, \omega) &= (\mu\xi^2 - \rho\omega^2)^2\{(\lambda + 2\mu)\xi^2 - \rho\omega^2\}, \\ \rho\omega_1^2 &= (\lambda + 2\mu)\xi^2, \\ \rho\omega_2^2 &= \rho\omega_3^2 = \mu\xi^2, \end{aligned} \tag{9.25}$$

and from (9.12)

$$\overset{1}{\phi}_i\overset{1}{\phi}_j = \xi_i\xi_j/\xi^2, \tag{9.26}$$

and from the second equation in (9.13)

$$\overset{2}{\phi}_i\overset{2}{\phi}_j + \overset{3}{\phi}_i\overset{3}{\phi}_j = \frac{\delta_{ij}\xi^2 - \xi_i\xi_j}{\xi^2}. \tag{9.27}$$

Substitution of the above result into (9.23) leads to

$$G_{ij}(\mathbf{x}, t) = \frac{-H(t)}{8\pi^2\rho}\frac{\partial}{\partial t}\left\{\int_{S^2}\delta\left(\sqrt{\frac{\lambda+2\mu}{\rho}}\,t - \bar{\boldsymbol{\xi}}\cdot\mathbf{x}\right)\bar{\xi}_i\bar{\xi}_j\left(\frac{\lambda+2\mu}{\rho}\right)^{-1} dS(\bar{\boldsymbol{\xi}})\right.$$

$$\left. + \int_{S^2}\delta\left(\sqrt{\frac{\mu}{\rho}}\,t - \bar{\boldsymbol{\xi}}\cdot\mathbf{x}\right)(\delta_{ij} - \bar{\xi}_i\bar{\xi}_j)\left(\frac{\mu}{\rho}\right)^{-1} dS(\bar{\boldsymbol{\xi}})\right\}. \tag{9.28}$$

The following integrals are used:

$$\int_{S^2}\delta\left(\sqrt{\frac{\lambda+2\mu}{\rho}}\,t - \bar{\boldsymbol{\xi}}\cdot\mathbf{x}\right)\bar{\xi}_i\bar{\xi}_j dS(\bar{\boldsymbol{\xi}})$$

$$= \frac{\sqrt{\frac{\lambda+2\mu}{\pi}}}{\sqrt{\frac{\lambda+2\mu}{\pi}}}\int_{c_2t-x}^{c_2t+x}\delta(u)\bar{\xi}_i\bar{\xi}_j x^{-1} du\int_0^{2\pi} d\phi$$

$$= \left(\frac{1}{x}\right)H\left(x - \sqrt{\frac{\lambda+2\mu}{\rho}}\,t\right)\int_0^{2\pi}\bar{\xi}_i\bar{\xi}_j d\phi, \tag{9.29}$$

where

$$\sqrt{\frac{\lambda+2\mu}{\rho}}\,t - \bar{\boldsymbol{\xi}}\cdot\mathbf{x}x = u, \quad c_2 = \sqrt{\frac{\lambda+2\mu}{\rho}},$$

$$\mathbf{x} = x\bar{\mathbf{x}}, \tag{9.30}$$

$$-d(\bar{\boldsymbol{\xi}}\cdot\bar{\mathbf{x}})x = du.$$

When $\sqrt{\frac{\lambda+2\mu}{\rho}}\,t - \bar{\boldsymbol{\xi}}\cdot\mathbf{x} = 0$ we have

$$\bar{\xi}_i = \sqrt{\frac{\lambda+2\mu}{\rho}}\,t\bar{x}_i/x + \eta_i,$$

$$\int_0^{2\pi}\eta_i\eta_j d\phi = \pi(\delta_{ij} - \bar{x}_i\bar{x}_j)\left(1 - \frac{\lambda+2\mu}{\rho}\frac{t^2}{x^2}\right). \tag{9.31}$$

Therefore, when $\sqrt{\frac{\lambda+2\mu}{\rho}}\,t - \bar{\boldsymbol{\xi}}\cdot\mathbf{x} = 0$,

$$\int_0^{2\pi}\bar{\xi}_i\bar{\xi}_j d\phi = 2\pi\left(\sqrt{\frac{\lambda+2\mu}{\rho}}\,t/x\right)^2\bar{x}_i\bar{x}_j + \pi(\delta_{ij} - \bar{x}_i\bar{x}_j)\left(1 - \frac{\lambda+2\mu}{\rho}\frac{t^2}{x^2}\right). \tag{9.32}$$

Finally, we have

$$G_{ij}(\mathbf{x}, t) = \frac{-H(t)}{8\pi\rho x}\frac{\partial}{\partial t}\left[H\left(x - \sqrt{\frac{\lambda + 2\mu}{\rho}}\, t\right)\right.$$
$$\times\left\{(\delta_{ij} - \bar{x}_i\bar{x}_j)\Big/\frac{\lambda + 2\mu}{\rho} - \frac{t^2}{x^2}(\delta_{ij} - 3\bar{x}_i\bar{x}_j)\right\}$$
$$\left. + H\left(x - \sqrt{\frac{\mu}{\rho}}\, t\right)\left\{(\delta_{ij} + \bar{x}_i\bar{x}_j)\Big/\frac{\mu}{\rho} + \frac{t^2}{x^2}(\delta_{ij} - 3\bar{x}_i\bar{x}_j)\right\}\right] \tag{9.33}$$

or

$$G_{ij}(\mathbf{x}, t) = \frac{H(t)}{4\pi\rho x}\left[\delta\left(x - \sqrt{\frac{\lambda + 2\mu}{\rho}}\, t\right)\bar{x}_i\bar{x}_j\Big/\sqrt{\frac{\lambda + 2\mu}{\rho}}\right.$$
$$+ \delta\left(x - \sqrt{\frac{\mu}{\rho}}\, t\right)(\delta_{ij} - \bar{x}_i\bar{x}_j)\Big/\sqrt{\frac{\mu}{\rho}}$$
$$\left. + (\delta_{ij} - 3\bar{x}_i\bar{x}_j)(t/x^2)\left\{H\left(x - \sqrt{\frac{\lambda + 2\mu}{\rho}}\, t\right) - H\left(x - \sqrt{\frac{\mu}{\rho}}\, t\right)\right\}\right]. \tag{9.34}$$

The above result was first obtained by Stokes (1849). The static Green's function (5.8) can be obtained as

$$G_{ij}(\mathbf{x}) = \int_{-\infty}^{\infty} G_{ij}(\mathbf{x}, t - t')dt'$$
$$= \int_{-\infty}^{\infty} \frac{H(\bar{t})}{8\pi\rho x}\frac{\partial}{\partial t'}\left[H\left(x - \sqrt{\frac{\lambda + 2\mu}{\rho}}\, \bar{t}\right)\right.$$
$$\times\left\{(\delta_{ij} - \bar{x}_i\bar{x}_j)\Big/\frac{\lambda + 2\mu}{\rho} - \frac{\bar{t}^2}{x^2}(\delta_{ij} - 3\bar{x}_i\bar{x}_j)\right\}$$
$$\left. + H\left(x - \sqrt{\frac{\mu}{\rho}}\, \bar{t}\right)\left\{(\delta_{ij} + \bar{x}_i\bar{x}_j)\Big/\frac{\mu}{\rho} + \frac{\bar{t}^2}{x^2}(\delta_{ij} - 3\bar{x}_i\bar{x}_j)\right\}\right]dt'$$
$$= \frac{1}{8\pi\rho x}\left\{(\delta_{ij} - \bar{x}_i\bar{x}_j)\Big/\frac{\lambda + 2\mu}{\rho} + (\delta_{ij} + \bar{x}_i\bar{x}_j)\Big/\frac{\mu}{\rho}\right\}, \tag{9.35}$$

where $\bar{t} = t - t'$, $\partial/\partial t = -\partial/\partial t'$. The last result agrees with (5.8).

Steady-state

If a unit body force in the x_m-direction with time dependence $\exp(-i\omega t)$ is applied at the origin of coordinates, the ith component of a displacement field at point $\mathbf{x}$ can be expressed by $g_{im}(\mathbf{x})\exp(-i\omega t)$. $g_{im}(\mathbf{x})$ is called the steady-state elastic wave Green's function. It satisfies the equation of motion

$$C_{ijkl}g_{km,lj}(\mathbf{x}) + \rho\omega^2 g_{im}(\mathbf{x}) + \delta_{im}\delta(\mathbf{x}) = 0. \tag{9.36}$$

The solution can be written as

$$g_{ij}(\mathbf{x}) = (2\pi)^{-3}\int_{-\infty}^{\infty} \bar{G}_{ij}(\boldsymbol{\xi}, \omega)\exp(i\boldsymbol{\xi}\cdot\mathbf{x})d\boldsymbol{\xi}, \tag{9.37}$$

where $\bar{G}_{ij}(\boldsymbol{\xi}, \omega)$ has been defined by (9.6) or (9.11). For isotropic materials, (9.26) and (9.27) yield

$$\bar{G}_{ij}(\boldsymbol{\xi}, \omega) = \frac{\xi_i\xi_j/\xi^2}{(\lambda + 2\mu)(\xi^2 - \alpha^2)} + \frac{\delta_{ij} - \xi_i\xi_j/\xi^2}{\mu(\xi^2 - \beta^2)} \tag{9.38}$$

with

$$\alpha^2 = \rho\omega^2/(\lambda + 2\mu), \quad \beta^2 = \rho\omega^2/\mu. \tag{9.39}$$

Then, we have

$$g_{ij}(\mathbf{x}) = \frac{1}{4\pi\rho\omega^2 x}\left[\beta^2\delta_{ij}\exp(i\beta x) - x\frac{\partial^2}{\partial x_i\partial x_j}\left\{\frac{\exp(i\alpha x)}{x} - \frac{\exp(i\beta x)}{x}\right\}\right], \tag{9.40}$$

where $x = (\delta_{ij}x_ix_j)^{1/2}$.

10. Incompatibility

Eigenstresses are caused by the incompatibility of the eigenstrains. Uniform or linear distributions of eigenstrains throughout a free body (compatible eigenstrains) do not introduce any eigenstress in the material.

The incompatibility of ϵ_{ij}^* has been discussed by Reissner (1931), and Neményi (1931). Dislocations due to the incompatibility have been studied by Volterra (1907), Weingarten (1901), Cesáro (1906), and Moriguti (1947) from the viewpoint of elasticity theory in connection with the multiple-valueness of displacements and rotations. A plasticity theory viewpoint has been developed by Kondo (1955), Bilby (1960), and Kröner (1958). In this section we follow the work of Moriguti.

Let us consider an elastic body D which is free from external force but is subjected to a distribution of eigenstrain ϵ_{ij}^*. Imagine a slender rod AP inside the elastic body (see Fig. 10.1). If the rod were isolated from its surroundings and set free, it would change its form slightly and take on a form, such as AP', where one end of the rod, A, is fixed (see Fig.

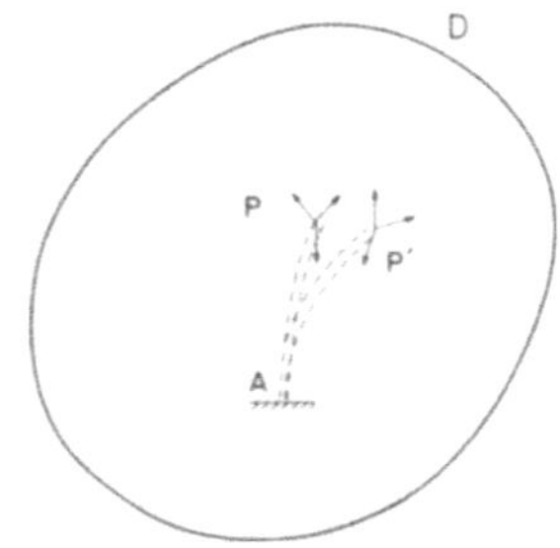

Fig. 10.1. Liberation of an imaginary slender rod AP in a continuum D gives deformation AP'.

10.1). If a Cartesian coordinate is attached at point P, it will rotate in a different direction. Both displacement $\bar{u}_i$ and the rotation $\bar{\omega}_i$ at P are caused by the release of elastic strain e_{ij} along AP.

Let us try to express these components of displacement and rotation as functions of e_{ij} in the form of line integrals along AP. Since the deformation of AP to AP' is completely elastic with displacement $\bar{u}_i$, we can write

$$e_{ij} = -\frac{1}{2}\left(\frac{\delta\bar{u}_i}{\delta x_j} + \frac{\delta\bar{u}_j}{\delta x_i}\right)$$
$$\bar{\omega}_{ij} = \epsilon_{ijk}\bar{\omega}_k = -\frac{1}{2}\left(\frac{\delta\bar{u}_j}{\delta x_i} - \frac{\delta\bar{u}_i}{\delta x_j}\right) \tag{10.1}$$

where $\delta\bar{u}_i$ is the change of $\bar{u}_i$ along a small line element $ds = (\delta x_j \delta x_j)^{1/2}$ along AP, ($\delta\bar{u}_i$ is caused by the release of ds). Since $\bar{u}_i$ is path-dependent, it does not have the usual derivatives $\bar{u}_{i,j}$ but $\delta\bar{u}_i/\delta x_j$ which are defined only along AP. The elastic strain components e_{ij} have been caused by ϵ^*_{ij}, which are piece-wise continuous functions with derivatives. The equations in (10.1) easily lead to

$$\delta\bar{u}_i = -(e_{ij} - \bar{\omega}_{ij})\delta x_j. \tag{10.2}$$

We have also

$$e_{kj,i} = -\frac{1}{2}\left(\frac{\partial}{\partial x_i}\frac{\delta\bar{u}_k}{\delta x_j} + \frac{\partial}{\partial x_i}\frac{\delta\bar{u}_j}{\delta x_k}\right),$$
$$e_{ki,j} = -\frac{1}{2}\left(\frac{\partial}{\partial x_j}\frac{\delta\bar{u}_k}{\delta x_i} + \frac{\partial}{\partial x_j}\frac{\delta\bar{u}_i}{\delta x_k}\right) \tag{10.3}$$

and therefore

$$(e_{kj,i} - e_{ki,j}) = -\frac{1}{2}\left(\frac{\partial}{\partial x_i}\frac{\delta\bar{u}_j}{\delta x_k} - \frac{\partial}{\partial x_j}\frac{\delta\bar{u}_i}{\delta x_k}\right), \tag{10.4}$$

where

$$\frac{\partial}{\partial x_i}\frac{\delta \bar{u}_k}{\delta x_j} = \frac{\partial}{\partial x_j}\frac{\delta \bar{u}_k}{\delta x_i} \tag{10.5}$$

is used.

Multiplying (10.4) by δx_k, we have

$$(e_{kj,i} - e_{ki,j})\delta x_k = \delta\bar{\omega}_{ij} \tag{10.6}$$

or

$$\delta\bar{\omega}_{ij} = \epsilon_{hij}\epsilon_{hlm}e_{km,l}\delta x_k. \tag{10.7}$$

From (10.7) and (10.2),

$$\begin{aligned} \bar{\omega}_{ij} &= \int_A^P \epsilon_{hij}\epsilon_{hlm}e_{km,l}dx_k, \\ \bar{u}_i &= -\int_A^P (e_{ij} - \bar{\omega}_{ij})dx_j, \end{aligned} \tag{10.8}$$

with the initial conditions

$$\bar{u}_i = \bar{\omega}_{ij} = 0 \quad \text{at } A. \tag{10.9}$$

Consider a special case where the end point P coincides with the starting point A (see Fig. 10.2). Then we have strain release along a closed curve c. Let us call the displacement and the rotation at the end point due to the strain release a dislocation and a disclination, respectively, as shown in Fig. 10.3. The end point shifts to A', and the coordinate system x_i (fixed to the end point) rotates to the coordinate

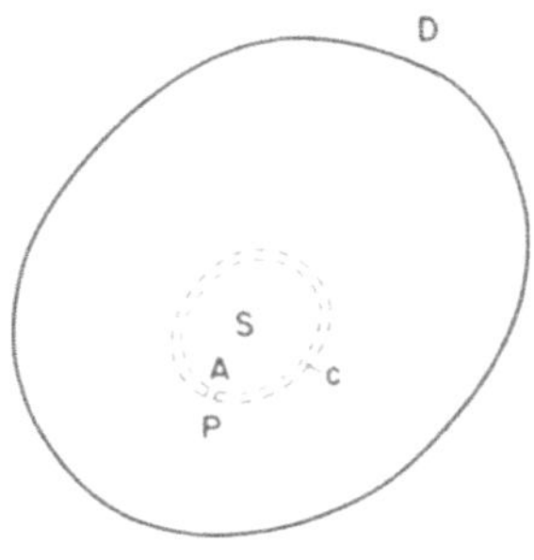

Fig. 10.2. End point P in Fig. 10.1 coincides with the starting point A. The closed loop and the surface bounded by the loop are denoted by c and S respectively.

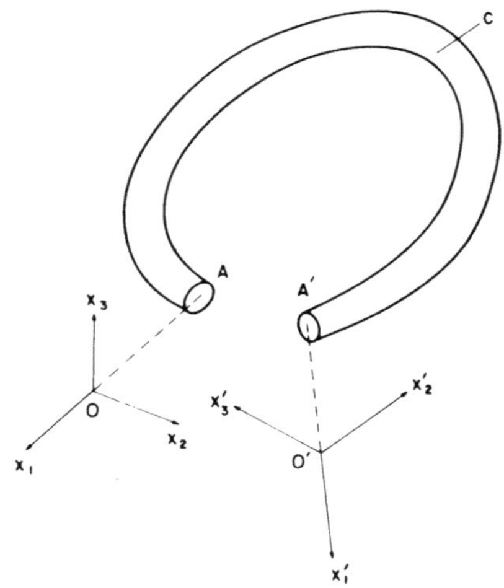

*Fig. 10.3. Dislocation and disclination are relative orientations between the **x** and **x'** coordinate systems.*

system x_i' after the release. The disclination is

$$\bar{\omega}_{ij} = \oint_c \epsilon_{hij}\epsilon_{hlm}e_{km,l}dx_k \tag{10.10}$$

and the dislocation is

$$\bar{u}_i = -\oint_c (e_{ij} - \bar{\omega}_{ij})dx_j. \tag{10.11}$$

The closed curve c is called the Burgers circuit. The integrations are defined on the closed curve c before the release. Applying Stokes' theorem to (10.10), we have

$$\bar{\omega}_{ij} = \iint_S \epsilon_{hij}R_{hq}n_q dS, \tag{10.12}$$

where n_q is the unit normal on dS, and

$$R_{hq} = \epsilon_{hlm}\epsilon_{qpk}e_{km,lp}. \tag{10.13}$$

The symmetric tensor R_{hq} is called the incompatibility tensor.

Before we apply Stokes' theorem, expression (10.11) is slightly modified. This is necessary since $\bar{\omega}_{ij}$ in the integrand (10.11) has no derivatives. Integration by parts of (10.11) and the definition of (10.7) lead to

$$\begin{aligned}\bar{u}_i &= -\oint_c e_{ij}dx_j + [\bar{\omega}_{ij}x_j]_A^P - \oint_c x_j\delta\bar{\omega}_{ij} \\ &= -\oint_c e_{ij}dx_j + \bar{\omega}_{ij}x_j^0 - \oint_c x_j\epsilon_{hij}\epsilon_{hlm}e_{km,l}dx_k,\end{aligned} \tag{10.14}$$

where x_j^0 is the coordinate of point $P(=A)$ in the x_i-coordinate before the release. The initial conditions (10.9) have been considered. The last expression of (10.14) is called Cesáro's integral (1906). By the use of Stokes' theorem, (10.14) can be simplified to

$$\bar{u}_i = \bar{\omega}_{ij}x_j^0 - \iint_S x_j\epsilon_{hij}R_{hq}n_q dS. \tag{10.15}$$

Since the components of the vector $\overrightarrow{00'}$ are the sum of $\bar{u}_i$ and the effect of the rotation $\bar{\omega}_{ij}$ at A', they can be expressed as

$$\bar{\xi}_i = \bar{u}_i - \bar{\omega}_{ij}x_j^0 = \bar{u}_i + \epsilon_{ijk}\bar{\omega}_j x_k^0. \tag{10.16}$$

Then,

$$\bar{\xi}_i = -\iint_S x_j\epsilon_{hij}R_{hq}n_q dS. \tag{10.17}$$

The result (10.16) corresponds to (7.19). $[u_i]$, $\epsilon_{ipq}\omega_p x_q$, $-\epsilon_{ipq}\omega_p x_q^0$ in (7.19) are equivalent to $\bar{\xi}_i$, $\epsilon_{ijk}\bar{\omega}_j x_k^0$, and $\bar{u}_i$ in (10.16), respectively, and c is the Burgers circuit.

An elastic body subjected to an applied load has a compatible elastic strain and $e_{ij} = \frac{1}{2}(u_{i,j} + u_{j,i})$. In this case, R_{hq} becomes zero when it is substituted into (10.13). Compatible elastic deformations do not cause any dislocation or disclination. If an elastic body has an internal stress (eigenstress) field due to an eigenstrain ϵ_{ij}^*, the dislocation and disclination defined by (10.17) and (10.12) are not zero. The incompatibility caused by ϵ_{ij}^* can be calculated by substitution of

$$e_{ij} = \tfrac{1}{2}(u_{i,j} + u_{j,i}) - \epsilon_{ij}^* \tag{10.18}$$

into (10.13). Then, we have

$$R_{hq} = -\epsilon_{hlm}\epsilon_{qpk}\epsilon_{km,lp}^*. \tag{10.19}$$

Equations (10.12) and (10.17) are expressions for disclination and dislocation as functions of the eigenstrains.

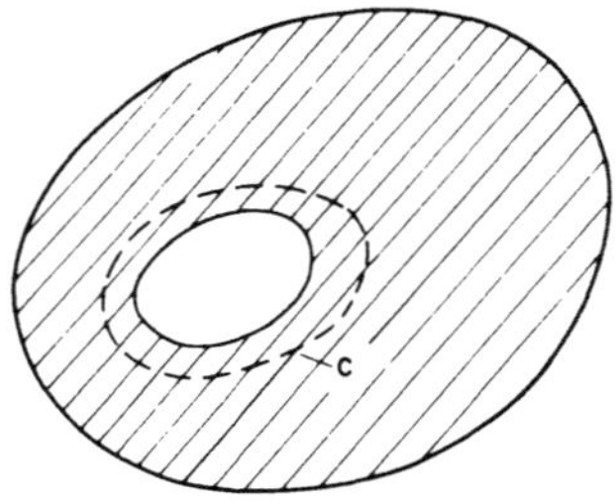

Fig. 10.4. In a multiply-connected material, the Burgers circuit c may contain a cavity.

The surface S in (10.12) and (10.17) should be bounded by the closed curve c and should consist of material points. When a multiply-connected material is considered and a closed curve contains a cavity (see Fig. 10.4), the line integrals (10.10) and (10.14) defined on c cannot be transformed into (10.12) and (10.17). Therefore, dislocations and disclinations can exist even if $R_{hq} = 0$. It is easily shown that any two closed curves surrounding the same cavity define the same dislocation and disclinations if $R_{hq} = 0$. The proof can be shown by applying Stokes' theorem to the domain bounded by the two curves. Examples are Volterra's dislocation (Fig. 10.5b) and disclination (Fig. 10.6b) which are made by welding from Figs. 10.5a and 10.6a.

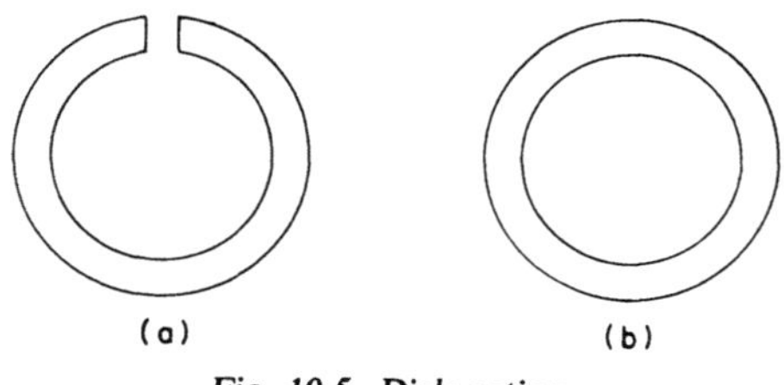

Fig. 10.5. Dislocation.

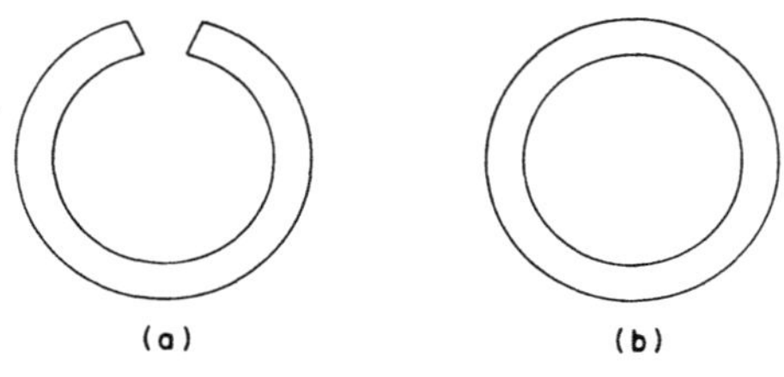

Fig. 10.6. Disclination.

∗Riemann-Christoffel curvature tensor

When the Burgers circuit is taken as an infinitesimal small loop, formula (10.17) can be written as

$$\bar{\xi}_i = -x_j \epsilon_{hij} R_{hq} n_q dS. \tag{10.20}$$

Furthermore, we put

$$n_q = \epsilon_{qmn} f_{mn}. \tag{10.21}$$

Then, (10.20) can be written as

$$\bar{\xi}_i = -x_j R_{mnji} f_{mn} dS, \tag{10.22}$$

where

$$R_{mnji} = \epsilon_{hij} \epsilon_{qmn} R_{hq}. \tag{10.23}$$

R_{mnji} is called the Riemann–Christoffel curvature tensor and R_{hq} the Ricci tensor. Although the eigenstrain ϵ^*_{ij} in (10.18) is incompatible in the three-dimensional Euclidean space when $R_{hq} \neq 0$, it might be compatible in the three-dimensional Riemann space. The curvature of the Riemann space is expressed by R_{mnji}. ϵ^*_{ij} can be considered as a strain in the natural state in the Riemann space with the fundamental metric tensor g_{ij}, that is,

$$\epsilon^*_{ij} = \tfrac{1}{2}(g_{ij} - \delta_{ij}). \tag{10.24}$$

When (10.19) is substituted into (10.23) by use of (10.24), we have

$$R_{mnji} = \tfrac{1}{2}(g_{mi,nj} + g_{nj,im} - g_{mj,ni} - g_{ni,mj}) \tag{10.25}$$

which is nothing more than the definition of the Riemann–Christoffel curvature tensor for small strains (see Sokolnikoff 1964, pps. 89 and 91).

In order to explain the physical meaning of the above result, let us consider a two-dimensional case for simplicity. Figure 10.7(a) shows a two-dimensional Euclidean space subjected to an eigenstress field due to a constraint eigenstrain ϵ^*_{ij}. This is the usual dislocated state of an imperfect crystal. Figure 10.7(*b*) is a two-dimensional Riemannian space with the fundamental metric tensor $2\epsilon^*_{ij} + \delta_{ij}$. This Riemannian space is an imaginary state of the material and is stress-free; it is called the natural state. The associate elastic stress and strain are necessary to make the natural state shown in Fig. 10.7(b) conform to the physical state shown in Fig. 10.7(a). The eigenstrain is compatible in Fig. 10.7(b), but it is constrained and incompatible in Fig. 10.7(a). The sum of the elastic strain and the eigenstrain is, of course, compatible in Fig. 10.7(a).

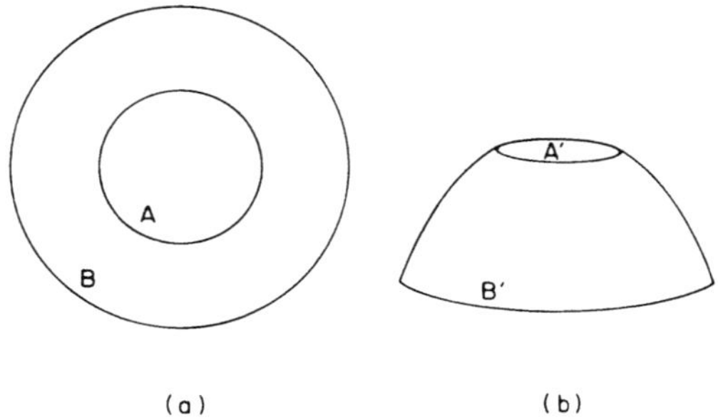

Fig. 10.7. Two-dimensional Euclidian space (a) and two-dimensional Riemannian space (b). Plastic deformation can be characterized by the transformation from (a) to (b).

It is important to note that the two-dimensional Riemannian space is embedded in the three-dimensional Euclidean space. Similarly, a three-dimensional Riemannian space can be embedded in a six-dimensional Euclidean space. If a deformation is defined from the strained state (Fig. 10.7(a) for the two-dimensional case) into the natural state (Fig. 10.7(b) for the two-dimensional case), the deformation can be characterized by a deviation from one space into another higher dimensional space, as seen in the buckling phenomena of elastic plates. Kondo (1949, 1955) used this buckling analogy in his explanation of the yielding phenomena of mild steel. He postulated that yielding is a buckling phenomenon of a three-dimensional Euclidean space into a six-dimensional Euclidean space (three-dimensional Riemannian space). The Riemann–Christoffel curvature tensor or the Euler–Schouten curvature tensor is the geometric object in Kondo's theory of plasticity. On the other hand, Bilby and his collaborators (1955, 1960) and Kröner and his collaborator (1956, 1958) adopted the distant parallelism in a linear connexion and described the Burgers vector by Schouten's torsion tensor. A more general expression for the Burgers vector has been given by Kondo (1953).

2

Isotropic inclusions

Explicit formulae are derived for elastic fields caused by inclusions. Most of the cases considered in this chapter are ellipsoidal inclusions in an isotropic infinite body, and the elastic moduli are the same for inclusions and matrices. The case when inclusions and matrices have different elastic moduli is treated in Chapter 4.

Eigenstrains in the inclusions are given by constants or by polynomials of coordinates. A special emphasis is placed on the case of constant eigenstrains.

11. Eshelby's solution

An ellipsoidal inclusion Ω is considered in an isotropic infinite body. Eigenstrains given in the ellipsoidal domain are assumed to be uniform (constant). The solution of the problem has been investigated by Goodier (1937) in the case when eigenstrains are thermal expansion strains. For general eigenstrains, the solution has been given by Eshelby (1957, 1959, 1961). Expressions for the solution are different for interior points (points inside the inclusion) and exterior points (points outside the inclusion). Eshelby's most valuable result is that the strain and stress fields become uniform for the interior points.

From (6.1) we have

$$u_i(\mathbf{x}) = -C_{jkmn}\epsilon^*_{mn}\int_\Omega G_{ij,k}(\mathbf{x}-\mathbf{x}')d\mathbf{x}', \tag{11.1}$$

where Ω is given by (see Fig. 11.1)

$$x_1^2/a_1^2 + x_2^2/a_2^2 + x_3^2/a_3^2 \leq 1 \tag{11.2}$$

and $G_{ij}(\mathbf{x}-\mathbf{x}')$ is, from (5.8),

$$G_{ij}(\mathbf{x}-\mathbf{x}') = \frac{1}{16\pi\mu(1-\nu)|\mathbf{x}-\mathbf{x}'|}\left[(3-4\nu)\delta_{ij} + \frac{(x_i - x_i')(x_j - x_j')}{|\mathbf{x}-\mathbf{x}'|^2}\right]. \tag{11.3}$$

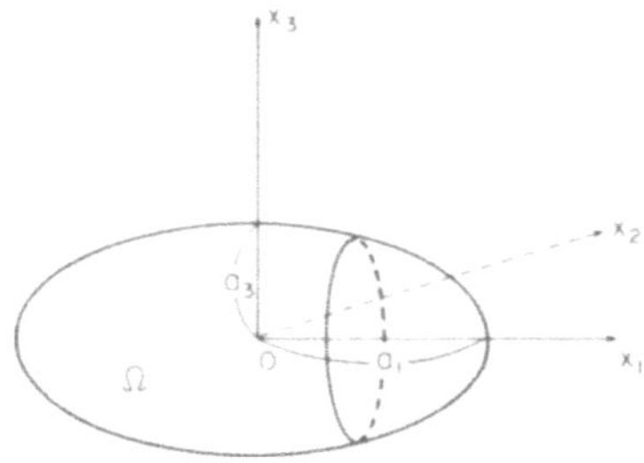

Fig. 11.1. An ellipsoidal inclusion with principal half axes a_1, a_2, and a_3.

After some manipulation we obtain

$$u_i(\mathbf{x}) = \frac{-\epsilon^*_{jk}}{8\pi(1-\nu)} \int_\Omega g_{ijk}(\mathbf{l}) \frac{d\mathbf{x}'}{|\mathbf{x}'-\mathbf{x}|^2}, \tag{11.4}$$

where

$$g_{ijk}(\mathbf{l}) = (1-2\nu)(\delta_{ij}l_k + \delta_{ik}l_j - \delta_{jk}l_i) + 3l_il_jl_k. \tag{11.5}$$

The vector $\mathbf{l}$ is a unit vector $(\mathbf{x}'-\mathbf{x})/|\mathbf{x}'-\mathbf{x}|$.

Interior points

When point $\mathbf{x}$ is located inside the inclusion, the integral in (11.4) is explicitly performed. As shown in Fig. 11.2, the volume element $d\mathbf{x}'$ in (11.4) can be written as

$$d\mathbf{x}' = drdS = drr^2d\omega, \tag{11.6}$$

where $r = |\mathbf{x}'-\mathbf{x}|$ and $d\omega$ is a surface element of a unit sphere Σ centered at point $\mathbf{x}$. Upon integration with respect to r, we have

$$u_i(\mathbf{x}) = \frac{-\epsilon^*_{jk}}{8\pi(1-\nu)} \int_\Sigma r(\mathbf{l})g_{ijk}(\mathbf{l})d\omega, \tag{11.7}$$

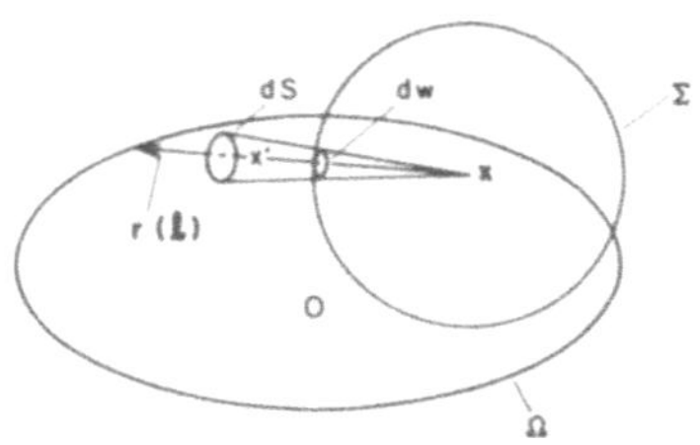

Fig. 11.2. Ω is an ellipsoidal inclusion, Σ is a unit sphere centered at point **x**.

where $r(\mathbf{l})$ is the positive root of

$$(x_1 + rl_1)^2/a_1^2 + (x_2 + rl_2)^2/a_2^2 + (x_3 + rl_3)^2/a_3^2 = 1, \tag{11.8}$$

that is,

$$r(\mathbf{l}) = -f/g + (f^2/g^2 + e/g)^{1/2}, \tag{11.9}$$

where

$$\begin{aligned} g &= l_1^2/a_1^2 + l_2^2/a_2^2 + l_3^2/a_3^2, \\ f &= l_1x_1/a_1^2 + l_2x_2/a_2^2 + l_3x_3/a_3^2, \\ e &= 1 - x_1^2/a_1^2 - x_2^2/a_2^2 - x_3^2/a_3^2. \end{aligned} \tag{11.10}$$

When (11.9) is inserted in (11.7), the term $(f^2/g^2 + e/g)^{1/2}$ can be omitted, since it is even in $\mathbf{l}$, while g_{ijk} is odd. To retain the advantages of suffix notation we introduce the vector

$$\lambda_1 = l_1/a_1^2, \quad \lambda_2 = l_2/a_2^2, \quad \lambda_3 = l_3/a_3^2. \tag{11.11}$$

Then,

$$u_i(\mathbf{x}) = \frac{x_m \epsilon_{jk}^*}{8\pi(1-\nu)} \int_\Sigma \frac{\lambda_m g_{ijk}}{g}\, d\omega \tag{11.12}$$

and the strain components become

$$\epsilon_{ij}(\mathbf{x}) = \frac{\epsilon_{mn}^*}{16\pi(1-\nu)} \int_\Sigma \frac{\lambda_i g_{jmn} + \lambda_j g_{imn}}{g}\, d\omega. \tag{11.13}$$

The integral in (11.13) is independent of $\mathbf{x}$. Therefore, we have an attractive conclusion in that the strain (and therefore the stress) is uniform inside the inclusion. The surface integrals of the type $\int_\Sigma l_1^m l_2^n l_3^k g^{-1} d\omega$ can be reduced to simple integrals according to the work by Routh (1895),

$$\begin{aligned} I_1 &= \int_\Sigma \frac{l_1^2 d\omega}{a_1^2 g} = 2\pi a_1 a_2 a_3 \int_0^\infty \frac{ds}{(a_1^2 + s)\Delta(s)}, \\ I_{11} &= \int_\Sigma \frac{l_1^4 d\omega}{a_1^4 g} = 2\pi a_1 a_2 a_3 \int_0^\infty \frac{ds}{(a_1^2 + s)^2\Delta(s)}, \\ I_{12} &= 3\int_\Sigma \frac{l_1^2 l_2^2 d\omega}{a_1^2 a_2^2 g} = 2\pi a_1 a_2 a_3 \int_0^\infty \frac{ds}{(a_1^2 + s)(a_2^2 + s)\Delta(s)}, \end{aligned} \tag{11.14}$$

with $\Delta(s) = (a_1^2 + s)^{1/2}(a_2^2 + s)^{1/2}(a_3^2 + s)^{1/2}$. The remaining coefficients are found by the simultaneous cyclic interchange of (1, 2, 3), (a_1, a_2, a_3) and (l_1, l_2, l_3).

It is convenient to write (11.13) as

$$\epsilon_{ij} = S_{ijkl}\epsilon^*_{kl}. \tag{11.15}$$

Then,

$$
\begin{aligned}
S_{ijkl} &= S_{jikl} = S_{ijlk},\\
S_{1111} &= \frac{3}{8\pi(1-\nu)}a_1^2 I_{11} + \frac{1-2\nu}{8\pi(1-\nu)}I_1,\\
S_{1122} &= \frac{1}{8\pi(1-\nu)}a_2^2 I_{12} - \frac{1-2\nu}{8\pi(1-\nu)}I_1,\\
S_{1133} &= \frac{1}{8\pi(1-\nu)}a_3^2 I_{13} - \frac{1-2\nu}{8\pi(1-\nu)}I_1,\\
S_{1212} &= \frac{a_1^2 + a_2^2}{16\pi(1-\nu)}I_{12} + \frac{1-2\nu}{16\pi(1-\nu)}(I_1 + I_2).
\end{aligned}
\tag{11.16}
$$

All other non-zero components are obtained by the cyclic permutation of $(1,2,3)$. The components which cannot be obtained by the cyclic permutation are zero; for instance, $S_{1112} = S_{1223} = S_{1232} = 0$. S_{ijkl} is called as Eshelby's tensor.

The integrals in (11.14) are expressed by the standard elliptic integrals (Gradshteyn and Ryzhik, 1965). Assuming $a_1 > a_2 > a_3$,

$$
\begin{aligned}
I_1 &= \frac{4\pi a_1 a_2 a_3}{(a_1^2 - a_2^2)(a_1^2 - a_3^2)^{1/2}}\{F(\theta,k) - E(\theta,k)\},\\
I_3 &= \frac{4\pi a_1 a_2 a_3}{(a_2^2 - a_3^2)(a_1^2 - a_3^2)^{1/2}}\left\{\frac{a_2(a_1^2 - a_3^2)^{1/2}}{a_1 a_3} - E(\theta,k)\right\},
\end{aligned}
\tag{11.17}
$$

where

$$
\begin{aligned}
&F(\theta,k) = \int_0^\theta \frac{dw}{(1-k^2\sin^2 w)^{1/2}}, \quad E(\theta,k) = \int_0^\theta (1-k^2\sin^2 w)^{1/2}dw,\\
&\theta = \sin^{-1}(1 - a_3^2/a_1^2)^{1/2}, \quad k = \{(a_1^2 - a_2^2)/(a_1^2 - a_3^2)\}^{1/2}.
\end{aligned}
\tag{11.18}
$$

The definitions for g and l and (11.14) yield the following formulae:

$$
\begin{aligned}
&I_1 + I_2 + I_3 = 4\pi,\\
&3I_{11} + I_{12} + I_{13} = 4\pi/a_1^2,\\
&3a_1^2 I_{11} + a_2^2 I_{12} + a_3^2 I_{13} = 3I_1,\\
&I_{12} = (I_2 - I_1)/(a_1^2 - a_2^2).
\end{aligned}
\tag{11.19}
$$

The last formula in (11.19) is obtained when we split the factor $(a_1^2+s)^{-1}(a_2^2+s)^{-1}$ in the integral (11.14) for I_{12} into partial fractions. Equations in (11.19) and their cyclic counterparts give sufficient relations to express I_i, I_{ij} in terms of I_1 and I_3.

Finally, the stress components become

$$
\begin{aligned}
\sigma_{11}/2\mu = &\left[\frac{a_1^2}{8\pi(1-\nu)}\left\{\frac{1-\nu}{1-2\nu}3I_{11}+\frac{\nu}{1-2\nu}(I_{21}+I_{31})\right\}\right.\\
&\left.+\frac{1-2\nu}{8\pi(1-\nu)}\left\{\frac{1-\nu}{1-2\nu}I_1-\frac{\nu}{1-2\nu}(I_2+I_3)\right\}-\frac{1-\nu}{1-2\nu}\right]\epsilon_{11}^*\\
&+\left[\frac{a_2^2}{8\pi(1-\nu)}\left\{\frac{1-\nu}{1-2\nu}I_{12}+\frac{\nu}{1-2\nu}(3I_{22}+I_{32})\right\}\right.\\
&\left.-\frac{1-2\nu}{8\pi(1-\nu)}\left\{\frac{1-\nu}{1-2\nu}I_1-\frac{\nu}{1-2\nu}(I_2-I_3)\right\}-\frac{\nu}{1-2\nu}\right]\epsilon_{22}^*\\
&+\left[\frac{a_3^2}{8\pi(1-\nu)}\left\{\frac{1-\nu}{1-2\nu}I_{13}+\frac{\nu}{1-2\nu}(3I_{33}+I_{23})\right\}\right.\\
&\left.-\frac{1-2\nu}{8\pi(1-\nu)}\left\{\frac{1-\nu}{1-2\nu}I_1-\frac{\nu}{1-2\nu}(I_3-I_2)\right\}-\frac{\nu}{1-2\nu}\right]\epsilon_{33}^*,\\
\sigma_{12}/2\mu = &\left\{\frac{a_1^2+a_2^2}{8\pi(1-\nu)}I_{12}+\frac{1-2\nu}{8\pi(1-\nu)}(I_1+I_2)-1\right\}\epsilon_{12}^*,
\end{aligned}
\tag{11.20}
$$

and other components are obtained by the cyclic permutation of (1, 2, 3). It is remarkable that the stress components depend only on Poisson's ration and the ratio of a_1, a_2, a_3 in addition to ϵ_{ij}^*.

The integrals (11.17) become elementary functions for special shapes of inclusions, as listed below.

Sphere $(a_1 = a_2 = a_3 = a)$

$$
\begin{aligned}
&I_1 = I_2 = I_3 = 4\pi/3,\\
&I_{11} = I_{22} = I_{33} = I_{12} = I_{23} = I_{31} = 4\pi/5a^2,\\
&S_{1111} = S_{2222} = S_{3333} = \frac{7-5\nu}{15(1-\nu)},\\
&S_{1122} = S_{2233} = S_{3311} = S_{1133} = S_{2211} = S_{3322} = \frac{5\nu-1}{15(1-\nu)},\\
&S_{1212} = S_{2323} = S_{3131} = \frac{4-5\nu}{15(1-\nu)}.
\end{aligned}
\tag{11.21}
$$

From (11.20) we have

$$\sigma_{11} = -\mu \frac{16}{15(1-\nu)} \epsilon^*_{11} - 2\mu \frac{5\nu+1}{15(1-\nu)} \epsilon^*_{22} - 2\mu \frac{5\nu+1}{15(1-\nu)} \epsilon^*_{33},$$

$$\sigma_{12} = -2\mu \frac{7-5\nu}{15(1-\nu)} \epsilon^*_{12}. \tag{11.21a}$$

All other stress components are obtained by the cyclic permutation of (1, 2, 3).

Elliptic cylinder ($a_3 \to \infty$)

$$I_1 = 4\pi a_2/(a_1+a_2), \quad I_2 = 4\pi a_1/(a_1+a_2), \quad I_3 = 0,$$
$$I_{12} = 4\pi/(a_1+a_2)^2, \quad 3I_{11} = 4\pi/a_1^2 - I_{12},$$
$$3I_{22} = 4\pi/a_2^2 - I_{12}, \quad I_{13} = I_{23} = I_{33} = 0,$$
$$a_3^2 I_{13} = I_1, \quad a_3^2 I_{23} = I_2, \quad a_3^2 I_{33} = 0.$$

$$S_{1111} = \frac{1}{2(1-\nu)} \left\{ \frac{a_2^2 + 2a_1a_2}{(a_1+a_2)^2} + (1-2\nu)\frac{a_2}{a_1+a_2} \right\},$$

$$S_{2222} = \frac{1}{2(1-\nu)} \left\{ \frac{a_1^2 + 2a_1a_2}{(a_1+a_2)^2} + (1-2\nu)\frac{a_1}{a_1+a_2} \right\}, \quad S_{3333} = 0,$$

$$S_{1122} = \frac{1}{2(1-\nu)} \left\{ \frac{a_2^2}{(a_1+a_2)^2} - (1-2\nu)\frac{a_2}{a_1+a_2} \right\}, \tag{11.22}$$

$$S_{2233} = \frac{1}{2(1-\nu)} \frac{2\nu a_1}{a_1+a_2}, \quad S_{3311} = 0,$$

$$S_{1133} = \frac{1}{2(1-\nu)} \frac{2\nu a_2}{a_1+a_2}, \quad S_{2211} = \frac{1}{2(1-\nu)} \left\{ \frac{a_1^2}{(a_1+a_2)^2} - (1-2\nu)\frac{a_1}{a_1+a_2} \right\},$$

$$S_{3322} = 0, \quad S_{1212} = \frac{1}{2(1-\nu)} \left\{ \frac{a_1^2 + a_2^2}{2(a_1+a_2)^2} + \frac{1-2\nu}{2} \right\},$$

$$S_{2323} = \frac{a_1}{2(a_1+a_2)}, \quad S_{3131} = \frac{a_2}{2(a_1+a_2)},$$

and

$$\sigma_{11} = \frac{\mu}{1-\nu} \left\{ -2 + \frac{a_2^2 + 2a_1a_2}{(a_1+a_2)^2} + \frac{a_2}{a_1+a_2} \right\} \epsilon^*_{11}$$
$$+ \frac{\mu}{1-\nu} \left\{ \frac{a_2^2}{(a_1+a_2)^2} - \frac{a_2}{a_1+a_2} \right\} \epsilon^*_{22} - \frac{2\mu\nu}{1-\nu} \frac{a_1}{a_1+a_2} \epsilon^*_{33}, \tag{11.22a}$$

$$\sigma_{22} = \frac{\mu}{1-\nu}\left\{-2 + \frac{a_1^2 + 2a_1a_2}{(a_1+a_2)^2} + \frac{a_1}{a_1+a_2}\right\}\epsilon^*_{22}$$

$$+\frac{\mu}{1-\nu}\left\{\frac{a_1^2}{(a_1+a_2)^2} - \frac{a_1}{a_1+a_2}\right\}\epsilon^*_{11} - \frac{2\mu\nu}{1-\nu}\frac{a_2}{a_1+a_2}\epsilon^*_{33},$$

$$\sigma_{33} = \frac{-2\mu\nu}{1-\nu}\frac{a_1}{a_1+a_2}\epsilon^*_{11} - \frac{2\mu\nu}{1-\nu}\frac{a_2}{a_1+a_2}\epsilon^*_{22} - \frac{2\mu}{1-\nu}\epsilon^*_{33},$$

$$\sigma_{12} = -\frac{2\mu}{1-\nu}\frac{a_1a_2}{(a_1+a_2)^2}\epsilon^*_{12}, \quad \sigma_{23} = -2\mu\frac{a_2}{a_1+a_2}\epsilon^*_{23},$$

$$\sigma_{31} = -2\mu\frac{a_1}{a_1+a_2}\epsilon^*_{31}.$$

Penny-shape $(a_1 = a_2 \gg a_3)$

$$I_1 = I_2 = \pi^2 a_3/a_1, \quad I_3 = 4\pi - 2\pi^2 a_3/a_1,$$

$$I_{12} = I_{21} = 3\pi^2 a_3/4a_1^3,$$

$$I_{13} = I_{23} = I_{31} = I_{32} = 3(\tfrac{4}{3}\pi - \pi^2 a_3/a_1)/a_1^2,$$

$$I_{11} = I_{22} = 3\pi^2 a_3/4a_1^3, \quad I_{33} = \frac{4\pi}{3}\Big/ a_3^2;$$

$$\begin{aligned}
&S_{1111} = S_{2222} = \frac{13-8\nu}{32(1-\nu)}\pi\frac{a_3}{a_1}, \quad S_{3333} = 1 - \frac{1-2\nu}{1-\nu}\frac{\pi}{4}\frac{a_3}{a_1},\\
&S_{1122} = S_{2211} = \frac{8\nu-1}{32(1-\nu)}\pi\frac{a_3}{a_1}, \quad S_{1133} = S_{2233} = \frac{2\nu-1}{8(1-\nu)}\pi\frac{a_3}{a_1},\\
&S_{3311} = S_{3322} = \frac{\nu}{1-\nu}\left(1 - \frac{4\nu+1}{8\nu}\pi\frac{a_3}{a_1}\right),\\
&S_{1212} = \frac{7-8\nu}{32(1-\nu)}\pi\frac{a_3}{a_1}, \quad S_{1313} = S_{2323} = \frac{1}{2}\left(1 + \frac{\nu-2}{1-\nu}\frac{\pi}{4}\frac{a_3}{a_1}\right),\\
&S_{kk11} = S_{kk22} = \frac{1-2\nu}{1-\nu}\frac{\pi}{4}\frac{a_3}{a_1} + \frac{\nu}{1-\nu}, \quad S_{kk33} = 1 - \frac{1-2\nu}{1-\nu}\frac{\pi}{2}\frac{a_3}{a_1};
\end{aligned} \tag{11.23}$$

and

$$\sigma_{11}/2\mu = \frac{-\nu}{1-\nu}(\epsilon^*_{11} + \epsilon^*_{22}) - \epsilon^*_{11} + \frac{13}{32(1-\nu)}\frac{\pi a_3}{a_1}\epsilon^*_{11} + \frac{16\nu-1}{32(1-\nu)}\frac{\pi a_3}{a_1}\epsilon^*_{22}$$

$$-\frac{2\nu+1}{8(1-\nu)}\frac{\pi a_3}{a_1}\epsilon^*_{33}$$

$$\sigma_{22}/2\mu = \frac{-\nu}{1-\nu}(\epsilon^*_{11}+\epsilon^*_{22}) - \epsilon^*_{22} + \frac{16\nu-1}{32(1-\nu)}\frac{\pi a_3}{a_1}\epsilon^*_{11} + \frac{13}{32(1-\nu)}\frac{\pi a_3}{a_1}\epsilon^*_{22}$$
$$-\frac{2\nu+1}{8(1-\nu)}\frac{\pi a_3}{a_1}\epsilon^*_{33} \tag{11.23a}$$

$$\sigma_{33}/2\mu = -\frac{2\nu+1}{8(1-\nu)}\frac{\pi a_3}{a_1}\epsilon^*_{11} - \frac{2\nu+1}{8(1-\nu)}\frac{\pi a_3}{a_1}\epsilon^*_{22} - \frac{1}{4(1-\nu)}\frac{\pi a_3}{a_1}\epsilon^*_{33}$$

$$\sigma_{23}/2\mu = \frac{\nu-2}{4(1-\nu)}\frac{\pi a_3}{a_1}\epsilon^*_{23}, \quad \sigma_{31}/2\mu = \frac{\nu-2}{4(1-\nu)}\frac{\pi a_3}{a_1}\epsilon^*_{31},$$

$$\sigma_{12}/2\mu = -\epsilon^*_{12} + \frac{7-8\nu}{16(1-\nu)}\frac{\pi a_3}{a_1}\epsilon^*_{12};$$

when $a_3 \to 0$

$$\begin{aligned}
&I_1 = I_2 = 0, \quad I_3 = 4\pi,\\
&I_{12} = 0, \quad I_{23} = 4\pi/a_2^2, \quad I_{31} = 4\pi/a_1^2,\\
&I_{11} = I_{22} = 0, \quad a_3^2 I_{33} = 4\pi/3,\\
&S_{2323} = S_{3131} = 1/2,\\
&S_{3311} = S_{3322} = \nu/(1-\nu),\\
&S_{3333} = 1, \quad \text{and all other } S_{ijkl} = 0,
\end{aligned} \tag{11.23b}$$

and

$$\begin{aligned}
&\sigma_{11} = -2\mu\left\{\frac{\nu}{1-\nu}(\epsilon^*_{11}+\epsilon^*_{22}) + \epsilon^*_{11}\right\},\\
&\sigma_{22} = -2\mu\left\{\frac{\nu}{1-\nu}(\epsilon^*_{11}+\epsilon^*_{22}) + \epsilon^*_{22}\right\}, \quad \sigma_{12} = -2\mu\epsilon^*_{12},\\
&\sigma_{33} = 0, \quad \sigma_{31} = 0, \quad \sigma_{32} = 0.
\end{aligned} \tag{11.23c}$$

Flat ellipsoid $(a_1 > a_2 \gg a_3)$

$$\begin{aligned}
&I_1 = 4\pi a_2 a_3\{F(k) - E(k)\}/(a_1^2 - a_2^2),\\
&I_2 = 4\pi a_3 E(k)/a_2 - 4\pi a_2 a_3\{F(k) - E(k)\}/(a_1^2 - a_2^2),\\
&I_3 = 4\pi - 4\pi a_3 E(k)/a_2,\\
&I_{12} = [4\pi a_3 E(k)/a_2 - 8\pi a_2 a_3\{F(k) - E(k)\}/(a_1^2 - a_2^2)]/(a_1^2 - a_2^2),\\
&I_{23} = [4\pi - 8\pi a_3 E(k)/a_2 + 4\pi a_2 a_3\{F(k) - E(k)\}/(a_1^2 - a_2^2)]/a_2^2,
\end{aligned}$$

$$I_{31} = [4\pi - 4\pi a_2 a_3\{F(k) - E(k)\}/(a_1^2 - a_2^2) - 4\pi a_3 E(k)/a_2]/a_1^2,$$

$$I_{33} = 4\pi/3a_3^2, \tag{11.24}$$

where $F(k)$ and $E(k)$ are the complete elliptic integrals of the first and the second kind, respectively,

$$E(k) = \int_0^{\pi/2} (1 - k^2 \sin^2 \phi)^{1/2} d\phi,$$

$$F(k) = \int_0^{\pi/2} (1 - k^2 \sin^2 \phi)^{-1/2} d\phi, \tag{11.25}$$

$$k^2 = (a_1^2 - a_2^2)/a_1^2.$$

For $\epsilon_{ij}^* = \epsilon_{33}^*$, (11.20) becomes

$$\sigma_{33}/2\mu = -a_3\epsilon_{33}^* E(k)/2(1-\nu)a_2. \tag{11.26}$$

For $\epsilon_{ij}^* = \epsilon_{13}^*$,

$$\sigma_{31}/2\mu = -a_3\epsilon_{13}^*\left[\frac{\nu}{1-\nu} a_2\{F(k) - E(k)\}/(a_1^2 - a_2^2) + E(k)/a_2\right]. \tag{11.27}$$

Oblate spheroid ($a_1 = a_2 > a_3$)

$$I_1 = I_2 = \frac{2\pi a_1^2 a_3}{(a_1^2 - a_3^2)^{3/2}}\left\{\cos^{-1}\frac{a_3}{a_1} - \frac{a_3}{a_1}\left(1 - \frac{a_3^2}{a_1^2}\right)^{1/2}\right\}, \quad I_3 = 4\pi - 2I_1,$$

$$I_{11} = I_{22} = I_{12}, \tag{11.28}$$

$$I_{12} = \pi/a_1^2 - I_{13}/4 = \pi/a_1^2 - \frac{(I_1 - I_3)}{4(a_3^2 - a_1^2)},$$

$$I_{13} = I_{23} = (I_1 - I_3)/(a_3^2 - a_1^2), \quad 3I_{33} = 4\pi/a_3^2 - 2I_{13}.$$

Prolate spheroid ($a_1 > a_2 = a_3$)

$$I_2 = I_3 = \frac{2\pi a_1 a_3^2}{(a_1^2 - a_3^2)^{3/2}}\left\{\frac{a_1}{a_3}\left(\frac{a_1^2}{a_3^2} - 1\right)^{1/2} - \cosh^{-1}\frac{a_1}{a_3}\right\},$$

$$I_1 = 4\pi - 2I_2, \quad I_{12} = (I_2 - I_1)/(a_1^2 - a_2^2), \quad 3I_{11} = 4\pi/a_1^2 - 2I_{12},$$

$$I_{22} = I_{33} = I_{23}, \tag{11.29}$$

$$3I_{22} = 4\pi/a_2^2 - I_{23} - (I_2 - I_1)/(a_1^2 - a_2^2),$$

$$I_{23} = \pi/a_2^2 - (I_2 - I_1)/4(a_1^2 - a_2^2).$$

Exterior points

When point $\mathbf{x}$ is located outside the inclusion Ω, a slightly different approach is employed. When expression (5.9) for G_{ij} is substituted into (11.1), we have

$$u_i(\mathbf{x}) = \frac{1}{8\pi(1-\nu)}\{\Psi_{jl,jli} - 2\nu\Phi_{mm,i} - 4(1-\nu)\Phi_{il,l}\}, \tag{11.30}$$

where

$$\Psi_{ij}(\mathbf{x}) = \epsilon^*_{ij}\int_\Omega |\mathbf{x}-\mathbf{x}'|d\mathbf{x}', \quad \Phi_{ij}(\mathbf{x}) = \epsilon^*_{ij}\int_\Omega 1/|\mathbf{x}-\mathbf{x}'|d\mathbf{x}'. \tag{11.31}$$

From (11.30) the induced strain field can be written as

$$\epsilon_{ij}(\mathbf{x}) = \frac{1}{8\pi(1-\nu)}\{\Psi_{kl,klij} - 2\nu\Phi_{kk,ij} - 2(1-\nu)(\Phi_{ik,kj} + \Phi_{jk,ki})\} \tag{11.32}$$

or

$$\epsilon_{ij}(\mathbf{x}) = D_{ijkl}(\mathbf{x})\epsilon^*_{kl}, \tag{11.33}$$

where

$$\begin{aligned} 8\pi(1-\nu)D_{ijkl}(\mathbf{x}) = {} & \psi_{,klij} - 2\nu\delta_{kl}\phi_{,ij} \\ & - (1-\nu)[\phi_{,kj}\delta_{il} + \phi_{,ki}\delta_{jl} + \phi_{,lj}\delta_{ik} + \phi_{,li}\delta_{jk}], \end{aligned} \tag{11.34}$$

and

$$\begin{aligned} \psi(\mathbf{x}) &= \int_\Omega |\mathbf{x}-\mathbf{x}'|d\mathbf{x}', \\ \phi(\mathbf{x}) &= \int_\Omega 1/|\mathbf{x}-\mathbf{x}'|d\mathbf{x}'. \end{aligned} \tag{11.35}$$

As will be shown in the next section, Ferrers (1877) and Dyson (1891) expressed the above integrals in terms of the following elliptic integrals:

$$\begin{aligned} I(\lambda) &= 2\pi a_1a_2a_3\int_\lambda^\infty \frac{ds}{\Delta(s)}, \\ I_i(\lambda) &= 2\pi a_1a_2a_3\int_\lambda^\infty \frac{ds}{(a_i^2+s)\Delta(s)}, \\ I_{ij}(\lambda) &= 2\pi a_1a_2a_3\int_\lambda^\infty \frac{ds}{(a_i^2+s)(a_j^2+s)\Delta(s)}, \end{aligned} \tag{11.36}$$

where $\Delta(s) = \{(a_1^2+s)(a_2^2+s)(a_3^2+s)\}^{1/2}$, and λ is the largest positive root of the equation

$$x_1^2/(a_1^2+\lambda)+x_2^2/(a_2^2+\lambda)+x_3^2/(a_3^2+\lambda)=1 \tag{11.37}$$

for an exterior point $\mathbf{x}$. $\lambda=0$ for points interior to Ω. The result is

$$\begin{aligned}\phi &= \tfrac{1}{2}[I(\lambda)-x_n x_n I_N(\lambda)],\\ \psi_{,i} &= \tfrac{1}{2}x_i\{I(\lambda)-x_n x_n I_N(\lambda)-a_I^2[I_I(\lambda)-x_n x_n I_{IN}(\lambda)]\},\end{aligned} \tag{11.38}$$

where the following summation convention has been used: repeated lower case indices are summed over from 1 to 3; upper case indices take on the same numbers as the corresponding lower case ones but are not summed. For example,

$$x_i x_i/(a_I^2+\lambda)=x_1^2/(a_1^2+\lambda)+x_2^2/(a_2^2+\lambda)+x_3^2/(a_3^2+\lambda), \tag{11.39}$$

$$\begin{aligned} x_k I_K &= x_1 I_1 \quad \text{where } k=1,\\ &= x_2 I_2 \quad \text{where } k=2,\\ \delta_{ij} I_J &= I_2 \quad\;\; \text{if } i=j=2,\\ &=0 \quad\;\;\;\; \text{if } i\neq j.\end{aligned}$$

Equation (11.34) requires higher derivates of ϕ and ψ. Since the lower bound of the integral limits in (11.36) is only a function of $\mathbf{x}$, the derivatives of I, I_i, I_{ij} can be reduced to the derivative of λ. From (11.37) we have

$$\lambda_{,l}=\frac{2x_l}{a_L^2+\lambda}\Big/\frac{x_i x_i}{(a_I^2+\lambda)^2}. \tag{11.40}$$

The following obvious relations are useful for further calculations: Consider the derivative

$$I_{i\cdots jk,p}(\lambda)=\frac{-2\pi a_1 a_2 a_3}{(a_i^2+\lambda)\cdots(a_j^2+\lambda)(a_k^2+\lambda)\Delta(\lambda)}\lambda_{,p}, \tag{11.40.1}$$

from which it follows that

$$I_{i\cdots jk,p}(\lambda)=\frac{1}{a_k^2+\lambda}I_{i\cdots j,p}(\lambda). \tag{11.40.2}$$

Using (11.39), it can be shown that

$$x_k x_k I_{i\cdots jK,p}(\lambda)=\frac{x_k x_k}{a_K^2+\lambda}I_{i\cdots j,p}(\lambda)=I_{i\cdots j,p}(\lambda). \tag{11.40.3}$$

It can be easily seen from (11.40.3) that

$$\frac{\partial}{\partial x_q}\{I_{ij\cdots k}(\lambda)-x_rx_rI_{Rij\cdots k}(\lambda)\}=-2x_qI_{Qij\cdots k}(\lambda). \tag{11.40.4}$$

The above result (11.40.4) is used for derivatives of (11.38). Then, we have

$$\phi_{,ij}=-\delta_{ij}I_I(\lambda)-x_iI_{I,J}(\lambda), \tag{11.40.5}$$

$$\begin{aligned}\psi_{,ijkl}=&-\delta_{ij}\delta_{kl}[I_K(\lambda)-a_I^2I_{IK}(\lambda)]-(\delta_{ik}\delta_{jl}+\delta_{jk}\delta_{il})[I_J(\lambda)\\&-a_I^2I_{IJ}(\lambda)]-\delta_{ij}x_k[I_K(\lambda)-a_I^2I_{IK}(\lambda)]_{,l}\\&-(\delta_{ik}x_j+\delta_{jk}x_i)[I_J(\lambda)-a_I^2I_{IJ}(\lambda)]_{,l}\\&-(\delta_{il}x_j+\delta_{jl}x_i)[I_J(\lambda)-a_I^2I_{IJ}(\lambda)]_{,k}\\&-x_ix_j[I_J(\lambda)-a_I^2I_{IJ}(\lambda)]_{,kl}.\end{aligned} \tag{11.40.6}$$

Finally, we have, from (11.34),

$$\begin{aligned}8\pi(1-\nu)D_{ijkl}(\mathbf{x})=&\,8\pi(1-\nu)S_{ijkl}(\lambda)+2\nu\delta_{kl}x_iI_{I,j}(\lambda)\\&+(1-\nu)\{\delta_{il}x_kI_{K,j}(\lambda)+\delta_{jl}x_kI_{K,i}(\lambda)+\delta_{ik}x_lI_{L,j}(\lambda)+\delta_{jk}x_lI_{L,i}(\lambda)\}\\&-\delta_{ij}x_k[I_K(\lambda)-a_I^2I_{KI}(\lambda)]_{,l}-(\delta_{ik}x_j+\delta_{jk}x_i)[I_J(\lambda)-a_I^2I_{IJ}(\lambda)]_{,l}\\&-(\delta_{il}x_j+\delta_{jl}x_i)[I_J(\lambda)-a_I^2I_{IJ}(\lambda)]_{,k}-x_ix_j[I_J(\lambda)-a_I^2I_{IJ}(\lambda)]_{,lk},\end{aligned} \tag{11.41}$$

where

$$\begin{aligned}8\pi(1-\nu)S_{ijkl}(\lambda)=&\,\delta_{ij}\delta_{kl}[2\nu I_I(\lambda)-I_K(\lambda)+a_I^2I_{KI}(\lambda)]\\&+(\delta_{ik}\delta_{jl}+\delta_{jk}\delta_{il})\{a_I^2I_{IJ}(\lambda)-I_J(\lambda)+(1-\nu)[I_K(\lambda)+I_L(\lambda)]\}.\end{aligned} \tag{11.42}$$

The above result holds for both exterior and interior points. For interior points, $\lambda=0$ and all derivatives of I_i and I_{ij} vanish. Then D_{ijkl} becomes $S_{ijkl}(0)$, which can be found to be equal to (11.16), where the relation (11.19) is used to establish this equivalency.

Recently, Tanaka and Mura (1981) have proposed an alternative method to evaluate the elastic field for exterior points. First, obtain the stress field for the points interior to Ω and denote it by σ_{ij}(in). Next, find the stress field for the exterior points, assuming that Ω is a void and the applied stress is $-\sigma_{ij}$(in). Denote the stress field by σ_{ij}(out). Then, the stress field for the exterior point in the original problem is the sum of σ_{ij}(out) and σ_{ij}(in).

Thermal expansion with central symmetry

The thermal stress problem in an isotropic infinite body under a temperature T symmetrical with respect to a point, has been well-known and its solution is found to be

$$u = \frac{1+\nu}{1-\nu} r^{-2} \int_0^r \alpha T r^2 dr \tag{11.43}$$

(e.g., Timoshenko and Goodier 1934, p. 417) where r is the distance from the center of symmetry, u is the displacement in the r direction, and α is the thermal expansion coefficient.

When T is constant inside a spherical domain Ω with radius a, and zero outside the sphere, we have

$$\epsilon_r = \epsilon_t = \frac{1+\nu}{3(1-\nu)} \alpha T \tag{11.44}$$

for interior points of Ω, and

$$\epsilon_r = -\frac{2}{3} \frac{1+\nu}{1-\nu} \frac{a^3}{r^3} \alpha T, \quad \epsilon_t = \frac{1}{3} \frac{1+\nu}{1-\nu} \frac{a^3}{r^3} \alpha T, \tag{11.45}$$

for exterior points, where ϵ_r and ϵ_t are the radial and tangential strains.

It will be shown that the same result can be obtained from (11.33) with (11.41), where $\epsilon^*_{ij} = \delta_{ij} \alpha T$.

The following calculation is straightforward. From (11.37), $\lambda = r^2 - a^2$. From (12.19) in the next section, $I_1(\lambda) = I_2(\lambda) = I_3(\lambda) = 4\pi a^3/3r^3$, $I_{ij}(\lambda) = 4\pi a^3/5r^5$. For interior points, $\lambda = 0$ and $I_1 = I_2 = I_3 = 4\pi/3$, $I_{ij} = 4\pi/5a^2$. When these results are substituted into (11.41) and (11.42), we have (11.44) or (11.45), depending on the position of $\mathbf{x}$.

*12. Ellipsoidal inclusions with polynomial eigenstrains

An ellipsoidal inclusion is considered in an isotropic infinite body. Eigenstrains in the ellipsoidal domain are given in the form of polynomials of coordinates such as

$$\epsilon^*_{ij}(\mathbf{x}) = B_{ijk} x_k + B_{ijkl} x_k x_l + \cdots, \tag{12.1}$$

where B_{ijk}, $B_{ijkl}, \ldots$ are constants symmetric with respect to the free indices i, j and $B_{ijkl} = B_{ijlk}$, $B_{ijklm} = B_{ijkml}$, etc. The constant term is excluded in (12.1) since it has been discussed already in the last section.

In this section, we follow to a large extent the Ph.D. theses of Sendeckyj (1967) and Moschovidis (1975).

Both displacement fields for interior and exterior points of Ω can be expressed by (11.30),

$$u_i(\mathbf{x}) = \frac{1}{8\pi(1-\nu)}\{\Psi_{jl,jli} - 2\nu\Phi_{mm,i} - 4(1-\nu)\Phi_{il,l}\}, \tag{12.2}$$

where

$$\Psi_{ij}(\mathbf{x}) = \int_\Omega |\mathbf{x}-\mathbf{x}'|\epsilon^*_{ij}(\mathbf{x}')d\mathbf{x}' \tag{12.3}$$

and

$$\Phi_{ij}(\mathbf{x}) = \int_\Omega \frac{\epsilon^*_{ij}(\mathbf{x}')}{|\mathbf{x}-\mathbf{x}'|}d\mathbf{x}'. \tag{12.4}$$

$\Phi_{ij}(\mathbf{x})$ and $\Psi_{ij}(\mathbf{x})$ are the harmonic and biharmonic potentials due to a body Ω of density $\epsilon^*_{ij}(\mathbf{x}')$, and have the following properties:

$$\Psi_{ij,kk} = 2\Phi_{ij} \tag{12.5}$$

and

$$\Psi_{ij,kkll} = 2\Phi_{ij,ll} = \begin{cases} -8\pi\epsilon^*_{ij}(\mathbf{x}) & \text{for } \mathbf{x} \text{ in } \Omega, \\ 0 & \text{for } \mathbf{x} \text{ out of } \Omega. \end{cases} \tag{12.6}$$

The strain field can be expressed by (11.32). When (12.1) is substituted into (12.3) and (12.4), we have

$$\Psi_{ij} = B_{ijk}\psi_k + B_{ijkl}\psi_{kl} + \cdots \tag{12.7}$$

and

$$\Phi_{ij} = B_{ijk}\phi_k + B_{ijkl}\phi_{kl} + \cdots, \tag{12.8}$$

where the following definitions have been made:

$$\begin{aligned} \psi_{ij\cdots k}(\mathbf{x}) &= \int_\Omega x'_i x'_j \cdots x'_k |\mathbf{x}-\mathbf{x}'| d\mathbf{x}', \\ \phi_{ij\cdots k}(\mathbf{x}) &= \int_\Omega \frac{x'_i x'_j \cdots x'_k d\mathbf{x}'}{|\mathbf{x}-\mathbf{x}'|}. \end{aligned} \tag{12.9}$$

ψ and ϕ have the same definitions as (11.35). Substitution of (12.7) and (12.8) into (11.32) leads to

$$\epsilon_{ij}(\mathbf{x}) = D_{ijklq}(\mathbf{x})B_{klq} + D_{ijklqr}(\mathbf{x})B_{klqr} + \cdots, \tag{12.10}$$

where

$$8\pi(1-\nu)D_{ijklq} = \psi_{q,klij} - 2\nu\delta_{kl}\phi_{q,ij} - (1-\nu)[\phi_{q,kj}\delta_{il} + \phi_{q,ki}\delta_{jl} + \phi_{q,lj}\delta_{ik} + \phi_{q,li}\delta_{jk}], \qquad (12.11)$$

etc.

The expression for D_{ijklqr} is obtained from the above by replacing ψ_q and ϕ_q by ψ_{qr} and ϕ_{qr}, respectively.

As seen in the next subsection, the harmonic potentials in (12.9) can be written in terms of the following elliptic integrals:

$$V(\mathbf{x}) = \pi a_1 a_2 a_3 \int_\lambda^\infty \frac{U(s)}{\Delta(s)}\,ds,$$

$$V_i(\mathbf{x}) = \pi a_1 a_2 a_3 \int_\lambda^\infty \frac{U(s)}{(a_I^2+s)\Delta(s)}\,ds, \qquad (12.12)$$

$$V_{ij}(\mathbf{x}) = \pi a_1 a_2 a_3 \int_\lambda^\infty \frac{U(s)}{(a_I^2+s)(a_J^2+s)\Delta(s)}\,ds,$$

etc.,

where $\Delta(s) = \{(a_1^2+s)(a_2^2+s)(a_3^2+s)\}^{1/2}$, $U(s) = 1 - \{x_1^2/(a_1^2+s) + x_2^2/(a_2^2+s) + x_3^2/(a_3^2+s)\}$, and λ is the largest positive root of (11.37) for an exterior point $\mathbf{x}$. $\lambda = 0$ for points interior to Ω. According to Dyson (1891), (12.9) can be written as

$$\phi = V,$$

$$\phi_n = a_N^2 x_n V_N, \qquad (12.13)$$

$$\phi_{mn} = a_M^2\{x_m x_n a_N^2 V_{MN} + \tfrac{1}{4}\delta_{mn}[V - x_r x_r V_R - a_M^2(V_M - x_r x_r V_{RM})]\},$$

and

$$\psi_{n,i} = -\tfrac{1}{4}\delta_{in}a_N^2\{(V - x_r x_r V_R) - a_N^2(V_N - x_r x_r V_{RN})\} + a_N^2 x_n x_i (V_I - a_N^2 V_{IN}), \qquad (12.14)$$

$$\psi_{mn,i} = a_M^2 a_N^2\{-\tfrac{1}{4}(\delta_{mi}x_n + \delta_{ni}x_m)[V_M - x_r x_r V_{RM} - a_N^2(V_{NM} - x_r x_r V_{RMN})] + x_m x_n x_i (V_{MN} - a_I^2 V_{IMN})\} + \tfrac{1}{4}a_M^2\delta_{mn}x_i\{(V - x_r x_r V_R) - (a_I^2 + a_M^2)(V_M - x_r x_r V_{RM}) + a_I^4(V_{MI} - x_r x_r V_{RMI})\}.$$

In the above expressions, the summation convention demonstrated by (11.39) had been used. By definition, the V-integrals in (12.12) and the

I-integrals in (11.36) are related,

$$\begin{aligned}
V(\mathbf{x}) &= \tfrac{1}{2}\{I(\lambda) - x_r x_r I_R(\lambda)\},\\
V_i(\mathbf{x}) &= \tfrac{1}{2}\{I_i(\lambda) - x_r x_r I_{Ri}(\lambda)\},\\
V_{ij}(\mathbf{x}) &= \tfrac{1}{2}\{I_{ij}(\lambda) - x_r x_r I_{Rij}(\lambda)\},
\end{aligned} \tag{12.15}$$

etc.

When $\mathbf{x}$ is an interior point of Ω, then $\lambda = 0$, I, I_i, $I_{ij}, \ldots$ are constants, and all their derivatives vanish. Thus, the first derivatives of the V-integrals are linear functions of $\mathbf{x}$, and the second derivatives are constants. Therefore, $D_{ijklq}(\mathbf{x})$ becomes a homogeneous linear function of $\mathbf{x}$ and $D_{ijklqr}(\mathbf{x})$ becomes a sum of constants and quadratic functions of $\mathbf{x}$. That is, for interior points $\mathbf{x}$,

$$\begin{aligned}
D_{ijklq}(\mathbf{x}) &= D_{ijklq,m}(0)x_m,\\
D_{ijklqr}(\mathbf{x}) &= D_{ijklqr}(0) + \tfrac{1}{2}D_{ijklqr,mn}(0)x_m x_n.
\end{aligned} \tag{12.16}$$

The $\mathbf{x}$ dependency of higher order $D_{ijkl\ldots}$ can be similarly investigated, and it is concluded that $\epsilon_{ij}(\mathbf{x})$ for interior points becomes an inhomogeneous polynomial of $\mathbf{x}$ whose terms are of degree $n, (n-2), (n-4), \ldots$, if $\epsilon^*_{kl}(\mathbf{x})$ is a homogeneous polynomial of $\mathbf{x}$ with degree n. This general conclusion was first pointed out by Eshelby (1961). Asaro and Barnett (1975) reached the same conclusion for anisotropic materials.

**The I-integrals*

The I-integrals in (11.36) can be expressed by the first and second elliptic integrals defined by (11.18); for $a_1 > a_2 > a_3$,

$$\begin{aligned}
I(\lambda) &= 4\pi a_1 a_2 a_3 (a_1^2 - a_3^2)^{-1/2} F(\theta, k),\\
I_1(\lambda) &= 4\pi a_1 a_2 a_3 (a_1^2 - a_2^2)^{-1} (a_1^2 - a_3^2)^{-1/2} \{F(\theta, k) - E(\theta, k)\},\\
I_2(\lambda) &= 4\pi a_1 a_2 a_3 \{(a_1^2 - a_3^2)^{1/2} (a_1^2 - a_2^2)^{-1} (a_2^2 - a_3^2)^{-1} E(\theta, k)\\
&\quad - (a_1^2 - a_2^2)^{-1} (a_1^2 - a_3^2)^{-1/2} F(\theta, k)\\
&\quad - (a_2^2 - a_3^2)^{-1} (a_3^2 + \lambda)^{1/2} (a_1^2 + \lambda)^{-1/2} (a_2^2 + \lambda)^{-1/2}\},\\
I_3(\lambda) &= 4\pi a_1 a_2 a_3 (a_2^2 - a_3^2)^{-1} (a_1^2 - a_3^2)^{-1/2} \{(a_2^2 + \lambda)^{1/2} (a_1^2 - a_3^2)^{1/2}\\
&\quad \times (a_3^2 + \lambda)^{-1/2} (a_1^2 + \lambda)^{-1/2} - E(\theta, k)\},
\end{aligned} \tag{12.17}$$

where F, E, θ and k have been defined in (11.18).

The other higher order I-integrals can be expressed by I_i, using the following relations:

$$
\begin{aligned}
&I_1(\lambda)+I_2(\lambda)+I_3(\lambda)=4\pi a_1a_2a_3/\Delta(\lambda),\\
&I_{12}(\lambda)=[I_2(\lambda)-I_1(\lambda)]/(a_1^2-a_2^2),\\
&3I_{11}(\lambda)=4\pi a_1a_2a_3/(a_1^2+\lambda)\Delta(\lambda)-I_{12}(\lambda)-I_{13}(\lambda),\\
&I_{123}(\lambda)=[I_{23}(\lambda)-I_{13}(\lambda)]/(a_1^2-a_2^2),\\
&I_{112}(\lambda)=[I_{12}(\lambda)-I_{11}(\lambda)]/(a_1^2-a_2^2),\\
&I_{113}(\lambda)=[I_{13}(\lambda)-I_{11}(\lambda)]/(a_1^2-a_3^2),\\
&5I_{111}(\lambda)=4\pi a_1a_2a_3/(a_1^2+\lambda)^2\Delta(\lambda)-I_{112}(\lambda)-I_{113}(\lambda),
\end{aligned}
\tag{12.18}
$$

where

$$
\begin{aligned}
&2\frac{d}{ds}\{1/\Delta(s)\}=-\{(a_1^2+s)^{-1}+(a_2^2+s)^{-1}+(a_3^2+s)^{-1}\}/\Delta(s),\\
&\Delta(\lambda)=\{(a_1^2+\lambda)(a_2^2+\lambda)(a_3^2+\lambda)\}^{1/2}.
\end{aligned}
$$

The cyclic permutation with respect to (1, 2, 3) can be applied in (12.18).

The I-integrals become elementary functions for special cases as shown in the following.

**Sphere* ($a_1=a_2=a_3=a$)

$$
\begin{aligned}
&I(\lambda)=4\pi a^3/(a^2+\lambda)^{1/2},\\
&I_i(\lambda)=4\pi a^3/3(a^2+\lambda)^{3/2},\\
&\underbrace{I_{ij\cdots k}}_{n}(\lambda)=\frac{4\pi a^3}{(2n+1)(a^2+\lambda)^{n+1/2}}.
\end{aligned}
\tag{12.19}
$$

**Elliptic cylinder* ($a_3\to\infty$)

$$
\begin{aligned}
&I_1(\lambda)=\frac{4\pi a_1a_2}{a_2^2-a_1^2}\left\{\frac{(a_2^2+\lambda)^{1/2}}{(a_1^2+\lambda)^{1/2}}-1\right\},\\
&I_2(\lambda)=\frac{4\pi a_1a_2}{a_1^2-a_2^2}\left\{\frac{(a_1^2+\lambda)^{1/2}}{(a_2^2+\lambda)^{1/2}}-1\right\},\\
&I_3(\lambda)=0.
\end{aligned}
\tag{12.20}
$$

$*$Oblate spheroid ($a_1 = a_2 > a_3$)

$$I(\lambda) = \frac{4\pi a_1^2 a_3}{\sqrt{a_1^2 - a_3^2}}\left[\frac{\pi}{2} - \text{arc tan}\left(\frac{\sqrt{a_3^2+\lambda}}{\sqrt{a_1^2-a_3^2}}\right)\right] = \frac{4\pi a_1^2 a_3}{\sqrt{a_1^2 - a_3^2}} \text{arc cos}\, b,$$

$$I_1(\lambda) = I_2(\lambda) = 2\pi a_1^2 a_3(\text{arc cos}\, b - bd)/(a_1^2 - a_3^2)^{3/2}, \tag{12.21}$$

$$I_3(\lambda) = \frac{4\pi a_1^2 a_3}{\Delta(\lambda)} - 2I_1(\lambda) = 4\pi a_1^2 a_3\left(\frac{d}{b} - \text{arc cos}\, b\right)\Big/(a_1^2 - a_3^2)^{3/2},$$

where $b = \sqrt{(a_3^2 + \lambda)/(a_1^2 + \lambda)}$ and $d = \sqrt{(a_1^2 - a_3^2)/(a_1^2 + \lambda)}$.

$*$*Prolate spheroid* ($a_1 > a_2 = a_3$)

$$I(\lambda) = -\frac{2\pi a_1 a_3^2}{\sqrt{a_1^2 - a_3^2}} \ln \frac{\sqrt{a_1^2+\lambda} - \sqrt{a_1^2 - a_3^2}}{\sqrt{a_1^2+\lambda} + \sqrt{a_1^2 - a_3^2}} = \frac{4\pi a_1 a_3^2}{\sqrt{a_1^2 - a_3^2}} \text{arc cosh}\, \bar{b},$$

$$I_1(\lambda) = 4\pi a_1 a_3^2(\text{arc cosh}\, \bar{b} - \bar{d}/\bar{b})/(a_1^2 - a_3^2)^{3/2}, \tag{12.22}$$

$$I_2(\lambda) = I_3(\lambda) = 2\pi a_1 a_3^2(\bar{b}\bar{d} - \text{arc cosh}\, \bar{b})/(a_1^2 - a_3^2)^{3/2},$$

where $\bar{b} = \sqrt{(a_1^2 + \lambda)/(a_3^2 + \lambda)}$ and $\bar{d} = \sqrt{(a_1^2 - a_3^2)/(a_3^2 + \lambda)}$.

$*$*Elliptical plate* ($a_1 \geq a_2 > a_3$, $a_3 \to 0$)

$$\lim_{a_3 \to 0} I_{ij\ldots}(\lambda) = 0 \quad \text{for } i, j, \ldots \neq 3. \tag{12.23}$$

Other cases are listed in Table 12.1.

$*$*The Ferrers and Dyson formula*

Ferrers (1877) and Dyson (1891) investigated integration of the harmonic potential of an ellipsoid,

$$P(\mathbf{x}) = \int_\Omega \frac{\rho(\mathbf{x}')d\mathbf{x}'}{|\mathbf{x} - \mathbf{x}'|}. \tag{12.24}$$

Their results have already been used in (11.38), (12.13), and (12.14).

A general result, given in Dyson's paper, states that if

$$\rho(\mathbf{x}) = \frac{m}{\pi a_1 a_2 a_3}\left(1 - \frac{x_i x_i}{a_I^2}\right)^{m-1} f\left(\frac{x_1}{a_1}, \frac{x_2}{a_2}, \frac{x_3}{a_3}\right) \quad \text{for } m > 0, \tag{12.25}$$

then

$$P(\mathbf{x}) = \int_{\lambda}^{\infty} U^{m}(s) \sum_{n=0}^{\infty} \frac{s^{n} U^{n}(s)}{2^{2n} n!(n+m)!/m!} L^{n} f\left(\frac{a_1 x_1}{a_1^2+s}, \frac{a_2 x_2}{a_2^2+s}, \frac{a_3 x_3}{a_3^2+s}\right) \frac{ds}{\Delta(s)}, \tag{12.26}$$

where

$$U(s) = 1 - \frac{x_i x_i}{a_I^2 + s}, \quad \Delta(s) = \sqrt{(a_1^2+s)(a_2^2+s)(a_3^2+s)},$$
$$L = \frac{a_I^2 + s}{a_I^2} \frac{\partial}{\partial x_i} \frac{\partial}{\partial x_i}. \tag{12.27}$$

Table 12.1 Limits of the I-integrals as $a_3 \to 0$ for $\lambda \neq 0$ and $\lambda = 0$.

$\lim_{a_3 \to 0} (\cdots)$	$\lambda \neq 0$	$\lambda = 0$
$I_{ij\cdots}(\lambda)$ with $i, j, \ldots \neq 3$	0	0
$I_3(\lambda)$	0	4π
$I_{13}(\lambda)$	0	$4\pi/a_1^2$
$I_{23}(\lambda)$	0	$4\pi/a_2^2$
$I_{33}(\lambda)$	0	∞
$a_3^2 I_{33}(\lambda)$	0	$4\pi/3$
$I_{113}(\lambda)$	0	$4\pi/a_1^4$
$I_{223}(\lambda)$	0	$4\pi/a_2^4$
$I_{123}(\lambda)$	0	$4\pi/a_1^2 a_2^2$
$I_{133}(\lambda)$	0	∞
$I_{233}(\lambda)$	0	∞
$a_3^2 I_{133}(\lambda)$	0	$4\pi/3a_1^2$
$a_3^2 I_{233}(\lambda)$	0	$4\pi/3a_2^2$
$I_{333}(\lambda)$	0	∞
$a_3^4 I_{333}(\lambda)$	0	$4\pi/5$
$I_{1113}(\lambda)$	0	$4\pi/a_1^6$
$I_{2223}(\lambda)$	0	$4\pi/a_2^6$
$I_{1123}(\lambda)$	0	$4\pi/a_1^4 a_2^2$
$I_{1223}(\lambda)$	0	$4\pi/a_1^2 a_2^4$
...	...	...

Furthermore, λ is the largest positive root of

$$\frac{x_i x_i}{a_I^2 + \lambda} = 1; \quad \text{i.e., } U(\lambda) = 0. \tag{12.28}$$

The summation convention in (11.39) has been used in the above equations.

All integrals in (12.9) can be expressed in a type of integral

$$\phi_{ij\cdots}(\mathbf{x}) = \int_{\Omega} \frac{x_i' x_j' \cdots d\mathbf{x}'}{|\mathbf{x}-\mathbf{x}'|}, \tag{12.29}$$

since $|\mathbf{x}-\mathbf{x}'| = (x_i x_i - 2x_i x_i' + x_i' x_i')/|\mathbf{x}-\mathbf{x}'|$. Comparing (12.25) with $x_i x_j \cdots$, we have $m = 1$ and

$$\phi_{ij\cdots}(\mathbf{x}) = \pi a_1 a_2 a_3 a_I^2 a_J^2 \cdots \int_{\lambda}^{\infty} U(s) \sum_{n=0}^{\infty} \frac{s^n U^n(s)}{2^{2n} n!(n+1)!}$$
$$\times L^n \left[\frac{x_i}{a_I^2+s} \frac{x_j}{a_J^2+s} \cdots \right] \frac{ds}{\Delta(s)}. \tag{12.30}$$

When the differential operator L^n in (12.30) is performed, the required integrals shown in (11.38), (12.13) and (12.14) can easily be obtained.

13. Energies of inclusions

Consider a finite body D containing inclusions Ω. The material is homogeneous, and anisotropic or isotropic. Ω is the sum of domains occupied by the inclusions. The volume of Ω is denoted by V.

Elastic strain energy

First we consider the case when the body D is free from any external force and surface constraint, but eigenstrains ϵ_{ij}^* are prescribed in Ω. The (elastic) strain energy is

$$W^* = (1/2) \int_D \sigma_{ij} e_{ij} dD, \tag{13.1}$$

where $e_{ij} = \epsilon_{ij} - \epsilon_{ij}^*$ and $\epsilon_{ij} = \frac{1}{2}(u_{i,j} + u_{j,i})$. Integrating by parts we obtain

$$\int_D \sigma_{ij} u_{i,j} dD = \int_S \sigma_{ij} u_i n_j dS - \int_D \sigma_{ij,j} u_i dD = 0, \tag{13.2}$$

since $\sigma_{ij} n_j = 0$ (free surface condition) on S and $\sigma_{ij,j} = 0$ (equilibrium condition) in D, where S is the boundary of D and n_i is the exterior unit vector normal to S. Therefore, we have

$$W^* = -(1/2) \int_{\Omega} \sigma_{ij} \epsilon_{ij}^* dD. \tag{13.3}$$

If Ω is an ellipsoidal inclusion and ϵ^*_{ij} is uniform, σ_{ij} in Ω is also uniform. Then, (13.3) is written as

$$W^* = -(1/2)V\sigma_{ij}\epsilon^*_{ij}. \tag{13.4}$$

σ_{ij} has been obtained in Section 11 for isotropic, and in Section 17 for anisotropic materials.

Consider now the case when the body D, containing inclusions Ω, is additionally subjected to external surface tractions F_i. The displacement field is the sum of u^0_i and u_i, where u^0_i is the displacement if F_i act alone in the absence of eigenstrains, and u_i is due to the eigenstrains prescribed in the inclusions. Then, the (elastic) strain energy is

$$W^* = (1/2)\int_D (\sigma^0_{ij} + \sigma_{ij})(u^0_{i,j} + u_{i,j} - \epsilon^*_{ij})dD, \tag{13.5}$$

where $\sigma^0_{ij} = C_{ijkl}u^0_{k,l}$. Since $\sigma_{ij,j} = 0$ in D and $\sigma_{ij}n_j = 0$ on S, the integration by parts gives

$$\int_D \sigma_{ij}(u^0_{i,j} + u_{i,j})dD = 0. \tag{13.6}$$

It is also seen that

$$\int_D \sigma^0_{ij}(u_{i,j} - \epsilon^*_{ij})dD = 0, \tag{13.7}$$

since $\sigma^0_{ij}(u_{i,j} - \epsilon^*_{ij}) = C_{ijkl}u^0_{k,l}(u_{i,j} - \epsilon^*_{ij}) = u^0_{k,l}\sigma_{kl} = \sigma_{ij}u^0_{i,j}$ and $\int_D \sigma_{ij}u^0_{i,j}dD = 0$.

The elastic strain energy, therefore, becomes

$$W^* = (1/2)\int_D \sigma^0_{ij}u^0_{i,j}dD - (1/2)\int_\Omega \sigma_{ij}\epsilon^*_{ij}dD. \tag{13.8}$$

It is interesting to note that the elastic strain energy is the sum of the two energies caused respectively by F_i and ϵ^*_{ij}.

Interaction energy

The total potential energy of the system subjected to F_i and eigenstrains due to its inclusions is defined by

$$W = W^* - \int_S F_i(u^0_i + u_i)dS, \tag{13.9}$$

where $\sigma^0_{ij}n_j = F_i$ on S, and W^* is defined by (13.5). If $\epsilon^*_{ij} = 0$ in D, W becomes

$$W_0 = (1/2)\int_D \sigma_{ij}^0 u_{i,j}^0 dD - \int_S F_i u_i^0 ds. \tag{13.10}$$

If $F_i = 0$ on S, W becomes

$$W_1 = (1/2)\int_D \sigma_{ij}(u_{i,j} - \epsilon_{ij}^*)dD = -(1/2)\int_\Omega \sigma_{ij}\epsilon_{ij}^* dD. \tag{13.11}$$

The interaction energy between ϵ_{ij}^* and F_i is defined by

$$\Delta W = W - W_0 - W_1. \tag{13.12}$$

Then, we have

$$\Delta W = -\int_S F_i u_i dS = -\int_\Omega \sigma_{ij}^0 \epsilon_{ij}^* dD, \tag{13.13}$$

since $\int_S F_i u_i dS = \int_S \sigma_{ij}^0 n_j u_i dS = \int_D \sigma_{ij}^0 u_{i,j} dD = \int_D \sigma_{ij}^0(u_{i,j} - \epsilon_{ij}^*)dD + \int_D \sigma_{ij}^0 \epsilon_{ij}^* dD = \int_D \sigma_{ij}^0 \epsilon_{ij}^* dD$ due to (13.7).

If Ω is an ellipsoidal inclusion and ϵ_{ij}^* is uniform,

$$\Delta W = -V\sigma_{ij}^0 \epsilon_{ij}^*. \tag{13.14}$$

The total potential energy of the system, (13.9), is the free energy, either the Gibbs free energy or the Helmholtz free energy, depending on a variable whose variations are kept constant. When an inclusion is a source of the stress field, the corresponding free energy is (13.11). When an external force is a source of the stress field, the free energy is (13.10). Therefore, the interaction energy (13.12) is a part of the free energy caused by the interaction between ϵ_{ij}^* and F_i.

Suppose that a body D is subjected to F_i alone. If eigenstrains due to inclusions are introduced in Ω, the free energy will increase by

$$\Delta \bar{W} = W - W_0. \tag{13.15}$$

This $\Delta \bar{W}$ is an interesting quantity, particularly in fracture mechanics, as explained in Chapter 5. Equation (13.15) can be written as

$$\Delta \bar{W} = -(1/2)\int_\Omega \sigma_{ij}\epsilon_{ij}^* dD - \int_\Omega \sigma_{ij}^0 \epsilon_{ij}^* dD. \tag{13.16}$$

It should be emphasized that energies (13.3), (13.8), (13.11), (13.13) and (13.16) are expressed by the quantities defined inside Ω. Energy calculations, in view of this, become easier.

Elastic strain energies due only to constant eigenstrains prescribed in inclusions of indicated geometries, in an isotropic infinite body are presented below.

Strain energy due to a spherical inclusion

$$W^* = \frac{8\pi a^3\mu}{45(1-\nu)}\{4(\epsilon_{11}^{*2}+\epsilon_{22}^{*2}+\epsilon_{33}^{*2})+(5\nu+1)(\epsilon_{11}^{*}\epsilon_{22}^{*}+\epsilon_{22}^{*}\epsilon_{33}^{*}+\epsilon_{33}^{*}\epsilon_{11}^{*})$$
$$+(7-5\nu)(\epsilon_{12}^{*2}+\epsilon_{23}^{*2}+\epsilon_{31}^{*2})\}, \qquad (13.17)$$

where a is the radius of the sphere.

Elliptic cylinder $(a_3\to\infty)$

$$W^* = \frac{\pi a_1 a_2\mu}{2(1-\nu)}\left[\left\{2-\frac{a_2^2+2a_1a_2}{(a_1+a_2)^2}-\frac{a_2}{a_1+a_2}\right\}\epsilon_{11}^{*2}\right.$$
$$+\left\{2-\frac{a_1^2+2a_1a_2}{(a_1+a_2)^2}-\frac{a_1}{a_1+a_2}\right\}\epsilon_{22}^{*2}$$
$$+2\epsilon_{33}^{*2}+\frac{2a_1a_2}{(a_1+a_2)^2}\epsilon_{11}^{*}\epsilon_{22}^{*}+\frac{4\nu a_2}{a_1+a_2}\epsilon_{22}^{*}\epsilon_{33}^{*}+\frac{4\nu a_1}{a_1+a_2}\epsilon_{33}^{*}\epsilon_{11}^{*} \qquad (13.18)$$
$$\left.+\frac{4a_1a_2}{(a_1+a_2)^2}\epsilon_{12}^{*2}+4(1-\nu)\frac{a_2}{a_1+a_2}\epsilon_{23}^{*2}+4(1-\nu)\frac{a_1}{a_1+a_2}\epsilon_{31}^{*2}\right]$$

which is defined per unit length of the cylinder.

Penny-shaped flat ellipsoid $(a_1=a_2=a,\ a_3\ll a)$

$$W^* = (4/3)\pi a^2 a_3\mu\{(\epsilon_{11}^{*2}+\epsilon_{22}^{*2})/(1-\nu)+2\nu\epsilon_{11}^{*}\epsilon_{22}^{*}/(1-\nu)+2\epsilon_{12}^{*2}\}. \qquad (13.19)$$

Spheroid

For $a_1=a_2=a,\ \beta=a_3/a$,

$$W^* = (4/3)\pi a^2 a_3\mu\{b_1(\epsilon_{11}^{*2}+\epsilon_{22}^{*2})+b_2\epsilon_{11}^{*}\epsilon_{22}^{*}+b_3(\epsilon_{11}^{*}+\epsilon_{22}^{*})\epsilon_{33}^{*}$$
$$+b_4\epsilon_{33}^{*2}+b_5(\epsilon_{32}^{*2}+\epsilon_{31}^{*2})+b_6\epsilon_{12}^{*}\}. \qquad (13.20)$$

For $a_3=a_1=a,\ \beta=a_2/a$,

$$W^* = (4/3)\pi a^2 a_2\mu\{b_1(\epsilon_{11}^{*2}+\epsilon_{33}^{*2})+b_2\epsilon_{11}^{*}\epsilon_{33}^{*}+b_3(\epsilon_{11}^{*}+\epsilon_{33}^{*})\epsilon_{22}^{*}$$
$$+b_4\epsilon_{22}^{*2}+b_5(\epsilon_{23}^{*2}+\epsilon_{21}^{*2})+b_6\epsilon_{13}^{*2}\}. \qquad (13.20')$$

Numerical values of b_i have been calculated by Shibata and Ono (1975) as functions of β for oblate spheroids as shown in Fig. 13.1.

It is interesting to note that for dilatational eigenstrains $\epsilon_{ij}^* = \delta_{ij}\epsilon^*$, the elastic strain energy per unit volume of inclusions is a constant independent of the shape of inclusions,

$$W^*/V = 2\mu(\epsilon^*)^2(1+\nu)/(1-\nu), \tag{13.21}$$

where $V = (4/3)\pi a_1 a_2 a_3$.

The above result (13.21) holds for any shape of inclusion Ω with volume V. This is because when $\epsilon_{ij}^* = \delta_{ij}\epsilon^*$, (2.7) and (11.33) lead to

$$\sigma_{ii} = \{(2\mu + 3\lambda)D_{iijj} - (6\mu + 9\lambda)\}\epsilon^*. \tag{13.22}$$

From (11.34) we have

$$D_{iijj} = (1+\nu)/(1-\nu), \tag{13.23}$$

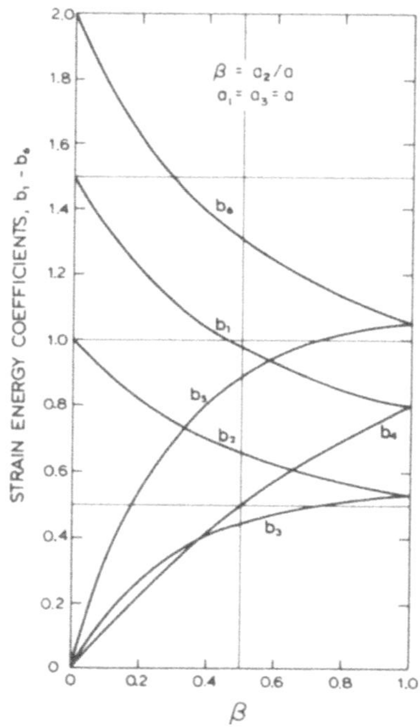

Fig. 13.1. Strain energy coefficients for an oblate spheroidal inclusion.

where

$$\psi_{,jj} = 2\phi,$$
$$\phi_{,ii} = -4\pi \text{ in } \Omega, \qquad (13.24)$$
$$= 0 \text{ outside } \Omega,$$

are used. Equation (13.24) follows directly from (12.5) and (12.6). Thus, it can be said that the hydrostatic pressure $(1/3)\sigma_{ii}$ is uniform for any shape of inclusion when the eigenstrain is dilatational. It becomes

$$\sigma_{ii} = -4\mu \frac{1+\nu}{1-\nu} \epsilon^* \text{ in } \Omega,$$
$$= 0 \text{ outside } \Omega. \qquad (13.25)$$

The elastic strain energy (13.3) becomes

$$W^* = -(1/2)\sigma_{ii}\epsilon^* V \qquad (13.26)$$

which leads to (13.21).

The above results (13.25) hold also when $\epsilon^*_{ij} = \delta_{ij}\epsilon^*$ is a function of $\mathbf{x}$. From (2.8) and (12.2) we have

$$\sigma_{ii} = \frac{2\mu(1+\nu)}{1-2\nu}(\epsilon_{ii} - \epsilon^*_{ii}),$$
$$\epsilon_{ii} = \frac{-(1+\nu)}{12\pi(1-\nu)} \Phi_{mm,ll}, \qquad (13.27)$$

where (12.5) and $\Phi_{ij} = \delta_{ij}\Phi_{mm}/3$ are used. Furthermore (12.6) leads to

$$\sigma_{ii} = \frac{-4\mu(1+\nu)}{1-\nu} \epsilon^*(\mathbf{x}) \text{ in } \Omega$$
$$= 0 \text{ outside } \Omega. \qquad (13.28)$$

Then, the elastic strain energy (13.3) becomes

$$W^* = \frac{2\mu(1+\nu)}{1-\nu} \int_\Omega \{\epsilon^*(\mathbf{x})\}^2 dD. \qquad (13.29)$$

In view of (13.28), two separate inclusions, Ω_1 and Ω_2, with pure dilatational eigenstrains, $\epsilon^*_{ij} = \delta_{ij}\epsilon^*$, do not interact. For anisotropic materials, (13.28) does not hold generally and, therefore, there exists an interaction between two dilatational inclusions. This problem is discussed by considering specific examples, in Section 26, see also Sines and Kikuchi (1958).

If the inclusion is an inhomogeneous inclusion with dilatational eigenstrain $\epsilon_{ij}^{p} = \delta_{ij}\epsilon^{p}$, the elastic strain energy depends on the shape of the inclusion even if the elastic moduli of the matrix and inhomogeneity are isotropic. As seen in (22.29) or (22.30), the equivalent eigenstrain ϵ_{ij}^{**} defined in (22.14) and (22.15) is not purely dilatational. Because of this, the hydrostatic stress in the inhomogeneous inclusion depends on its shape and (13.26) is not valid. These facts should be kept in mind when one discusses the elastic state of precipitates.

14. Cuboidal inclusions

The elastic fields due to a cuboidal inclusion with uniform eigenstrains have been calculated by Faivre (1964), Sankaran and Laird (1976), Lee and Johnson (1977), and Y. P. Chiu (1977). The solid is assumed to be an infinite isotropic elastic medium. The calculation is interesting, because it shows how the stress distribution for interior points deviates from the result for the ellipsoidal inclusion where the stress distribution is uniform.

The cuboidal inclusion is shown in Fig. 14.1. Expression (6.2) is used, and ϵ_{ij}^{*} is constant in the cuboidal domain Ω. The integral in (6.2) with respect to $\mathbf{x}'$ is easily performed and we have

$$u_{i,q}(\mathbf{x}) = (i/8\pi^3)\epsilon_{mn}^{*} \sum_{p=1}^{8} (-1)^p \int_{-\infty}^{\infty} C_{jlmn}\xi_l\xi_q N_{ij}(\boldsymbol{\xi}) \times \{\xi_1\xi_2\xi_3 D(\boldsymbol{\xi})\}^{-1} \exp(i\boldsymbol{\xi}\cdot\mathbf{c}_p)d\boldsymbol{\xi}, \tag{14.1}$$

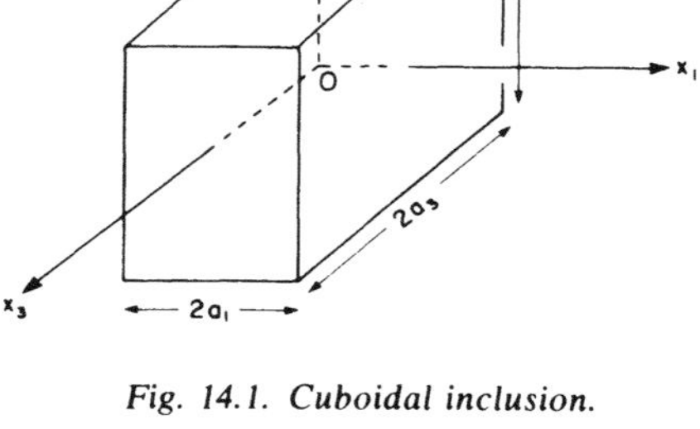

Fig. 14.1. Cuboidal inclusion.

where

$$
\begin{aligned}
\mathbf{c}_1 &= (x_1 - a_1, x_2 - a_2, x_3 - a_3),\\
\mathbf{c}_2 &= (x_1 + a_1, x_2 - a_2, x_3 - a_3),\\
\mathbf{c}_3 &= (x_1 + a_1, x_2 + a_2, x_3 - a_3),\\
\mathbf{c}_4 &= (x_1 - a_1, x_2 + a_2, x_3 - a_3),\\
\mathbf{c}_5 &= (x_1 - a_1, x_2 + a_2, x_3 + a_3),\\
\mathbf{c}_6 &= (x_1 - a_1, x_2 - a_2, x_3 + a_3),\\
\mathbf{c}_7 &= (x_1 + a_1, x_2 - a_2, x_3 + a_3),\\
\mathbf{c}_8 &= (x_1 + a_1, x_2 + a_2, x_3 + a_3).
\end{aligned}
\tag{14.2}
$$

The following integrals are used for the integrations in (14.1), a detailed calculation is given by Chiu (1977):

$$
\begin{aligned}
&\int_{-\infty}^{\infty} \frac{\xi_1\xi_2 \exp(i\boldsymbol{\xi}\cdot\mathbf{x})d\boldsymbol{\xi}}{(\xi_1^2+\xi_2^2+\xi_3^2)^2} = -\frac{\pi^2 x_1x_2}{R^3},\\
&i\int_{-\infty}^{\infty} \frac{\xi_1\xi_2 \exp(i\boldsymbol{\xi}\cdot\mathbf{x})d\boldsymbol{\xi}}{\xi_3(\xi_1^2+\xi_2^2+\xi_3^2)^2} = \frac{\pi^2 x_1x_2x_3}{(x_1^2+x_2^2)R},\\
&\int_{-\infty}^{\infty} \frac{\xi_1^2 \exp(i\boldsymbol{\xi}\cdot\mathbf{x})d\boldsymbol{\xi}}{(\xi_1^2+\xi_2^2+\xi_3^2)^2} = \pi^2\frac{x_2^2+x_3^2}{R^3},\\
&-i\int_{-\infty}^{\infty} \frac{\xi_1^2 \exp(i\boldsymbol{\xi}\cdot\mathbf{x})d\boldsymbol{\xi}}{\xi_3(\xi_1^2+\xi_2^2+\xi_3^2)^2} = \pi^2\left[\operatorname{sgn}(x_3)\log\frac{R+|x_3|}{(x_1^2+x_2^2)^{1/2}} - \frac{x_1^2x_3}{(x_1^2+x_2^2)R}\right],\\
&i\int_{-\infty}^{\infty} \frac{\xi_1^3 \exp(i\boldsymbol{\xi}\cdot\mathbf{x})d\boldsymbol{\xi}}{\xi_2\xi_3(\xi_1^2+\xi_2^2+\xi_3^2)^2} = 2\pi^2\left\{\tan^{-1}\frac{x_2x_3}{x_1R} - \frac{x_1x_2x_3}{2R}\left(\frac{1}{x_1^2+x_2^2}+\frac{1}{x_1^2+x_3^2}\right)\right\},
\end{aligned}
\tag{14.3}
$$

where $R = (x_1^2 + x_2^2 + x_3^2)^{1/2}$.

The expression (14.1) can be reduced to

$$
\begin{aligned}
2\mu u_{i,q}(\mathbf{x}) = (\mu/4\pi^3)\sum_{n=1}^{8}(-1)^n\Bigg[&\frac{\nu}{1-\nu}\epsilon_{kk}^{*}D_{iqmm}(\mathbf{c}_n) + 2\epsilon_{ij}^{*}D_{jqmm}(\mathbf{c}_n)\\
&-\frac{1}{1-\nu}\epsilon_{mj}^{*}D_{iqmj}(\mathbf{c}_n)\Bigg],
\end{aligned}
\tag{14.4}
$$

where

$$
D_{ijkl}(\mathbf{c}) = (\partial^4/\partial x_i\partial x_j\partial x_k\partial x_l)\int_{-\infty}^{\infty}\frac{i\exp(i\boldsymbol{\xi}\cdot\mathbf{c})}{\xi_1\xi_2\xi_3(\xi_1^2+\xi_2^2+\xi_3^2)^2}\,d\boldsymbol{\xi}. \tag{14.5}
$$

Integrations with respect to $\boldsymbol{\xi}$ lead to

$$
\begin{aligned}
D_{1111}(\mathbf{c}) &= 2\pi^2\left\{\tan^{-1}\frac{c_2c_3}{c_1R} - \frac{c_1c_2c_3}{2R}\left(\frac{1}{c_1^2+c_2^2} + \frac{1}{c_1^2+c_3^2}\right)\right\},\\
D_{1112}(\mathbf{c}) &= -\pi^2\left\{\operatorname{sgn}(c_3)\log\frac{R+|c_3|}{(c_1^2+c_2^2)^{1/2}} - \frac{c_1^2c_3}{(c_1^2+c_2^2)R}\right\},\\
D_{1122}(\mathbf{c}) &= \frac{\pi^2c_1c_2c_3}{(c_1^2+c_2^2)R},\\
D_{1123}(\mathbf{c}) &= -\frac{\pi^2c_1}{R},
\end{aligned}
\tag{14.6}
$$

where $\mathbf{c} = (c_1, c_2, c_3)$ and other components are obtained by cyclic permutation of $(1, 2, 3)$.

It can be seen that there are logarithmic singularities in certain edges and corners of the cuboid, depending upon the type of eigenstrains in the cuboid; i.e., the singularities only occur in shear stress components when ϵ_{ij}^* has no shear strain components.

The stress components are obtained from $\sigma_{ij}(\mathbf{x}) = C_{ijkl}\{u_{k,l}(\mathbf{x}) - \epsilon_{kl}^*(\mathbf{x})\}$. Figures 14.2 ~ 14.4 are taken from Chiu (1977). Poisson's ratio $\nu = 0.3$ is used in all calculations. For simplicity it is assumed that all components of the eigenstrain, except ϵ_{11}^*, vanish and that a_2 is equal to a_3. Figure 14.2 shows variations of the normal stresses along the x_1-axis for the three types of cuboidal domains. The curves show that the normal stress components are continuous across the

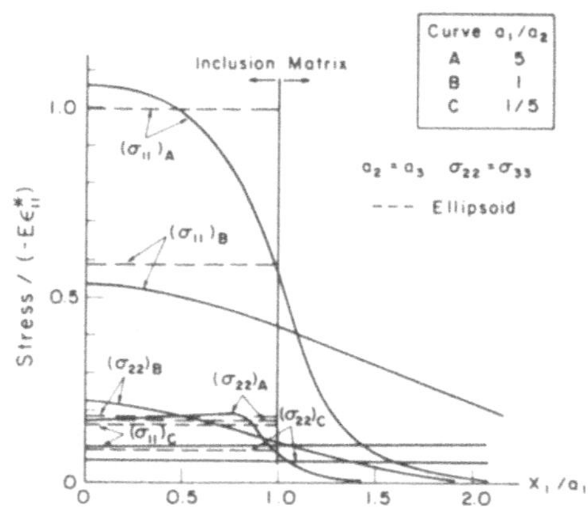

Fig. 14.2. Variation of normal stresses along the x_1-axis for three different a_1/a_2-values, where $a_2 = a_3$. (Dotted lines denote stress values for ellipsoidal inclusions.

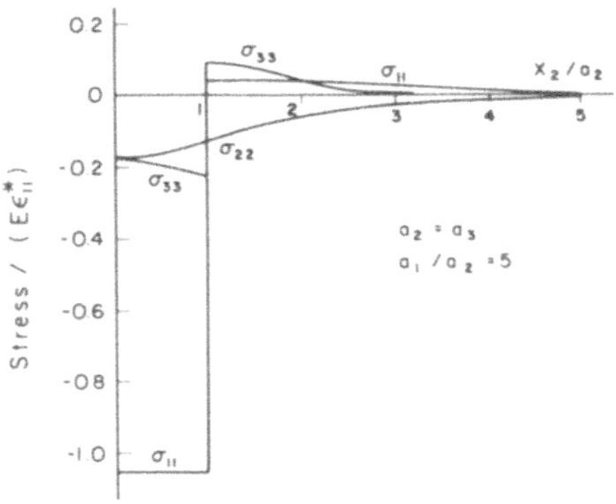

Fig. 14.3. *Variation of normal stresses along the* x_2*-axis for* $a_1/a_2 = 5$ *and* $a_2 = a_3$.

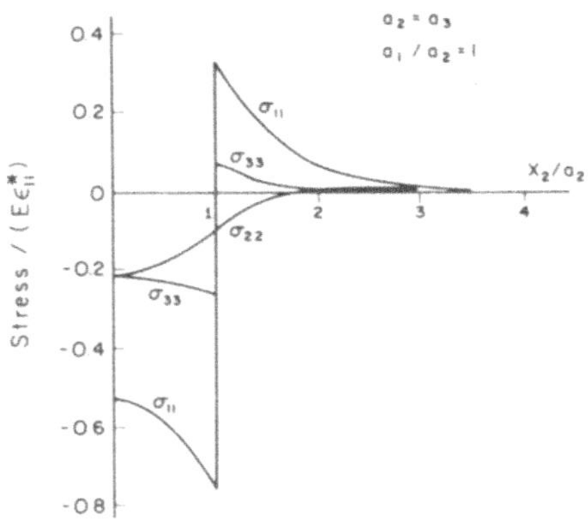

Fig. 14.4. *Variation of normal stresses along the* x_2*-axis for a cubical inclusion* $a_1 = a_2 = a_3$.

boundary of the cuboid at $x_1 = a_1$. The dotted lines are the stress components in an ellipsoidal inclusion with the semi-axes a_1, a_2, $a_3 = a_2$. Figures 14.3 and 14.4 show that σ_{22} is continuous across the boundary of the cuboid at $x_2 = a_2$, while the other two normal components of the stress are discontinuous at $x_2 = a_2$. It can be seen from (6.11) that the discontinuities are local quantities which are independent of the overall shape of the inclusion (independent of a_1/a_2 in the present calculation). It is found that $[\sigma_{11}] = 1.099\epsilon^*_{11}E$ and $[\sigma_{33}] = 0.329\epsilon^*_{11}E$, where E is Young's modulus.

Lee and Johnson (1977) have performed an alternative calculation by using (11.30) and (11.31). The $\psi(\mathbf{x})$ and $\phi(\mathbf{x})$ defined by (11.35) have been obtained by MacMillan (1930) when Ω is a cuboidal domain. They are

$$\psi(\mathbf{x}) = \sum_{n=1}^{8} (-1)^n D(\mathbf{c}_n),$$
$$\phi(\mathbf{x}) = \sum_{n=1}^{8} (-1)^n E(\mathbf{c}_n), \tag{14.7}$$

where

$$\begin{aligned} D(\mathbf{c}) = {} & (1/4)c_1c_2c_3R + (1/6)\{(R^2 - c_1^2)c_2c_3 \log(R + c_1) \\ & + (R^2 - c_2^2)c_3c_1 \log(R + c_2) + (R^2 - c_3^2)c_1c_2 \log(R + c_3)\} \\ & - (1/12)\{c_1^4 \tan^{-1}(c_2c_3/c_1R) + c_2^4 \tan^{-1}(c_3c_1/c_2R) \\ & + c_3^4 \tan^{-1}(c_1c_2/c_3R)\}, \end{aligned} \tag{14.8}$$

$$\begin{aligned} E(\mathbf{c}) = {} & c_1c_2 \log(R + c_3) + c_2c_3 \log(R + c_1) + c_3c_1 \log(R + c_2) \\ & - (1/2)\{c_1^2 \tan^{-1}(c_2c_3/c_1R) + c_2^2 \tan^{-1}(c_3c_1/c_2R) \\ & + c_3^2 \tan^{-1}(c_1c_2/c_3R)\}, \end{aligned}$$

where $\mathbf{c} = (c_1, c_2, c_3)$; $R = (x_1^2 + x_2^2 + x_3^2)^{1/2}$ and $\mathbf{c}_n$ is defined in (14.2). The strain can be derived from (11.32) where $\Psi_{ij} = \epsilon_{ij}^*\psi$ and $\Phi_{ij} = \epsilon_{ij}^*\phi$.

Lee and Johnson (1977) further have calculated the elastic strain energy W^* of a single cuboidal inclusion and the interaction energy between two cuboidal inclusions, where $\epsilon_{ij}^* = \epsilon_{31}^*$, and other components are zero.

From (13.3) and σ_{ij} calculated from (14.7) or (14.4), Lee and Johnson obtain the elastic strain energy, for $a_1 = a_2 = a_3/\beta$,

$$W^*/8a_1a_2a_3 = 2\mu(\epsilon_{31}^*)^2\{1 - f(\beta)/4\pi(1-\nu)\beta\}, \tag{14.9}$$

where

$$\begin{aligned} f(\beta) = {} & 8(1-\nu)\beta[\tan^{-1}\{1/\beta(2+\beta^2)^{1/2}\} + \tan^{-1}\{\beta/(2+\beta^2)^{1/2}\}] \\ & + 4\{\nu + (2-\nu)\beta^2\} \tanh^{-1}\{1/(2+\beta^2)^{1/2}\} \\ & - 4(2-\nu)\beta^2 \tanh^{-1}\{1/(1+\beta^2)^{1/2}\} \\ & - (4/3)\{\nu + (3-\nu)\beta^2\}(2+\beta^2)^{1/2} + (4/3)\{\nu + 2(3-\nu)\beta^2\}(1+\beta^2)^{1/2} \\ & - (4/3)(3-\nu)\beta^3 + \nu[2\log(3-2\sqrt{2}) + (4/3)(\sqrt{2}-1)]. \end{aligned} \tag{14.10}$$

The elastic strain energy per unit volume of a precipitate is smaller for a cuboid than for an ellipsoid when $\beta > 0.35$. This relation is reversed when $\beta < 0.35$.

Sankaran and Laird (1976) employed another method of calculation.

Starting from (11.1) we can write

$$u_i(\mathbf{x}) = C_{jkmn}\epsilon^*_{mn}\int_{|\Omega|} G_{ij}(\mathbf{x}-\mathbf{x}')n_k dS(\mathbf{x}') \qquad (14.11)$$

where Gauss' theorem is used, and where $|\Omega|$ is the boundary of Ω, $\mathbf{n}$ is the outward normal vector on the boundary, and $G_{ij,k} = -(\partial/\partial x'_k)G_{ij}$.

Sankaran and Laird have calculated the stress field around misfitting precipitates of plate morphology, like θ' in Al–Cu and η in Al–Au (see Aaronson et al. 1970, Sankaran and Laird 1974). The stresses are shown to attain very high values near the interfaces at the edges. This can significantly influence the generation of dislocations to accommodate the mismatch.

Argon (1976) has obtained approximate solutions for the stresses around slender finite elastic rods and platelets with moduli different from those of the infinite matrix.

15. Inclusions in a half space

Green's functions

A semi-infinite domain is defined by $x_3 \geq 0$ as shown in Fig. 15.1. The surface $x_3 = 0$ is free from external tractions. The objective is to express the elastic field when the eigenstrain $\epsilon^*_{ij}(\mathbf{x})$ is given in the half space.

Green's functions $G_{ij}(\mathbf{x}, \mathbf{x}')$ for the half space are defined by the following equations: For $x_3 \geq 0$,

$$C_{ijkl}G_{km,lj}(\mathbf{x}, \mathbf{x}') + \delta_{im}\delta(\mathbf{x}, \mathbf{x}') = 0 \qquad (15.1)$$

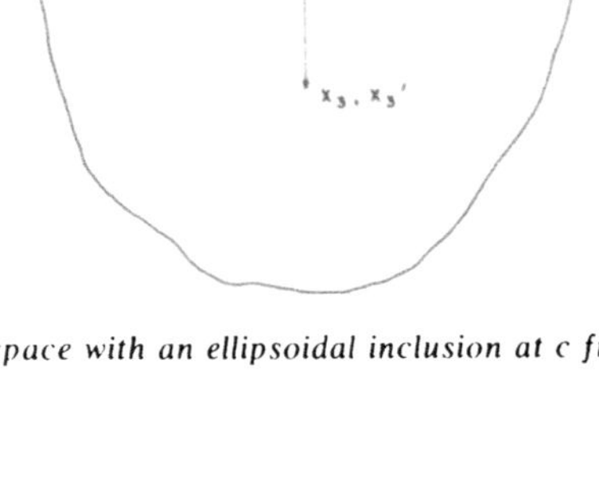

Fig. 15.1. Half-space with an ellipsoidal inclusion at c from the surface.

and on $x_3=0$,

$$C_{ijkl}G_{km,l}(\mathbf{x},\mathbf{x}')n_j = \delta_{im}\delta_S(\mathbf{x},\mathbf{x}'), \tag{15.2}$$

where $G_{km,l}(\mathbf{x},\mathbf{x}')=(\partial/\partial x_l)G_{km}(\mathbf{x},\mathbf{x}')$ and n_j is the unit normal on surface $x_3=0$. $\delta(\mathbf{x},\mathbf{x}')$ and $\delta_S(\mathbf{x},\mathbf{x}')$ are the three and two-dimensional Dirac delta functions, respectively, having the following properties:

$$\int_0^\infty f(\mathbf{x}')\delta(\mathbf{x},\mathbf{x}')d\mathbf{x}' = f(\mathbf{x}), \tag{15.3}$$

and

$$\int_S f(\mathbf{x}')\delta_S(\mathbf{x},\mathbf{x}')dS(x') = f(\mathbf{x}) \quad \text{at } x_3=0, \tag{15.4}$$

where S is the plane $x_3=0$ and $\delta(\mathbf{x},\mathbf{x}')=\delta_S(\mathbf{x},\mathbf{x}')=0$ when $\mathbf{x}\neq\mathbf{x}'$. $G_{ij}(\mathbf{x},\mathbf{x}')$ is the x_i-component of displacement at point $\mathbf{x}$ when a unit force (body force, when $\mathbf{x}'$ is an interior point and surface force when $\mathbf{x}'$ is on S) in the x_j-direction is applied at point $\mathbf{x}'$.

The displacement $u_i(\mathbf{x})$ caused by $\epsilon^*_{ij}(\mathbf{x})$ is expressed in a form similar to (3.23),

$$u_i(\mathbf{x}) = \int_0^\infty C_{jlmn}\epsilon^*_{mn}(\mathbf{x}')\frac{\partial}{\partial x'_l}G_{ij}(\mathbf{x},\mathbf{x}')d\mathbf{x}'. \tag{15.5}$$

It can be shown that (15.5) satisfies the equations of equilibrium (2.12),

$$C_{ijkl}u_{k,lj}(\mathbf{x}) = C_{ijkl}\epsilon^*_{kl,j}(\mathbf{x}), \tag{15.6}$$

and the boundary condition (2.13),

$$C_{ijkl}u_{k,l}(\mathbf{x})n_j = C_{ijkl}\epsilon^*_{kl}(\mathbf{x})n_j. \tag{15.7}$$

In view of the properties of Green's function, we have

$$\begin{aligned} C_{ijkl}u_{k,lj}(\mathbf{x}) &= C_{ijkl}\int_0^\infty C_{pqmn}\epsilon^*_{mn}(\mathbf{x}')\frac{\partial}{\partial x'_q}G_{kp,lj}(\mathbf{x},\mathbf{x}')d\mathbf{x}' \\ &= -\int_0^\infty C_{pqmn}\epsilon^*_{mn}(\mathbf{x}')\frac{\partial}{\partial x'_q}\delta_{ip}\delta(\mathbf{x},\mathbf{x}')d\mathbf{x}' \\ &= -\int_0^\infty C_{ijkl}\epsilon^*_{kl}(\mathbf{x}')\frac{\partial}{\partial x'_j}\delta(\mathbf{x},\mathbf{x}')d\mathbf{x}' \\ &= \int_0^\infty C_{ijkl}\epsilon^*_{kl,j}(\mathbf{x}')\delta(\mathbf{x},\mathbf{x}')d\mathbf{x}' = C_{ijkl}\epsilon^*_{kl,j}(\mathbf{x}) \end{aligned} \tag{15.8}$$

and on the plane $x_3 = 0$

$$
\begin{aligned}
C_{ijkl}u_{k,l}(\mathbf{x})n_j &= C_{ijkl}\int_0^\infty C_{pqmn}\epsilon^*_{mn}(\mathbf{x}')\frac{\partial}{\partial x'_q}G_{kp,l}(\mathbf{x},\mathbf{x}')n_j d\mathbf{x}' \\
&= \int_0^\infty C_{pqmn}\epsilon^*_{mn}(\mathbf{x}')\frac{\partial}{\partial x'_q}C_{ijkl}G_{kp,l}(\mathbf{x},\mathbf{x}')n_j d\mathbf{x}' \\
&= \int_S C_{pqmn}\epsilon^*_{mn}(\mathbf{x}')\delta_{ip}\delta_S(\mathbf{x},\mathbf{x}')n_q dS(\mathbf{x}') \\
&= C_{ijkl}\epsilon^*_{kl}(\mathbf{x})n_j, \qquad (15.9)
\end{aligned}
$$

where S is the surface $x_3 = 0$.

Green's functions $G_{ij}(\mathbf{x},\mathbf{x}')$ for the semi-infinite isotropic medium have been found by Mindlin (1953); they are

$$
\begin{aligned}
G_{ij}(\mathbf{x},\mathbf{x}') &= \frac{1}{16\pi\mu(1-\nu)}\left[\frac{3-4\nu}{R_1}\delta_{ij} + \frac{1}{R_2}\delta_{ij} + \frac{(x_i - x'_i)(x_j - x'_j)}{R_1^3}\right. \\
&\quad + \frac{(3-4\nu)(x_i - x'_i)(x_j - x'_j)}{R_2^3} + \frac{2x_3x'_3}{R_2^3}\left\{\delta_{ij} - \frac{3(x_i - x'_i)(x_j - x'_j)}{R_2^2}\right\} \\
&\quad \left. + \frac{4(1-\nu)(1-2\nu)}{R_2 + x_3 + x'_3}\left\{\delta_{ij} - \frac{(x_i - x'_i)(x_j - x'_j)}{R_2(R_2 + x_3 + x'_3)}\right\}\right] = G_{ji}(\mathbf{x},\mathbf{x}'), \\
G_{3j}(\mathbf{x},\mathbf{x}') &= \frac{(x_j - x'_j)}{16\pi\mu(1-\nu)}\left[\frac{(x_3 - x'_3)}{R_1^3} + \frac{(3-4\nu)(x_3 - x'_3)}{R_2^3} - \frac{6x_3x'_3(x_3 + x'_3)}{R_2^5}\right. \\
&\quad \left. + \frac{4(1-\nu)(1-2\nu)}{R_2(R_2 + x_3 + x'_3)}\right], \qquad (15.10) \\
G_{i3}(\mathbf{x},\mathbf{x}') &= \frac{(x_i - x'_i)}{16\pi\mu(1-\nu)}\left[\frac{(x_3 - x'_3)}{R_1^3} + \frac{(3-4\nu)(x_3 - x'_3)}{R_2^3} + \frac{6x_3x'_3(x_3 + x'_3)}{R_2^5}\right. \\
&\quad \left. - \frac{4(1-\nu)(1-2\nu)}{R_2(R_2 + x_3 + x'_3)}\right], \\
G_{33}(\mathbf{x},\mathbf{x}') &= \frac{1}{16\pi\mu(1-\nu)}\left[\frac{3-4\nu}{R_1} + \frac{8(1-\nu)^2 - (3-4\nu)}{R_2} + \frac{(x_3 - x'_3)^2}{R_1^3}\right. \\
&\quad \left. + \frac{(3-4\nu)(x_3 + x'_3)^2 - 2x_3x'_3}{R_2^3} + \frac{6x_3x'_3(x_3 + x'_3)^2}{R_2^5}\right], \quad i, j = 1, 2
\end{aligned}
$$

where

$$
\begin{aligned}
R_1^2 &= (x_1 - x'_1)^2 + (x_2 - x'_2)^2 + (x_3 - x'_3)^2, \\
R_2^2 &= (x_1 - x'_1)^2 + (x_2 - x'_2)^2 + (x_3 + x'_3)^2, \qquad (15.11)
\end{aligned}
$$

μ = shear modulus, and ν = Poisson's ratio.

In these equations, the effect of the free surface appears in terms which contain R_2. (A minor sign error in the original expressions has been corrected later in a private communication; Mindlin 1975.)

The inclusion problem here occurs when ϵ^*_{ij} is given in domain Ω. If ϵ^*_{ij} is a constant, (15.5) becomes

$$u_i(\mathbf{x}) = C_{jlmn}\epsilon^*_{mn}\int_\Omega \frac{\partial}{\partial x'_l}G_{ij}(\mathbf{x},\mathbf{x}')d\mathbf{x}' \tag{15.12}$$

or

$$u_i(\mathbf{x}) = C_{jlmn}\epsilon^*_{mn}\int_{|\Omega|} G_{ij}(\mathbf{x},\mathbf{x}')n_l(\mathbf{x}')dS(\mathbf{x}'), \tag{15.13}$$

where Gauss' theorem is used, and $|\Omega|$ is the surface of Ω.

Recently, N.G. Lee (1979) and Pan and Chou (1979) have obtained explicit expressions for Green's function in a half space of a hexagonal continuum with its basal plane as surface. Pan and Chou (1979) further extended their solution to the two-phase transversely isotropic materials.

Ellipsoidal inclusion with a uniform dilatational eigenstrain

The elastic field is calculated when an ellipsoidal inclusion with a uniform dilatational eigenstrain

$$\epsilon^*_{ij} = \delta_{ij}\epsilon^* \tag{15.14}$$

is located near the surface of a half space. When (15.10) and (15.14) are substituted into (15.12) we obtain

$$u_i(\mathbf{x}) = \frac{\epsilon^*(1+\nu)}{4\pi(1-\nu)}\int_\Omega\left[-\left(\frac{1}{R_1}\right)_{,i} + (3-4\nu)(2\delta_{3i}-1)\left(\frac{1}{R_2}\right)_{,i} - 2x_3\left(\frac{1}{R_2}\right)_{,3i}\right]d\mathbf{x}', \tag{15.15}$$

where

$$\frac{\partial}{\partial x_i}\left(\frac{1}{R_1}\right) = \left(\frac{1}{R_1}\right)_{,i}$$

and the domain of integration is

$$\frac{x_1'^2}{a_1^2} + \frac{x_2'^2}{a_2^2} + \frac{(x_3'-c)^2}{a_3^2} \le 1.$$

If we set

$$x_1 = y_1,\ \ x_2 = y_2, x_3 = y_3 + c, x_1' = y_1', x_2' = y_2', x_3' = y_3' + c \tag{15.16}$$

and

$$x_1 = z_1, x_2 = z_2, x_3 = z_3 - c, x_1' = z_1', x_2' = z_2', x_3' = -z_3' + c \tag{15.17}$$

then R_1 and R_2 (see Fig. 15.2) become

$$\begin{aligned} R_1^2 &= (y_1 - y_1')^2 + (y_2 - y_2')^2 + (y_3 - y_3')^2, \\ R_2^2 &= (z_1 - z_1')^2 + (z_2 - z_2')^2 + (z_3 - z_3')^2. \end{aligned} \tag{15.18}$$

R is the distance between points $\mathbf{x}$ and $\mathbf{x}'$. R_2 is the distance between point $\mathbf{x}$ and the mirror image of $\mathbf{x}'$. In order to use formulae (12.9) ~ (12.13) in (15.15), domain Ω is transformed into its mirror image Ω_z when the integral $\int_\Omega d\mathbf{x}'/R_2$ is considered (this domain transformation is not necessary for $\int_\Omega d\mathbf{x}'/R_1$). Referring to (12.9) & (12.13), we write

$$\phi(\mathbf{y}) = \int_\Omega \frac{1}{R_1}\,d\mathbf{x}' = \int_\Omega \frac{1}{R_1}\,d\mathbf{y}' = \pi a_1 a_2 a_3 \int_\lambda^\infty \frac{U}{\Delta}\,ds = V(\mathbf{y}), \tag{15.19}$$

where

$$\begin{aligned} U &= 1 - \left(\frac{y_1^2}{a_1^2 + s} + \frac{y_2^2}{a_2^2 + s} + \frac{y_3^2}{a_3^2 + s}\right), \\ \Delta &= \{(a_1^2 + s) + (a_2^2 + s) + (a_3^2 + s)\}^{1/2}, \\ &\frac{y_1^2}{a_1^2 + \lambda} + \frac{y_2^2}{a_2^2 + \lambda} + \frac{y_3^2}{a_3^2 + \lambda} = 1 \text{ for exterior points } \mathbf{x} \text{ of } \Omega \\ &\lambda = 0 \text{ for interior points } \mathbf{x} \text{ of } \Omega \end{aligned} \tag{15.20}$$

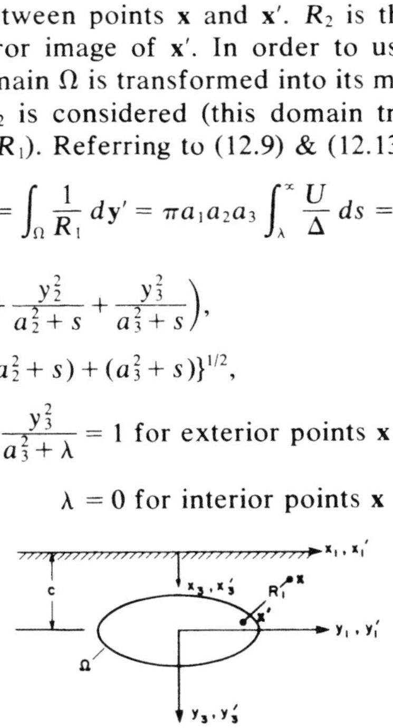

Fig. 15.2a. Coordinate systems for R_1 integral.

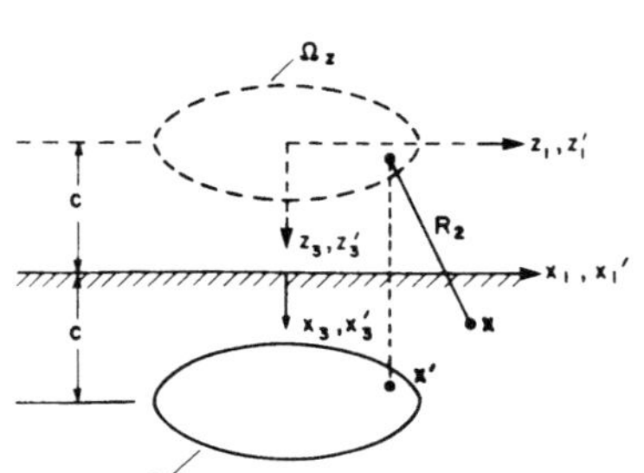

Fig. 15.2b. Coordinate systems for R_2 integral.

and

$$\phi(\mathbf{z}) = \int_\Omega \frac{1}{R_2}\,d\mathbf{x}' = \int_{\Omega_z} \frac{1}{R_2}\,dz' = \pi a_1 a_2 a_3 \int_\lambda^\infty \frac{U}{\Delta}\,ds = V(\mathbf{z}),$$

where

$$U = 1 - \left(\frac{z_1^2}{a_1^2+s} + \frac{z_2^2}{a_2^2+s} + \frac{z_3^2}{a_3^2+s}\right),$$

$$\Delta = \{(a_1^2+s)+(a_2^2+s)+(a_3^2+s)\}^{1/2},$$

$$\frac{z_1^2}{a_1^2+\lambda} + \frac{z_2^2}{a_2^2+\lambda} + \frac{z_3^2}{a_3^2+\lambda} = 1 \text{ for any point } \mathbf{x}.$$

Then (15.15) reduces to

$$u_i(\mathbf{x}) = \frac{\epsilon^*(1+\nu)}{4\pi(1-\nu)}[-\phi_{,i}(\mathbf{y}) + (3-4\nu)(2\delta_{3i}-1)\phi_{,i}(\mathbf{z}) - 2x_3\phi_{,3i}(\mathbf{z})] \quad (15.21)$$

and, using (11.39), we obtain

$$\begin{aligned}\epsilon_{ij}(\mathbf{x}) &= \tfrac{1}{2}(u_{i,j}+u_{j,i}) \\ &= \frac{\epsilon^*(1+\nu)}{4\pi(1-\nu)}[-\phi_{,ij}(\mathbf{y}) + (3-4\nu)(\delta_{3I}+\delta_{3J}-1)\phi_{,ij}(\mathbf{z}) \\ &\quad -(\delta_{3I}+\delta_{3J})\phi_{,ij}(\mathbf{z}) - 2x_3\phi_{,3ij}(\mathbf{z})]. \end{aligned} \quad (15.22)$$

The stress components are

$$\sigma_{ij}(\mathbf{x}) = C_{ijkl}[\epsilon_{ij}(\mathbf{x}) - \epsilon_{ij}^*(\mathbf{x})], \quad (15.23)$$

where $\epsilon_{ij}^* = 0$ for exterior points of Ω. For interior points, the stress components become

$$\begin{aligned}\sigma_{ij}(\mathbf{x}) = \frac{\epsilon^*(1+\nu)2\mu}{4\pi(1-\nu)}&[-4\pi\delta_{ij} - \phi_{,ij}(\mathbf{y}) + 4\nu\delta_{ij}\phi_{,33}(\mathbf{z}) \\ &+(3-4\nu)(\delta_{3I}+\delta_{3J}-1)\phi_{,ij}(\mathbf{z}) \\ &-(\delta_{3I}+\delta_{3J})\phi_{,ij}(\mathbf{z}) - 2x_3\phi_{,3ij}(\mathbf{z})] \end{aligned} \quad (15.24)$$

and for exterior points

$$\begin{aligned}\sigma_{ij}(\mathbf{x}) = \frac{\epsilon^*(1+\nu)2\mu}{4\pi(1-\nu)}&[-\phi_{,ij}(\mathbf{y}) + 4\nu\delta_{ij}\phi_{,33}(\mathbf{z}) + (3-4\nu)(\delta_{3I}+\delta_{3J}-1)\phi_{,ij}(\mathbf{z}) \\ &-(\delta_{3I}+\delta_{3J})\phi_{,ij}(\mathbf{z}) - 2x_3\phi_{,3ij}(\mathbf{z})]. \end{aligned} \quad (15.25)$$

Mindlin and Cheng (1950) have treated the same problem, except that their inclusion is a sphere ($a_1 = a_2 = a_3$).

The following numerical calculations are taken from the paper by Seo and Mura (1979), with $\nu = 0.3$ for all cases.

First consider the case when the inclusion with $a_1 = a_2 = 3a_3$ is located at depth $c = a_3$. The dimensionless stress distributions are shown in Figs. 15.3 ~ 15.4 by solid curves. For comparison, the results for a spherical inclusion ($a_1 = a_2 = a_3$) are shown by dotted curves. Figure 15.3 shows σ_{11}, σ_{33} components along the x_3-axis. For the ellipsoidal inclusion, σ_{11} is compressive at interior points and increases with x_3. It changes discontinuously to tension for exterior points and decays to zero at infinity. The spherical inclusion, however, causes tension for the interior points near the free surface. The stress component σ_{33} is continuous and takes a maximum tension at an interior point. The maximum compression appears at the interface point. Its magnitude is quite different for the two cases, and the compression region is broader in the ellipsoidal case than in the spherical. To see the effect of the free boundary surface, the stress distributions for an infinite medium are shown in Fig. 15.5. It can be seen that the stress uniformity inside the inclusion for the infinite medium is diminished by the existence of the free surface and some region near the surface undergoes tension. Note also that component σ_{33} in the matrix may be positive (tension in

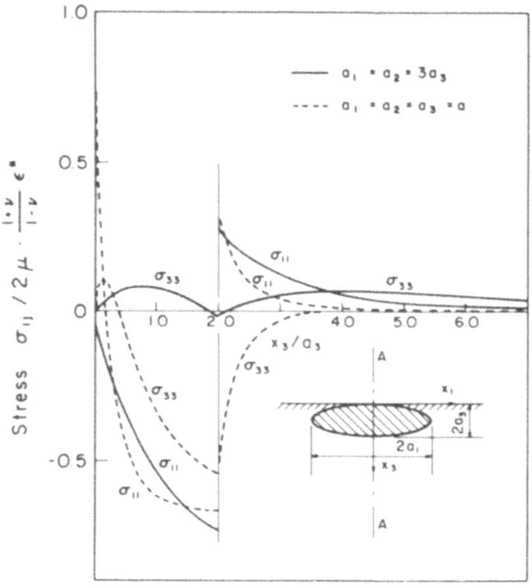

Fig. 15.3. Stress distribution on A-A section.

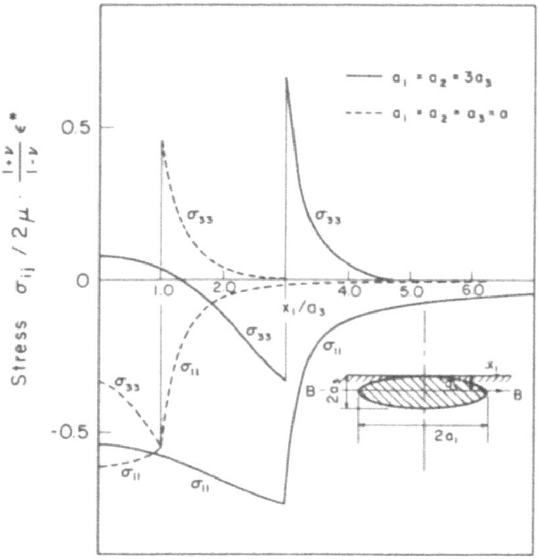

Fig. 15.4. Stress distribution on B-B section.

regions) far away from the inclusion, while σ_{33} for the infinite medium is compressive everywhere.

In Fig. 15.4 the stress distributions on the B–B section are shown. The stress jump at the interface is observed as expected for each

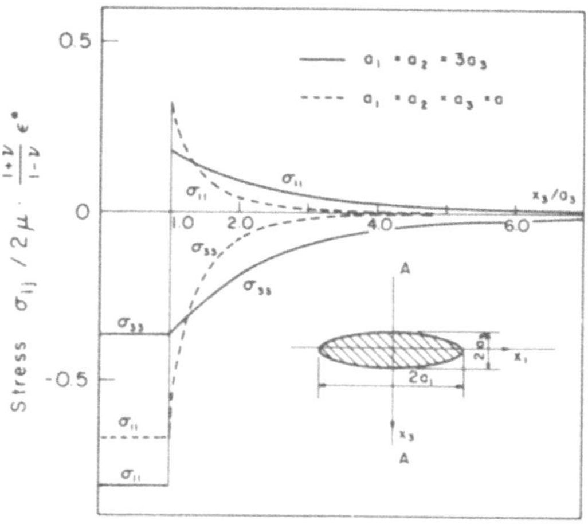

Fig. 15.5. Stress distribution on A-A section in case of infinite solid.

component of the stress. The effect of the free boundary surface appears in the σ_{33} component showing a tension region around the center of the ellipsoid. The effect of the inclusion shape on the stress distribution can be seen by comparing the interface point $x_1/a_3 = 1$ for the sphere with $x_1/a_3 = 3$ for the ellipsoid: $\sigma_{11} = \sigma_{22} = \sigma_{33}$ for sphere, but $\sigma_{11} = \sigma_{22} \neq \sigma_{33}$ for the ellipsoid.

The effect of depth c is shown by curves in Figs. 15.6 ~ 15.7. σ_{11} is tensile at the boundary surface and decreases with x_3. It jumps to compression for interior points of the inclusion and changes to tension again in the matrix. Due to the existence of the free surface, the compressive stress components for interior points of the inclusion increase with depth from the surface. This effect is less prevalent for the spherical inclusion. It may be said that the effect of the free surface disappears for the spherical inclusion, when $c > 2a$.

In Fig. 15.7 the effect of c on the stress component σ_{11} at $x_1 = x_2 = x_3 = 0$ is shown for various aspect ratios of the ellipsoid. For the spherical inclusion, σ_{11} shows large tension at the surface that decreases with c. It becomes almost zero when $c > 3a$. σ_{11} at the surface decreases when the ellipsoid becomes flatter. However, the effect of c is less prevalent in the flatter inclusion. It is seen that the eigenstress at the surface depends not only on the depth of the inclusion from the surface but also on the shape of the inclusion.

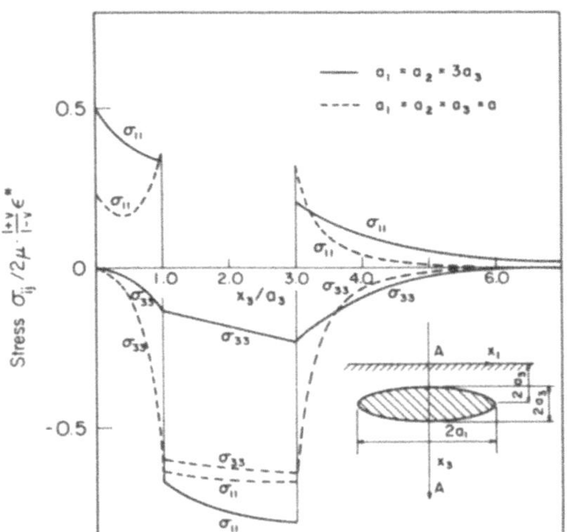

Fig. 15.6. Stress distribution on A-A section.

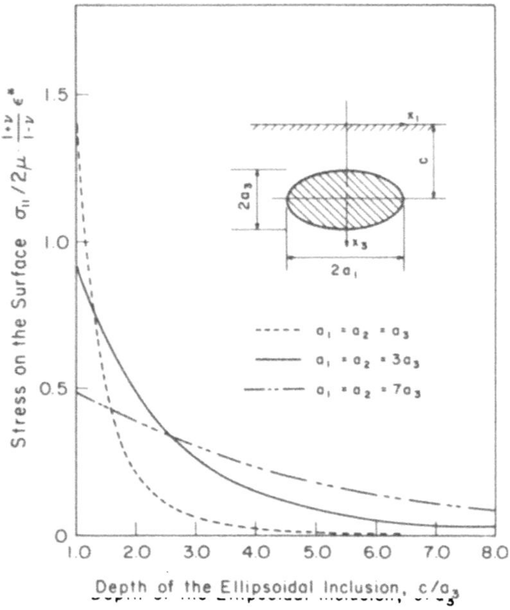

Fig. 15.7. Effect of the inclusion depth on the surface stress.

The elastic strain energy is, from (13.3),

$$W^* = -(1/2)\int_\Omega \sigma_{ii}\epsilon^* d\mathbf{x}. \tag{15.25.1}$$

Substituting from (15.24), we have

$$W^* = 2\mu\,\frac{1+\nu}{1-\nu}(\epsilon^*)^2 V\left(1 - \frac{1+\nu}{2\pi V}\int_\Omega \phi_{,33}(\mathbf{z})d\mathbf{x}\right), \tag{15.25.2}$$

where V is the volume of the inclusion. The factor $2\mu(1+\nu)(\epsilon^*)^2V/(1-\nu)$ is the elastic strain energy (13.21) for the case of an infinite space. The second term in (15.25.2) is the correction factor due to the free surface.

If the inclusion is a sphere, we have

$$\phi(z) = 4\pi a^3/3r = V/r, \tag{15.25.3}$$

where $r = (z_1^2 + z_2^2 + z_3^2)^{1/2}$. Since

$$\int_\Omega \left(\frac{1}{r}\right)_{,33} d\mathbf{x} = 2\pi a^3/6c^3, \tag{15.25.4}$$

the correction term becomes $-(1+\nu)a^3/6c^3$. Eshelby (1961) has given this correction term as $-(1+\nu)a^3/4c^3$ which involves a trivial numerical error.

The force acting on the inclusion is defined as

$$F = -\frac{\delta W^*}{\delta c} = -\mu\frac{(1+\nu)^2}{1-\nu}(\epsilon^*)^2 Va^3/c^2. \tag{15.25.5}$$

Thus, the free surface attracts the inclusion.

**Cuboidal inclusion with uniform eigenstrains*

When a cuboidal inclusion with a uniform eigenstrain is located in a half-space, the elastic field can be obtained from (15.12) ~ (15.13) by taking Ω as the cuboidal domain. The surface integrals (on $|\Omega|$) in (15.13) have been explicitly performed by T.H. Lin and T.K. Tung (1962) in their calculation of the stress field produced by a localized plastic slip at a free surface. Y.P. Chiu (1978) attempts to extend his solution for an infinite body to the case of a half-space by applying on the boundary plane, equal and opposite normal and shear stresses in order to obtain the free boundary conditions.

**Periodic distribution of eigenstrains*

When eigenstrain $\epsilon^*_{ij}(\mathbf{x})$ is given by

$$\epsilon^*_{ij}(\mathbf{x}) = \bar{\epsilon}^*_{ij}(\boldsymbol{\xi})\exp(i\boldsymbol{\xi}\cdot\mathbf{x}) \tag{15.26}$$

in a half-space, the displacement field is the sum of (3.10) and $u^s_i(\mathbf{x})$, where $u^s_i(\mathbf{x})$ is the displacement caused by applying on the boundary plane $x_3 = 0$ equal and opposite normal and shear stresses which would render this boundary stress-free. Since the stress field for the full space problem is (3.12), the x_i-component of the force to be applied on the boundary $x_3 = 0$ is

$$F_i = -C_{ijkl}\{C_{pqmn}\bar{\epsilon}^*_{mn}(\boldsymbol{\xi})\xi_q\xi_l N_{kp}(\boldsymbol{\xi})D^{-1}(\boldsymbol{\xi}) - \bar{\epsilon}^*_{kl}\}n_j\exp(i\boldsymbol{\xi}\cdot\mathbf{x}^0)$$

$$\equiv \bar{F}_i\exp(i\boldsymbol{\xi}\cdot\mathbf{x}^0), \tag{15.27}$$

where $\mathbf{x}^0$ is a point on the boundary, $\boldsymbol{\xi}\cdot\mathbf{x}^0 = \xi_1 x^0_1 + \xi_2 x^0_2$, and $\mathbf{n}$ is the outward unit normal on the boundary. If F_i is a unit force, the induced displacement field is given by (15.10) with $x'_3 = 0$. Thus

$$u^s_i(\mathbf{x}) = \iint_{-\infty}^{\infty} \bar{F}_j G_{ij}(\mathbf{x},\mathbf{x}^0)\exp(i\boldsymbol{\xi}\cdot\mathbf{x}^0)dx^0_1 dx^0_2. \tag{15.28}$$

The stress field of the problem is the sum of (3.12) and σ^s_{ij}, where

$$\sigma^s_{ij}(\mathbf{x}) = C_{ijkl}\frac{\partial}{\partial x_l}\int\!\!\int_{-\infty}^{\infty} \bar{F}_m G_{km}(\mathbf{x}, \mathbf{x}^0)\exp(i\boldsymbol{\xi}\cdot\mathbf{x}^0)dx^0_1 dx^0_2. \tag{15.29}$$

Owen and Mura (1967) have performed the above integrations and obtained

$$\sigma^s_{ij}(\mathbf{x}) = \frac{\bar{F}_m C_{ijkl}}{2\xi}\frac{\partial}{\partial x_l}[H_{km}\exp\{i(\xi_1 x_1 + \xi_2 x_2) - \xi x_3\}], \tag{15.30}$$

where

$$\begin{aligned} \xi &= (\xi_1^2 + \xi_2^2)^{1/2}, \\ H_{km} &= \mu^{-1}[\delta_{km} - ix_3\{\delta_{3k}\xi_m + \delta_{3m}\xi_k - \delta_{3k}\delta_{3m}(\xi_K + \xi_M + i\xi)\} \\ &\quad - i(\delta_{K1} + \delta_{K2})(\delta_{M1} + \delta_{M2})\xi_k\xi_m/\xi + i(\xi_2\delta_{k1} - \xi_1\delta_{k2})(\xi_2\delta_{m1} - \xi_1\delta_{m2})/\xi] \\ &\quad + (\lambda + \mu)^{-1}\{\delta_{k3}\delta_{m3} + (1 - \delta_{K3}\delta_{M3})\xi_k\xi_m/\xi^2 + i(\delta_{m3}\xi_k - \delta_{k3}\xi_m)/\xi\}. \end{aligned} \tag{15.31}$$

The solution in this subsection has many applications when the Fourier series and integral methods are employed. When eigenstrains are given by summations or integrations of (15.26), the corresponding solutions can be constructed from summations or integrations (with respect to $\boldsymbol{\xi}$) of the solution given here. Owen and Mura have calculated the stress field of a Frank dislocation network located near a free surface. Owen (1971) has further calculated by the present method the stress disturbance near the free surface of a fiber-reinforced material subjected to tensile traction. Saito et al. (1972) have considered the case of a periodic distribution of dislocations in a thin film.

Joined half-spaces

Let us consider the case when an infinite body consists of two half-spaces with different shear modulus and Poisson's ratio (see Fig. 15.8). When one of the half-spaces, e.g. region 1, contains an inclusion Ω with eigenstrain ϵ^*_{ij}, the displacement field can be expressed in the form similar to (15.5),

$$u_i(\mathbf{x}) = \int_{-\infty}^{\infty} C_{jlmn}\epsilon^*_{mn}(\mathbf{x}')\frac{\partial}{\partial x'_l}G_{ij}(\mathbf{x}, \mathbf{x}')d\mathbf{x}'. \tag{15.32}$$

It is easily proved that (15.32) satisfies (15.6); this done in the manner

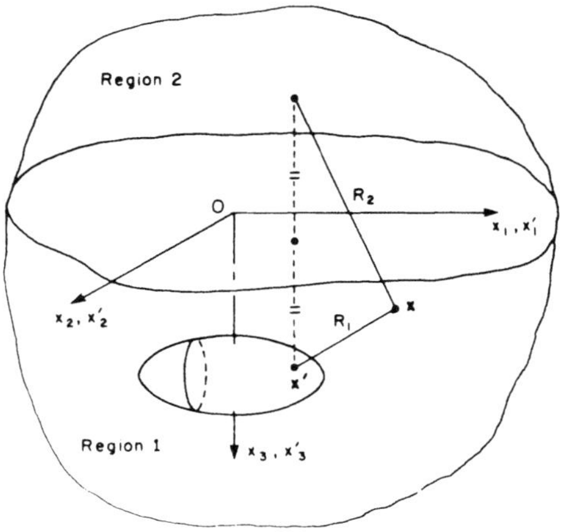

Fig. 15.8. Joint half-spaces.

outlined in (15.8). Since ϵ^*_{mn} is zero outside of Ω,

$$u_i(\mathbf{x}) = \int_\Omega C_{jlmn}\epsilon^*_{mn}(\mathbf{x}') \frac{\partial}{\partial x'_l} G_{ij}(\mathbf{x}, \mathbf{x}')d\mathbf{x}'. \tag{15.33}$$

$G_{ij}(\mathbf{x}, \mathbf{x}')$ in the above expressions is Green's function in the two semi-infinite solids; it is the x_i-direction displacement at point $\mathbf{x}$ caused by the x_j-direction unit body force applied at point $\mathbf{x}'$.

The elastic moduli in regions 1 and 2 are denoted by subscripts 1 and 2, respectively.

When the two half-spaces are perfectly bonded, continuity of all components of displacements and tractions at the interface is required. The Green's function for this case has been found by Rongved (1955). The solution is given in terms of the Papkovich-Neuber displacement functions (see Sokolnikoff 1956),

$$G_{ij}(\mathbf{x}, \mathbf{x}') = (1/2G)\{(\kappa + 1)\Phi_{ij} - (x_k\Phi_{kj} + \psi_j)_{,i}\}, \tag{15.34}$$

where $\kappa = 3 - 4\nu$ and point $\mathbf{x}'$ is in region 1. For $\mathbf{x}$ in region 1,

$$\Phi_{11} = \Phi_{22} = \frac{1}{2\pi(\kappa_1 + 1)}(1/R_1 + S/R_2),$$

$$\Phi_{21} = \Phi_{12} = 0,$$

$$\Phi_{3j} = \frac{1}{2\pi(\kappa_1 + 1)}\{(A\kappa_1 - S)(x_j - x'_j)/R_2(R_2 + x_3 + x'_3) \\ - 2Ax'_3(x_j - x'_j)/R_2^3\},$$

$$\psi_j = \frac{1}{2\pi(\kappa_1+1)}\{-(A\kappa_1^2 + B - 2S\kappa_1)(x_j - x_j')/2(R_2 + x_3 + x_3')$$
$$+ (A\kappa_1 - S)x_3'(x_j - x_j')/R_2(R_2 + x_3 + x_3')\}, \tag{15.35}$$
$$\Phi_{13} = \Phi_{23} = 0,$$
$$\Phi_{33} = \frac{1}{2\pi(\kappa_1+1)}\{1/R_1 + A\kappa_1/R_2 + 2Ax_3'(x_3 + x_3')/R_2^3\},$$
$$\psi_3 = \frac{1}{2\pi(\kappa_1+1)}\{-x_3'/R_1 - A\kappa_1 x_3'/R_2 + (A\kappa_1^2 - B)\log(R_2 + x_3' + x_3)^{1/2}\}.$$

For $\mathbf{x}$ in region 2,

$$\Phi_{11} = \Phi_{22} = \frac{T}{2\pi(\kappa_1+1)R_1},$$
$$\Phi_{21} = \Phi_{12} = 0,$$
$$\Phi_{3j} = -\frac{(1-B-T)}{2\pi(\kappa_1+1)}\frac{x_j - x_j'}{R_1(R_1 - x_3 + x_3')}, \tag{15.36}$$
$$\psi_j = \frac{1}{2\pi(\kappa_1+1)}[-\{(1-A)\kappa_1 + (1-B-2T)\kappa_2\}(x_j - x_j')/2(R_1 - x_3 + x_3')$$
$$+ (1-A-T)x_3'(x_j - x_j')/R_1(R_1 - x_3 + x_3')],$$
$$\Phi_{13} = \Phi_{23} = 0,$$
$$\Phi_{33} = \frac{(1-B)}{2\pi(\kappa_1+1)R_1},$$
$$\psi_3 = \frac{1}{2\pi(\kappa_1+1)}[-(1-A)x_3'/R_1$$
$$+ \{(1-A)\kappa_1 - (1-B)\kappa_2\}\log(R_1 - x_3 + x_3')^{1/2}],$$

where

$$i, j = 1, 2,$$
$$R_1^2 = (x_1 - x_1')^2 + (x_2 - x_2')^2 + (x_3 - x_3')^2,$$
$$R_2^2 = (x_1 - x_1')^2 + (x_2 - x_2')^2 + (x_3 + x_3')^2,$$
$$\Gamma = G_2/G_1, \tag{15.37}$$
$$A = (1-\Gamma)/(1+\kappa_1\Gamma), \quad B = (\kappa_2 - \kappa_1\Gamma)/(\kappa_2 + \Gamma),$$
$$S = (1-\Gamma)/(1+\Gamma), \quad T = 2(\kappa_1+1)\Gamma/(k_2+1)(1+\Gamma).$$

When the two half-spaces have smooth contact, the boundary

conditions for Green's function are continuity of the normal components of displacement and stress and a condition for vanishing shear components of stress at the interface. The Green function for this case has been found by Dundurs and Hetényi (1965). For $\mathbf{x}$ in region 1, ($\mathbf{x}'$ is in region 1),

$$\begin{aligned}
\Phi_{11} &= \Phi_{22} = \frac{1}{2\pi(\kappa_1+1)}(1/R_1 + 1/R_2),\\
\Phi_{21} &= \Phi_{12} = 0,\\
\Phi_{3j} &= \frac{1-D}{2\pi(\kappa_1+1)}\{(\kappa_1-1)(x_j-x_j')/R_2(R_2+x_3+x_3') - 2x_3'x_3/R_2^3\},\\
\psi_j &= \frac{(1-D)(\kappa_1-1)}{2\pi(\kappa_1+1)}\{-(\kappa_1-1)(x_j-x_j')/2(R_2+x_3+x_3')\\
&\quad + x_3'(x_j-x_j')/R_2(R_2+x_3+x_3')\},\\
\Phi_{13} &= \Phi_{23} = 0,\\
\Phi_{33} &= \frac{1}{2\pi(\kappa_1+1)}\{1/R_1 + [\kappa_1 - D(\kappa_1+1)]/R_2 + 2(1-D)x_3'(x_3+x_3')/R_2^3\},\\
\psi_3 &= \frac{1}{2\pi(\kappa_1+1)}\{-x_3'/R_1 - [\kappa_1 - D(\kappa_1-1)]x_3'/R_2\\
&\quad + (1-D)(\kappa_1^2-1)\log(R_2+x_3+x_3')^{1/2}\}.
\end{aligned} \tag{15.38}$$

For $\mathbf{x}$ in region 2,

$$\begin{aligned}
\Phi_{11} &= \Phi_{22} = \Phi_{12} = \Phi_{21} = 0,\\
\Phi_{3j} &= \frac{D}{2\pi(\kappa_1+1)}\{(\kappa_1-1)(x_j-x_j')/R_1(R_1-x_3+x_3') - 2x_3'(x_j-x_j')/R_1^3\},\\
\psi_j &= \frac{D(\kappa_2-1)}{2\pi(\kappa_1+1)}\{(\kappa_1-1)(x_j-x_j')/2(R_1-x_3+x_3')\\
&\quad - x_3'(x_j-x_j')/R_1(R_1-x_3+x_3')\},\\
\Phi_{13} &= \Phi_{23} = 0,\\
\Phi_{33} &= \frac{D}{2\pi(\kappa_1+1)}\{(\kappa_1+1)/R_1 - 2x_3'(x_3-x_3')/R_1^3\},\\
\psi_3 &= \frac{D(\kappa_2-1)}{2\pi(\kappa_1+1)}\{x_3'/R_1 - (\kappa_1+1)\log(R_1-x_3+x_3')^{1/2}\},
\end{aligned} \tag{15.39}$$

where

$$i, j = 1, 2, \qquad D = (\kappa_1 + 1)\Gamma/\{(\kappa_1 + 1)\Gamma + \kappa_2 + 1\}. \tag{15.40}$$

Special cases $\Gamma = 0$, 1 and ∞ correspond to a single half-space, a homogeneous whole space, and the smooth rigid base contact, respectively.

It should be noted from (15.35) ~ (15.40) that (15.34) can be derived from Green's function for an homogeneous infinite medium. The terms containing R_2 and $R_1 - x_3 + x_3'$ in (15.35) ~ (15.39) can be constructed from R_1 by proper elementary operations (changing arguments, multiplying by $(x_j - x_j')$, differentiating, and integrating). Aderogba (1977) has pointed out that an elastic solution for dissimilar joined half-spaces is constructed from the solution for an homogeneous whole-space.

Dundurs and Guell (1965) and Guell and Dundurs (1967) (see also Aderogba 1972) have calculated the stress field for a center of dilatation where

$$\epsilon_{ij}^*(\mathbf{x}) = \delta_{ij}\epsilon^*\delta(x_1, x_2, x_3 - c). \tag{15.41}$$

From (15.33) this case gives

$$u_i(\mathbf{x}) = C_{jlmn}\epsilon^* \left[\frac{\partial}{\partial x_l'} G_{ij}(x, x')\right]_{x_1'=0,\, x_2'=0,\, x_3'=c}. \tag{15.42}$$

According to Dundurs and Guell, the normal traction transmitted by the interface is

$$(\sigma_{33})_{x_3=0} = \frac{4DG_1}{c^3}\left(\frac{c}{R_1}\right)^3\left\{1 - 3\left(\frac{c}{R_1}\right)^2\right\} \tag{15.43}$$

for the smooth interface, and as

$$(\sigma_{33})_{x_3=0} = \frac{2(1-A)G_1}{c^3}\left(\frac{c}{R_1}\right)^3\left\{1 - 3\left(\frac{c}{R_1}\right)^2\right\} \tag{15.44}$$

for the perfectly bonded interface. It is interesting to compare the above results with the solution for the whole space (a center of dilatation is placed at point $(0, 0, c)$),

$$(\sigma_{33})_{x_3=0} = \frac{2G_1}{c^3}\left(\frac{c}{R_1}\right)^3\left\{1 - 3\left(\frac{c}{R_1}\right)^2\right\} \equiv \sigma_{33}^*. \tag{15.45}$$

Then, we have

$$\begin{aligned}(\sigma_{33})_{x_3=0} &= 2D\sigma_{33}^{*} \quad \text{for smooth surface,}\\ &= (1-A)\sigma_{33}^{*} \quad \text{for perfect bond.}\end{aligned} \tag{15.46}$$

Moon and Pao (1967) have calculated the interaction energy between a spherical inhomogeneity and a center of dilatation. The center of dilatation may be attracted to or repelled by the inhomogeneity, depending on the rigidity of the inhomogeneity and the matrix material.

3

Anisotropic inclusions

In this chapter we investigate elastic fields due to ellipsoidal inclusions in anistropic materials. Since explicit expressions for anisotropic Green's functions are not available, the technique used in the last chapter is not applicable. We start from the result in Section 6 and discuss how to integrate the integrals involved in the Fourier space and physical space.

The results are presented in general forms, which are applicable to any distribution of eigenstrains given in an ellipsoidal inclusion. Known solutions can be derived as special cases from these general results.

16. Elastic field of an ellipsoidal inclusion

When eigenstrain $\epsilon_{ij}^*(x)$ is distributed in an ellipsoidal sub-domain Ω of an infinitely extended anisotropic material, the displacement is expressed by (6.2),

$$u_i(\mathbf{x}) = -(2\pi)^{-3}\frac{\partial}{\partial x_l}\int_\Omega d\mathbf{x}'\int_{-\infty}^{\infty} C_{jlmn}\epsilon_{nm}^*(\mathbf{x}')N_{ij}(\boldsymbol{\xi})D^{-1}(\boldsymbol{\xi})\exp\{i\boldsymbol{\xi}\cdot(\mathbf{x}-\mathbf{x}')\}d\boldsymbol{\xi} \tag{16.1}$$

or, changing $\boldsymbol{\xi}$ to $-\boldsymbol{\xi}$, by

$$u_i(\mathbf{x}) = -(2\pi)^{-3}\frac{\partial}{\partial x_l}\int_\Omega d\mathbf{x}'\int_{-\infty}^{\infty} C_{jlmn}\epsilon_{nm}^*(\mathbf{x}')N_{ij}(\boldsymbol{\xi})D^{-1}(\boldsymbol{\xi}) \times\exp\{-i\boldsymbol{\xi}\cdot(\mathbf{x}-\mathbf{x}')\}d\boldsymbol{\xi}, \tag{16.2}$$

where the integral with respect to x_1', x_2', x_3' is defined in domain Ω, which is the ellipsoid

$$x_1^2/a_1^2 + x_2^2/a_2^2 + x_3^2/a_3^2 \le 1$$

(see Fig. 11.1).

The integration with respect to $\boldsymbol{\xi}$-space is considered first. The volume element in the $\boldsymbol{\xi}$-space, $d\boldsymbol{\xi}$, is

$$d\boldsymbol{\xi} = d\xi_1 d\xi_2 d\xi_3 = \xi^2 d\xi dS(\bar{\boldsymbol{\xi}}), \tag{16.3}$$

where

$$\begin{aligned} \xi &= (\xi_1^2 + \xi_2^2 + \xi_3^2)^{1/2}, \\ \bar{\boldsymbol{\xi}} &= \boldsymbol{\xi}/\xi, \end{aligned} \tag{16.4}$$

and $dS(\bar{\boldsymbol{\xi}})$ is a surface element on the unit sphere S^2 in the $\boldsymbol{\xi}$-space. Then, (16.1) is written as

$$\begin{aligned} u_i(\mathbf{x}) = & -(2\pi)^{-3} \frac{\partial}{\partial x_l} \int_\Omega d\mathbf{x}' \int_0^\infty \xi^2 d\xi \int_{S^2} C_{jlmn} \epsilon^*_{nm}(\mathbf{x}') N_{ij}(\boldsymbol{\xi}) D^{-1}(\boldsymbol{\xi}) \\ & \times \exp\{i\boldsymbol{\xi} \cdot (\mathbf{x} - \mathbf{x}')\} dS(\bar{\boldsymbol{\xi}}) \end{aligned} \tag{16.5}$$

and (16.2) leads to

$$\begin{aligned} u_1(\mathbf{x}) = & -(2\pi)^{-3} \frac{\partial}{\partial x_l} \int_\Omega d\mathbf{x}' \int_{-\infty}^0 \xi^2 d\xi \int_{S^2} C_{jlmn} \epsilon^*_{nm}(\mathbf{x}') N_{ij}(\boldsymbol{\xi}) D^{-1}(\boldsymbol{\xi}) \\ & \times \exp\{i\boldsymbol{\xi} \cdot (\mathbf{x} - \mathbf{x}')\} dS(\bar{\boldsymbol{\xi}}), \end{aligned} \tag{16.6}$$

where ξ is changed to $-\xi$. The sum of (16.5) and (16.6) is

$$\begin{aligned} u_i(\mathbf{x}) = & -\frac{1}{2}(2\pi)^{-3} \frac{\partial}{\partial x_l} \int_\Omega d\mathbf{x}' \int_{-\infty}^\infty d\xi \int_{S^2} C_{jlmn} \epsilon^*_{nm}(\mathbf{x}') N_{ij}(\bar{\boldsymbol{\xi}}) D^{-1}(\bar{\boldsymbol{\xi}}) \\ & \times \exp\{i\xi\bar{\boldsymbol{\xi}} \cdot (\mathbf{x} - \mathbf{x}')\} dS(\bar{\boldsymbol{\xi}}). \end{aligned} \tag{16.7}$$

In the above integrals the homogeneity of degree zero of $\xi^2 N_{ij}(\boldsymbol{\xi}) D^{-1}(\boldsymbol{\xi})$ has been used, namely,

$$\xi^2 N_{ij}(\boldsymbol{\xi}) D^{-1}(\boldsymbol{\xi}) = N_{ij}(\bar{\boldsymbol{\xi}}) D^{-1}(\bar{\boldsymbol{\xi}}). \tag{16.8}$$

Since $\int_{-\infty}^\infty \exp(i\xi\eta) d\xi = 2\pi\delta(\eta)$, (16.7) becomes

$$u_i(\mathbf{x}) = -\frac{1}{8\pi^2} \frac{\partial}{\partial x_l} \int_\Omega d\mathbf{x}' \int_{S^2} C_{jlmn} \epsilon^*_{nm}(\mathbf{x}') N_{ij}(\bar{\boldsymbol{\xi}}) D^{-1}(\bar{\boldsymbol{\xi}}) \delta(\bar{\boldsymbol{\xi}} \cdot (\mathbf{x} - \mathbf{x}')) dS(\bar{\boldsymbol{\xi}}). \tag{16.9}$$

The following transformations of variables are used

$$\begin{aligned} & x_1/a_1 = y_1, \quad x_2/a_2 = y_2, \quad x_3/a_3 = y_3, \\ & x_1'/a_1 = y_1', \quad x_2'/a_2 = y_2', \quad x_3'/a_3 = y_3', \\ & a_1\bar{\xi}_1 = \zeta_1, \quad a_2\bar{\xi}_2 = \zeta_2, \quad a_3\bar{\xi}_3 = \zeta_3, \\ & \zeta_1/\zeta = \bar{\zeta}_1, \quad \zeta_2/\zeta = \bar{\zeta}_2, \quad \zeta_3/\zeta = \bar{\zeta}_3, \\ & \zeta = (\zeta_1^2 + \zeta_2^2 + \zeta_3^2)^{1/2} = (a_1^2\bar{\xi}_1^2 + a_2^2\bar{\xi}_2^2 + a_3^2\bar{\xi}_3^2)^{1/2}, \\ & \bar{\boldsymbol{\zeta}} \cdot \mathbf{y} = \bar{\boldsymbol{\xi}} \cdot \mathbf{x}/\zeta. \end{aligned} \tag{16.10}$$

We have also

$$d\mathbf{x}' = dx_1'dx_2'dx_3' = a_1a_2a_3dy_1'dy_2'dy_3' = a_1a_2a_3rdrd\phi dz, \tag{16.11}$$

where r, ϕ and z are shown in Fig. 16.1 and

$$z = \bar{\boldsymbol{\zeta}} \cdot \mathbf{y}' = \bar{\boldsymbol{\xi}} \cdot \mathbf{x}'/\zeta. \tag{16.12}$$

Note that (16.10) transforms the ellipsoid into the unit sphere S^2. Since

$$\partial/\partial x_l \delta(\bar{\boldsymbol{\xi}} \cdot (\mathbf{x} - \mathbf{x}')) = \bar{\xi}_l \delta'(\bar{\boldsymbol{\xi}} \cdot (\mathbf{x} - \mathbf{x}')), \tag{16.13}$$

where δ' is the derivative of δ with respect to the argument, (16.9) is written as

$$u_i(\mathbf{x}) = -\frac{a_1a_2a_3}{8\pi^2}\int_{-1}^{1} dz \int_0^{2\pi} d\phi \int_0^R rdr \int_{S^2} C_{klmn}\epsilon^*_{nm}(\mathbf{x}')N_{ik}(\bar{\boldsymbol{\xi}}) \times D^{-1}(\bar{\boldsymbol{\xi}})\bar{\xi}_l\delta'(\zeta\bar{\boldsymbol{\zeta}} \cdot \mathbf{y} - \zeta z)dS(\bar{\boldsymbol{\xi}}), \tag{16.14}$$

where

$$R = (1 - z^2)^{1/2} \tag{16.15}$$

must be a real number.

Differentiation of (16.14) with respect to x_j leads to

$$u_{i,j}(\mathbf{x}) = -\frac{a_1a_2a_3}{8\pi^2}\int_{-1}^{1} dz \int_0^{2\pi} d\phi \int_0^R rdr \int_{S^2} C_{klmn}\epsilon^*_{nm}(\mathbf{x}')N_{ik}(\bar{\boldsymbol{\xi}})D^{-1}(\bar{\boldsymbol{\xi}}) \times \bar{\xi}_l\bar{\xi}_j\delta''(\zeta\bar{\boldsymbol{\zeta}} \cdot \mathbf{y} - \zeta z)dS(\bar{\boldsymbol{\xi}}). \tag{16.16}$$

The argument $\zeta(\bar{\boldsymbol{\zeta}} \cdot \mathbf{y} - z)$ of the derivatives of Dirac's delta function becomes zero when

$$\bar{\boldsymbol{\zeta}} \cdot \mathbf{y} = z. \tag{16.17}$$

When point $\mathbf{y}$ is inside the unit sphere, $z^2 < 1$ for any $\bar{\boldsymbol{\zeta}}$, and R becomes a real number. When point $\mathbf{y}$ is outside the unit sphere, R is real for only

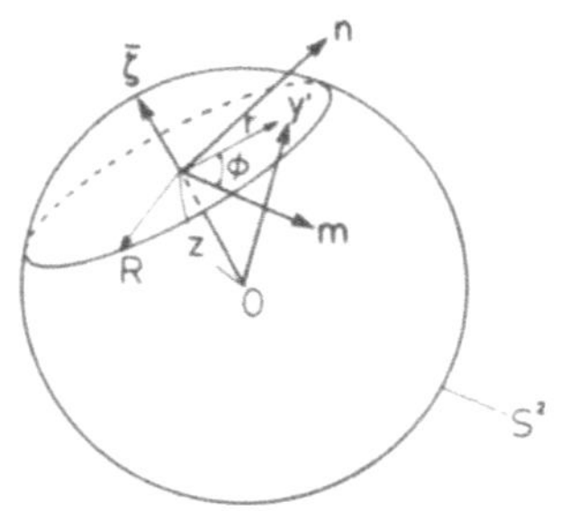

Fig. 16.1. Unit sphere and parameters.

certain $\bar{\zeta}$. In other words, for $\mathbf{x} \in \Omega$, the integration domain for $\bar{\zeta}$ is the whole surface of the unit sphere S^2, and for $\mathbf{x} \notin \Omega$ the integration domain becomes the subspace of S^2 satisfying the condition $z^2 < 1$. The subspace will be denoted by S^*, which is the unshaded domain seen in Fig. 16.2.

The expressions (16.14), (16.16) and the other modified expressions appearing in the later sections remain in the same form for any arbitrarily oriented coordinate system, except that ζ is taken as

$$\zeta = \{a_1^2(a_{j1}\bar{\xi}_j)^2 + a_2^2(a_{j2}\bar{\xi}_j)^2 + a_3^2(a_{j3}\bar{\xi}_j)^2\}^{1/2}, \tag{16.18}$$

where a_{jl} is the direction cosine between the x_j-axis and the lth principal direction of the ellipsoid.

17. Formulae for interior points

Let us consider the case when point $\mathbf{x}$ is inside the ellipsoid (interior point) or point $\mathbf{y}$ is inside the unit sphere. Since $\delta'(\zeta\bar{\zeta}\cdot\mathbf{y} - \zeta z) = -\zeta^{-1}(\partial/\partial z)\delta(\zeta\bar{\zeta}\cdot\mathbf{y} - \zeta z)$, the integration of (16.14) by parts with respect to z leads to

$$u_i(\mathbf{x}) = -\frac{a_1a_2a_3}{8\pi^2}\int_{-1}^{1} dz \int_0^{2\pi} d\phi\left[\int_0^R r dr \frac{\partial}{\partial z}\,\epsilon^*_{nm}(\mathbf{x}') - z\{\epsilon^*_{nm}(\mathbf{x}')\}_{r=R}\right]$$
$$\times \int_{S^2} C_{klmn}N_{ik}(\bar{\xi})D^{-1}(\bar{\xi})\bar{\xi}_l\zeta^{-1}\delta(\zeta\bar{\zeta}\cdot\mathbf{y} - \zeta z)dS(\bar{\xi}), \tag{17.1}$$

where R and $\mathbf{x}'$ are only functions of z for a fixed $\bar{\xi}$ through (16.15) and (16.12). The boundary values of the integration by parts vanish, since

$$\delta(\zeta\bar{\zeta}\cdot\mathbf{y} \pm \zeta) = 0 \tag{17.2}$$

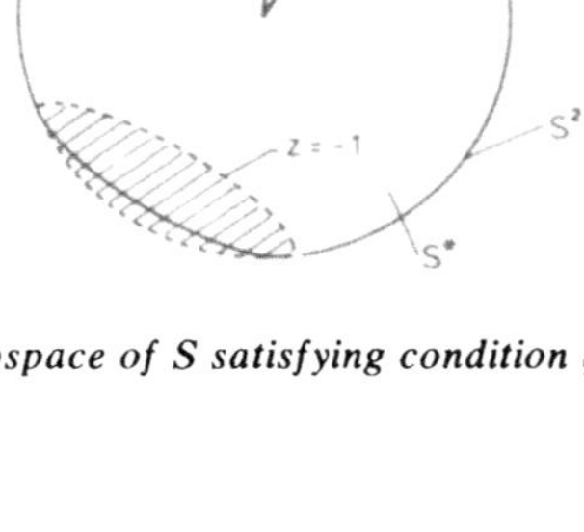

Fig. 16.2. S^* is a subspace of S satisfying condition $\bar{\zeta}\cdot\mathbf{y} \leq 1$ for a given $\mathbf{y}$.

for interior points. Furthermore, for the interior points there exists a z satisfying the condition

$$\bar{\boldsymbol{\zeta}} \cdot \mathbf{y} - z = 0, \quad -1 \leq z \leq 1, \tag{17.3}$$

for any $\bar{\boldsymbol{\zeta}}$. Then, (17.1) can be integrated with respect to z,

$$u_i(\mathbf{x}) = -\frac{a_1 a_2 a_3}{8\pi^2} \int_0^{2\pi} d\phi \left[\int_0^R r dr \frac{\partial}{\partial z} \epsilon^*_{nm}(\mathbf{x}') - (\bar{\boldsymbol{\zeta}} \cdot \mathbf{y}) \{\epsilon^*_{nm}(\mathbf{x}')\}_{r=R} \right]_{z=\bar{\zeta}\cdot\mathbf{y}}$$
$$\times \int_{S^2} C_{klmn} N_{ik}(\bar{\boldsymbol{\xi}}) D^{-1}(\bar{\boldsymbol{\xi}}) \bar{\xi}_l \zeta^{-2} dS(\bar{\boldsymbol{\xi}}), \tag{17.4}$$

where

$$R = \{1 - z^2\}^{1/2}. \tag{17.5}$$

The points $\mathbf{x}'$ in (17.4) must satisfy the condition

$$\zeta(\bar{\boldsymbol{\zeta}} \cdot \mathbf{y} - \bar{\boldsymbol{\zeta}} \cdot \mathbf{y}') = \bar{\boldsymbol{\xi}} \cdot (\mathbf{x} - \mathbf{x}') = 0 \tag{17.6}$$

or

$$z = \bar{\boldsymbol{\zeta}} \cdot \mathbf{y}. \tag{17.7}$$

Similarly, (16.16) becomes, after twice intergrating by parts with respect to z,

$$u_{i,j}(\mathbf{x}) = -\frac{a_1 a_2 a_3}{8\pi^2} \int_0^{2\pi} d\phi \left[\int_0^R r dr \frac{\partial^2}{\partial z^2} \epsilon^*_{nm}(\mathbf{x}') - z \left\{ \frac{\partial}{\partial z} \epsilon^*_{nm}(\mathbf{x}') \right\}_{r=R} \right.$$
$$- z \frac{\partial}{\partial z} \left\{ \epsilon^*_{nm}(\mathbf{x}') \right\}_{r=R}$$
$$\left. - \{\epsilon^*_{nm}(\mathbf{x}')\}_{r=R} \right]_{z=\bar{\zeta}\cdot\mathbf{y}} \int_{S^2} C_{klmn} N_{ik}(\bar{\boldsymbol{\xi}}) D^{-1}(\bar{\boldsymbol{\xi}}) \bar{\xi}_l \bar{\xi}_j \zeta^{-3} dS(\bar{\boldsymbol{\xi}}). \tag{17.8}$$

where $\mathbf{x}'$ is subject to condition (17.6), and R is defined by (17.5). The above result can also be derived by differentiating (17.4), where $(\bar{\boldsymbol{\zeta}} \cdot \mathbf{y})_{,j} = \bar{\xi}_j \zeta^{-1}$ and $R_{,j} = -(\bar{\boldsymbol{\zeta}} \cdot \mathbf{y}) \bar{\xi}_j \zeta^{-1} R^{-1}$ at $z = (\bar{\boldsymbol{\zeta}} \cdot \mathbf{y})$. It can be seen from (17.8) that if ϵ^*_{nm} is constant, $u_{i,j}$ is also constant, and if ϵ^*_{nm} is a linear function of coordinates, $u_{i,j}$ is also a linear function of coordinates.

The surface element $dS(\bar{\boldsymbol{\xi}})$ can be transformed to a new surface element defined at point $\bar{\boldsymbol{\zeta}}$ by use of (16.10). Let $dS(\bar{\boldsymbol{\xi}})$ be a surface element constructed by the vector product of $d\bar{\boldsymbol{\xi}}$ and $\delta\bar{\boldsymbol{\xi}}$, and $dS(\bar{\boldsymbol{\zeta}})$ by $d\bar{\boldsymbol{\zeta}}$

and $\delta\bar{\zeta}$, where $d\bar{\zeta}_1 = a_1 d\bar{\xi}_1/\zeta$, $\delta\bar{\zeta}_1 = a_1\delta\bar{\xi}_1/\zeta$, etc. (see Fig. 17.1). Then,

$$dS(\bar{\xi}) = \begin{vmatrix} \bar{\xi}_1 & \bar{\xi}_2 & \bar{\xi}_3 \\ d\bar{\xi}_1 & d\bar{\xi}_2 & d\bar{\xi}_3 \\ \delta\bar{\xi}_1 & \delta\bar{\xi}_2 & \delta\bar{\xi}_3 \end{vmatrix},$$

$$dS(\bar{\zeta}) = \begin{vmatrix} a_1\bar{\xi}_1/\zeta & a_2\bar{\xi}_2/\zeta & a_3\bar{\xi}_3/\zeta \\ a_1 d\bar{\xi}_1/\zeta & a_2 d\bar{\xi}_2/\zeta & a_3 d\bar{\xi}_3/\zeta \\ a_1\delta\bar{\xi}_1/\zeta & a_2\delta\bar{\xi}_2/\zeta & a_3\delta\bar{\xi}_3/\zeta \end{vmatrix}. \tag{17.9}$$

Therefore, we have the relation

$$dS(\bar{\zeta}) = a_1 a_2 a_3 \zeta^{-3} dS(\bar{\xi}). \tag{17.10}$$

Uniform eigenstrains

The displacement and strain fields for interior points become as follows when ϵ^*_{ij} is uniform:

$$u_i(\mathbf{x}) = \frac{a_1 a_2 a_3}{4\pi} C_{klmn}\epsilon^*_{nm} \int_{S^2} (\bar{\xi}\cdot\mathbf{x}) N_{ik}(\bar{\xi}) D^{-1}(\bar{\xi})\bar{\xi}_l \zeta^{-3} dS(\bar{\xi}) \tag{17.11}$$

and

$$u_{i,j}(\mathbf{x}) = \frac{a_1 a_2 a_3}{4\pi} C_{klmn}\epsilon^*_{nm} \int_{S^2} N_{ik}(\bar{\xi}) D^{-1}(\bar{\xi})\bar{\xi}_l\bar{\xi}_j \zeta^{-3} dS(\bar{\xi}), \tag{17.12}$$

or, using (17.10),

$$u_i(\mathbf{x}) = (1/4\pi) C_{klmn}\epsilon^*_{nm} \int_{S^2} (\bar{\xi}\cdot\mathbf{x}) N_{ik}(\bar{\xi}) D^{-1}(\bar{\xi})\bar{\xi}_l dS(\bar{\zeta}) \tag{17.13}$$

and

$$u_{i,j}(\mathbf{x}) = (1/4\pi) C_{klmn}\epsilon^*_{nm} \int_{S^2} N_{ik}(\bar{\xi}) D^{-1}(\bar{\xi})\bar{\xi}_l\bar{\xi}_j dS(\bar{\zeta}). \tag{17.14}$$

Equation (17.11) has been derived by Kinoshita and Mura (1971).

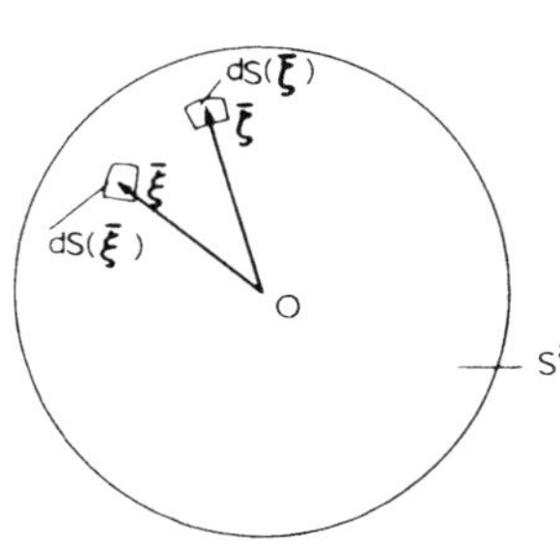

Fig. 17.1. Transformation of the surface element.

Furthermore, the surface element $dS(\bar{\boldsymbol{\zeta}})$ can be written as

$$dS(\bar{\boldsymbol{\zeta}}) = d\bar{\zeta}_3 d\theta, \tag{17.15}$$

where θ is measured counter-clockwise from the $\bar{\zeta}_1$-axis as shown in Fig. 17.2. Then, (17.13) is written as

$$u_i(\mathbf{x}) = (1/4\pi)C_{jlmn}\epsilon^*_{nm}x_k \int_{-1}^{1} d\bar{\zeta}_3 \int_0^{2\pi} G_{ijkl}(\bar{\boldsymbol{\xi}})d\theta, \tag{17.16}$$

where

$$G_{ijkl}(\bar{\boldsymbol{\xi}}) = \bar{\xi}_k\bar{\xi}_l N_{ij}(\bar{\boldsymbol{\xi}})/D(\bar{\boldsymbol{\xi}}). \tag{17.17}$$

If the notation introduced in (11.15) is used, we can write

$$\epsilon_{ij} = S_{ijmn}\epsilon^*_{mn}, \tag{17.18}$$

where

$$S_{ijmn} = (1/8\pi)C_{pqmn} \int_{-1}^{1} d\bar{\zeta}_3 \int_0^{2\pi} \{G_{ipjq}(\bar{\boldsymbol{\xi}}) + G_{jpiq}(\bar{\boldsymbol{\xi}})\}d\theta. \tag{17.19}$$

From (16.10)

$$\bar{\xi}_1 = \zeta\bar{\zeta}_1/a_1, \quad \bar{\xi}_2 = \zeta\bar{\zeta}_2/a_2, \quad \bar{\xi}_3 = \zeta\bar{\zeta}_3/a_3; \tag{17.20}$$

however, since $G_{ijkl}(\bar{\xi}_1, \bar{\xi}_2, \bar{\xi}_3)$ are homogeneous polynomials of degree 0, we can use

$$\bar{\xi}_1 = \bar{\zeta}_1/a_1, \quad \bar{\xi}_2 = \bar{\zeta}_2/a_2, \quad \bar{\xi}_3 = \bar{\zeta}_3/a_3, \tag{17.21}$$

for $\bar{\boldsymbol{\xi}}$ as the argument of $G_{ijkl}(\bar{\boldsymbol{\xi}})$. The components of $\bar{\boldsymbol{\zeta}}$ are

$$\begin{aligned} \bar{\zeta}_1 &= (1-\bar{\zeta}_3^2)^{1/2}\cos\theta, \\ \bar{\zeta}_2 &= (1-\bar{\zeta}_3^2)^{1/2}\sin\theta, \\ \bar{\zeta}_3 &= \bar{\zeta}_3; \end{aligned} \tag{17.22}$$

see Fig. 17.2.

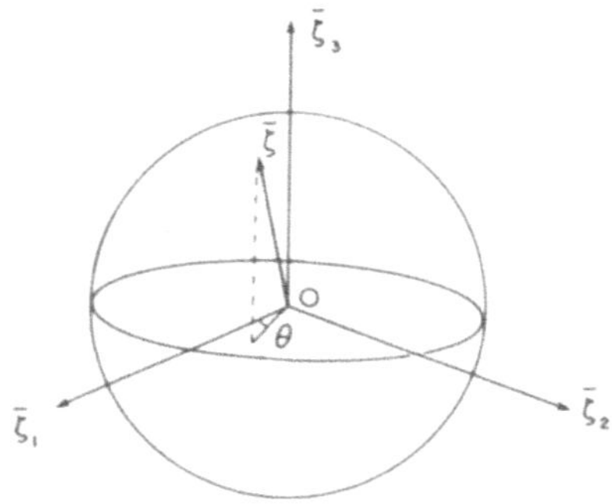

Fig. 17.2. New coordinate system $\bar{\zeta}_3$ and θ.

The integrals with respect to θ in (17.16) or (17.19) can be obtained by the residue calculation in a complex Z plane, where

$$\begin{aligned} \cos\theta &= (Z + Z^{-1})/2, \\ \sin\theta &= (Z - Z^{-1})/2i, \\ d\theta &= dZ/iZ. \end{aligned} \tag{17.23}$$

Then,

$$\int_0^{2\pi} G_{ijkl}(\bar{\xi})d\theta = 2\pi R[G_{ijkl}/Z], \tag{17.24}$$

where $R[G_{ijkl}/Z]$ is the sum of the residues of the function G_{ijkl}/Z existing within the unit circle $|Z| = 1$ under a fixed value of $\bar{\zeta}_3$. Then (17.16) is written as

$$u_i(\mathbf{x}) = \frac{1}{4\pi} C_{jlmn}\epsilon^*_{nm} x_k \bar{G}_{ijkl}, \tag{17.25}$$

where

$$\bar{G}_{ijkl} = 2\pi \int_{-1}^{1} R[G_{ijkl}/Z] d\bar{\zeta}_3. \tag{17.26}$$

Also, we have, from (17.19),

$$S_{ijmn} = (1/8\pi) C_{pqmn}(\bar{G}_{ipjq} + \bar{G}_{jpiq}). \tag{17.27}$$

Some special cases are considered below; see Lin and Mura (1973).

Spheroid $(a_1 = a_2, \quad a_1/a_3 = \rho)$

The crystalline directions are assumed to be coincident with the principal directions of the spheroid.

For cubic crystals, nonzero components of $\bar{G}_{ijkl}$ are given below:

$$\begin{aligned} \bar{G}_{1111} &= \bar{G}_{2222} \\ &= \frac{2\pi}{a}\int_0^1 \frac{(1-x^2)}{pq}(1 - x^2 + \rho^2 x^2)[\mu^2(1 - x^2 + \rho^2 x^2) + \beta\rho^2 x^2]dx \\ &\quad + \frac{\pi}{a}\int_0^1 \frac{(1-x^2)^2}{p(p+q)}[\beta(1 - x^2 + \rho^2 x^2) + \gamma\rho^2 x^2]dx, \\ \bar{G}_{3333} &= \frac{4\pi}{a}\int_0^1 \frac{\rho^2 x^2}{pq}(1 - x^2 + \rho^2 x^2)[\mu^2(1 - x^2 + \rho^2 x^2) + \beta(1 - x^2)]dx \\ &\quad + \frac{\pi\gamma}{a}\int_0^1 \frac{\rho^2 x^2(1-x^2)^2}{p(p+q)}dx, \end{aligned}$$

$$\begin{aligned}
\bar{G}_{1122} &= \bar{G}_{2211} \\
&= \frac{2\pi}{a}\int_0^1 \frac{(1-x^2)}{pq}\{(1-x^2+\rho^2x^2)[\mu^2(1-x^2+\rho^2x^2)+\beta\rho^2x^2] \\
&\quad + (1-x^2)[\beta(1-x^2+\rho^2x^2)+\gamma\rho^2x^2]\}dx - \frac{\pi}{a}\int_0^1 \frac{(1-x^2)^2}{p(p+q)} \\
&\quad \times [\beta(1-x^2+\rho^2x^2)+\gamma\rho^2x^2]dx,
\end{aligned} \tag{17.28}$$

$$\begin{aligned}
\bar{G}_{1133} &= \bar{G}_{2233} \\
&= \frac{2\pi}{a}\int_0^1 \frac{\rho^2x^2}{pq}\{2(1-x^2+\rho^2x^2)[\mu^2(1-x^2+\rho^2x^2)+\beta\rho^2x^2] \\
&\quad + (1-x^2)[\beta(1-x^2+\rho^2x^2)+\gamma\rho^2x^2]\}dx,
\end{aligned}$$

$$\bar{G}_{1212} = -\frac{\pi(\lambda+\mu)}{a}\int_0^1 \frac{(1-x^2)^2}{p(p+q)}[\mu(1-x^2+\rho^2x^2)+\mu'\rho^2x^2]dx,$$

$$\begin{aligned}
\bar{G}_{1313} &= \bar{G}_{2323} \\
&= -\frac{2\pi\mu(\lambda+\mu)}{a}\int_0^1 \frac{\rho^2x^2(1-x^2)(1-x^2+\rho^2x^2)}{pq}dx \\
&\quad - \frac{\pi\mu'(\lambda+\mu)}{a}\int_0^1 \frac{\rho^2x^2(1-x^2)^2}{p(p+q)}dx,
\end{aligned}$$

$$\begin{aligned}
\bar{G}_{3311} &= \bar{G}_{3322} \\
&= \frac{2\pi}{a}\int_0^1 \frac{(1-x^2)}{pq}(1-x^2+\rho^2x^2)[\mu^2(1-x^2+\rho^2x^2)+\beta(1-x^2)]dx \\
&\quad + \frac{\pi\gamma}{2a}\int_0^1 \frac{(1-x^2)^3}{p(p+q)}dx,
\end{aligned}$$

where

$$\begin{aligned}
\rho &= a_1/a_3, \\
a &= \mu^2(\lambda+2\mu+\mu'), \\
b &= a^{-1}\mu\mu'(2\lambda+2\mu+\mu'), \\
c &= a^{-1}\mu'^2(3\lambda+3\mu+\mu'), \\
\beta &= \mu(\lambda+\mu+\mu'), \\
\gamma &= \mu'(2\lambda+2\mu+\mu'), \\
p &= \{(1-x^2+\rho^2x^2)^3 + b\rho^2x^2(1-x^2)(1-x^2+\rho^2x^2) \\
&\quad + \frac{(1-x^2)^2}{4}[b(1-x^2+\rho^2x^2)+c\rho^2x^2]\}^{1/2}, \quad 0<x<1, \\
q &= \{(1-x^2+\rho^2x^2)^3 + b\rho^2x^2(1-x^2)(1-x^2+\rho^2x^2)\}^{1/2}, \quad 0<x<1.
\end{aligned} \tag{17.29}$$

It has been verified that p^2 and q^2 are positive for most of the cubic crystals. When $\rho = 1$, $\bar{G}_{1111}$ and $\bar{G}_{1212}$ agree with Kneer's results (1965). However, $\bar{G}_{1122}$ of Kneer's is incorrect, since it does not reduce to the isotropic case.

For hexagonal crystals, the elastic moduli are denoted by

$$\begin{aligned} C_{11} &= d, \\ (C_{11} - C_{12})/2 &= e, \\ C_{44} &= f, \\ C_{13} + C_{44} &= g, \\ C_{33} &= h, \end{aligned} \tag{17.30}$$

where C_{ij} are the Voigt constants. The nonzero components of $\bar{G}_{ijkl}$ are given below

$$\begin{aligned}
\bar{G}_{1111} = \bar{G}_{2222} &= \frac{\pi}{2}\int_0^1 \Delta(1-x^2)\{[f(1-x^2)+h\rho^2x^2] \\
&\quad \times [(3e+d)(1-x^2)+4f\rho^2x^2] - g^2\rho^2x^2(1-x^2)\}dx, \\
\bar{G}_{3333} &= 4\pi\int_0^1 \Delta\rho^2x^2[d(1-x^2)+f\rho^2x^2][e(1-x^2)+f\rho^2x^2]dx, \\
\bar{G}_{1122} = \bar{G}_{2211} &= \frac{\pi}{2}\int_0^1 \Delta(1-x^2)\{[f(1-x^2)+h\rho^2x^2] \\
&\quad \times [(e+3d)(1-x^2)+4f\rho^2x^2] - 3g^2\rho^2x^2(1-x^2)\}dx, \\
\bar{G}_{1133} = \bar{G}_{2233} &= 2\pi\int_0^1 \Delta\rho^2x^2\{[(d+e)(1-x^2)+2f\rho^2x^2] \\
&\quad \times [f(1-x^2)+h\rho^2x^2] - g^2\rho^2x^2(1-x^2)\}dx, \\
\bar{G}_{3311} &= \bar{G}_{3322} \\
&= 2\pi\int_0^1 \Delta(1-x^2)[d(1-x^2)+f\rho^2x^2][e(1-x^2)+f\rho^2x^2]dx, \\
\bar{G}_{1212} &= \frac{\pi}{2}\int_0^1 \Delta(1-x^2)^2\{g^2\rho^2x^2 - (d-e)[f(1-x^2)+h\rho^2x^2]\}dx, \\
\bar{G}_{1313} = \bar{G}_{2323} &= (-2\pi)\int_0^1 \Delta g\rho^2x^2(1-x^2)[e(1-x^2)+f\rho^2x^2]dx,
\end{aligned} \tag{17.31}$$

where

$$\begin{aligned}
\Delta^{-1} &= [e(1-x^2)+f\rho^2x^2]\{[d(1-x^2)+f\rho^2x^2][f(1-x^2)+h\rho^2x^2] \\
&\quad - g^2\rho^2x^2(1-x^2)\}.
\end{aligned}$$

When $\rho = 1$, all integrals except $\bar{G}_{3311}$ agree with Kneer's results (1965). Isotropic materials are obtained by a limiting process: $d \to \lambda + 2\mu$, $e \to \mu$, $f \to \mu$, $g \to \lambda + \mu$, and $h \to \lambda + 2\mu$.

Cylinder ($a_3 \to \infty$) (*elliptic inclusion*)

A rod or a needle-shaped inclusion can be approximated by a cylindrical inclusion. When $a_3 \to \infty$, (17.21) becomes

$$\bar{\xi}_1 = \bar{\zeta}_1/a_1, \quad \bar{\xi}_2 = \bar{\zeta}_2/a_2, \quad \bar{\xi}_3 = 0. \tag{17.32}$$

Since $G_{ijkl}(\bar{\zeta}_1/a_1, \bar{\zeta}_2/a_2, 0)$ are homogeneous functions of degree 0, the factor $(1 - \bar{\zeta}_3^2)^{1/2}$ in (17.22) can be dropped. Then G_{ijkl} are independent of $\bar{\zeta}_3$ and we have

$$u_i(\mathbf{x}) = (1/2\pi) C_{jlmn} \epsilon^*_{mn} x_k \int_0^{2\pi} G_{ijkl}(\bar{\zeta}_1/a_1, \bar{\zeta}_2/a_2, 0)\, d\theta \tag{17.33}$$

with

$$\bar{\zeta}_1 = \cos\theta, \quad \bar{\zeta}_2 = \sin\theta. \tag{17.34}$$

When the coordinate axes x_i are not taken along the directions of the principal axes of the cylinder, we have

$$u_i(\mathbf{x}) = (1/2\pi) C_{jlmn} \epsilon^*_{mn} x_k \int_0^{2\pi} G_{ijkl}(a_{1p}\bar{\zeta}_p/a_p, a_{2p}\bar{\zeta}_p/a_p, 0)\, d\theta, \tag{17.35}$$

where $p = 1, 2$ and a_{jp} are the direction cosines between the x_j-axis and the pth principal directions of the ellipsoid.

If the cylinder has a circular cross-section ($a_1 = a_2$), $\bar{G}_{ijkl}$ are obtained from (17.28) by taking $\rho = 0$.

Jaswon and Bhargava (1961) and List (1969) were able to obtain explicit solutions for interior and exterior points by using complex representation. Jaswon and Bhargava's analysis has been extended by Willis (1964) to the anisotropic case with cubic symmetry. Subsequent anisotropic analyses have been performed by Bhargava and Radhakrishna (1964) for an orthotropic medium and by Chen (1967) for a medium with one plane of symmetry. These treatments of two-dimensional inclusion problems are restricted to the state of plane strain or plane stress, based on the classical formulation for the plane theory of elasticity (e.g., Green and Zerna 1954). However, for a general anisotropic medium it is not always possible to treat the plane strain and antiplane strain fields independently. Yang and Chou (1976) have presented a general method for solving the generalized plane problems of inclusions in anisotropic solids. They have used Green's function (5.84) for the generalized plane problems developed by Eshelby, Read, and Shockley (1953).

The following is the result obtained by Yang and Chou for orthorhombic materials with elastic moduli given by (A2.17), where eigenstrain ϵ_{ij}^* is assumed to be uniform in the elliptic domain with the half-axes a_1 and a_2. For interior points,

$$\sigma_{11} = -\frac{A^2(\bar{C}_{12}^2 - C_{12}^2)}{2\bar{C}_{12}Q}\left[2\epsilon_{11}^* + \frac{e}{A\sin\alpha}(A^2\epsilon_{11}^* + \epsilon_{22}^*)\right],$$

$$\sigma_{22} = -\frac{(\bar{C}_{12}^2 - C_{12}^2)}{2\bar{C}_{12}Q}\left[\frac{e}{A\sin\alpha}(A^2\epsilon_{11}^* + \epsilon_{22}^*) + 2e^2\epsilon_{22}^*\right], \tag{17.35.1}$$

$$\sigma_{12} = -\frac{A(\bar{C}_{12}^2 - C_{12}^2)}{\bar{C}_{12}Q\sin\alpha}e\epsilon_{12}^*,$$

where

$$\begin{aligned} A &= (C_{11}/C_{22})^{1/4}, \quad \bar{C}_{12} = (C_{11}C_{22})^{1/2}, \quad e = a_2/a_1, \\ \alpha &= \cos^{-1}(\sqrt{-C}/2), \quad \text{for } -4 < C \le 0, \\ &= \cos^{-1}(i\sqrt{C}/2), \quad \text{for } C > 0, \\ C &= (\bar{C}_{12} + C_{12})(\bar{C}_{12} - C_{12} - 2C_{66})/\bar{C}_{12}C_{66}, \\ Q &= 1 + e^2A^2 + 2eA\sin\alpha. \end{aligned} \tag{17.35.2}$$

The component σ_{33} can be obtained from the condition $\epsilon_{33} = 0$. When isotropy is approached ($\lambda = 1$ and $C = 0$), equations (17.35.1) reduce to the results of Jaswon and Bhargava.

The strain energy is

$$W^* = \frac{\pi a_1 a_2(\bar{C}_{12}^2 - C_{12}^2)}{4\bar{C}_{12}Q}\left[2(A^2\epsilon_{11}^{*2} + e^2\epsilon_{22}^{*2}) + \frac{e}{A\sin\alpha}(A^2\epsilon_{11}^* + \epsilon_{22}^*)^2\right]$$

$$+\frac{\pi a_1 a_2 A(\bar{C}_{12}^2 - C_{12}^2)}{\bar{C}_{12}Q\sin\alpha}\epsilon_{12}^{*2}. \tag{17.35.3}$$

Yang and Chou (1977) further obtained solutions of antiplane strain problems of an elliptic inclusion in an anisotropic medium.

Flat ellipsoid ($a_3 \to 0$)

When $a_3 \to 0$, from (17.21),

$$\bar{\xi}_1, \bar{\xi}_2 \ll \bar{\xi}_3; \tag{17.36}$$

therefore, we can put $\bar{\xi}_1 = \bar{\xi}_2 = 0$ for the argument of $G_{ijkl}(\bar{\boldsymbol{\xi}})$. Then

$G_{ijkl}(0, 0, \bar{\xi}_3)$ are homogeneous of degree 0, and we set $\bar{\xi}_3 = 1$. Therefore, (17.16) becomes

$$u_i(\mathbf{x}) = C_{jlmn}\epsilon^*_{nm}x_kG_{ijkl}(0, 0, 1). \tag{17.37}$$

When the coordinate axes are not taken along the principal directions of the flat ellipsoid, we have

$$u_i(\mathbf{x}) = C_{jlmn}\epsilon^*_{nm}x_kG_{ijkl}(a_{13}, a_{23}, a_{33}), \tag{17.37'}$$

where a_{jp} is the direction cosine between the x_j-axis and the pth principal direction of the flat ellipsoid.

When the **x**-coordinate axes are taken in the principal directions of the flat ellipsoid, the displacement in the flat ellipsoid is identical to zero, since $G_{ijkl}(0, 0, 1)$ in (17.37) becomes $\delta_{k3}\delta_{l3}N_{ij}(0, 0, 1/D(0, 0, 1)$ and, therefore, $x_k\delta_{k3} = x_3 = 0$ on the flat ellipsoid. Then, we have

$$u_i = 0, \quad u_{i,1} = 0 \quad \text{and} \quad u_{i,2} = 0. \tag{17.38}$$

The stress becomes

$$\sigma_{pq} = C_{pqi3}C_{j3mn}\epsilon^*_{nm}N_{ij}(\bar{\boldsymbol{\xi}})D^{-1}(\bar{\boldsymbol{\xi}}) - C_{pqik}\epsilon^*_{ki}, \tag{17.39}$$

where $\bar{\boldsymbol{\xi}} = (0, 0, 1)$. By definition, we have from (5.5)

$$C_{pqik}\bar{\xi}_q\bar{\xi}_kN_{ij}(\bar{\boldsymbol{\xi}})D^{-1}(\bar{\boldsymbol{\xi}}) = \delta_{pj}. \tag{17.40}$$

Since $\bar{\xi}_1 = 0$, $\bar{\xi}_2 = 0$, $\bar{\xi}_3 = 1$, this is identical to

$$C_{p3i3}N_{ij}(\bar{\boldsymbol{\xi}})D^{-1}(\bar{\boldsymbol{\xi}}) = \delta_{pj}. \tag{17.41}$$

Then, (17.39) leads to

$$\sigma_{p3} = 0, \tag{17.42}$$

which should be expected for the case of a very flat inclusion. Furthermore, we have from (17.37) and (17.41)

$$u_{i,3} = C_{j3m1}\epsilon^*_{1m}N_{ij}(\bar{\boldsymbol{\xi}})D^{-1}(\bar{\boldsymbol{\xi}}) + C_{j3m2}\epsilon^*_{2m}N_{ij}(\bar{\boldsymbol{\xi}})D^{-1}(\bar{\boldsymbol{\xi}}) + \epsilon^*_{3i}. \tag{17.43}$$

If

$$\epsilon^*_{11} = \epsilon^*_{22} = \epsilon^*_{12} = 0, \tag{17.44}$$

it is shown that

$$\tfrac{1}{2}(u_{1,3} + u_{3,1}) = \epsilon^*_{13},\ \tfrac{1}{2}(u_{2,3} + u_{3,2}) = \epsilon^*_{23},\ u_{3,3} = \epsilon^*_{33} \tag{17.45}$$

and any stress component vanishes,

$$\sigma_{pq} = 0. \tag{17.46}$$

In short, when $\epsilon^*_{11} = \epsilon^*_{22} = \epsilon^*_{12} = 0$, all the stress components become zero, resulting in the vanishing elastic energy. Further, an arbitrary line element parallel to the flat surface does not change, since $u_{i,1} = u_{i,2} = 0$ from (17.37).

If the coordinate system is taken arbitrarily, (17.39) can be written as

$$\sigma_{pq} = C_{pqik}C_{jlmn}\epsilon^*_{mn}N_{ij}(\bar{\boldsymbol{\xi}})D^{-1}(\boldsymbol{\xi})\bar{\xi}_k\bar{\xi}_l - C_{pqik}\epsilon^*_{ki}, \tag{17.47}$$

where $\bar{\xi} = (a_{13}, a_{23}, a_{33})$.

Eigenstrains with polynomial variation

For this case the stresses at interior and exterior points are investigated simultaneously; see Section 19.

Eigenstrains with a periodic form

The elastic field inside an ellipsoidal inclusion is considered when the distribution of eigenstrains inside the inclusion is periodic,

$$\epsilon^*_{nm}(\mathbf{x}') = \bar{\epsilon}^*_{nm} \exp(ic_p x'_p/a_P) = \bar{\epsilon}^*_{nm} \exp(ic_p y'_p), \tag{17.48}$$

where $\bar{\epsilon}^*_{nm}$ are constants, $\mathbf{c}$ is a given vector, and the summation convention defined in (11.39), is used. From Fig. 16.1 we write

$$\mathbf{y}' = z\bar{\boldsymbol{\zeta}} + r\cos\phi\,\mathbf{m} + r\sin\phi\,\mathbf{n} \tag{17.49}$$

and, therefore,

$$c_p y'_p = (\mathbf{c}\cdot\bar{\boldsymbol{\zeta}})z + rc_0\cos(\phi - \alpha), \tag{17.50}$$

where (see Fig. 17.3)

$$(\mathbf{c}\cdot\mathbf{m})/c_0 = \cos\alpha,\ (\mathbf{c}\cdot\mathbf{n})/c_0 = \sin\alpha,\ c_0 = \{\mathbf{c}\cdot\mathbf{c} - (\mathbf{c}\cdot\bar{\boldsymbol{\zeta}})^2\}^{1/2}. \tag{17.51}$$

Then, we have

$$\begin{aligned} \frac{\partial}{\partial z}\epsilon^*_{nm}(\mathbf{x}') &= i(\mathbf{c}\cdot\bar{\boldsymbol{\zeta}})\bar{\epsilon}^*_{nm}\exp(ic_p y'_p), \\ \frac{\partial^2}{\partial z^2}\epsilon^*_{nm}(\mathbf{x}') &= -(\mathbf{c}\cdot\bar{\boldsymbol{\zeta}})^2\bar{\epsilon}^*_{nm}\exp(ic_p y'_p). \end{aligned} \tag{17.52}$$

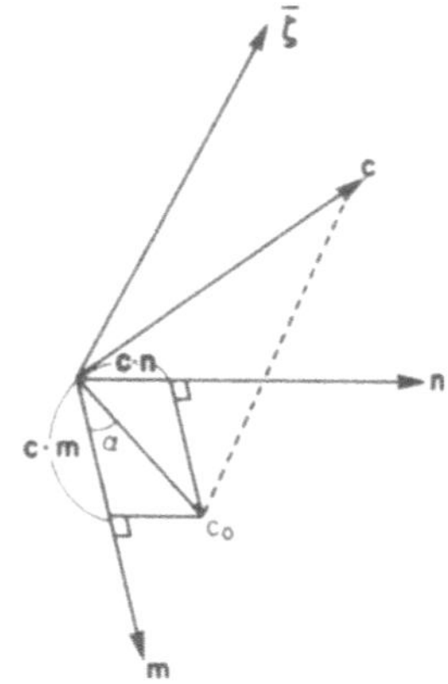

Fig. 17.3. Illustration of vector **c**.

Substituting (17.52) into (17.4), we obtain

$$\begin{aligned} u_i(\mathbf{x}) = \frac{-a_1a_2a_3}{8\pi^2}\int_0^{2\pi} d\phi\Bigg[\int_0^R & r\,dr\,i(\mathbf{c}\cdot\bar{\boldsymbol{\zeta}})\epsilon^*_{nm} \\ & \times\exp\{i(\bar{\boldsymbol{\zeta}}\cdot\mathbf{y})(\mathbf{c}\cdot\bar{\boldsymbol{\zeta}})+irc_0\cos(\phi-\alpha)\} \\ & -(\bar{\boldsymbol{\zeta}}\cdot\mathbf{y})\epsilon^*_{nm}\exp\{i(\bar{\boldsymbol{\zeta}}\cdot\mathbf{y})(\mathbf{c}\cdot\bar{\boldsymbol{\zeta}}) \\ & +iRc_0\cos(\phi-\alpha)\}\Bigg]\int_{S^2} C_{klmn}N_{ik}(\bar{\boldsymbol{\xi}})D^{-1}(\bar{\boldsymbol{\xi}})\bar{\xi}_l\zeta^{-2}dS(\bar{\boldsymbol{\xi}}). \end{aligned} \qquad (17.53)$$

From the properties of the Bessel functions,

$$\int_0^{2\pi}\exp\{irc_0\cos(\phi-\alpha)\}d\phi = 2\pi J_0(c_0r),\ \{zJ_1(z)\}' = zJ_0(z), \qquad (17.54)$$

and (17.53) is written as

$$\begin{aligned} u_i(\mathbf{x}) = \frac{-a_1a_2a_3}{4\pi}\int_{S^2} & [i(\mathbf{c}\cdot\bar{\boldsymbol{\zeta}})(R/c_0)J_1(c_0R)-(\bar{\boldsymbol{\zeta}}\cdot\mathbf{y})J_0(c_0R)]\epsilon^*_{nm} \\ & \times\exp\{i(\bar{\boldsymbol{\zeta}}\cdot\mathbf{y})(\mathbf{c}\cdot\bar{\boldsymbol{\zeta}})\}\bar{\xi}_lC_{klmn}N_{ik}(\bar{\boldsymbol{\xi}})D^{-1}(\bar{\boldsymbol{\xi}})\zeta^{-2}dS(\bar{\boldsymbol{\xi}}). \end{aligned} \qquad (17.55)$$

Similarly, (17.8) is written as

$$\begin{aligned} u_{i,j}(\mathbf{x}) = \frac{a_1a_2a_3}{4\pi}\int_{S^2} & [(R/c_0)J_1(c_0R)(\mathbf{c}\cdot\bar{\boldsymbol{\zeta}})^2+\{2i(\bar{\boldsymbol{\zeta}}\cdot\mathbf{y})(\mathbf{c}\cdot\bar{\boldsymbol{\zeta}})+1\}J_0(c_0R)] \\ & \times\epsilon^*_{nm}\exp\{i(\bar{\boldsymbol{\zeta}}\cdot\mathbf{y})(\mathbf{c}\cdot\bar{\boldsymbol{\zeta}})\}C_{klmn}N_{ik}(\bar{\boldsymbol{\xi}})D^{-1}(\bar{\boldsymbol{\xi}})\bar{\xi}_l\bar{\xi}_j\zeta^{-3}dS(\bar{\boldsymbol{\xi}}), \end{aligned} \qquad (17.56)$$

where

$$R = \{1-(\bar{\boldsymbol{\zeta}}\cdot\mathbf{y})^2\}^{1/2}, \quad \bar{\boldsymbol{\zeta}}\cdot\mathbf{y} = \bar{\boldsymbol{\xi}}\cdot\mathbf{x}\zeta^{-1}. \qquad (17.57)$$

The result (17.55) has been obtained by Mura, Mori, and Kato (1976). The above solution can be called the fundamental solution. When ϵ^*_{nm} is given as an arbitrary function, it can be expressed by the Fourier series or integrals in terms of (17.48). The corresponding solution $u_{i,j}(\mathbf{x})$ is the sum or integral of (17.56) with respect to various values of $\mathbf{c}$.

When the ellipsoid degenerates into a flat ellipsoid ($a_3 \to 0$), expression (17.55) simplifies. First the unit sphere S^2 is transformed into a cylinder as shown in Fig. 17.4, and the cylindrical surface element $d\Sigma$ reduces to $d\theta d\xi_3$ which is related to dS by

$$dS(\bar{\boldsymbol{\xi}}) = d\theta d\xi_3/(1+\xi_3^2)^{3/2}. \tag{17.58}$$

Also, from (16.10),

$$\begin{aligned} &\bar{\xi}_1 = \cos\theta/(1+\xi_3^2)^{1/2},\ \bar{\xi}_2 = \sin\theta/(1+\xi_3^2)^{1/2},\ \bar{\xi}_3 = \xi_3/(1+\xi_3^2)^{1/2}, \\ &\zeta^{-3} dS(\bar{\boldsymbol{\xi}}) = (a_1^2\cos^2\theta + a_2^2\sin^2\theta + a_3^2\xi_3^2)^{-3/2} d\theta d\xi_3, \end{aligned} \tag{17.59}$$

where $-\infty \le \xi_3 \le \infty$.

With use of the identity

$$\begin{aligned} &(a_1^2\cos^2\theta + a_2^2\sin^2\theta + a_3^2\xi_3^2)^{-3/2} \\ &\quad = (a_1^2\cos^2\theta + a_2^2\sin^2\theta)^{-1}\frac{\partial}{\partial\xi_3}\{\xi_3(a_1^2\cos^2\theta + a_2^2\sin^2\theta + a_3^2\xi_3^2)^{-1/2}\}, \end{aligned} \tag{17.60}$$

integration by parts with respect to ξ_3 is applied to (17.55). Then, we have

$$\begin{aligned} u_i(\mathbf{x}) = (a_1a_2/2\pi)C_{klmn}\bar{\epsilon}^*_{nm}\int_0^{2\pi} d\theta[\{(\mathbf{x}\cdot\bar{\boldsymbol{\xi}})J_0(c_0R) - ic_pa_p\bar{\xi}_PRJ_1(c_0R)/c_0\} \\ \times\exp\{i(\mathbf{x}\cdot\bar{\boldsymbol{\xi}})c_pa_p\bar{\xi}_P/\zeta^2\}N_{ik}(\bar{\boldsymbol{\xi}})\bar{\xi}_lD^{-1}(\bar{\boldsymbol{\xi}})]_{\xi_3=\infty}(a_1^2\cos^2\theta + a_2^2\sin^2\theta)^{-1} \end{aligned}$$

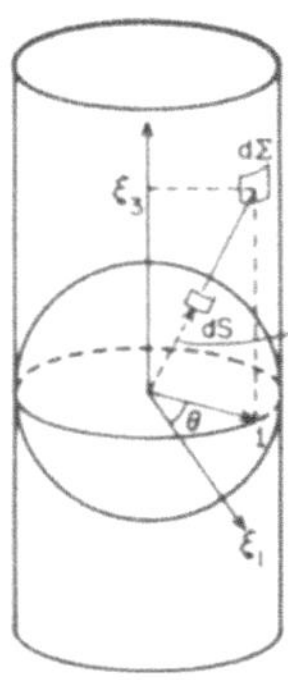

Fig. 17.4. The unit sphere S_2 is transformed into the infinitely long cylinder.

$$
\begin{aligned}
&-(a_1a_2a_3/4\pi)C_{klmn}\bar{\epsilon}^*_{nm}\int_0^{2\pi}d\theta\int_{-\infty}^{\infty}d\xi_3(a_1^2\cos^2\theta+a_2^2\sin^2\theta)^{-1}\\
&\times\xi_3(a_1^2\cos^2\theta+a_2^2\sin^2\theta+a_3^2\xi_3^2)^{-1/2}\frac{\partial}{\partial\xi_3}[\{(\mathbf{x}\cdot\bar{\boldsymbol{\xi}})J_0(c_0R) \qquad (17.61)\\
&-ic_pa_p\bar{\xi}_PRJ_1(c_0R)/c_0\}\exp\{i(\mathbf{x}\cdot\bar{\boldsymbol{\xi}})c_pa_p\bar{\xi}_P/\zeta^2\}N_{ik}(\bar{\boldsymbol{\xi}})\bar{\xi}_lD^{-1}(\bar{\boldsymbol{\xi}})],
\end{aligned}
$$

where (16.10) has been used.

For $\xi_3\to\infty$, from (17.59),

$$\bar{\xi}_1=0,\quad \bar{\xi}_2=0,\quad \bar{\xi}_3=1; \qquad (17.62)$$

therefore,

$$
\begin{aligned}
&(\mathbf{x}\cdot\bar{\boldsymbol{\xi}})=x_3,\quad \zeta=a_3,\\
&c_0=(c_1^2+c_2^2)^{1/2},\quad R=(1-x_3^2/a_3^2)^{1/2}.
\end{aligned} \qquad (17.63)
$$

Furthermore, for $a_3\to0$, the second integral in the right side of (17.61) vanishes. The factor [ ] in the integrand in the first integral is independent of θ and

$$\int_0^{2\pi}(a_1^2\cos^2\theta+a_2^2\sin^2\theta)^{-1}d\theta=2\pi/a_1a_2. \qquad (17.64)$$

Equation (17.61) becomes, for $a_3\to0$,

$$u_i(\mathbf{x})=C_{klmn}\bar{\epsilon}^*_{nm}x_3J_0(c_0R)\exp(ix_3c_3/a_3)N_{ik}(\bar{\boldsymbol{\xi}})\bar{\xi}_lD^{-1}(\bar{\boldsymbol{\xi}}), \qquad (17.65)$$

where $\bar{\boldsymbol{\xi}}=(0,0,1)$.

Since we are considering a point $\mathbf{x}$ inside the inclusion, we set $x_3=0$ before taking the limit $a_3\to0$. Then, (17.65) leads to

$$u_i(\mathbf{x})=0 \qquad (17.66)$$

for interior points. By taking the derivative of (17.65) with respect to x_j, we obtain the expressions for the displacement gradient and, therefore, the stress components throughout the inclusion,

$$
\begin{aligned}
&u_{i,j}(\mathbf{x})=C_{klmn}\bar{\epsilon}^*_{mn}J_0(c_0)\bar{\xi}_j\bar{\xi}_lN_{ik}(\bar{\boldsymbol{\xi}})D^{-1}(\bar{\boldsymbol{\xi}}),\\
&\sigma_{pq}(\mathbf{x})=C_{pqij}[C_{klmn}\bar{\epsilon}^*_{mn}J_0(c_0)\bar{\xi}_j\bar{\xi}_lN_{ik}(\bar{\boldsymbol{\xi}})D^{-1}(\bar{\boldsymbol{\xi}})\\
&\qquad\qquad -\bar{\epsilon}^*_{ij}\exp\{i(c_1x_1/a_1+c_2x_2/a_2)\}],
\end{aligned} \qquad (17.67)
$$

where $\bar{\boldsymbol{\xi}}$ is the unit normal to the surface of the flat ellipsoid.

*18. Formulae for exterior points

When point **x** is outside an ellipsoidal inclusion (point **y** is outside S^2), the boundary terms produced by the integration by parts with respect to z in (16.14) and (16.16) are not necessarily zero.

Since

$$\delta'(\zeta\bar{\boldsymbol{\zeta}}\cdot\mathbf{y}-\zeta z) = -\zeta^{-1}\frac{\partial}{\partial z}\delta(\zeta\bar{\boldsymbol{\zeta}}\cdot\mathbf{y}-\zeta z),$$
$$\delta''(\zeta\bar{\boldsymbol{\zeta}}\cdot\mathbf{y}-\zeta z) = \zeta^{-2}\frac{\partial^2}{\partial z^2}\delta(\zeta\bar{\boldsymbol{\zeta}}\cdot\mathbf{y}-\zeta z), \tag{18.1}$$

(16.14) with (17.10) can be written as

$$u_i(\mathbf{x}) = (8\pi^2)^{-1}\int_{-1}^{1}dz\int_0^{2\pi}d\phi\int_0^R r\,dr\int_{S^2}C_{klmn}\epsilon^*_{nm}(\mathbf{x}')N_{ik}(\boldsymbol{\xi})D^{-1}(\boldsymbol{\xi})\bar{\xi}_l\zeta^2 \times\frac{\partial}{\partial z}\delta(\zeta\bar{\boldsymbol{\zeta}}\cdot\mathbf{y}-\zeta z)dS(\bar{\boldsymbol{\zeta}}). \tag{18.2}$$

Integration by parts with respect to z leads to

$$\begin{aligned}
u_i(\mathbf{x}) &= \lim_{z\to 1}(8\pi^2)^{-1}\int_0^{2\pi}d\phi\int_0^R r\,dr\int_{S^2}C_{klmn}\epsilon^*_{nm}(\mathbf{x}')N_{ik}D^{-1}\bar{\xi}_l\zeta^2\delta(\zeta\bar{\boldsymbol{\zeta}}\cdot\mathbf{y}\\
&\quad-\zeta z)dS(\bar{\boldsymbol{\zeta}})\\
&\quad-\lim_{z\to -1}(8\pi^2)^{-1}\int_0^{2\pi}d\phi\int_0^R r\,dr\int_{S^2}C_{klmn}\epsilon^*_{nm}(\mathbf{x}')\\
&\quad\times N_{ik}D^{-1}\bar{\xi}_l\zeta^2\delta(\zeta\bar{\boldsymbol{\zeta}}\cdot\mathbf{y}-\zeta z)dS(\bar{\boldsymbol{\zeta}})\\
&\quad-(8\pi^2)^{-1}\int_{-1}^{1}dz\int_0^{2\pi}d\phi\int_0^R r\,dr\int_{S^2}C_{klmn}\frac{\partial}{\partial z}\epsilon^*_{nm}(\mathbf{x}')\\
&\quad\times N_{ik}D^{-1}\bar{\xi}_l\zeta^2\delta(\zeta\bar{\boldsymbol{\zeta}}\cdot\mathbf{y}-\zeta z)dS(\bar{\boldsymbol{\zeta}})\\
&\quad+(8\pi^2)^{-1}\int_{-1}^{1}dz\int_0^{2\pi}d\phi\int_{S^2}C_{klmn}z\{\epsilon^*_{nm}(\mathbf{x}')\}_{r=R}\\
&\quad\times N_{ik}D^{-1}\bar{\xi}_l\zeta^2\delta(\zeta\bar{\boldsymbol{\zeta}}\cdot\mathbf{y}-\zeta z)dS(\bar{\boldsymbol{\zeta}}).
\end{aligned}\tag{18.3}$$

The first two integrals in (18.3) become zero, since $R\to 0$ for $z\to\pm 1$ and $\int_{S^2}\delta(\zeta\bar{\boldsymbol{\zeta}}\cdot\mathbf{y}-\zeta z)dS(\bar{\boldsymbol{\zeta}})$ is finite. The integration with respect to z in the remaining integrals is performed explicitly by the use of the property of Dirac's delta function,

$$\int_{-1}^{1}\zeta\delta(\zeta\bar{\boldsymbol{\zeta}}\cdot\mathbf{y}-\zeta z)dz = 1,\qquad \int_{-1}^{1}z\zeta\delta(\zeta\bar{\boldsymbol{\zeta}}\cdot\mathbf{y}-\zeta z)dz = \bar{\boldsymbol{\zeta}}\cdot\mathbf{y}, \tag{18.4}$$

if $|\bar{\zeta}\cdot\mathbf{y}|<1$; otherwise, they are zero. Therefore, (18.3) becomes

$$u_i(\mathbf{x}) = -(8\pi^2)^{-1}\int_0^{2\pi} d\phi\left[\int_0^R rdr\frac{\partial}{\partial z}\epsilon^*_{nm}(\mathbf{x}') - (\bar{\zeta}\cdot\mathbf{y})\{\epsilon^*_{nm}(\mathbf{x}')\}_{r=R}\right]_{z=\bar{\zeta}\cdot\mathbf{y}}$$
$$\times C_{klmn}\int_{S^*} N_{ik}(\bar{\xi})D^{-1}(\bar{\xi})\bar{\xi}_l\zeta dS(\bar{\zeta}), \tag{18.5}$$

where S^* is a subdomain of S^2 satisfying condition

$$|\bar{\zeta}\cdot\mathbf{y}|<1 \tag{18.5.1}$$

and S^* is shown in Fig. 16.2 by the unshaded surface on S^2. The notation defined in (16.10) and Fig. 16.1 are used in (18.5), and $R = (1-z^2)^{1/2}$.

Similarly, after repeating the integration by parts with respect to z, Mura and Cheng (1977) rewrite (16.16) as

$$u_{i,j}(\mathbf{x}) = -(8\pi^2)^{-1}\int_0^{2\pi} d\phi\left[\int_0^R rdr\frac{\partial^2}{\partial z^2}\epsilon^*_{mn}(\mathbf{x}') - z\left\{\frac{\partial}{\partial z}\epsilon^*_{mn}(\mathbf{x}')\right\}_{r=R}\right.$$
$$\left. - z\frac{\partial}{\partial z}\{\epsilon^*_{mn}(\mathbf{x}')\}_{r=R} - \{\epsilon^*_{mn}(\mathbf{x}')\}_{r=R}\right]_{z=\bar{\zeta}\cdot\mathbf{y}}$$
$$\times\int_{S^*} C_{klmn}N_{ik}(\bar{\xi})D^{-1}(\bar{\xi})\bar{\xi}_l\bar{\xi}_j dS(\bar{\zeta}) - (2\pi)^{-1}\oint_{L^+}\{\epsilon^*_{mn}(\mathbf{x}')\}_{z=\bar{\zeta}\cdot\mathbf{y}=1}$$
$$\times y^{-1}C_{klmn}N_{ik}(\bar{\xi})D^{-1}(\bar{\xi})\bar{\xi}_l\bar{\xi}_j d\theta(\bar{\zeta}). \tag{18.6}$$

The line integrals in (18.6) are defined on L^+ in Fig. 18.1 where $\bar{\zeta}\cdot\mathbf{y}=1$. The surface integrals on S^* in (18.6) can be obtained directly from (18.5) by differentiation with respect to x_j. $\partial/\partial x_j = \bar{\xi}_j\zeta^{-1}\partial/\partial z$ from (16.12).

The above result can be derived directly by differentiating (18.5)

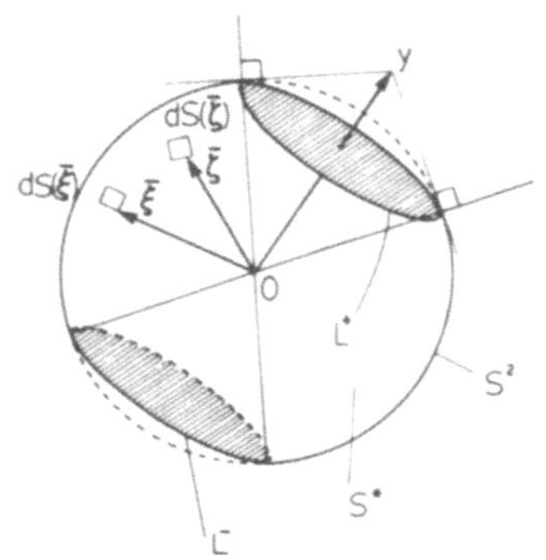

Fig. 18.1. L^+ is the boundary of S^* in the side of **y**.

with respect to x_j. Since $R_{,j} = -(\bar{\boldsymbol{\zeta}} \cdot \mathbf{y})\bar{\xi}_j/R\zeta$ and $z_{,j} = \bar{\xi}_j/\zeta$ at $z = \bar{\boldsymbol{\zeta}} \cdot \mathbf{y}$, the differentiation gives

$$u_{i,j}(\mathbf{x}) = -(8\pi^2)^{-1}\int_0^{2\pi} d\phi\left[\int_0^R rdr\,\frac{\partial^2}{\partial z^2}\epsilon^*_{nm}(\mathbf{x}') - (\bar{\boldsymbol{\zeta}}\cdot\mathbf{y})\left\{\frac{\partial}{\partial z}\epsilon^*_{nm}(\mathbf{x}')\right\}_{r=R}\right.$$
$$\left. - (\bar{\boldsymbol{\zeta}}\cdot\mathbf{y})\frac{\partial}{\partial z}\{\epsilon^*_{nm}(\mathbf{x}')\}_{r=R} - \epsilon^*_{nm}(\mathbf{x}')\right]_{z=\bar{\zeta}\cdot\mathbf{y}}$$
$$\times C_{klmn}\int_{S^*} N_{ik}(\bar{\boldsymbol{\xi}})D^{-1}(\bar{\boldsymbol{\xi}})\bar{\xi}_l\bar{\xi}_j dS(\bar{\boldsymbol{\zeta}})$$
$$-(8\pi^2)^{-1}\int_0^{2\pi} d\phi\left[\int_0^R rdr\,\frac{\partial}{\partial z}\epsilon^*_{nm}(\mathbf{x}') - (\bar{\boldsymbol{\zeta}}\cdot\mathbf{y})\{\epsilon^*_{nm}(\mathbf{x}')\}_{r=R}\right]_{z=\bar{\zeta}\cdot\mathbf{y}}$$
$$\times C_{klmn}N_{ik}(\bar{\boldsymbol{\xi}})D^{-1}(\bar{\boldsymbol{\xi}})\bar{\xi}_l\zeta\,\frac{\partial S}{\partial x_j}. \tag{18.6.1}$$

The last term comes from the change in S^* when $\mathbf{x}$ changes by $\delta\mathbf{x}$. It is clear that this term is considered only along L^+ and L^{-1}, where L^- is the circle $\bar{\boldsymbol{\zeta}} \cdot \mathbf{y} = -1$. From Figs. 18.2 and 18.3,

$$\frac{\partial S}{\partial x_j} = -\frac{\bar{\xi}_j}{\zeta y}\oint_{L^+} d\theta \tag{18.6.2}$$

along L^+. To understand this equation we note, in view of Fig. 18.3,

$$\begin{aligned} dS &= -\sin\alpha d\theta d\alpha,\\ \sin\alpha &= l/y,\\ ld\alpha &= \delta\mathbf{y}\cdot\bar{\boldsymbol{\zeta}},\\ (\delta\mathbf{y}\cdot\bar{\boldsymbol{\zeta}}) &= (\delta\mathbf{x}\cdot\bar{\boldsymbol{\xi}})/\zeta. \end{aligned} \tag{18.6.3}$$

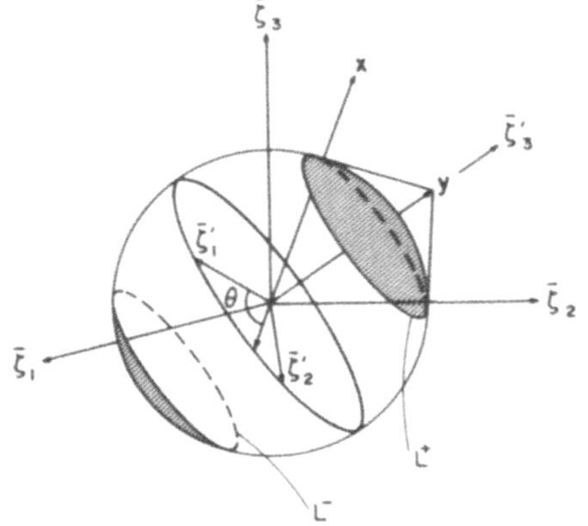

Fig. 18.2. New coordinate system $\bar{\zeta}'$.

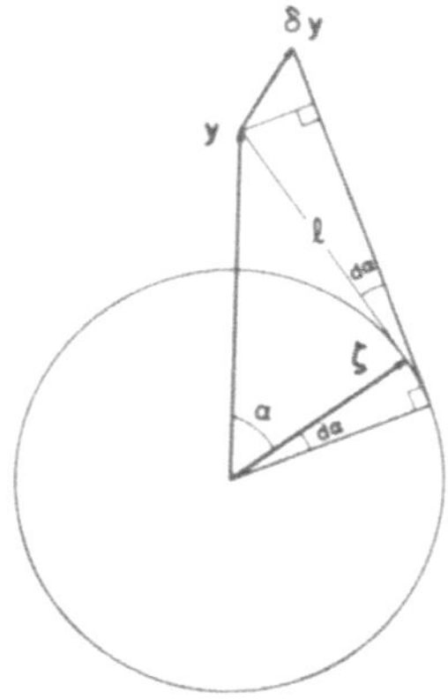

Fig. 18.3. Vectors in (18.6.3).

The last relation is obtained from (16.10). Similarly, $\partial S/\partial x_j$ along L^- is given by

$$\frac{\partial S}{\partial x_j} = \frac{\xi_j}{\zeta y} \oint_{L^-} d\theta. \tag{18.6.4}$$

Thus, (18.6.1) is written as

$$\begin{aligned} u_{i,j}(\mathbf{x}) = & -(8\pi^2)^{-1} \int_0^{2\pi} d\phi \Big[\int_0^R r dr \frac{\partial^2}{\partial z^2} \epsilon^*_{nm}(\mathbf{x}') - (\bar{\boldsymbol{\zeta}} \cdot \mathbf{y}) \Big\{ \frac{\partial}{\partial z} \epsilon^*_{nm}(\mathbf{x}') \Big\}_{r=R} \\ & - (\bar{\boldsymbol{\zeta}} \cdot \mathbf{y}) \frac{\partial}{\partial z} \{\epsilon^*_{nm}(\mathbf{x}')\}_{r=R} - \epsilon^*_{nm}(\mathbf{x}') \Big]_{z=\bar{\zeta} \cdot \mathbf{y}} \\ & \times C_{klmn} \int_{S^*} N_{ik}(\bar{\boldsymbol{\xi}}) D^{-1}(\bar{\boldsymbol{\xi}}) \bar{\xi}_l \bar{\xi}_j dS(\bar{\boldsymbol{\zeta}}) \\ & + (8\pi^2)^{-1} \int_0^{2\pi} d\phi \Big[\int_0^R r dr \frac{\partial}{\partial z} \epsilon^*_{nm}(\mathbf{x}') - (\bar{\boldsymbol{\zeta}} \cdot \mathbf{y}) \{\epsilon^*_{nm}(\mathbf{x}')\} \Big]_{z=\bar{\zeta} \cdot \mathbf{y}=1} \\ & \times \oint_{L^+} d\theta(\bar{\boldsymbol{\zeta}}) C_{klmn} N_{ik}(\bar{\boldsymbol{\xi}}) D^{-1}(\bar{\boldsymbol{\xi}}) \bar{\xi}_l \bar{\xi}_j y^{-1} \\ & - (8\pi^2)^{-1} \int_0^{2\pi} d\phi \Big[\int_0^R r dr \frac{\partial}{\partial z} \epsilon^*_{nm}(\mathbf{x}') - (\bar{\boldsymbol{\zeta}} \cdot \mathbf{y}) \{\epsilon^*_{nm}(\mathbf{x}')\} \Big]_{z=\bar{\zeta} \cdot \mathbf{y}=-1} \\ & \times \oint_{L^-} d\theta(\bar{\zeta}) C_{klmn} N_{ik}(\bar{\boldsymbol{\xi}}) D^{-1}(\bar{\boldsymbol{\xi}}) \bar{\xi}_l \bar{\xi}_j y^{-1}, \end{aligned} \tag{18.6.5}$$

when $z \to \pm 1$, $R \to 0$ and $\oint_0^{2\pi} d\phi \to 2\pi$. Furthermore, note from (17.49) that $[\epsilon^*_{nm}(\mathbf{x}')]_{z=-1}$ for $[\bar{\boldsymbol{\zeta}}]_{L^-}$ on L^{-1} is equal to $[\epsilon^*_{nm}(\mathbf{x}')]_{z=1}$ for $[\bar{\boldsymbol{\zeta}}]_{L^+}$(= −

$[\bar{\zeta}]_{L^-}$) on L^+, and $N_{ik}(\bar{\xi})D^{-1}(\bar{\xi})\bar{\xi}_l\bar{\xi}_j$ is an even function of $\bar{\xi}$. Thus, one obtains

$$u_{i,j}(\mathbf{x}') = -(8\pi^2)^{-1}\int_0^{2\pi} d\phi\left[\left[\int_0^R rdr\frac{\partial^2}{\partial z^2}\epsilon^*_{nm}(\mathbf{x}') - (\bar{\zeta}\cdot\mathbf{y})\left\{\frac{\partial}{\partial z}\epsilon^*_{nm}(\mathbf{x}')\right\}_{r=R}\right.\right.$$

$$\left.-(\bar{\zeta}\cdot\mathbf{y})\frac{\partial}{\partial z}\{\epsilon^*_{nm}(\mathbf{x}')\}_{r=R} - \epsilon^*_{nm}(\mathbf{x}')\right]_{z=\bar{\zeta}\cdot\mathbf{y}}$$

$$\times C_{klmn}\int_{S^*} N_{ik}(\bar{\xi})D^{-1}(\bar{\xi})\bar{\xi}_l\bar{\xi}_j dS(\bar{\zeta})$$

$$-(2\pi)^{-1}\oint_{L^+}\{\epsilon^*_{nm}(\mathbf{x}')\}_{z=\bar{\zeta}\cdot\mathbf{y}=1}C_{klmn}N_{ik}(\bar{\xi})D^{-1}(\bar{\xi})\bar{\xi}_l\bar{\xi}_j y^{-1}d\theta(\bar{\zeta}),$$

which is identical to (18.6).

If point $\mathbf{x}$ is inside the inclusion, the line integrals in (18.6) can be dropped, and S^* in the surface integrals is replaced by S^2. The resulting equation is identical to (17.8).

We note that the stress field has a discontinuity across the boundary of the inclusion. When $\mathbf{x}$ approaches the inclusion boundary from exterior points, the integral domain S^* in (18.6) becomes S^2, and $\bar{\zeta}$, satisfying the condition $\bar{\zeta}\cdot\mathbf{y} = 1$, approaches $\mathbf{y}$; that is, $a_1\bar{\xi}_1/\zeta \to x_1/a_1$, $a_2\bar{\xi}_2/\zeta \to x_2/a_2$, and $a_3\bar{\xi}_3/\zeta \to x_3/a_3$. It is easily shown that $\bar{\xi}_i \to n_i$, where n_i is the outward normal on the inclusion boundary. The integrals along L^+ in (18.6) are reduced to $-C_{klmn}\epsilon^*_{nm}(\mathbf{x})N_{ik}(\mathbf{n})D^{-1}(\mathbf{n})n_jn_l$. Thus, the stress jump on the inclusion boundary is

$$[\sigma_{ij}] = \sigma_{ij}(\text{out}) - \sigma_{ij}(\text{in})$$

$$= C_{ijkl}\{-C_{pqmn}\epsilon^*_{nm}(\mathbf{x})N_{kp}(\mathbf{n})D^{-1}(\mathbf{n})n_qn_l + \epsilon^*_{kl}(\mathbf{x})\} \tag{18.7}$$

which agrees with (6.11).

The stress concentration of an inclusion or inhomogeneity is obtained from σ_{ij}(out) which can be evaluated from (18.7) when σ_{ij}(in) and $[\sigma_{ij}]$ are known (see Laws 1977). The stress concentration factor of a lens-shaped void and its relation to the stress intensity factor of a crack ($a_3 \to 0$) has been discussed by Mura and Cheng (1977). The dynamic stress concentration factor of a spherical cavity is investigated by Moon and Pao (1967).

The integrals on S^* in (18.6) can be further modified as follows. A new coordinate system $\bar{\zeta}'_i$ is introduced as shown in Fig. 18.2, where the $\bar{\zeta}'_3$-axis is taken in the direction of $\mathbf{y}$, and the $\bar{\zeta}'_1$, $\bar{\zeta}'_2$-axes are lying on the plane perpendicular to $\mathbf{y}$. Then we have

$$dS(\bar{\zeta}) = d\bar{\zeta}'_3 d\theta, \tag{18.8}$$

where θ is measured counter-clockwise from the $\bar{\zeta}'_1$-axis, and, therefore,

$$\int_{S^*} [\ ]dS(\bar{\boldsymbol{\zeta}}) = \int_{-1/y}^{1/y} d\bar{\zeta}'_3 \int_0^{2\pi} [\ ]d\theta, \tag{18.9}$$

since S^* is determined from $\bar{\boldsymbol{\zeta}} \cdot \mathbf{y} = \bar{\zeta}'_3 y = \pm 1$.

The integrands in (18.6) are functions of $\bar{\boldsymbol{\xi}}$ and therefore must be expressed in terms of $\bar{\zeta}'_3$ and θ. The necessary transformation is as follows (see (16.10)):

$$\begin{aligned}
&\bar{\xi}_1 = \bar{\zeta}_1\zeta/a_1, \quad \bar{\xi}_2 = \bar{\zeta}_2\zeta/a_2, \quad \bar{\xi}_3 = \bar{\zeta}_3\zeta/a_3, \quad \bar{\boldsymbol{\zeta}} \cdot \mathbf{y} = \bar{\boldsymbol{\xi}} \cdot \mathbf{x}\zeta^{-1}, \\
&\bar{\zeta}_i = a_{ij}\bar{\zeta}'_j, \quad \zeta = (\zeta_i\zeta_i)^{1/2} = (a_1^2\bar{\xi}_1^2 + a_2^2\bar{\xi}_2^2 + a_3^2\bar{\xi}_3^2)^{1/2}, \\
&\bar{\zeta}'_1 = (1 - \bar{\zeta}'^2_3)^{1/2}\cos\theta, \\
&\bar{\zeta}'_2 = (1 - \bar{\zeta}'^2_3)^{1/2}\sin\theta, \\
&\bar{\zeta}'_3 = \bar{\xi}'_3,
\end{aligned} \tag{18.10}$$

where a_{ij} is the direction cosine between the $\bar{\zeta}_i$-and $\bar{\zeta}'_j$-axes.

Examples

It will now be shown that (18.5) and (18.6) lead to well-known solutions. Let us consider a sphere with radius a and with a uniform dilatational distribution of eigenstrain,

$$\epsilon^*_{ij} = \delta_{ij}\epsilon^*. \tag{18.11}$$

Consider (18.5) with (18.9) at point $(x_1, 0, 0)$. Then, the $\bar{\zeta}'_3$-axis coincides with the $\bar{\zeta}_1$-axis, and $\bar{\xi}_i = \bar{\zeta}_i$. Equation (18.5) becomes

$$u_1(x_1, 0, 0) = \frac{x_1\epsilon^*}{4\pi} C_{klmm} \int_0^{2\pi} d\theta \int_{-a/x_1}^{a/x_1} G_{1k1l}(\bar{\boldsymbol{\xi}})d\bar{\xi}_1, \tag{18.12}$$

where $G_{ijkl}(\bar{\boldsymbol{\xi}}) = N_{ij}(\bar{\boldsymbol{\xi}})\bar{\xi}_k\bar{\xi}_l D^{-1}(\bar{\boldsymbol{\xi}})$.

For an isotropic material, we have

$$u_1(x_1, 0, 0) = \frac{x_1\epsilon^*}{4\pi}\frac{3\lambda + 2\mu}{\lambda + 2\mu}\int_0^{2\pi} d\theta \int_{-a/x_1}^{a/x_1} \bar{\xi}_1 d\bar{\xi}_1 = \frac{a^3\epsilon^*}{3x_1^2}\frac{1+\nu}{1-\nu}. \tag{18.13}$$

Because of the symmetry of the problem, the above result can be written as $u = a^3\epsilon^*(1+\nu)/3r^2(1-\nu)$, where u is the displacement in the radial direction. Then $\epsilon_r = \partial u/\partial r = -2a^3\epsilon^*(1+\nu)/3r^3(1-\nu)$, which agrees with (11.45).

Next, consider the case when a centrally symmetric eigenstrain is prescribed in a sphere,

$$\epsilon^*_{ij}(\mathbf{x}) = \delta_{ij}\epsilon^*|\mathbf{x}|. \tag{18.14}$$

Then, the use of the notation in Fig. 16.1 and (16.10), leads to

$$\begin{aligned} &\epsilon^*_{ij}(\mathbf{x}') = \delta_{ij}\epsilon^* a(z^2 + r^2)^{1/2}, \\ &\left[\int_0^R r \frac{\partial}{\partial z} \epsilon^*_{nm}(\mathbf{x}')dr\right]_{z=\bar{\zeta}\cdot\mathbf{y}} = a\delta_{nm}\epsilon^*(\bar{\zeta}\cdot\mathbf{y})(1 - |\bar{\zeta}\cdot\mathbf{y}|). \end{aligned} \tag{18.15}$$

Consider point $(x_1, 0, 0)$. Equation (18.5) yields, for an isotropic medium,

$$\begin{aligned} u_1(x_1, 0, 0) &= \frac{x_1\epsilon^*}{4\pi} \int_0^{2\pi} d\theta \int_{-a/x_1}^{a/x_1} \frac{1+\nu}{1-\nu} \bar{\xi}_1^2 |x_1 \cdot \bar{\xi}_1| d\bar{\xi}_1 \\ &= \frac{a^4\epsilon^*}{4x_1^2} \frac{1+\nu}{1-\nu} \end{aligned} \tag{18.16}$$

or

$$u = \frac{a^4\epsilon^*}{4r^2} \frac{1+\nu}{1-\nu} \tag{18.16'}$$

which is identical to u obtained from (11.43) by taking $\alpha T = \epsilon^* r$. Equation (18.6) becomes

$$\begin{aligned} u_{1,1}(x_1, 0, 0) = \frac{a\epsilon^*}{4\pi} \frac{1+\nu}{1-\nu} &\left\{\int_0^{2\pi} d\theta \int_{-a/x_1}^{a/x_1} 2|\bar{\zeta}\cdot\mathbf{y}|\bar{\zeta}_1^2 d\bar{\zeta}_1 - 2 \right. \\ &\left. \times \int_{L^+} [\bar{\zeta}_1^2]_{\bar{\zeta}_1 = a/x_1} y^{-1} d\theta \right\} = -\frac{1}{2}\frac{1+\nu}{1-\nu}\frac{a^4\epsilon^*}{x_1^3} \end{aligned} \tag{18.17}$$

or

$$\frac{\partial u}{\partial r} = -\frac{1}{2}\frac{1+\nu}{1-\nu}\frac{a^4\epsilon^*}{r^3}. \tag{18.18}$$

It can also be shown that (18.6) leads to (11.33) when ϵ^*_{ij} is uniform and the material is isotropic. This result agrees with Eshelby's solution (1959).

In cases where analytical results are obtained from (18.5), equation (18.6) does not necessarily have to be used for $u_{i,j}$, as seen in the considered examples. However, (18.6) is necessary for numerical calculations when no analytical expression is available from (18.5).

19. Ellipsoidal inclusions with polynomial eigenstrains in anisotropic media

We consider the case when eigenstrains inside the inclusion are given in the following polynomial form of $\mathbf{x}'$ or $\mathbf{y}'$:

$$\epsilon^*_{nm}(\mathbf{x}') = \epsilon_{nm}(y_1')^{\alpha_1}(y_2')^{\alpha_2}(y_3')^{\alpha_3}, \tag{19.1}$$

where ϵ_{nm} are constants and α_1, α_2 and α_3 are integers. As shown in Fig. 16.1, we can write

$$y_p' = z\bar{\zeta}_p + r(m_p \cos\phi + n_p \sin\phi), \tag{19.2}$$

where $\mathbf{m}, \mathbf{n}, \bar{\zeta}$ are orthogonal unit vectors. The binomial theorem leads to

$$(y_1')^{\alpha_1} = \sum_{\beta_1=0}^{\alpha_1} \frac{\alpha_1!}{\beta_1!(\alpha_1-\beta_1)!}(z\bar{\zeta}_1)^{\alpha_1-\beta_1} r^{\beta_1}(m_1\cos\phi + n_1\sin\phi)^{\beta_1}. \tag{19.3}$$

The integrals with respect to ϕ in (18.5) have the form

$$\int_0^{2\pi} (m_1\cos\phi + n_1\sin\phi)^{\beta_1}(m_2\cos\phi + n_2\sin\phi)^{\beta_2} \times (m_3\cos\phi + n_3\sin\phi)^{\beta_3} d\phi. \tag{19.4}$$

A complex number $Z = \exp(i\phi)$ is introduced here. By defining

$$W_p = (m_p + in_p)/2, \quad \bar{W}_p = (m_p - in_p)/2, \tag{19.5}$$

where $i = \sqrt{-1}$, we can write (19.4) as

$$\oint_{|Z|=1} \sum_{q_1=0}^{\beta_1} \frac{\beta_1!}{q_1!(\beta_1-q_1)!} \bar{W}_1^{\beta_1-q_1} W_1^{q_1} Z^{\beta_1-2q_1} \sum_{q_2=0}^{\beta_2} \frac{\beta_2!}{q_2!(\beta_2-q_2)!} \times \bar{W}_2^{\beta_2-q_2} W_2^{q_2} Z^{\beta_2-2q_2} \sum_{q_3=0}^{\beta_3} \frac{\beta_3!}{q_3!(\beta_3-q_3)!} \bar{W}_3^{\beta_3-q_3} W_3^{q_3} Z^{\beta_3-2q_3} \frac{dZ}{iZ}, \tag{19.6}$$

where $|Z| = 1$ is a unit circle in the complex plane. The residue calculation is applied. The constant terms in the product of the power series of Z give the residues. Then, (19.6) is written as

$$2\pi \sum^{\beta_1} \sum^{\beta_2} \sum^{\beta_3}_{q_1+q_2+q_3=(\beta_1+\beta_2+\beta_3)/2} \frac{\beta_1!}{q_1!(\beta_1-q_1)!} \frac{\beta_2!}{q_2!(\beta_2-q_2)!} \frac{\beta_3!}{q_3!(\beta_3-q_3)!} \times \bar{W}_1^{\beta_1-q_1} \bar{W}_2^{\beta_2-q_2} \bar{W}_3^{\beta_3-q_3} W_1^{q_1} W_2^{q_2} W_3^{q_3}. \tag{19.7}$$

The integral (19.6) becomes zero when $(\beta_1+\beta_2+\beta_3)/2$ is not integer, that is, (19.6) reduces to (19.7) when $\beta_1+\beta_2+\beta_3$ is even; otherwise it is zero.

Therefore equation (18.5), after integration with respect to r, becomes

$$u_i(\mathbf{x}) = -(4\pi)^{-1}\int_{S^*} C_{klmn}\epsilon_{nm}N_{ik}(\bar{\boldsymbol{\xi}})D^{-1}(\bar{\boldsymbol{\xi}})\bar{\xi}_l\zeta\alpha_1!\alpha_2!\alpha_3!$$
$$\times\sum_{\beta_1=0}^{\alpha_1}\sum_{\beta_2=0}^{\alpha_2}\sum_{\beta_3=0}^{\alpha_3}\left\{(\bar{\boldsymbol{\zeta}}\cdot\mathbf{y})^{\alpha-\beta-1}(1-(\bar{\boldsymbol{\zeta}}\cdot\mathbf{y})^2)^{(\beta+2)/2}\frac{\alpha-\beta}{\beta+2}\right.$$
$$\left.-(\bar{\boldsymbol{\zeta}}\cdot\mathbf{y})^{\alpha-\beta+1}(1-(\bar{\boldsymbol{\zeta}}\cdot\mathbf{y})^2)^{\beta/2}\right\}\frac{\bar{\zeta}_1^{\alpha_1-\beta_1}}{(\alpha_1-\beta_1)!}\frac{\bar{\zeta}_2^{\alpha_2-\beta_2}}{(\alpha_2-\beta_2)!}\frac{\bar{\zeta}_3^{\alpha_3-\beta_3}}{(\alpha_3-\beta_3)!}$$
$$\times\sum_{q_1+q_2+q_3=\beta/2}\frac{\bar{W}_1^{\beta_1-q_1}W_1^{q_1}}{q_1!(\beta_1-q_1)!}\frac{\bar{W}_2^{\beta_2-q_2}W_2^{q_2}}{q_2!(\beta_2-q_2)!}\frac{\bar{W}_3^{\beta_3-q_3}W_3^{q_3}}{q_3!(\beta_3-q_3)!}\,dS(\bar{\boldsymbol{\zeta}}) \tag{19.8}$$

where $\alpha=\alpha_1+\alpha_2+\alpha_3$, $\beta=\beta_1+\beta_2+\beta_3$ and the summation with respect to $\beta_1, \beta_2, \beta_3$ is taken only when β is even. The $\mathbf{x}$ dependency of u_i can be seen from terms containing $(\bar{\boldsymbol{\zeta}}\cdot\mathbf{y}) = (\bar{\boldsymbol{\xi}}\cdot\mathbf{x})/\zeta$. If α equals N, then u_i is the sum of the polynomials of degrees $N+1, N-1, N-2, \ldots$, for interior points, since $S^* = S^2$ for these points.

The displacement gradient can be obtained as the sum of the direct derivative of the intergrand of (19.8) and the line integral on L^+, as seen in (18.6), where

$$\{\epsilon^*_{nm}(\mathbf{x}')\}_{z=1} = \epsilon_{nm}\bar{\zeta}_1^{\alpha_1}\bar{\zeta}_2^{\alpha_2}\bar{\zeta}_3^{\alpha_3}, \tag{19.9}$$

since in Fig. 16.1 $\mathbf{y}'\to\bar{\boldsymbol{\zeta}}$ when $z\to 1$. Equation (19.8) has been given by Mura and Kinoshita (1978).

Asaro and Barnett (1975) have obtained essentially the same expression.

The elastic field for exterior points of the inclusion becomes important when the interaction between two or more inclusions is considered. An approximation method has been proposed by Moschovidis and Mura (1975) for two isotropic inhomogeneous inclusions. The elastic field in the second inclusion caused by the first inclusion is approximated by its Taylor's expansion at the center of the second inclusion. The sum of the two elastic fields caused by individual inclusions can satisfy the required conditions for the equivalent inclusion problem at the second inclusion. A similar procedure is applied for points located in the first inclusion.

Special cases

When a given eigenstrain is linear,

$$\epsilon^*_{nm}(\mathbf{x}') = \epsilon^p_{nm}x'_p/a_p = \epsilon^p_{nm}y'_p, \tag{19.10}$$

we have from (19.2)

$$\frac{\partial}{\partial z} \epsilon^*_{nm}(\mathbf{x}') = \epsilon^p_{nm} \bar{\zeta}_p, \tag{19.11}$$

and, therefore,

$$u_i(\mathbf{x}) = -(4\pi)^{-1} \int_{S^*} \frac{1}{2}\{1 - 3(\bar{\boldsymbol{\zeta}} \cdot \mathbf{y})^2\} \epsilon^p_{nm} \bar{\zeta}_p C_{klmn} N_{ik}(\bar{\boldsymbol{\xi}}) D^{-1}(\bar{\boldsymbol{\xi}}) \bar{\xi}_l \zeta dS(\bar{\boldsymbol{\zeta}}). \tag{19.12}$$

When

$$\epsilon^*_{nm}(\mathbf{x}') = \epsilon^p_{nm}(y'_p)^2 \tag{19.13}$$

we have, similarly,

$$u_i(\mathbf{x}) = (4\pi)^{-1} \int_{S^*} [-(\bar{\boldsymbol{\zeta}} \cdot \mathbf{y})\bar{\zeta}_p^2 + 2(\bar{\boldsymbol{\zeta}} \cdot \mathbf{y})^3 \bar{\zeta}_p^2 + \{(\bar{\boldsymbol{\zeta}} \cdot \mathbf{y}) - (\bar{\boldsymbol{\zeta}} \cdot \mathbf{y})^3\}(1 - \bar{\zeta}_p^2)/2]$$
$$\times \epsilon^p_{nm} C_{klmn} N_{ik}(\bar{\boldsymbol{\xi}}) D^{-1}(\bar{\boldsymbol{\xi}}) \bar{\xi}_l \zeta dS(\bar{\boldsymbol{\zeta}}). \tag{19.14}$$

When the ellipsoid degenerates into a flat ellipsoid ($a_3 \to 0$), an argument similar to (17.36) can be employed. For interior points we have, from (19.12),

$$u_{i,j}(\mathbf{x}) = (4\pi)^{-1} C_{klmn} \epsilon^p_{nm} \int_{S^2} 3(\bar{\boldsymbol{\zeta}} \cdot \mathbf{y}) N_{ik}(\bar{\boldsymbol{\xi}}) D^{-1}(\bar{\boldsymbol{\xi}}) \bar{\xi}_j \bar{\xi}_l \bar{\zeta}_p dS(\bar{\boldsymbol{\zeta}}), \tag{19.15}$$

where $\bar{\xi}_1 = \zeta\bar{\zeta}_1/a_1$, $\bar{\xi}_2 = \zeta\bar{\zeta}_2/a_2$, $\bar{\xi}_3 = \zeta\bar{\zeta}_3/a_3$, $y_1 = x_1/a_1$, $y_2 = x_2/a_2$, $y_3 = x_3/a_3$. $G_{ikjl}(\bar{\xi}_1, \bar{\xi}_2, \bar{\xi}_3) \equiv N_{ik}(\bar{\boldsymbol{\xi}}) D^{-1}(\bar{\boldsymbol{\xi}}) \bar{\xi}_j \bar{\xi}_l$ is a homogeneous function of $\bar{\boldsymbol{\xi}}$ of degree zero and independent of the geometry of the ellipsoid. When $a_3 \to 0$, $\bar{\xi}_1, \bar{\xi}_2 \ll \bar{\xi}_3$, it can be written as $G_{ikjl}(0, 0, 1)$. The integral in (19.15), therefore, can be reduced to the integral

$$\int_{S^2} (\bar{\boldsymbol{\zeta}} \cdot \mathbf{y}) \bar{\zeta}_p dS(\bar{\boldsymbol{\zeta}}) = \int_{S^2} (\bar{\boldsymbol{\zeta}} \cdot \mathbf{y}) n_p dS(\bar{\boldsymbol{\zeta}}) = \int_V y_p dV = (4\pi/3) y_p, \tag{19.16}$$

where V is the volume of S^2 and n_p is the normal vector on S^2. Finally, (19.15) is written as

$$u_{i,j}(\mathbf{x}) = C_{klmn} \epsilon^p_{nm} y_p G_{ikjl}(0, 0, 1). \tag{19.17}$$

Similarly, we have, from (19.14),

$$u_{i,j}(\mathbf{x}) = C_{klmn} \epsilon^p_{mn} y_p^2 G_{ikjl}(0, 0, 1), \tag{19.18}$$

which is the displacement gradient inside a flat ellipsoid when an eigenstrain is given by (19.13).

*20. Harmonic eigenstrains

Let us consider the case where eigenstrain component ϵ_{ij}^* is a solid harmonic function of $\mathbf{y}$ in an ellipsoidal inclusion Ω,

$$\epsilon_{ij}^*(\mathbf{x}) = \bar{\epsilon}_{ij}^*(\boldsymbol{\omega})y^n P_n(\mathbf{y}\cdot\boldsymbol{\omega}/y), \tag{20.1}$$

where $y_1 = x_1/a_1$, $y_2 = x_2/a_2$, $y_3 = x_3/a_3$, $y = (y_iy_i)^{1/2}$; $\boldsymbol{\omega}$ is an arbitrary vector on the unit sphere S^2 and P_n is the Legendre polynomial of degree n. It is known that any function $f(\mathbf{y})$ of class $\mathbf{C}'$ defined on S^2 can be expanded in a uniformly convergent series,

$$f(\mathbf{y}) = \sum_{n=0}^{\infty}\frac{2n+1}{4\pi}\int_{S^2} f(\boldsymbol{\omega})P_n(\mathbf{y}\cdot\boldsymbol{\omega})dS(\boldsymbol{\omega}) \tag{20.2}$$

(see e.g., Hobson 1931). Therefore, the eigenstrain given by (20.1) may have a broad application, particularly in an interaction problem of two ellipsoidal inhomogeneities subjected to an applied stress. Note that (20.1) is not a solid harmonic function of $\mathbf{x}$.

When (20.1) is substituted into (16.14), we have

$$u_i(\mathbf{x}) = -\frac{a_1a_2a_3}{8\pi^2}\int_{-1}^{1}dz\int_0^{2\pi}d\phi\int_0^R rdr\int_{S^2} C_{klpq}\bar{\epsilon}_{pq}^*(\boldsymbol{\omega})N_{ik}(\bar{\boldsymbol{\xi}})$$
$$\times D^{-1}(\bar{\boldsymbol{\xi}})\bar{\xi}_l(y')^n P_n(\mathbf{y}'\cdot\boldsymbol{\omega}/y')\delta'(\zeta\bar{\boldsymbol{\zeta}}\cdot\mathbf{y}-\zeta z)dS(\bar{\boldsymbol{\xi}}), \tag{20.3}$$

where $y_1' = x_1'/a_1$, $y' = (y_i'y_i')^{1/2}$, $\bar{\zeta}_1 = a_1\bar{\xi}_1/\zeta$, $\zeta = (a_1^2\bar{\xi}_1^2 + a_2^2\bar{\xi}_2^2 + a_3^2\bar{\xi}_3^2)^{1/2}$, and z, ϕ, r and R are shown in Fig. 20.1. We have

$$\mathbf{y}' = z\bar{\boldsymbol{\zeta}} + r(\mathbf{m}\cos\phi + \mathbf{n}\sin\phi). \tag{20.4}$$

The addition theorem of the Legendre polynomial leads to

$$P_n(\mathbf{y}'\cdot\boldsymbol{\omega}/y') = P_n(\mathbf{y}'\cdot\bar{\boldsymbol{\zeta}}/y')P_n(\bar{\boldsymbol{\zeta}}\cdot\boldsymbol{\omega})$$
$$+2\sum_{m=1}^{n}\frac{(n-m)!}{(n+m)!}P_n^m(\mathbf{y}'\cdot\bar{\boldsymbol{\zeta}}/y')P_n^m(\bar{\boldsymbol{\zeta}}\cdot\boldsymbol{\omega})\cos(m\phi). \tag{20.5}$$

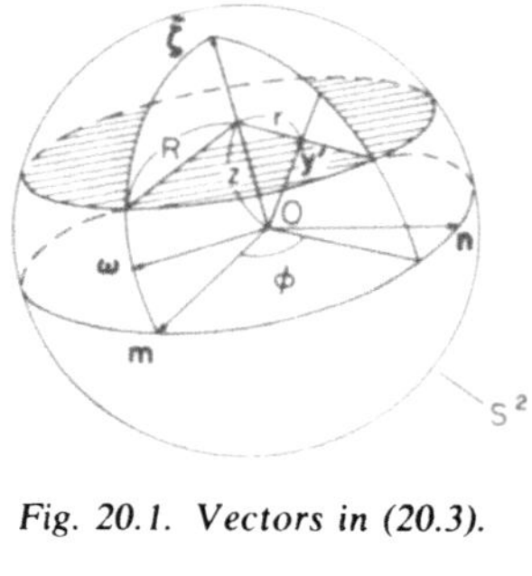

Fig. 20.1. Vectors in (20.3).

Since $\int_0^{2\pi} \cos m\phi d\phi = 0$, only the first term in (20.5) is used for P_n in (20.3). Furthermore, we have

$$\int_0^{2\pi} d\phi \int_0^R rdr(y')^n P_n(\mathbf{y}' \cdot \bar{\boldsymbol{\zeta}}/y') = 2\pi \int_0^{\sqrt{1-z^2}} rdr(z^2 + r^2)^{n/2} P_n(z/\sqrt{z^2+r^2})$$
$$= 2\pi(1 - z^2)P'_{n+1}(z)/(n+1)(n+2) \tag{20.6}$$

and

$$\int_{-1}^{1} (1 - z^2)P'_{n+1}(z)\delta'(\zeta\bar{\boldsymbol{\zeta}} \cdot \mathbf{y} - \zeta z)dz$$
$$= \int_{-1}^{1} \{(1 - z^2)P'_{n+1}(z)\}'\zeta^{-1}\delta(\zeta\bar{\boldsymbol{\zeta}} \cdot \mathbf{y} - \zeta z)dz$$
$$= \int_{-1}^{1} -(n+1)(n+2)P_{n+1}(z)\zeta^{-1}\delta(\zeta\bar{\boldsymbol{\zeta}} \cdot \mathbf{y} - \zeta z)dz$$
$$= -(n+1)(n+2)P_{n+1}(\bar{\boldsymbol{\zeta}} \cdot \mathbf{y})\zeta^{-2} \quad \text{for } |\bar{\boldsymbol{\zeta}} \cdot \mathbf{y}| < 1$$
$$= 0 \quad \text{for } |\bar{\boldsymbol{\zeta}} \cdot \mathbf{y}| > 1. \tag{20.7}$$

Therefore, (20.3) can be written as

$$u_i(\mathbf{x}) = \frac{a_1 a_2 a_3}{4\pi} C_{klpq}\epsilon^*_{pq}(\boldsymbol{\omega}) \int_{S^*} N_{ik}(\bar{\boldsymbol{\xi}})D^{-1}(\bar{\boldsymbol{\xi}})\bar{\xi}_l P_n(\bar{\boldsymbol{\zeta}} \cdot \boldsymbol{\omega})$$
$$\times P_{n+1}(\bar{\boldsymbol{\zeta}} \cdot \mathbf{y})\zeta^{-2}dS(\bar{\boldsymbol{\xi}}), \tag{20.8}$$

where S^* is the subspace of S^2 satisfying $|\bar{\boldsymbol{\zeta}} \cdot \mathbf{y}| \le 1$. For interior points of Ω, S^* is taken as S^2. When $n = 0$, then $P_0 = 1$, $P_1 = \bar{\boldsymbol{\zeta}} \cdot \mathbf{y} = \bar{\boldsymbol{\xi}} \cdot \mathbf{x}\zeta^{-1}$, and (20.8) becomes identical to (17.11).

The above result, (20.8), can be further modified, since $N_{ik}(\bar{\boldsymbol{\xi}})D^{-1}(\bar{\boldsymbol{\xi}})\bar{\xi}_l$ are homogeneous functions of degree -1. Using (17.10), we have

$$u_i(\mathbf{x}) = (1/4\pi) \int_{S^*} Q_{ni}(\bar{\boldsymbol{\zeta}}, \boldsymbol{\omega})P_{n+1}(\bar{\boldsymbol{\zeta}} \cdot \mathbf{y})dS(\bar{\boldsymbol{\zeta}}), \tag{20.9}$$

where

$$Q_{ni}(\bar{\boldsymbol{\zeta}}, \boldsymbol{\omega}) = C_{klpq}\epsilon^*_{pq}(\boldsymbol{\omega})N_{ik}(\boldsymbol{\xi})D^{-1}(\boldsymbol{\xi})\xi_l P_n(\bar{\boldsymbol{\zeta}} \cdot \boldsymbol{\omega}),$$
$$\xi_1 = \bar{\zeta}_1/a_1, \quad \xi_2 = \bar{\zeta}_2/a_2, \quad \xi_3 = \bar{\zeta}_3/a_3. \tag{20.10}$$

Q_{ni} can be expanded in a series of surface harmonic functions,

$$Q_{ni}(\bar{\boldsymbol{\zeta}}, \boldsymbol{\omega}) = \sum_{m=0}^{\infty} \frac{2m+1}{4\pi} \int_{S^2} Q_{ni}(\bar{\boldsymbol{\eta}}, \boldsymbol{\omega})P_m(\bar{\boldsymbol{\zeta}} \cdot \bar{\boldsymbol{\eta}})dS(\bar{\boldsymbol{\eta}}). \tag{20.11}$$

The above expression is substituted into (20.9) and the integration on S^*

is performed. The addition theorem

$$P_m(\bar{\boldsymbol{\zeta}}\cdot\bar{\boldsymbol{\eta}}) = P_m(\bar{\boldsymbol{\zeta}}\cdot\bar{\mathbf{y}})P_m(\bar{\boldsymbol{\eta}}\cdot\bar{\mathbf{y}}) + 2\sum_{l=1}^{m}\frac{(m-l)!}{(m+l)!}P_m^l(\bar{\boldsymbol{\zeta}}\cdot\bar{\mathbf{y}})P_m^l(\bar{\boldsymbol{\eta}}\cdot\bar{\mathbf{y}})\cos(l\theta),$$

yields

$$\int_{S^*} P_m(\bar{\boldsymbol{\zeta}}\cdot\bar{\boldsymbol{\eta}})P_{n+1}(\bar{\boldsymbol{\zeta}}\cdot\mathbf{y})dS(\bar{\boldsymbol{\zeta}}) = 2\pi P_m(\bar{\boldsymbol{\eta}}\cdot\bar{\mathbf{y}})\Phi_{nm}(y), \tag{20.12}$$

where $\bar{\mathbf{y}} = \mathbf{y}/y$, $y = |\mathbf{y}|$, $dS(\bar{\boldsymbol{\zeta}}) = d\theta dt$, $t = \bar{\boldsymbol{\zeta}}\cdot\bar{\mathbf{y}}$, and

$$\Phi_{nm}(y) = \int_{-1}^{1} P_m(t)P_{n+1}(yt)dt \tag{20.13}$$

for interior points $\mathbf{x}$ of Ω, since $S^* = S^2$, and

$$\Phi_{nm}(y) = \int_{-1/y}^{1/y} P_m(t)P_{n+1}(yt)dt = \frac{1}{y}\int_{-1}^{1} P_m(t/y)P_{n+1}(t)dt \tag{20.14}$$

for exterior points $\mathbf{x}$, since S^* is determined from the condition

$$|\bar{\boldsymbol{\zeta}}\cdot\mathbf{y}| = |ty| \leq 1. \tag{20.15}$$

The integrals (20.13) and (20.14) can be performed by the use of Rodrigues' formula (Whittaker and Watson, 1962),

$$P_m(t) = \frac{1}{2^m m!}\frac{d^m}{dt^m}(t^2-1)^m. \tag{20.16}$$

By repeated integration by parts, (20.13) becomes

$$\Phi_{nm}(y) = (-1)^{(n+1-m)/2}\sum_{l=0}^{(n+1-m)/2}\frac{(-1)^l\Gamma\left(\frac{n+m+2l+2}{2}\right)}{l!\,\Gamma\left(\frac{n-m-2l+3}{2}\right)\Gamma\left(\frac{2m+2l+3}{2}\right)}y^{m+2l}, \tag{20.17}$$

where $y \leq 1$, $m \leq n+1$, and Γ is the Gamma-function. When $m > n+1$, $\Phi_{nm} = 0$. Similarly, (20.14) becomes

$$\Phi_{nm}(y) = \sum_{l=0}^{(m-n-1)/2}\frac{(-1)^l\Gamma\left(\frac{2m-2l+1}{2}\right)}{l!\,\Gamma\left(\frac{m-n-2l+1}{2}\right)\Gamma\left(\frac{m+n-2l+4}{2}\right)}y^{-m-1+2l}, \tag{20.18}$$

where $y \geq 1$, $n+1 \leq m$. When $n+1 > m$, $\Phi_{nm} = 0$. In addition,

$$\begin{aligned}\Phi_{nm}(1) &= 0 \quad \text{for } m \neq n+1\\ &= \frac{2}{2n+3} \quad \text{for } m = n+1.\end{aligned} \tag{20.19}$$

Finally, (20.9) can be written as

$$u_i(\mathbf{x}) = (1/2) \sum_{m=0}^{\infty} \frac{2m+1}{4\pi} \Phi_{nm}(y) \int_{S^2} Q_{ni}(\bar{\boldsymbol{\eta}}, \boldsymbol{\omega}) P_m(\bar{\boldsymbol{\eta}} \cdot \bar{\mathbf{y}}) dS(\bar{\boldsymbol{\eta}}). \qquad (20.20)$$

From the definition (20.10), terms for which $n+m$ is even, are zero. It should be noticed that (20.20) is a uniformly convergent series and the integral domain S^2 is independent of $\mathbf{x}$ or $\mathbf{y}$. Therefore, the displacement gradient $u_{i,j}$ can be obtained directly from (20.20) by differentiation, where $\partial y/\partial x_1 = y_1/a_1 y$ and $\partial(\bar{\boldsymbol{\eta}} \cdot \mathbf{y})/\partial x_1 = \bar{\eta}_1/a_1$, etc.

21. Periodic distribution of spherical inclusions

Let us consider a periodic distribution of spherical inclusions with a period $2L$ in all three directions in the Cartesian coordinates (Fig. 21.1a). The distribution of eigenstrains can be expressed in the Fourier series as

$$\epsilon^*_{ij}(\mathbf{x}) = \sum{}' \sum_{\nu_1, \nu_2, \nu_3 = -\infty}^{\infty}{}' \sum{}' \bar{\epsilon}^*_{ij}(\boldsymbol{\xi}) \exp(i\boldsymbol{\xi} \cdot \mathbf{x}), \qquad (21.1)$$

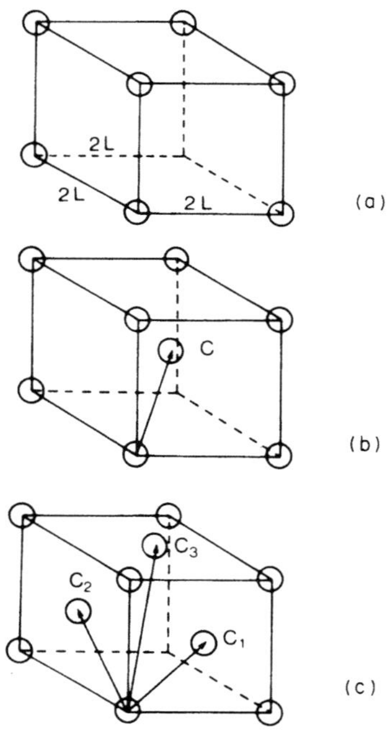

Fig. 21.1. Periodic distributions of spherical inclusions in a unit cell.

where $\xi_1 = \pi\nu_1/L$, $\xi_2 = \pi\nu_2/L$, $\xi_3 = \pi\nu_3/L$ (excluding $\nu_1 = \nu_2 = \nu_3 = 0$, since the uniform eigenstrain in a whole space gives a divergent solution), and

$$\bar{\epsilon}^*_{ij}(\boldsymbol{\xi}) = \frac{1}{8L^3}\int\int\int_{-L}^{L} \epsilon^*_{ij}(\mathbf{x}) \exp(i\boldsymbol{\xi}\cdot\mathbf{x})d\mathbf{x}. \tag{21.2}$$

Similarly, the displacement $u_i(\mathbf{x})$ can also be expressed in the Fourier series form:

$$u_i(\mathbf{x}) = \sum{}'\sum{}'\sum_{\nu_1,\nu_2,\nu_3=-\infty}^{\infty}{}' \bar{u}_i(\boldsymbol{\xi}) \exp(i\boldsymbol{\xi}\cdot\mathbf{x}). \tag{21.3}$$

According to (3.14), the solution can be written as

$$u_i(\mathbf{x}) = -i\sum{}'\sum{}'\sum_{\nu_1,\nu_2,\nu_3=-\infty}^{\infty}{}' C_{jlmn}\bar{\epsilon}^*_{mn}\xi_l N_{ij}(\boldsymbol{\xi})D^{-1}(\boldsymbol{\xi}) \exp(i\boldsymbol{\xi}\cdot\mathbf{x}). \tag{21.4}$$

If $\epsilon^*_{ij}(\mathbf{x})$ is uniform inside the spherical domains and zero outside, then (21.2) becomes

$$\bar{\epsilon}_{ij}(\boldsymbol{\xi}) = \frac{\epsilon^*_{ij}}{8L^3}\int_\Omega \exp(i\boldsymbol{\xi}\cdot\mathbf{x})d\mathbf{x}. \tag{21.5}$$

The integral has been carried out by Asaro (1975) and gives

$$\bar{\epsilon}^*_{ij} = (a/2L)^3(2\pi)^{3/2}\epsilon^*_{ij}\eta^{-3/2}J_{3/2}(\eta), \tag{21.6}$$

where a is the radius of the sphere,

$$\eta = a\xi = a(\xi_1^2+\xi_2^2+\xi_3^2)^{1/2} = \frac{\pi a}{L}(\nu_1^2+\nu_2^2+\nu_3^2)^{1/2} = \frac{\pi a}{L}\nu, \tag{21.7}$$

and

$$J_{3/2}(\eta) = (2/\pi\eta)^{1/2}(\eta^{-1}\sin\eta - \cos\eta). \tag{21.8}$$

Finally, the displacement field $u_i(\mathbf{x})$ becomes

$$u_i(\mathbf{x}) = -i\left(\frac{a}{2L}\right)^{3/2} C_{jlmn}\epsilon^*_{mn}\sum{}'\sum{}'\sum_{\nu_1,\nu_2,\nu_3=-\infty}^{\infty}{}' N_{ij}(\boldsymbol{\xi})D^{-1}(\boldsymbol{\xi})\xi_l\nu^{-3/2} \times J_{3/2}\left(\frac{\pi a\nu}{L}\right)\exp(i\boldsymbol{\xi}\cdot\mathbf{x}). \tag{21.9}$$

If the spherical inclusions are distributed in a body-centered cubic lattice (Fig. 21.1b), the solution can be the sum of (21.9) and the modified (21.9) where $\mathbf{x}$ is replaced by $\mathbf{x}-\mathbf{c}$. If the spherical inclusions are distributed in a face-centered cubic lattice (see Fig. 21.1c), the solution is obtained as the sum of four expressions similar to (21.9), having exponential factors

$\exp(i\boldsymbol{\xi}\cdot\mathbf{x})$, $\exp\{i\boldsymbol{\xi}\cdot(\mathbf{x}-\mathbf{c}_1)\}$, $\exp\{i\boldsymbol{\xi}\cdot(\mathbf{x}-\mathbf{c}_2)\}$ and $\exp\{i\boldsymbol{\xi}\cdot(\mathbf{x}-\mathbf{c}_3)\}$. The distortions corresponding to these cases can be summarized by

$$u_{i,k}(\mathbf{x}) = \left(\frac{a}{2L}\right)^{3/2} C_{jlmn}\epsilon^*_{mn} \sum{}' \sum_{\nu_1,\nu_2,\nu_3=-\infty}^{\infty}{}' \sum{}' G_{ijkl}(\boldsymbol{\xi})\nu^{-3/2} \times J_{3/2}\left(\frac{\pi a\nu}{L}\right) f(\nu)\exp(i\boldsymbol{\xi}\cdot\mathbf{x}) \tag{21.10}$$

where

$$f(\boldsymbol{\nu}) = \begin{cases} 1, & \text{for simple cubic} \\ 1+\exp(-i\boldsymbol{\xi}\cdot\mathbf{c}), & \text{for b.c.c.} \\ 1+\exp(-i\boldsymbol{\xi}\cdot\mathbf{c}_1)+\exp(-i\boldsymbol{\xi}\cdot\mathbf{c}_2)+\exp(-i\boldsymbol{\xi}\cdot\mathbf{c}_3), & \text{for f.c.c.} \end{cases} \tag{21.11}$$

The elastic strain energy assigned to a single inclusion is obtained from (13.3) as

$$W^* = -\frac{1}{2}\int_{V_0} \sigma_{ij}\epsilon^*_{ij}\,dD, \tag{21.12}$$

where V_0 is the volume of a single inclusion, or $V_0 = 4\pi a^3/3$. σ_{ij} in V_0 can be calculated from (21.10) and (2.4). Then, (21.12) becomes

$$W^* = \frac{2}{3}\pi a^3 C_{ijkl}\epsilon^*_{kl}\epsilon^*_{ij} - \frac{1}{2}a^3 C_{ijkl}C_{pqmn}\epsilon^*_{mn}\epsilon^*_{ij} \times \sum{}' \sum_{\nu_1,\nu_2,\nu_3=-\infty}^{\infty}{}' \sum{}' G_{kplq}(\boldsymbol{\nu})\nu^{-3} J^2_{3/2}\left(\frac{\pi a\nu}{L}\right) F(\boldsymbol{\nu}), \tag{21.13}$$

where $F(\boldsymbol{\nu})$ is the real part of $f(\boldsymbol{\nu})$.

Khachaturyan (1969, 1970) and Khachaturyan and Airapetyan (1974) have studied a spatial distribution of the inclusions by a different method and concluded that the periodic distribution proves to be stable.

In some alloys, clusters or precipitates are formed in a periodic manner (Ardell et al. 1966, Eurin et al. 1973), which is usually interpreted in terms of the interaction of two neighboring clusters or precipitates. However, the elastic strain energy in such a situation can be directly calculated from (21.13). Mori et al. (1978) have calculated the elastic strain energy for periodic arrangements of nitrogen atoms in iron. It is assumed that a nitrogen atom occupies a spherical region, V_0, the volume of which is equal to that of an iron atom and has eigenstrains of $\epsilon^*_{33} = 0.86$, $\epsilon^*_{11} = \epsilon^*_{22} = -0.08$ (Nowick and Berry 1972, misfitting sphere model). There are numerous ways of arranging the nitrogen atoms in periodic arrays. Mori et al. (1978) have calculated three cases: simple cubic, body-centered cubic, and face-centered cubic arrangements, all of which have symmetry identical to that of the host iron atoms. The elastic constants are assumed to be $G_{1111} = 2.37 \times 10^{11}$ N/m^2, $C_{1122} = 1.41 \times 10^{11}$ N/m^2, $C_{1212} = 1.16 \times 10^{11}$ N/m^2. The calculated energy is plotted against $2L$ in Fig. 21.2. First, it is seen that the lowest elastic strain energy is attained when the nitrogen atoms are arranged in body-centered cubic arrays. Secondly, in the simple and face-centered cubic arrays, the elastic strain energy decreases monotonously to the asymptotic value as the period $2L$ increases. The difference between the asymptotic value and the elastic strain energy at any specified arrangement is the interaction energy among all the nitrogen atoms. Contrary to the cases of the simple cubic and face-centered cubic, the elastic strain energy in the body-centered cubic array first decreases and then increases toward the asymptotic value as the separation increases. The lowest energy state is attained when $2L/a \cong 4.0$. It is noted that this situation corresponds approximately to the arrangement of nitrogen atoms in the α'' precipitate, $Fe_{16}N_2$, observed as an intermediate phase in supersaturated iron-nitrogen alloys. Although one cannot examine all the possible arrangements of the nitrogen atoms, it is clear from the above results that the formation of the α'' intermediate phase on tempering of supersaturated iron-nitrogen alloys can be understood on the basis of the elastic energy associated with the clustering of nitrogen atoms.

Instead of taking the approach described above, it is possible to use a direct method which utilizes the summation of the interaction energy between nitrogen atoms arranged in a periodic manner. Although Brown et al. (1973) have reported that the summation converges very slowly,

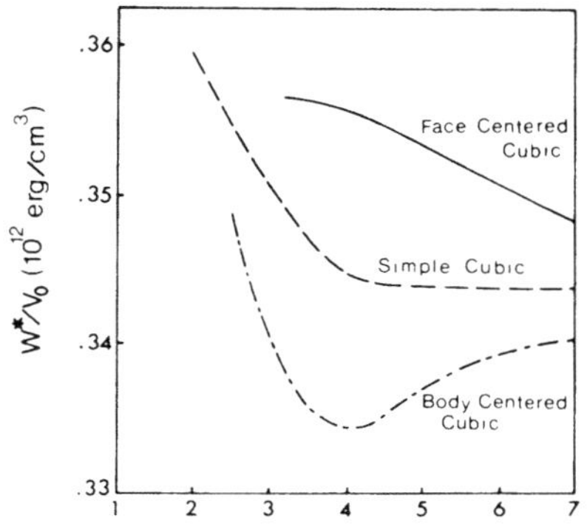

Fig. 21.2. Elastic energy per nitrogen atom.

Furuhashi et al. (1981) recently have proved that the summation of the stress fields of individual inclusions does not converge.

If a bounded inclusion Ω_P with eigenstrain ϵ^{*p}_{mn} is given in an infinite three-dimensional space D, the stress σ^p_{ij} due to the inclusion can be obtained from (6.1).

If an infinite number of inclusions $\Omega_p (p = 1, 2, \ldots)$ are distributed in D, the stress σ_{ij} may be the sum of σ^p_{ij},

$$\sigma_{ij}(\mathbf{x}) = \sum_{p=1}^{\infty} \sigma^p_{ij}(\mathbf{x}). \tag{21.14}$$

The convergency of this series, however, is not obvious. Furuhashi et al. showed that the convergency depends on the dimensions of distributions (line, plane, or three-dimensional) and types of eigenstrains (uniform or linear function of coordinates, etc.). The summation (21.14) is replaced by a new summation when it does not converge.

Let us assume that the material is isotropic and $\epsilon^{*p}_{ij} = \delta_{ij}$ for all indices p. According to Timoshenko and Goodier (1951), $\sigma^p_{ij}(\sigma^p_r$ or $\sigma^p_t)$ has a form of

$$1/\{(\bar{x}_1)^2 + (\bar{x}_2)^2 + (\bar{x}_3)^2\}^{3/2} \tag{21.15}$$

for exterior points of Ω_p, where $\bar{x}_1$, $\bar{x}_2$, $\bar{x}_3$ are the Cartesian coordinates measured from the center of spherical inclusion Ω_p.

The following Eisenstein's theorem (see Whittaker and Watson, 1962) is used,

$$\sum 1/(q_1^2 + q_2^2 + \ldots + q_n^2)^s \tag{21.16}$$

in which the summation extends over all positive and negative integral values and zero values of $q_1, q_2, \ldots, q_n$, except the set of simultaneous zero values, is absolutely convergent if $s > n/2$ and divergent if $s \leq n/2$.

Suppose that the distribution of Ω_p is defined in the region $x_1 > 0, x_2 > 0, x_3 > 0$ as

$$\begin{aligned} &\Omega_p = \{\mathbf{x}; |\mathbf{x} - 3\mathbf{p}| < 1\}, \\ &\mathbf{p} = (p_1, p_2, p_3), p_1, p_2, p_3\text{: positive integers.} \end{aligned} \tag{21.17}$$

In order to investigate the convergency of (21.14), it is sufficient to investigate the convergency of the series

$$S = \sum_{p_1, p_2, p_3} 1/\{(x_1 - 3p_1)^2 + (x_2 - 3p_2)^2 + (x_3 - 3p_3)^2\}^{3/2}. \tag{21.18}$$

Since it is always possible for a given set of x_i and p_i to choose the smallest integers q_i such that

$$|x_i - 3p_i| \leq q_i, \tag{21.19}$$

then, for such q_i,

$$S \geq \sum_{q_1, q_2, q_3} 1/(q_1^2 + q_2^2 + q_3^2)^{3/2}. \tag{21.20}$$

Since $n = 3$, $s = 3/2$ in Eisenstein's theorem, the series on the right hand side in (21.20) is divergent and, therefore, series (21.18) diverges.

Although q_1, q_2, q_3 in the series (21.20) do not take all integers used for the series (21.16), Eisenstein's theorem still holds. Consider the series $\Sigma_{p_i \geq n_i} 1/|\mathbf{p} - (1/3)\mathbf{x}|^3$ instead of (21.18), where n_i are arbitrary positive integers such that $p_i - (1/3)x_i \geq 1$ for $p_i \geq n_i$ and let q_i be the smallest integers such as $p_i - (1/3)x_i \leq q_i$. Then the inequality

$$\sum_{p_i \geqq n_i} 1/|\mathbf{p} - (1/3)\mathbf{x}|^3 \geqq \sum_{q_i \geqq q_i'} 1/(q_1^2 + q_2^2 + q_3^2)^{3/2} \tag{21.21}$$

holds for some positive integers q_i'.

It is concluded that the resulting stress caused by a three-dimensional periodic distribution of spherical inclusions with uniform dilatational eigenstrains cannot be expressed as (21.14). The expression for the stress field must be modified as shown later.

Let us consider the following plane distribution of inclusions:

$$\begin{aligned} &\Omega_p = \{\mathbf{x}; |\mathbf{x} - 3\mathbf{p}| < 1\}, \\ &\quad \mathbf{p} = (p_1, p_2, 0),\ p_1, p_2\text{: all positive and negative integers.} \end{aligned} \tag{21.22}$$

In order to investigate the convergency of (21.14) it is sufficient to investigate the convergency of the series

$$S = \sum_{p_1, p_2} 1/\{(x_1 - 3p_1)^2 + (x_2 - 3p_2)^2 + x_3^2\}^{3/2}. \tag{21.23}$$

It is always possible to choose the largest integers q_1, q_2 for given values of x_i and p_i such that

$$|x_1 - 3p_1| \geq q_1,\ |x_2 - 3p_2| \geq q_2. \tag{21.24}$$

Then

$$S \leq \sum_{q_1, q_2} 1/(q_1^2 + q_2^2)^{3/2}. \tag{21.25}$$

When Eisenstein's theorem is applied to the right hand side in (21.25), the series converges because $n = 2$ and $s = 3/2$. An argument similar to the following sentences after (21.20) can also be applied if one wants strictness of argument on the convergency. It is concluded that for any plane or linear periodic distribution of inclusions the resulting stress can be expressed as (21.14).

Now let us construct a suitable expression for the resulting stress caused by a three-dimensional periodic distribution of inclusions. In order to have a convergent series, we consider

$$\sum_{p} (\sigma_{ij}^{p}(\mathbf{x}) - \sigma_{ij}^{p}(\mathbf{0})). \tag{21.26}$$

Since

$$\left|\sum_{p} (1/|\mathbf{x}-\mathbf{p}|^3 - 1/|\mathbf{p}|^3)\right| \le M \sum_{p} 1/|\mathbf{x}-\mathbf{p}|^3|\mathbf{p}| + C\sum_{p} 1/|\mathbf{x}-\mathbf{p}|^2|\mathbf{p}|^2, \tag{21.27}$$

where M and C are finite constants, the series (21.26) is convergent ($n = 3$, $s = 4/2$). An arbitrary constant can be added to (21.26) when (21.14) is replaced by (21.26). The arbitrary constant must be determined from the condition of periodicity.

Let us define v_0 as a unit cell of the periodic distribution of inclusions. For distribution (21.17), v_0 is a hexahedron defined by $-(3/2) \le x_i \le (3/2)$, $i = 1, 2, 3$. The integration σ_{ij} over v_0 becomes

$$\begin{aligned}\int_{v_0} \sigma_{ij}(\mathbf{x})d\mathbf{x} &= C_{ijkl}\int_{v_0} \{u_{k,l}(\mathbf{x}) - \epsilon_{kl}^{*}(\mathbf{x})\}d\mathbf{x}\\ &= C_{ijkl}\int_{|v_0|} u_k(\mathbf{x})n_l dS - C_{ijkl}\int_{v_0} \epsilon_{kl}^{*}(\mathbf{x})d\mathbf{x},\end{aligned} \tag{21.28}$$

where $|v_0|$ is the boundary of v_0 and C_{ijkl} are the elastic moduli. The condition of periodicity gives

$$\int_{|v_0|} u_k(\mathbf{x})n_l dS = 0. \tag{21.29}$$

Then, (21.28) becomes

$$\int_{v_0} \sigma_{ij}(\mathbf{x})d\mathbf{x} = -C_{ijkl}\int_{v_0} \epsilon_{kl}^{*}(\mathbf{x})d\mathbf{x}. \tag{21.30}$$

In order to satisfy the above condition, the stress expression must have the form

$$\sigma_{ij}(\mathbf{x}) = \sum_{p} (\sigma_{ij}^{p}(\mathbf{x}) - \sigma_{ij}^{p}(0)) + \alpha_{ij}, \tag{21.31}$$

and

$$\alpha_{ij} = -(1/v_0)\left\{C_{ijkl}\int_{v_0} \epsilon^*_{kl}(\mathbf{x})d\mathbf{x} + \int_{v_0}\sum_p (\sigma^p_{ij}(\mathbf{x}) - \sigma^p_{ij}(\mathbf{0}))d\mathbf{x}\right\}. \tag{21.32}$$

Expression (21.14) can be recovered from (21.32) when the summation (21.14) is convergent. It can also be observed that expression (21.31) is independent of the choice of the origin of coordinates.

To generalize the last result for eigenstrains prescribed by polynomials of the coordinates, we start our discussion from equation (12.2). The displacement field for interior and exterior points of Ω is expressed as

$$u_i(x) = \frac{1}{8\pi(1-\nu)}\{\Psi_{jl,jli} - 2\nu\Phi_{mm,i} - 4(1-\nu)\Phi_{il,l}\}, \tag{21.33}$$

where

$$\Psi_{ij}(\mathbf{x}) = \int_\Omega |\mathbf{x}-\mathbf{x}'|\epsilon^*_{ij}(\mathbf{x}')d\mathbf{x}' \tag{21.34}$$

and

$$\Phi_{ij}(\mathbf{x}) = \int_\Omega \frac{\epsilon^*_{ij}(\mathbf{x}')}{|\mathbf{x}-\mathbf{x}'|}d\mathbf{x}'. \tag{21.35}$$

$\Phi_{ij}(\mathbf{x})$ and $\Psi_{ij}(\mathbf{x})$ are the harmonic and biharmonic potentials due to a body Ω of density $\epsilon^*_{ij}(\mathbf{x}')$, and have the properties described by (12.5) and (12.6).

Let $\epsilon^*_{ij}(\mathbf{x})$ be any homogeneous polynomial of degree l,

$$\epsilon^*_{ij}(\mathbf{x}) = B_{ijs_1\ldots s_l}x_{s_1}x_{s_2}\ldots x_{s_l}. \tag{21.36}$$

Then, the structure of the elastic field can be determined by the property of the integral

$$\phi_{s_1s_2\cdots s_l} = \int_\Omega \frac{x'_{s_1}x'_{s_2}\cdots x'_{s_l}}{|\mathbf{x}-\mathbf{x}'|}d\mathbf{x}'. \tag{21.37}$$

According to Dyson (1891), when Ω is a sphere with radius a, the integral (21.37) for exterior point $\mathbf{x}$ of Ω is explicitly written as

$$\phi_{s_1s_2\cdots s_l} = \sum_{n=0}^{[l/2]} |\mathbf{x}|^{-2(l-n)-1}\left\{\left(\frac{\partial}{\partial x_j}\frac{\partial}{\partial x_j}\right)^n x_{s_1}x_{s_2}\cdots x_{s_l}\right\}\sum_{\alpha=0}^{n} C(n,\alpha)|\mathbf{x}|^{2\alpha}, \tag{21.38}$$

where [l/2] is Gauss' symbol (a maximum integer not exceeding l/2) and

$$C(n, \alpha) = \frac{\pi a^{l+3-2\alpha}(-1)^{\alpha+n+1}}{2^{2n} n!(n+1)!} \sum_{r=0}^{n+1} \frac{(-1)^r {}_nC_r}{(\alpha + n - l - r - 1/2)}. \tag{21.39}$$

The finite series (21.38) can be summarized in the form

$$\{f_0^l(\mathbf{x})/|\mathbf{x}|^l + f_1^l(\mathbf{x})/|\mathbf{x}|^{l-2} + \ldots + f_{(l+2)/2}^l(\mathbf{x})\}/|\mathbf{x}|^{l+1} \tag{21.40}$$

if l is even, and in the form

$$\{f_0^l(\mathbf{x})/|\mathbf{x}|^{l-1} + f_1^l(\mathbf{x})/|\mathbf{x}|^{l-3} + \ldots + f_{(l+1)/2}^l(\mathbf{x})\}/|\mathbf{x}|^{l+2} \tag{21.41}$$

if l is odd, where f_j^l are homogeneous polynomials of $\mathbf{x}$ of degree l for all indices j. Accordingly, the displacement expression (21.33) can be summarized in the form

$$\{f_0^{l+3}(\mathbf{x})/|\mathbf{x}|^{l+2} + f_1^{l+3}(\mathbf{x})/|\mathbf{x}|^l + \ldots + f_{(l+2)/2}^{l+3}(\mathbf{x})\}/|\mathbf{x}|^{l+5} \tag{21.42}$$

if l is even, and in the form

$$\{f_0^{l+3}(\mathbf{x})/|\mathbf{x}|^{l+1} + f_1^{l+3}(\mathbf{x})/|\mathbf{x}|^{l-1} + \ldots + f_{(l+1)/2}^{l+3}(\mathbf{x})\}/|\mathbf{x}|^{l+6} \tag{21.43}$$

if l is odd.

The corresponding expression for stress becomes

$$\{f_0^{l+4}(\mathbf{x})/|\mathbf{x}|^{l+2} + f_1^{l+4}(\mathbf{x})/|\mathbf{x}|^l + \ldots + f_{(l+2)/2}^{l+4}(\mathbf{x})\}/|\mathbf{x}|^{l+7} \tag{21.44}$$

if l is even, and

$$\{f_0^{l+4}(\mathbf{x})/|\mathbf{x}|^{l+1} + f_1^{l+4}(\mathbf{x})/|\mathbf{x}|^{l-1} + \ldots + f_{(l+1)/2}^{l+4}(\mathbf{x})\}/|\mathbf{x}|^{l+8} \tag{21.45}$$

if l is odd.

If an infinite number of inclusions are periodically distributed in the three-dimensional space as shown by (21.17), the resulting stress can be obtained by the sum of (21.43) or (21.44), depending on l. These summations are taken for all $\Omega_p (p = 1, 2, \ldots)$. The coordinates appearing in (21.44) and (21.45) are measured from the center of each Ω_p. Convergency of the resulting series can be investigated by Eisenstein's theorem. The theorem is applied to the last terms in (21.42) ~ (21.45), since they have the lowest $0(1/|\mathbf{x}|)$. Since $n = 3$ and $s = 4/2$ for (21.45), the series constructed from (21.45) is convergent. It is concluded, therefore, that the stress field can be expressed as (21.14) when the eigenstrain is a homogeneous polynomial with degree of an odd number l. The other series representing displacement fields and the stress field for an even number of l are generally divergent. However, the series may converge when the last few terms in (21.42) ~ (21.44) are zero. It is always possible, however, to write the stress field in the form of (21.31).

We show that expression (21.31) with (21.32) is valid even for arbitrary-shaped inclusions in an anistropic medium with an arbitrary eigenstrain, where ϵ^{*p}_{ij} is integrable in Ω_p. Ω_p are identical for all p.

For Green's function of elasticity $G_{sk}(\mathbf{x})$, we have (5.61),

$$G_{sk,rl}(\mathbf{x}) = V(\mathbf{x})/|\mathbf{x}|^3, \tag{21.46}$$

where $V(\mathbf{x})$ is analytic for $\mathbf{x} \neq 0$ and positively homogeneous of degree 0. Then

$$V(\mathbf{x}+\mathbf{p}) = V(\mathbf{p}) + V_{,i}(\mathbf{p}+\theta\mathbf{x})x_i, \; (0 < \theta < 1) \tag{21.47}$$

and

$$\frac{V(\mathbf{x}+\mathbf{p})}{|\mathbf{x}+\mathbf{p}|^3} - \frac{V(\mathbf{p})}{|\mathbf{x}|^3} = V(\mathbf{p})\left(\frac{1}{|\mathbf{x}+\mathbf{p}|^3} - \frac{1}{|\mathbf{x}|^3}\right) + \frac{V_{,i}(\mathbf{p}+\theta\mathbf{x})x_i}{|\mathbf{x}+\mathbf{p}|^3}. \tag{21.48}$$

From (21.27) the series constructed from the first term in the right hand side in (21.48) converges. Furthermore,

$$|V_{,i}(\mathbf{x})| \leq \frac{V_1}{|\mathbf{x}|} \tag{21.49}$$

and

$$|V_{,i}(\mathbf{p}+\theta\mathbf{x})| \leq \frac{V_1}{|\mathbf{p}+\theta\mathbf{x}|} \leq \frac{V_1}{|\mathbf{p}|-|\mathbf{x}|}, \; (|\mathbf{p}| > |\mathbf{x}|). \tag{21.50}$$

Therefore, the series constructed from the last term in the right hand side in (21.48) is also convergent. From (6.1) we have

$$\sigma^p_{ij}(\mathbf{x}) = -C_{ijrs}\int_{\Omega_p} G_{sk,rl}(\mathbf{x}-\mathbf{p}-\boldsymbol{\xi})C_{klmn}\epsilon^*_{mn}(\boldsymbol{\xi})d\boldsymbol{\xi} - C_{ijmn}\epsilon^*_{mn}(\mathbf{x}-\mathbf{p}) \tag{21.51}$$

and ϵ^*_{mn} is integrable in Ω_p, therefore, the series in (21.31) and (21.32) is convergent.

4

Ellipsoidal inhomogeneities

When the elastic moduli of an ellipsoidal subdomain of a material differ from those of the remainder (matrix), the subdomain is called an ellipsoidal inhomogeneity. Voids, cracks and precipitates are examples of the inhomogeneity which might also be called inclusion. However, the term "inclusion" has been used in a different context in this book. Here, an inclusion has the same elastic moduli as the matrix, but it contains eigenstrains.

A material containing inhomogeneities is free from any stress field unless a load is applied. On the other hand, a material containing inclusions is subjected to an internal stress (eigenstress) field even if it is free from any external tractions.

If an inhomogeneity contains an eigenstrain, it is called an inhomogeneous inclusion. Most of the precipitates in alloys and martensites in phase transformation are inhomogeneous inclusions. Eigenstrains inside these inhomogeneous inclusions are misfit and phase transformation strains.

Eshelby (1957) first pointed out that the stress disturbance in an applied stress due to the presence of an inhomogeneity can be simulated by an eigenstress caused by an inclusion when the eigenstrain is chosen properly. This equivalency will be called the equivalent inclusion method.

This book emphasizes the equivalent inclusion method, since it provides a consistent method where the results of inclusions in homogeneous media obtained in Chapters 2 and 3 are used. Exact methods of analysis for stress concentration around ellipsoidal cavities in isotropic media subjected to a uniform tension or torsion usually involve harmonic or bi-harmonic displacement potentials. Particular solutions of this kind have been developed by Lamé (1866), Boussinesq (1885), Galerkin (1930), Love (1927), Papkovich (1932), Neuber (1934) and Mindlin (1936). A brief survey of this approach has been given by Sternberg (1958). The solution for a spherical cavity under an arbitrary uniform field at infinity is attributed to Neuber (1937) to Leon (1908). It has also been given by Southwell and Gough (1926), Goodier (1933), Sternberg et al. (1952) and Ling and Yang (1951). Sadowsky and Sternberg (1947), Shapiro (1947), Miyamoto (1950, 1951, 1953), Edward (1951) and Das (1954) have given solutions for a spheroidal cavity or inhomogeneity. For an ellipsoidal inhomogeneity or inclusion, solutions have been given by Sadowsky and Sternberg (1949), Lurie (1952), Robinson (1951) and Niesel (1953). The problem of a spheroidal inhomogeneity in a transversely isotropic material has been solved by Bose (1956) and W.T. Chen (1966, 1968). In connexion with a geotechnical study, Selvadurai (1976, 1978) investigated the load-deflexion characteristics of deep rigid anchors embedded in cohesive soil or rock media.

22. Equivalent inclusion method

Consider an infinitely extended material with the elastic moduli C_{ijkl} containing an ellipsoidal domain Ω with the elastic moduli C^*_{ijkl}. Ω is called an ellipsoidal inhomogeneity. We investigate the disturbance in an applied stress caused by the presence of this inhomogeneity. Let us denote the applied stress at infinity by σ^0_{ij} and the corresponding strain by $\frac{1}{2}(u^0_{i,j} + u^0_{j,i})$. The stress disturbance and the displacement disturbance are denoted by σ_{ij} and u_i, respectively. The total stress (actual stress) is $\sigma^0_{ij} + \sigma_{ij}$, and the total displacement is $u^0_i + u_i$. Stress components σ_{ij} are in self-equilibrium, that is,

$$\sigma_{ij,j} = 0 \tag{22.1}$$

and $\sigma_{ij} = 0$ at infinity. When a finite body is considered,

$$\sigma_{ij}n_j = 0 \tag{22.2}$$

on the boundary, where n_i is the unit normal on the boundary.

Hooke's law is written as

$$\begin{aligned} \sigma^0_{ij} + \sigma_{ij} &= C^*_{ijkl}(u^0_{k,l} + u_{k,l}) \quad \text{in } \Omega, \\ \sigma^0_{ij} + \sigma_{ij} &= C_{ijkl}(u^0_{k,l} + u_{k,l}) \quad \text{in } D - \Omega. \end{aligned} \tag{22.3}$$

The equivalent inclusion method is used to simulate the stress disturbance using the eigenstress resulting from an inclusion which occupies the space Ω.

Consider an infinitely extended homogeneous material with the elastic moduli C_{ijkl} everywhere, containing domain Ω with an eigenstrain

ϵ^*_{ij}. ϵ^*_{ij} has been introduced here arbitrarily in order to simulate the inhomogeneity problem by use of the inclusion method. Such an eigenstrain is called an equivalent eigenstrain. When this homogeneous material is subjected to the applied strain $\epsilon^0_{ij} = \frac{1}{2}(u^0_{i,j} + u^0_{j,i})$ at infinity, the resulting total stress, distortion, and elastic distortion, respectively, are $\sigma^0_{ij} + \sigma_{ij}$, $u^0_{i,j} + u_{i,j}$, and $u^0_{i,j} + u_{i,j} - \epsilon^*_{ij}$ in Ω. Then, Hooke's law yields

$$\begin{aligned} \sigma^0_{ij} + \sigma_{ij} &= C_{ijkl}(u^0_{k,l} + u_{k,l} - \epsilon^*_{kl}) \text{ in } \Omega, \\ \sigma^0_{ij} + \sigma_{ij} &= C_{ijkl}(u^0_{k,l} + u_{k,l}) \qquad \text{in } D - \Omega, \end{aligned} \tag{22.4}$$

where $\sigma^0_{ij} = C_{ijkl}u^0_{k,l}$.

The necessary and sufficient condition for the equivalency of the stresses and strains in the above two problems of inhomogeneity and inclusion is,

$$C^*_{ijkl}(u^0_{k,l} + u_{k,l}) = C_{ijkl}(u^0_{k,l} + u_{k,l} - \epsilon^*_{kl}) \text{ in } \Omega. \tag{22.5}$$

or

$$C^*_{ijkl}(\epsilon^0_{kl} + \epsilon_{kl}) = C_{ijkl}(\epsilon^0_{kl} + \epsilon_{kl} - \epsilon^*_{kl}) \qquad \text{in } \Omega. \tag{22.5$'$}$$

As mentioned in the preceding sections, ϵ_{kl} in the above equation can be obtained as a known function of ϵ^*_{kl} when the eigenstrain problem in the homogeneous material is solved. Thus, (22.5) determines ϵ^*_{kl} for a given ϵ^0_{kl}, in such a manner that equivalency holds. After obtaining ϵ^*_{kl}, the stress $\sigma^0_{ij} + \sigma_{ij}$ can be found from (22.3) or (22.4).

If σ^0_{ij} is a uniform stress, ϵ^*_{ij} is also uniform in Ω. Then, from (11.15) or (17.18)

$$\epsilon_{kl} = \tfrac{1}{2}(u_{k,l} + u_{l,k}) = S_{klmn}\epsilon^*_{nm}, \tag{22.6}$$

where S_{klmn} are given by (11.16) for the isotropic materials and by (17.27) for anisotropic materials.

Substitution of (22.6) into (22.5′) leads to

$$C^*_{ijkl}(\epsilon^0_{kl} + S_{klmn}\epsilon^*_{mn}) = C_{ijkl}(\epsilon^0_{kl} + S_{klmn}\epsilon^*_{mn} - \epsilon^*_{kl}) \tag{22.7}$$

from which the six unknowns, ϵ^*_{ij}, are determined.

Sometimes the inhomogeneity may involve its own eigenstrain. Then, it is called the inhomogeneous inclusion. In other words, a domain which has different elastic constants from those of the matrix may also have a given distribution of eigenstrain ϵ^p_{ij}. An example is the formation of martensite blades in quenched carbon steel and precipitations in alloys.

Let a material D, containing an ellipsoidal inhomogeneous inclusion Ω, be under a stress field $\sigma_{ij}^0 + \sigma_{ij}$. σ_{ij}^0 is the applied stress if the material is homogeneous, i.e. having no inclusions. σ_{ij} is the sum of the two stress disturbances, one caused by the inhomogeneity, and the other by the eigenstress associated with eigenstrain ϵ_{ij}^p in Ω.

Denoting the elastic constants of $D - \Omega$ by C_{ijkl} and those of Ω by C_{ijkl}^*, we have Hooke's law

$$\begin{aligned} \sigma_{ij}^0 + \sigma_{ij} &= C_{ijkl}^*(u_{k,l}^0 + u_{k,l} - \epsilon_{kl}^p) \text{ in } \Omega, \\ \sigma_{ij}^0 + \sigma_{ij} &= C_{ijkl}(u_{k,l}^0 + u_{k,l}) \qquad \text{in } D - \Omega, \end{aligned} \tag{22.8}$$

where

$$\begin{aligned} &\sigma_{ij}^0 = C_{ijkl}u_{k,l}^0 \text{ in } D, \\ &\sigma_{ij,j}^0 = 0 \qquad \text{in } D, \\ &\sigma_{ij}^0 n_j = F_i \qquad \text{on } |D|, \\ &\sigma_{ij,j} = 0 \qquad \text{in } D, \\ &\sigma_{ij}n_j = 0 \qquad \text{on } |D|. \end{aligned} \tag{22.9}$$

The inhomogeneous inclusion is simulated by an inclusion in the homogeneous material with eigenstrain ϵ_{ij}^p plus equivalent eigenstrain ϵ_{ij}^*,

$$\begin{aligned} \sigma_{ij}^0 + \sigma_{ij} &= C_{ijkl}(u_{k,l}^0 + u_{k,l} - \epsilon_{kl}^p - \epsilon_{kl}^*) \text{ in } \Omega, \\ \sigma_{ij}^0 + \sigma_{ij} &= C_{ijkl}(u_{k,l}^0 + u_{k,l}) \qquad \text{in } D - \Omega. \end{aligned} \tag{22.10}$$

Eigenstrain ϵ_{ij}^* is a fictitious one, introduced for this simulation. The equivalency between (22.8) and (22.10) holds when

$$C_{ijkl}^*(\epsilon_{kl}^0 + \epsilon_{kl} - \epsilon_{kl}^p) = C_{ijkl}(\epsilon_{kl}^0 + \epsilon_{kl} - \epsilon_{kl}^p - \epsilon_{kl}^*) \text{ in } \Omega. \tag{22.11}$$

If σ_{ij}^0 is a given uniform stress field and ϵ_{ij}^p is a given uniform eigenstrain in Ω, (22.11) is satisfied by taking ϵ_{ij} as the solution of the inclusion problem with a uniform eigenstrain $\epsilon_{ij}^p + \epsilon_{ij}^*$. Instead of (22.6), we now have

$$\epsilon_{kl} = \tfrac{1}{2}(u_{k,l} + u_{l,k}) = S_{klmn}(\epsilon_{mn}^p + \epsilon_{mn}^*) \equiv S_{klmn}\epsilon_{mn}^{**}, \tag{22.12}$$

where

$$\epsilon_{ij}^{**} = \epsilon_{ij}^p + \epsilon_{ij}^*. \tag{22.12a}$$

Substituting (22.12) into (22.11), we have the equation to determine ϵ_{ij}^{**}:

$$\begin{aligned}\sigma_{ij}^0 + \sigma_{ij} &= C_{ijkl}^*(\epsilon_{kl}^0 + S_{klmn}\epsilon_{mn}^{**} - \epsilon_{kl}^p)\\ &= C_{ijkl}(\epsilon_{kl}^0 + S_{klmn}\epsilon_{mn}^{**} - \epsilon_{kl}^{**}).\end{aligned} \tag{22.13}$$

In the absence of the applied stress, this becomes

$$\sigma_{ij} = C_{ijkl}^*(S_{klmn}\epsilon_{mn}^{**} - \epsilon_{kl}^p) = C_{ijkl}(S_{klmn}\epsilon_{mn}^{**} - \epsilon_{kl}^{**}). \tag{22.13a}$$

Isotropic materials

Let us illustrate in more detail the equivalent inclusion method described above by assuming the material to be isotropic. When both the matrix and inhomogeneity are isotropic, (22.11) can be written as

$$\begin{aligned}&2\mu^*(\epsilon_{ij}^0 + \epsilon_{ij} - \epsilon_{ij}^p) + \lambda^*\delta_{ij}(\epsilon_{kk}^0 + \epsilon_{kk} - \epsilon_{kk}^p)\\ &\quad = 2\mu(\epsilon_{ij}^0 + \epsilon_{ij} - \epsilon_{ij}^{**}) + \lambda\delta_{ij}(\epsilon_{kk}^0 + \epsilon_{kk} - \epsilon_{kk}^{**}),\end{aligned} \tag{22.14}$$

where λ^*, μ^* and λ, μ are the Lamé constants in the inhomogeneity and the matrix, respectively, and

$$\begin{aligned}\epsilon_{ij}^0 &= \tfrac{1}{2}(u_{i,j}^0 + u_{j,i}^0),\\ \epsilon_{ij}^{**} &= \epsilon_{ij}^* + \epsilon_{ij}^p.\end{aligned} \tag{22.15}$$

It is convenient to introduce the following deviatoric (reduced) strains:

$$\begin{aligned}'\epsilon_{ij}^0 &= \epsilon_{ij}^0 - \delta_{ij}\epsilon_{kk}^0/3,\\ '\epsilon_{ij}^p &= \epsilon_{ij}^p - \delta_{ij}\epsilon_{kk}^p/3,\\ '\epsilon_{ij}^{**} &= \epsilon_{ij}^{**} - \delta_{ij}\epsilon_{kk}^{**}/3,\\ '\epsilon_{ij} &= \epsilon_{ij} - \delta_{ij}\epsilon_{kk}/3.\end{aligned} \tag{22.16}$$

Then (22.14) is equivalent to

$$\begin{aligned}2\mu^*('\epsilon_{ij}^0 + '\epsilon_{ij} - '\epsilon_{ij}^p) &= 2\mu('\epsilon_{ij}^0 + '\epsilon_{ij} - '\epsilon_{ij}^{**}),\\ K^*(\epsilon_{kk}^0 + \epsilon_{kk} - \epsilon_{kk}^p) &= K(\epsilon_{kk}^0 + \epsilon_{kk} - \epsilon_{kk}^{**}),\end{aligned} \tag{22.17}$$

where

$$\begin{aligned}K &= \lambda + 2\mu/3 = E/3(1-2\nu)\\ E &= 2(1+\nu)\mu\\ \lambda &= 2\mu\nu/(1-2\nu),\end{aligned} \tag{22.18}$$

and K, E, ν are the bulk modulus, Young's modulus, and Poisson's ratio of the matrix, respectively. Similar relations hold among the elastic constants of the inhomogeneity.

If σ_{ij}^0 is a given uniform stress field and ϵ_{ij}^p is a given uniform strain in Ω, we can use (22.12) for solving ϵ_{ij}^* from (22.17). The shear components of ϵ_{ij}^{**} are obtained directly from the first equation in (22.17):

$$\begin{aligned}
\epsilon_{12}^{**} &= \{2(\mu - \mu^*)\epsilon_{12}^0 + 2\mu^*\epsilon_{12}^p\}/\{4(\mu^* - \mu)S_{1212} + 2\mu\}, \\
\epsilon_{23}^{**} &= \{2(\mu - \mu^*)\epsilon_{23}^0 + 2\mu^*\epsilon_{23}^p\}/\{4(\mu^* - \mu)S_{2323} + 2\mu\}, \\
\epsilon_{31}^{**} &= \{2(\mu - \mu^*)\epsilon_{31}^0 + 2\mu^*\epsilon_{31}^p\}/\{4(\mu^* - \mu)S_{3131} + 2\mu\}.
\end{aligned} \tag{22.19}$$

The determination of ϵ_{11}^{**}, ϵ_{22}^{**} and ϵ_{33}^{**} is not a simple matter. After ϵ_{ij}^{**} are calculated, the stress field in Ω is obtained as

$$\sigma_{ij}^0 + \sigma_{ij} = 2\mu(\epsilon_{ij}^0 + S_{ijmn}\epsilon_{mn}^{**} - \epsilon_{ij}^{**}) + \lambda\delta_{ij}(\epsilon_{kk}^0 + S_{kkmn}\epsilon_{mn}^{**} - \epsilon_{kk}^{**}). \tag{22.20}$$

The strain outside Ω is given by (11.33) and (11.41).

Let us consider the case where applied stress σ_{ij}^0 is uniform at infinity (see Fig. 22.1), and the inhomogeneity is an inhomogeneous inclusion with a uniform plastic strain field ϵ_{11}^p, ϵ_{22}^p, ϵ_{33}^p. When σ_{ij}^0 and ϵ_{ij}^p are shear, the solution has already been given by (22.19) and (22.20).

Sphere

From (11.15) and (11.21), ϵ_{ij} inside the spherical inclusion with uniform

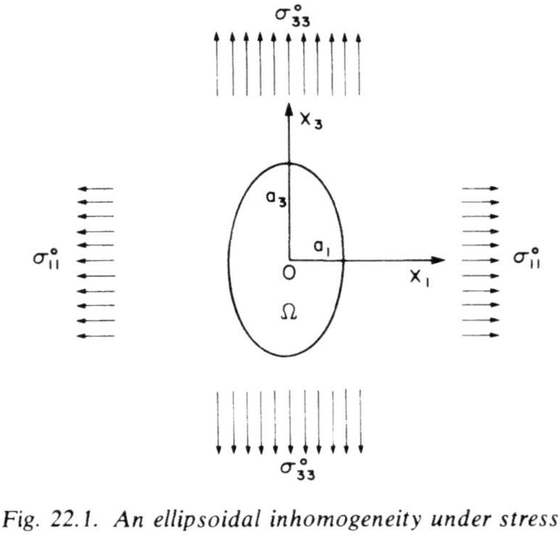

Fig. 22.1. An ellipsoidal inhomogeneity under stress σ_{ij}^0.

eigenstrains ϵ_{11}^{**}, ϵ_{22}^{**} and ϵ_{33}^{**} is given by

$$\begin{aligned}15(1-\nu)\epsilon_{11} &= (7-5\nu)\epsilon_{11}^{**}+(5\nu-1)\epsilon_{22}^{**}+(5\nu-1)\epsilon_{33}^{**},\\ 15(1-\nu)\epsilon_{22} &= (5\nu-1)\epsilon_{11}^{**}+(7-5\nu)\epsilon_{22}^{**}+(5\nu-1)\epsilon_{33}^{**},\\ 15(1-\nu)\epsilon_{33} &= (5\nu-1)\epsilon_{11}^{**}+(5\nu-1)\epsilon_{22}^{**}+(7-5\nu)\epsilon_{33}^{**},\end{aligned} \tag{22.21}$$

and

$$\epsilon_{kk} = \epsilon_{kk}^{**}(1+\nu)/3(1-\nu). \tag{22.22}$$

Then,

$$'\epsilon_{ij} = {}'\epsilon_{ij}^{**}(8-10\nu)/15(1-\nu). \tag{22.23}$$

With the aid of this last result, it follows from the first equation in (22.17) that

$$'\epsilon_{ij}^{**} = 15\{'\epsilon_{ij}^{0}(\mu^*-\mu)-{}'\epsilon_{ij}^{p}\mu^*\}(1-\nu)/\{(5\nu-7)\mu-(8-10\nu)\mu^*\}. \tag{22.24}$$

The second equations in (22.17) and (22.22) yield

$$\epsilon_{kk}^{**} = 3\{(K^*-K)\epsilon_{kk}^{0}-K^*\epsilon_{kk}^{p}\}(1-\nu)/\{(4\nu-2)K-(1+\nu)K^*\}. \tag{22.25}$$

Thus, we have

$$\begin{aligned}\epsilon_{ij}^{**} = {} & 15\{'\epsilon_{ij}^{0}(\mu^*-\mu)-{}'\epsilon_{ij}^{p}\mu^*\}(1-\nu)/\{(5\nu-7)\mu-(8-10\nu)\mu^*\}\\ & +\delta_{ij}\{(K^*-K)\epsilon_{kk}^{0}-K^*\epsilon_{kk}^{p}\}(1-\nu)/\{(4\nu-2)K-(1+\nu)K^*\}.\end{aligned} \tag{22.26}$$

The stress components in Ω can be calculated from (22.20). For the case when tension σ is applied in the x_3-direction, $'\epsilon_{11}^{0} = {}'\epsilon_{22}^{0} = -\sigma/6\mu$, $'\epsilon_{33}^{0} = \sigma/3\mu$ and $\epsilon_{kk}^{0} = \sigma/3K$.

Penny shape

When Ω is a penny-shaped inhomogeneity, $a_1 = a_2 = a$ and $a_3 = 0$. From (11.23), ϵ_{ij} inside Ω, for given ϵ_{11}^{**}, ϵ_{22}^{**} and ϵ_{33}^{**} is

$$\epsilon_{11} = \epsilon_{22} = 0, \qquad \epsilon_{33} = \epsilon_{11}^{**}\nu/(1-\nu)+\epsilon_{22}^{**}\nu/(1-\nu)+\epsilon_{33}^{**}. \tag{22.27}$$

Then, (22.14) becomes

$$\begin{aligned}&(2\mu^*-2\mu)\epsilon_{11}^{0}-2\mu^*\epsilon_{11}^{p}+(\lambda^*-\lambda)\epsilon_{kk}^{0}-\lambda^*\epsilon_{kk}^{p}+\lambda^*\epsilon_{33}\\ &\quad = -2\mu(\epsilon_{11}^{**}+\nu\epsilon_{22}^{**})/(1-\nu),\\ &(2\mu^*-2\mu)\epsilon_{22}^{0}-2\mu^*\epsilon_{22}^{p}+(\lambda^*-\lambda)\epsilon_{kk}^{0}-\lambda^*\epsilon_{kk}^{p}+\lambda^*\epsilon_{33}\\ &\quad = -2\mu(\epsilon_{22}^{**}+\nu\epsilon_{11}^{**})/(1-\nu),\end{aligned}$$

$$(2\mu^* - 2\mu)\epsilon^0_{33} - 2\mu^*\epsilon^p_{33} + (\lambda^* - \lambda)\epsilon^0_{kk} - \lambda^*\epsilon^p_{kk} = -(\lambda^* + 2\mu^*)\epsilon_{33}. \tag{22.28}$$

It should be noted that σ_{33} is zero. When (22.28) and (22.27) are solved for ϵ^{**}_{11}, ϵ^{**}_{22} and ϵ^{**}_{33}, one obtains

$$\begin{aligned}
2\mu(1+\nu)\epsilon^{**}_{11} &= \{(2\mu - 2\mu^*)\epsilon^0_{11} + 2\mu^*\epsilon^p_{11}\} - \nu\{(2\mu^* - 2\mu)\epsilon^0_{22} + 2\mu^*\epsilon^p_{22}\} \\
&\quad - \{(2\mu - 2\mu^*)\epsilon^0_{33} + 2\mu^*\epsilon^p_{33}\}(1-\nu)\lambda^*/(\lambda^* + 2\mu^*) \\
&\quad + \{(\lambda - \lambda^*)\epsilon^0_{kk} + \lambda^*\epsilon^p_{kk}\}(1-\nu)2\mu^*/(\lambda^* + 2\mu^*), \\
2\mu(1+\nu)\epsilon^{**}_{22} &= \{(2\mu - 2\mu^*)\epsilon^0_{22} + 2\mu^*\epsilon^p_{22}\} - \nu\{(2\mu^* - 2\mu)\epsilon^0_{11} + 2\mu^*\epsilon^p_{11}\} \\
&\quad - \{(2\mu - 2\mu^*)\epsilon^0_{33} + 2\mu^*\epsilon^p_{33}\}(1-\nu)\lambda^*/(\lambda^* + 2\mu^*) \\
&\quad + \{(\lambda - \lambda^*)\epsilon^0_{kk} + \lambda^*\epsilon^p_{kk}\}(1-\nu)2\mu^*/(\lambda^* + 2\mu^*), \\
2\mu(1+\nu)\epsilon^{**}_{33} &= -\{(2\mu - 2\mu^*)\epsilon^0_{11} + 2\mu^*\epsilon^p_{11}\}\nu \\
&\quad - \{(2\mu - 2\mu^*)\epsilon^0_{22} + 2\mu^*\epsilon^p_{22}\}\nu \\
&\quad + \{(2\mu - 2\mu^*)\epsilon^0_{33} + 2\mu^*\epsilon^p_{33}\} \\
&\quad \times \{2\mu(1+\nu) + 2\lambda^*\nu\}/(\lambda^* + 2\mu^*) \\
&\quad + \{(\lambda - \lambda^*)\epsilon^0_{kk} + \lambda^*\epsilon^p_{kk}\}\{2\mu(1+\nu) - 4\mu^*\nu\}/(\lambda^* + 2\mu^*).
\end{aligned} \tag{22.29}$$

The stress components are calculated from (22.20).

Rod

A rod-shaped geometry results when $a_1 = a_2 = a$, and $a_3 = \infty$. From (11.22), ϵ_{ij} inside Ω, for given ϵ^{**}_{11}, ϵ^{**}_{22}, ϵ^{**}_{33}, is

$$\begin{aligned}
8(1-\nu)\epsilon_{11} &= (5 - 4\nu)\epsilon^{**}_{11} + (4\nu - 1)\epsilon^{**}_{22} + 4\nu\epsilon^{**}_{33} \\
8(1-\nu)\epsilon_{22} &= (4\nu - 1)\epsilon^{**}_{11} + (5 - 4\nu)\epsilon^{**}_{22} + 4\nu\epsilon^{**}_{33} \\
\epsilon_{33} &= 0.
\end{aligned} \tag{22.30}$$

Substitution of (22.30) into (22.14), after some modification gives

$$\begin{aligned}
&\{\mu^*(\tfrac{5}{4} - \nu) + \mu(\tfrac{3}{4} - \nu)\}\epsilon^{**}_{11} + \{\mu^*(\nu - \tfrac{1}{4}) + \mu(\tfrac{1}{4} - \nu)\}\epsilon^{**}_{22} + \{\mu^*\nu + \mu(\nu - 2)\}\epsilon^{**}_{33} \\
&\quad = \{(2\mu - 2\mu^*)(\epsilon^0_{11} - \epsilon^0_{33}) + 2\mu^*(\epsilon^p_{11} - \epsilon^p_{33})\}(1-\nu), \\
&\{\mu^*(\nu - \tfrac{1}{4}) + \mu(\tfrac{1}{4} - \nu)\}\epsilon^{**}_{11} + \{\mu^*(\tfrac{5}{4} - \nu) + \mu(\tfrac{3}{4} - \nu)\}\epsilon^{**}_{22} + \{\mu^*\nu + \mu(\nu - 2)\}\epsilon^{**}_{33} \\
&\quad = \{(2\mu - 2\mu^*)(\epsilon^0_{22} - \epsilon^0_{33}) + 2\mu^*(\epsilon^p_{22} - \epsilon^p_{33})\}(1-\nu), \\
&(\lambda^*/2 + \mu\nu)\epsilon^{**}_{11} + (\lambda^*/2 + \mu\nu)\epsilon^{**}_{22} + (\lambda^*\nu + 2\mu)\epsilon^{**}_{33} \\
&\quad = \{(2\mu - 2\mu^*)\epsilon^0_{33} + 2\mu^*\epsilon^p_{33} + (\lambda - \lambda^*)\epsilon^0_{kk} + \lambda^*\epsilon^p_{kk}\}(1-\nu).
\end{aligned} \tag{22.31}$$

ϵ^{**}_{ij} are obtained from these linear algebraic equations.

If the problem is symmetric, so that $\epsilon_{11}^p = \epsilon_{22}^p$, $\epsilon_{11}^0 = \epsilon_{22}^0$ and $\epsilon_{11}^{**} = \epsilon_{22}^{**}$, we have

$$\begin{aligned}
\epsilon_{11}^{**} = \frac{1-\nu}{\Delta} & [(\lambda^*\nu + 2\mu)\{(2\mu - 2\mu^*)(\epsilon_{11}^0 - \epsilon_{33}^0) + 2\mu^*(\epsilon_{11}^p - \epsilon_{33}^p)\} \\
& - \{\mu^*\nu + \mu(\nu - 2)\} \\
& \times \{(2\mu - 2\mu^*)\epsilon_{33}^0 + 2\mu^*\epsilon_{33}^p + (\lambda - \lambda^*)\epsilon_{kk}^0 + \lambda^*\epsilon_{kk}^p\}], \\
\epsilon_{33}^{**} = \frac{1-\nu}{\Delta} & [-(\lambda^* + 2\mu\nu)\{(2\mu - 2\mu^*)(\epsilon_{11}^0 - \epsilon_{33}^0) + 2\mu^*(\epsilon_{11}^p - \epsilon_{33}^p)\} \\
& + \{-\mu^* + \mu(1 - 2\nu)\} \\
& \times \{(2\mu - 2\mu^*)\epsilon_{33}^0 + 2\mu^*\epsilon_{33}^p + (\lambda - \lambda^*)\epsilon_{kk}^0 + \lambda^*\epsilon_{kk}^p\}],
\end{aligned} \tag{22.32}$$

where

$$\Delta = -2\lambda^*\mu^*\nu + 2(1-\nu)\mu(\lambda^* + \mu) - 2\mu^*\mu(1+\nu^2). \tag{22.33}$$

It should be noted that the stress components are independent of the size of the inhomogeneity, but are dependent on its shape.

Anisotropic inhomogeneities in isotropic matrices

When an anisotropic inhomogeneous inclusion is contained in an isotropic matrix, C_{ijkl} and S_{klmn} in (22.13a) are taken for the isotropic material. The calculation then follows that of the isotropic inhomogeneity case.

Stress field for exterior points

Displacement and stress fields outside of an ellipsoidal inhomogeneity, Ω, can be obtained from the formulae in Sections 12 and 18 when the equivalent eigenstrain is determined. However, these formulae are complicated for practice. Recently, Tanaka and Mura (1981) have proposed an alternative method to evaluate the elastic fields for points exterior to Ω.

Consider an ellipsoidal inhomogeneity Ω with elastic moduli C_{ijkl}^* in a matrix with elastic moduli C_{ijkl} and an applied stress σ_{ij}^0 at infinity. First, obtain the stress field for the points interior to Ω and denote it by $\sigma_{ij}^0 + \sigma_{ij}(\text{in})$. Next, find the stress field for the points exterior to Ω if Ω is a void and the applied stress is $-\sigma_{ij}(\text{in})$. Denote the stress field by $\sigma_{ij}(\text{out})$. Then, the stress field for the points exterior to Ω in the original problem is the sum of $\sigma_{ij}^0 + \sigma_{ij}(\text{in})$ and $\sigma_{ij}(\text{out})$ (see Fig. 22.2). The proof is obvious. $\sigma_{ij}(\text{out})$ becomes $-\sigma_{ij}(\text{in})$ at infinity and zero in Ω. There-

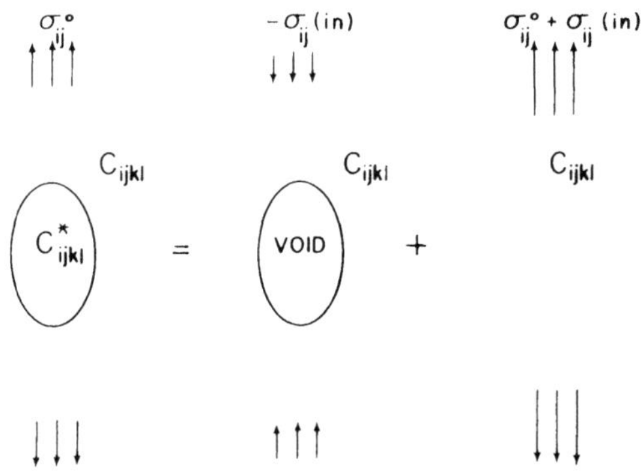

Fig. 22.2. The method to obtain the exterior elastic field.

fore, $\sigma_{ij}^0 + \sigma_{ij}(\text{in}) + \sigma_{ij}(\text{out})$ is σ_{ij}^0 at infinity and $\sigma_{ij}^0 + \sigma_{ij}(\text{in})$ in Ω, which are the required conditions in the original problem. The method is useful since solutions for the void problem under various stress fields are usually available in the literature (e.g. Neuber 1937, Muskhelishvili 1953, Savin 1961, Peterson 1962, Miyamoto 1967).

23. Numerical calculations

The equivalent inclusion method can be extended to cases where applied stresses are not uniform. If an applied stress is linear with respect to the coordinates, the equivalent eigenstrain is also linear with respect to these coordinates. Generally, when the applied stress is a polynomial of degree n, the equivalent eigenstrain is also chosen as a polynomial of degree n. This arises from the fact that for a polynomial eigenstrain of degree n the corresponding eigenstress of the inclusion problem is the sum of polynomials of degrees $n, n-2, n-4, \ldots$ for interior points (see the paragraph following (19.8)).

Let us consider an ellipsoidal inhomogeneity Ω in an isotropic infinite body (Fig. 23.1). The stress field is calculated when an applied strain (or stress) before the disturbance due to inhomogeneity is given by

$$\epsilon_{ij}^0(\mathbf{x}) = E_{ij} + E_{ijk}x_k + E_{ijkl}x_kx_l + \cdots, \tag{23.1}$$

where the coefficients are constants. The eigenstrain in (22.5′) is assumed to be

$$\epsilon_{ij}^*(\mathbf{x}) = B_{ij} + B_{ijk}x_k + B_{ijkl}x_kx_l + \cdots \tag{23.2}$$

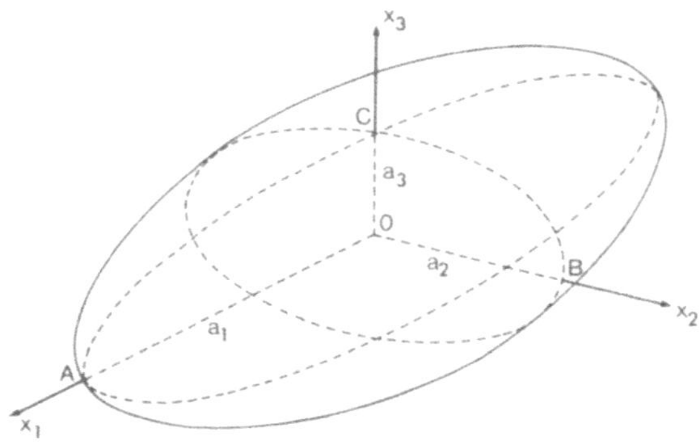

Fig. 23.1. An ellipsoidal inhomogeneity, indicating points A, B and C.

The strain associated with (23.2) is given by (12.10) and (11.33),

$$\epsilon_{ij}(\mathbf{x}) = D_{ijkl}(\mathbf{x})B_{kl} + D_{ijklq}(\mathbf{x})B_{klq} + D_{ijklqr}(\mathbf{x})B_{klqr} + \cdots. \tag{23.3}$$

It should be remembered that for points interior to Ω, D_{ijkl} is constant, D_{ijklq} is linear in $\mathbf{x}$, and D_{ijklqr} is a sum of constants and quadratic functions of $\mathbf{x}$. Using expression (12.16), substituting (23.1), (23.2) and (23.3) into (22.5′), and comparing coefficients in the power series, we obtain

$$\begin{aligned}
&(C_{stmn} - C^*_{stmn})\{D_{mnij}(0)B_{ij} + D_{mnijkl}(0)B_{ijkl} + \cdots\} - C_{stmn}B_{mn} \\
&\quad = (C^*_{stmn} - C_{stmn})E_{mn}, \qquad\qquad (23.4)\\
&(C_{stmn} - C^*_{stmn})D_{mnijk,p}(0)B_{ijk} + \cdots - C_{stmn}B_{mnp} = (C^*_{stmn} - C_{stmn})E_{mnp},\\
&(1/2!)(C_{stmn} - C^*_{stmn})D_{mnijkl,pq}(0)B_{ijkl} + \cdots - C_{stmn}B_{mnpq}\\
&\quad = (C^*_{stmn} - C_{stmn})E_{mnpq},\\
&\text{etc.,}
\end{aligned}$$

where C^*_{ijkl} and C_{ijkl} are the elastic moduli of the inhomogeneity and the matrix respectively. The system of equations (23.4) determines the unknown B constants. The stress and strain fields are determined from (23.3) and these constants.

Numerical calculations are performed by Moschovidis (1975). Figures 23.2 and 23.3 show $\sigma^T_{33} = \sigma^0_{33} + \sigma_{33}$ at A and B in Fig. 23.1 when $\sigma^0_{33} = 1$ is applied to an infinite isotropic body containing an ellipsoidal void. The curves in these figures show that the normal stress σ^T_{33} at point B is always larger than that at point A when $a_1 > a_2$.

Figures 23.4 and 23.5 show $\sigma^T_{33} = \sigma^0_{33} + \sigma_{33}$ at A and B when $\sigma^0_{33} = 1$ is applied to an infinite isotropic body containing spheroidal in-

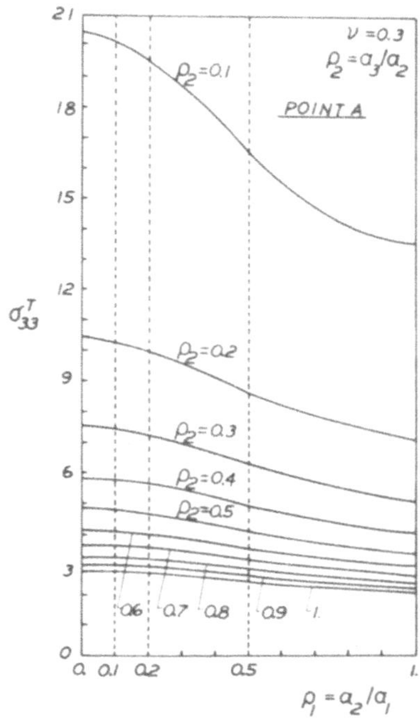

Fig. 23.2. Ellipsoidal void under $\sigma^0_{33} = 1$. The stress component σ^T_{33} at point A in Fig. 23.1.

homogeneity with shear modulus μ^*. These curves show the dependency of the normal stress on shear modulus and geometry. The shear modulus of the matrix is μ, and Poisson's ratios ν and ν^* are equal to 0.3. For large values of μ^*, σ^T_{33} shows large compression at point A. It is also found that σ^T_{11} and σ^T_{22} are everywhere much smaller than σ^T_{33} except at point A where σ^T_{22} is comparable to σ^T_{33}.

Figure 23.6 shows $\sigma^T_{23} = \sigma^0_{23} + \sigma_{23}$ along the x_1-axis when σ^0_{23} is applied linearly (as shown by the dotted line) to an infinite isotropic body containing a spherical inhomogeneity. The axis of torsion is the x_3-axis, and $\gamma = \mu^*/\mu$. It is important to note that the system of algebraic equations (23.4), to be solved for B_{ijk}, becomes singular when the inhomogeneity reduces to a void ($\gamma = 0$). It is easy to see this from equation (22.5′) which, for $\gamma = 0$, yields

$$C_{ijkl}(\epsilon_{kl} - \epsilon^*_{kl}) = -C_{ijkl}\epsilon^0_{kl}. \tag{23.5}$$

As a solution of the inclusion problem, $\sigma_{ij} = C_{ijkl}(\epsilon_{kl} - \epsilon^*_{kl})$ always satisfies the equations of equilibrium. Therefore, the right-hand side in (23.5)

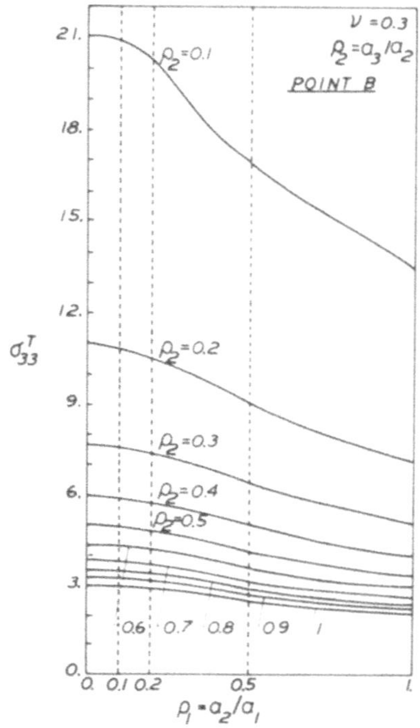

Fig. 23.3. Ellipsoidal void under $\sigma_{33}^0 = 1$. σ_{33}^T at point B.

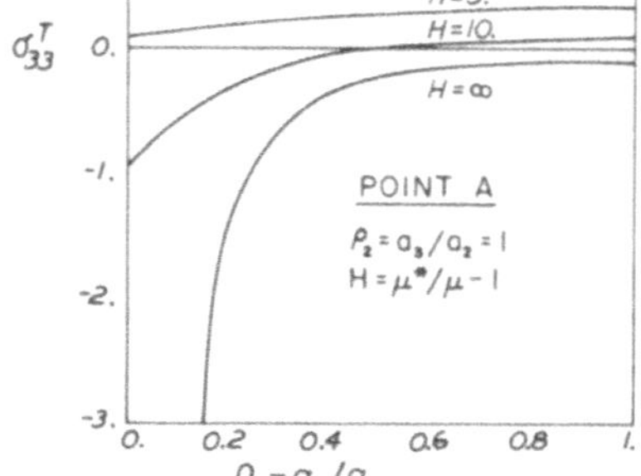

Fig. 23.4. Ellipsoidal inhomogeneity under $\sigma_{33}^0 = 1$. σ_{33}^T at point A. μ^* is the shear modulus of the inhomogeneity.

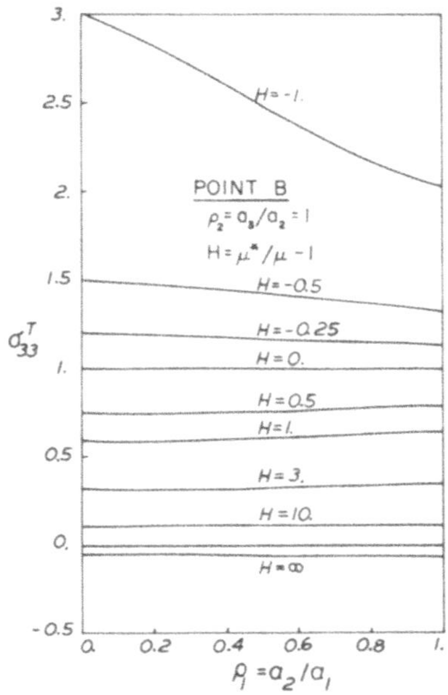

Fig. 23.5. Ellipsoidal inhomogeneity under $\sigma_{33}^0 = 1$. σ_{33}^T at point B.

must also satisfy the equations of equilibrium. In other words, the system of algebraic equations has a non-trivial solution only when terms in the right-hand side of the equations are proper. This situation leads to

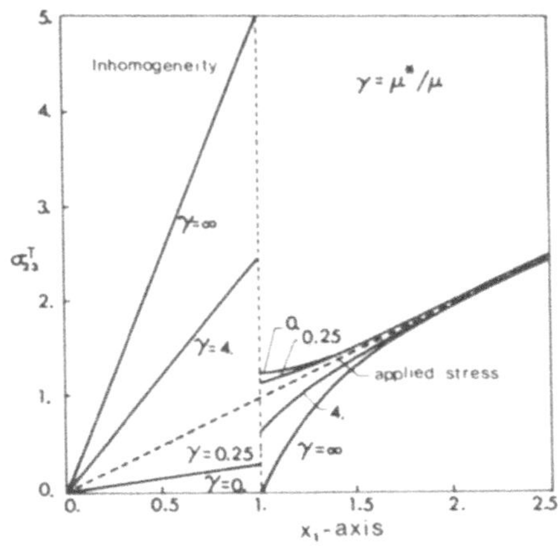

Fig. 23.6. Spherical inhomogeneity under linear torsion about the x_3 axis.

the conclusion that the system of algebraic equations (23.4) must be singular for $\gamma = 0$. This singularity will be discussed in Section 24.

Two ellipsoidal inhomogeneities

Let us consider two inhomogeneities, Ω_1 and Ω_2, as shown in Fig. 23.7, under the applied stress σ^0_{ij}. Applying the equivalent inclusion method in domains Ω_1 and Ω_2, the equivalency equations become, from (22.5′),

$$\begin{aligned} C^I_{ijkl}(\epsilon^0_{kl} + \epsilon_{kl}) &= C_{ijkl}(\epsilon^0_{kl} + \epsilon_{kl} - \beta^I_{kl}) \text{ in } \Omega_1, \\ C^{II}_{ijkl}(\epsilon^0_{kl} + \epsilon_{kl}) &= C_{ijkl}(\epsilon^0_{kl} + \epsilon_{kl} - \beta^{II}_{kl}) \text{ in } \Omega_2, \end{aligned} \tag{23.6}$$

where C^I_{ijkl} and β^I_{kl} represent respectively the elastic moduli and the equivalent eigenstrains, defined in Ω_1. The corresponding quantities in Ω_2 have the superscript *II*. It should be emphasized here that ϵ_{kl} is no longer uniform even if ϵ^0_{kl} is uniform, since interior points of Ω_1 are exterior points of Ω_2 and the stress in Ω_1 is disturbed by Ω_2.

Let us assume that the material is isotropic and infinitely extended. The applied strain before disturbance is given by

$$\epsilon^0_{ij}(\mathbf{x}) = E_{ij} + E_{ijk}x_k + E_{ijkl}x_kx_l + \cdots, \tag{23.7}$$

where the x_i coordinate system is taken at the center of Ω_1. If the $\bar{x}_i$

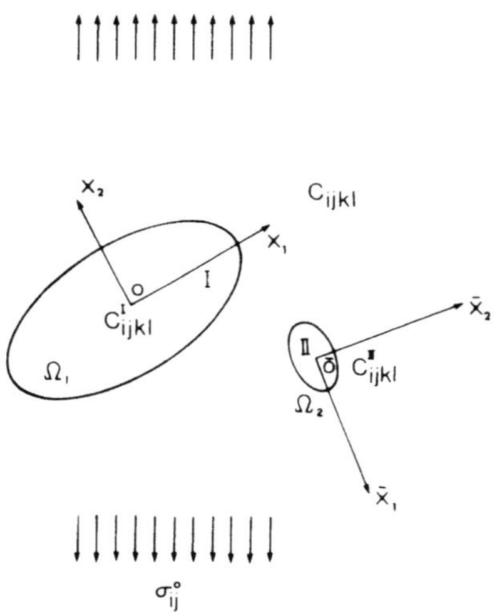

Fig. 23.7. Two ellipsoidal inhomogeneities under applied stress σ^0_{ij}.

coordinate system is taken at the center of Ω_2, the corresponding expression becomes

$$\bar{\epsilon}^0_{ij}(\mathbf{x}) = \bar{E}_{ij} + \bar{E}_{ijk}\bar{x}_k + \bar{E}_{ijkl}\bar{x}_k\bar{x}_l + \cdots. \tag{23.8}$$

Hereafter, tensor components referred to the $\bar{x}_i$ coordinate system are denoted by bars.

The two coordinate systems are related by

$$x_i - c_i = a_{ij}\bar{x}_j, \quad \bar{x}_i = a_{ji}(x_j - c_j), \tag{23.9}$$

where a_{ij} is the direction cosine between the x_i-axis and the $\bar{x}_j$-axis, and c_i is the x_i coordinate of the origin $\bar{0}$ in Ω_2.

We assume that the equivalent eigenstrains are

$$\begin{aligned} \beta^I_{ij}(\mathbf{x}) &= B^I_{ij} + B^I_{ijk}x_k + B^I_{ijkl}x_kx_l + \cdots, \\ \bar{\beta}^{II}_{ij}(\mathbf{x}) &= B^{II}_{ij} + B^{II}_{ijk}\bar{x}_k + B^{II}_{ijkl}\bar{x}_k\bar{x}_l + \cdots. \end{aligned} \tag{23.10}$$

From (11.33) and (12.10) we can write

$$\begin{aligned} \epsilon^I_{ij}(\mathbf{x}) &= D^I_{ijkl}(\mathbf{x})B^I_{kl} + D^I_{ijklq}(\mathbf{x})B^I_{klq} + D^I_{ijklqr}(\mathbf{x})B^I_{klqr} + \cdots, \\ \bar{\epsilon}^{II}_{ij}(\mathbf{x}) &= D^{II}_{ijkl}(\bar{\mathbf{x}})B^{II}_{kl} + D^{II}_{ijklq}(\bar{\mathbf{x}})B^{II}_{klq} + D^{II}_{ijklqr}(\bar{\mathbf{x}})B^{II}_{klqr} + \cdots, \end{aligned} \tag{23.11}$$

where ϵ^I_{ij} and $\bar{\epsilon}^{II}_{ij}$ are the respective strains caused by β^I_{ij} and $\bar{\beta}^{II}_{ij}$. The strain ϵ_{ij} in (23.6) is the sum of ϵ^I_{ij} and ϵ^{II}_{ij}. Interior points of Ω_1 are exterior points of Ω_2, and functions D in (23.11) depend on position $\mathbf{x}$ as has been discussed following (12.15). As seen in (12.16), $D^I(\mathbf{x})$ are polynomials of $\mathbf{x}$ in Ω_1; however, $D^{II}(\bar{\mathbf{x}})$ are not polynomials of $\mathbf{x}$ in Ω_1. For $\mathbf{x}$ in Ω_1, $D^{II}(\bar{\mathbf{x}})$ are approximated by a Taylor's expansion around point 0. Similarly, $D^I(\mathbf{x})$ are also expanded in a Taylor's series of $\mathbf{x}$ in Ω_2. $D^{II}(\bar{\mathbf{x}})$ are polynomials of $\bar{\mathbf{x}}$ in Ω_2 according to (12.16). After these considerations the system of equations (23.6) is solved for B. The coefficients of power series in the left- and right-hand sides in (23.6) are equated. Consequently, in Ω_1,

$$\begin{aligned} &\Delta C^I_{stmn}\{[D^I_{mnij}[0]B^I_{ij} + D^I_{mnijkl}[0]B^I_{ijkl} + \cdots] + a_{mc}a_{nh}[D^{II}_{chij}[0]B^{II}_{ij} \\ &\quad + D^{II}_{chijk}[0]B^{II}_{ijk} + D^{II}_{chijkl}[0]B^{II}_{ijkl} + \cdots]\} - C_{stmn}B^I_{mn} = -\Delta C^I_{stmn}E_{mn}, \end{aligned}$$

$$\begin{aligned} &\Delta C^I_{stmn}\left\{\left[\frac{\partial}{\partial x_p} D^I_{mnijk}[0]B^I_{ijk} + \cdots\right] + a_{mc}a_{nh}a_{pf}\left[\frac{\partial}{\partial \bar{x}_f} D^{II}_{chij}[0]B^{II}_{ij}\right.\right. \\ &\quad \left.\left. + \frac{\partial}{\partial \bar{x}_f} D^{II}_{chijk}[0]B^{II}_{ijk} + \frac{\partial}{\partial \bar{x}_f} D^{II}_{chijkl}[0]B^{II}_{ijkl} + \cdots\right]\right\} - C_{stmn}B^I_{mnp} \\ &= -\Delta C^I_{stmn}E_{mnp}, \end{aligned}$$

$$\frac{1}{2!}\Delta C^{I}_{stmn}\left\{\left[\frac{\partial^2}{\partial x_p \partial x_q} D^{I}_{mnijkl}[0]B^{I}_{ijkl} + \cdots\right] + a_{mc}a_{nh}a_{pf}a_{qg}\right.$$
$$\times\left[\frac{\partial^2}{\partial \bar{x}_f \partial \bar{x}_g} D^{II}_{chij}[0]B^{II}_{ij}\right.$$
$$\left.\left. + \frac{\partial^2}{\partial \bar{x}_f \partial \bar{x}_g} D^{II}_{chijk}[0]B^{II}_{ijk} + \frac{\partial^2}{\partial \bar{x}_f \partial \bar{x}_g} D^{II}_{chijkl}[0]B^{II}_{ijkl} + \cdots\right]\right\} - C_{stmn}B^{I}_{mnpq}$$
$$= -\Delta C^{I}_{stmn}E_{mnpq}, \tag{23.12}$$

etc.

Similarly, in Ω_2,

$$\Delta C^{II}_{stmn}\{a_{cm}a_{hn}[D^{I}_{chij}[\bar{0}]B^{I}_{ij} + D^{I}_{chijk}[\bar{0}]B^{I}_{ijk} + D^{I}_{chijkl}[\bar{0}]B^{I}_{ijkl} + \cdots]$$
$$+ [D^{II}_{mnij}[\bar{0}]B^{II}_{ij} + D^{II}_{mnijkl}[\bar{0}]B^{II}_{ijkl} + \cdots]\} - C_{stmn}B^{II}_{mn} = -\Delta C^{II}_{stmn}\bar{E}_{mn},$$

$$\Delta C^{II}_{stmn}\left\{a_{cm}a_{hn}a_{fp}\left[\frac{\partial}{\partial x_f} D^{I}_{chij}[\bar{0}]B^{I}_{ij} + \frac{\partial}{\partial x_f} D^{I}_{chijk}[\bar{0}]B^{I}_{ijk}\right.\right.$$
$$\left.\left. + \frac{\partial}{\partial x_f} D^{I}_{chijkl}[\bar{0}]B^{I}_{ijkl} + \cdots\right] + \left[\frac{\partial}{\partial \bar{x}_p} D^{II}_{mnijk}[\bar{0}]B^{II}_{ijk} + \cdots\right]\right\} - C_{stmn}B^{II}_{mnp}$$
$$= -\Delta C^{II}_{stmn}\bar{E}_{mnp}, \tag{23.13}$$

$$\frac{1}{2!}\Delta C^{II}_{stmn}\left\{a_{cm}a_{hn}a_{fp}a_{gq}\left[\frac{\partial^2}{\partial x_f \partial x_g} D^{I}_{chij}[\bar{0}]B^{I}_{ij} + \frac{\partial^2}{\partial x_f \partial x_g} D^{I}_{chijk}[\bar{0}]B^{I}_{ijk}\right.\right.$$
$$\left.\left. + \frac{\partial^2}{\partial x_f \partial x_g} D^{I}_{chijkl}[\bar{0}]B^{I}_{ijkl} + \cdots\right] + \left[\frac{\partial^2}{\partial \bar{x}_p \partial \bar{x}_q} D^{II}_{mnijkl}[\bar{0}]B^{II}_{ijkl} + \cdots\right]\right\}$$
$$- C_{stmn}B^{II}_{mnpq} = -\Delta C^{II}_{stmn}\bar{E}_{mnpq},$$

etc.,

where

$$\Delta C^{I}_{ijkl} = C_{ijkl} - C^{I}_{ijkl}, \quad \Delta C^{II}_{ijkl} = C_{ijkl} - C^{II}_{ijkl}. \tag{23.14}$$

Equations (23.12) and (23.13) determine the unknowns B^I and B^{II}. The number of unknowns is the same as the number of equations.

Numerical calculations are performed by Moschovidis (1975). Table 23.1 shows the stresses at A, C' and C (see Fig. 23.8) for the two spherical cavities in the uniform tension field $\sigma^0_{11} = \sigma^0_{22} = \sigma^0_{33} = 1$. The first row of the table shows the exact solution obtained by Sternberg and Sadowsky (1952). The second and third rows of the table show the current approximation calculated from the equivalent inclusion method when the equivalent eigenstrains are assumed to be polynomials of

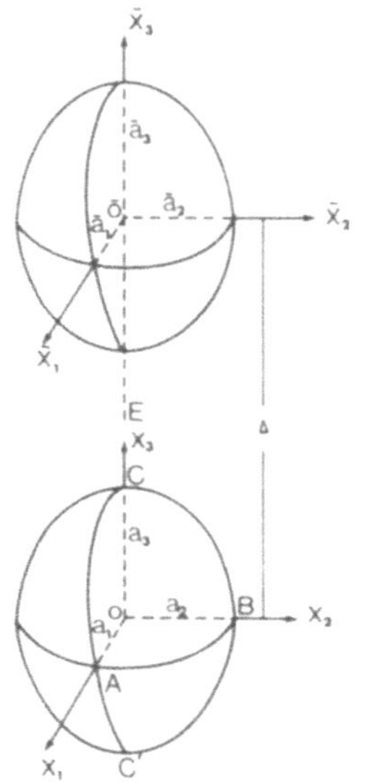

Fig. 23.8. Two ellipsoidal inhomogeneities under σ^0_{ij}, indicating points A, B and C.

degree zero and of degree one in the position coordinates. Δ is the distance between 0 and $\bar{0}$.

It is found that the interaction between the two cavities becomes negligible when $\Delta > 4a_1$.

Figure 23.9 shows the stress distributions $\sigma^T_{ij} = \sigma^0_{ij} + \sigma_{ij}$ for two

Table 23.1. Two spherical cavities of unit radii in uniform applied tension $\sigma^0_{11} = \sigma^0_{22} = \sigma^0_{33} = 1$ at ∞; stresses at equators and poles. $\nu = 0.25$, $\Delta = 4$.

	Point A			Point C′			Point C		
	σ^T_{11}	σ^T_{22}	σ^T_{33}	σ^T_{11}	σ^T_{22}	σ^T_{33}	σ^T_{11}	σ^T_{22}	σ^T_{33}
Sternberg and Sadowsky (1952)	0.0	1.50	1.47	1.51	1.51	0.0	1.57	1.57	0.0
Zeroth degree polynomial	0.011	1.508	1.467	1.519	1.519	0.01	1.534	1.534	−0.017
1st degree polynomial	0.001	1.5079	1.4676	1.503	1.503	−0.0008	1.5495	1.5495	−0.006

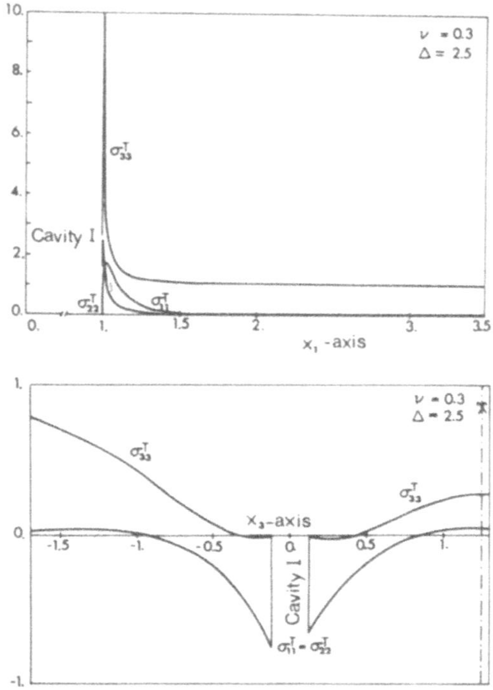

Fig. 23.9. Two spheroidal cavities ($a_1 = \bar{a}_1 = a_2 = \bar{a}_2 = 1$, $a_3 = \bar{a}_3 = 0.125$) under $\sigma_{33}^0 = 1$.

spheroidal cavities ($a_1 = a_2 = \bar{a}_1 = \bar{a}_2 = 1$, and $a_3 = \bar{a}_3 = 0.125$) under $\sigma_{33}^0 = 1$. The distance between cavities is relatively small ($\Delta = 2.5a_1$), and the interaction effect is apparent. The solutions have been approximated by using linear eigenstrains, and the cavities have been regarded as weak inhomogeneities ($\mu^I = \mu^{II} = 10^{-7}\mu$).

Figure 23.10 shows σ_{33}^T for two identical spheroidal inhomogeneities ($a_1 = a_2 = 1$, $a_3 = 0.5$, $\gamma = \mu^I/\mu = \mu^{II}/\mu$, $\nu = \nu^I = \nu^{II} = 0.3$) under an applied stress $\sigma_{33}^0 = 1$. It can be seen that the stress field inside the inhomogeneities is practically constant.

The accurate elastic interaction between two finite gas bubbles, each of arbitrary size under some excess gas pressure, has been investigated by Willis and Bullough (1969). They found that two such bubbles always attract one another in an elastically isotropic medium. The Taylor series method has been recently used by Johnson et al. (1980) to study cuboidal precipitates.

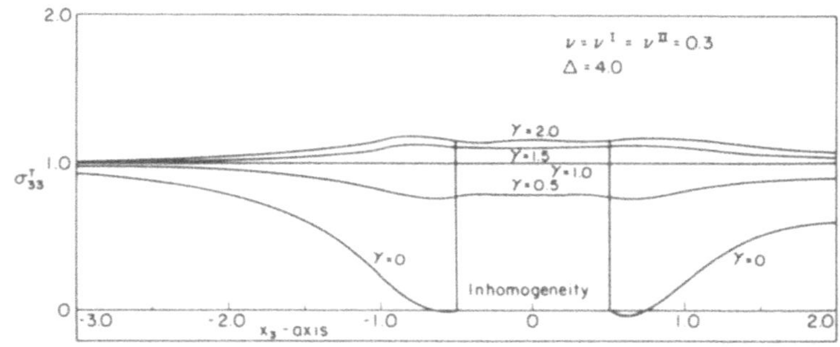

Fig. 23.10. Two spheroidal inhomogeneities ($a_1 = \bar{a}_1 = a_2 = \bar{a}_2 = 1$, $a_3 = \bar{a}_3 = 0.5$) in uniaxial tension $\sigma^0_{33} = 1$ at ∞, σ^T_{33} along x_3 axis for various values of $\gamma = \mu^I/\mu = \mu^{II}/\mu$.

*24. Impotent eigenstrains

As mentioned in the paragraph preceding (23.5), the system of equations (22.5′) or (23.4) becomes singular when the inhomogeneity is a void, and the applied stress, before the disturbance due to the inhomogeneity, is in the form of degree one polynomial. The solution of this system of equivalency equations is not unique. In other words, there is an infinite number of equivalent eigenstrains which do not generate any stress field within the material. Such eigenstrains are called impotent eigenstrains.

In this section analytical expressions for solutions of the system of homogeneous equivalency equations are presented for a spherical void in an isotropic material. The result of this special case is extended to the general case of arbitrarily shaped inclusions in anisotropic materials. Most parts of this section follow the work of Furuhashi and Mura (1979).

Let Ω be an inhomogeneity in D having elastic moduli C^*_{ijmn}, and let C_{ijmn} be the elastic moduli of the remainder $D - \Omega$. Let σ^0_{ij} and ϵ^0_{ij} be the stress and strain when the inhomogeneity is absent. Then, we have $\sigma^0_{ij} = C_{ijmn}\epsilon^0_{mn}$. Due to the presence of the inhomogeneity, the stress and the strain are changed to $\sigma^0_{ij} + \sigma_{ij}$, and $\epsilon^0_{ij} + \epsilon_{ij}$, where

$$\epsilon_{ij} = (1/2)(u_{i,j} + u_{j,i}). \tag{24.1}$$

The equivalency equation which transforms an inhomogeneity problem to an equivalent inclusion problem is expressed by (22.5′) or

$$C_{ijmn}\epsilon^*_{mn} - \Delta C_{ijmn}\epsilon_{mn} = \sigma^0_{ij} - C^*_{ijmn}\epsilon^0_{mn}, \tag{24.2}$$

where ϵ^*_{mn} is the equivalent eigenstrain in Ω and

$$\Delta C_{ijmn} = C_{ijmn} - C^*_{ijmn}.$$

When Ω has an eigenstrain ϵ^*_{rq}, the displacement u_m is expressed by (6.1),

$$u_m(\mathbf{x}) = -\int_\Omega G_{mk,l}(\mathbf{x}-\mathbf{x}')C_{klrq}\epsilon^*_{rq}(\mathbf{x}')d\mathbf{x}'. \tag{24.3}$$

Substitution of (24.3) into (24.2) leads to the following fundamental integro-differential equation to give the equivalent eigenstrain ϵ^*_{ij}:

$$\sigma^*_{ij}(\mathbf{x}) + \Delta C_{ijmn}\frac{\partial^2}{\partial x_n \partial x_l}\int_\Omega G_{km}(\mathbf{x}-\mathbf{x}')\sigma^*_{kl}(\mathbf{x}')d\mathbf{x}' = q_{ij}(\mathbf{x}), \tag{24.4}$$

where

$$\sigma^*_{ij} = C_{ijmn}\epsilon^*_{mn}, \quad q_{ij} = \sigma^0_{ij} - C^*_{ijmn}\epsilon^0_{mn}. \tag{24.5}$$

We can calculate the stress σ_{ij} from the formula

$$\sigma_{ij}(\mathbf{x}) = -C_{ijmn}\frac{\partial^2}{\partial x_n \partial x_l}\int_\Omega G_{mk}(\mathbf{x}-\mathbf{x}')\sigma^*_{kl}(\mathbf{x}')d\mathbf{x}' - \sigma^*_{ij}(\mathbf{x}). \tag{24.6}$$

Let us consider a special case in which the inhomogeneity Ω is a spherical isotropic inhomogeneity with shear modulus μ^* and Poisson's ratio ν^* and the matrix is an infinitely extended isotropic material with shear modulus μ and Poisson's ratio ν.

The applied stress field before the inhomogeneity disturbance, is assumed to be a homogeneous polynomial of degree one,

$$q_{ij}(\mathbf{x}) = B_{ijp}x_p, \quad B_{ijp} = B_{jip}, \tag{24.7}$$

where B_{ijp} are given constants.

It is well known that when Ω is an ellipsoid, the integral

$$\frac{\partial^2}{\partial x_n \partial x_l}\int_\Omega G_{km}(\mathbf{x}-\mathbf{x}')x'_p d\mathbf{x}'$$

becomes a polynomial of degree one without the constant term for interior points, as discussed following (12.16). Therefore, we can write

$$\Delta C_{ijmn}\frac{\partial^2}{\partial x_n \partial x_l}\int_\Omega G_{km}(\mathbf{x}-\mathbf{x}')x'_p d\mathbf{x}' = \omega^{klp}_{ijq}x_q. \tag{24.8}$$

We seek a solution of the integro-differential equation (24.4) in the form of

$$\sigma^*_{ij}(\mathbf{x}) = A_{ijp}x_p, \quad A_{ijp} = A_{jip}, \tag{24.9}$$

where A_{ijp} are constants to be determined. Substituting this expression into (24.5) and using (24.8), we obtain

$$A_{ijq}x_q + \omega_{ijq}^{klp} A_{klp}x_q = B_{ijq}x_q.$$

Hence, A_{ijq} must be a solution of the following linear system of equations:

$$A_{ijq} + \omega_{ijq}^{klp} A_{klp} = B_{ijq}. \tag{24.10}$$

The number of equations in (24.10) is 18. If Ω is not a void, the independent number of B_{ijq} is also 18. From (24.7) and (24.5), we have

$$q_{ij,j} = B_{ijj} = -C^*_{ijmn}\epsilon^0_{mn,j}, \tag{24.11}$$

since $\sigma^0_{ij,j} = 0$.

When Ω is a void, we get the three relations $B_{ijj} = 0$. Therefore, when Ω is a void, the maximum number of independent coefficients B_{ijp} is 15. The system of 18 equations (24.10) consists of four groups of equations with coefficients (A_{111}, A_{221}, A_{331}, A_{122}, A_{133}), (A_{222}, A_{112}, A_{332}, A_{121}, A_{233}), (A_{333}, A_{113}, A_{223}, A_{131}, A_{232}) and (A_{123}, A_{132}, A_{231}). Each of the first three groups is made of a 5×5 matrix, and the last group is made of a 3×3 matrix. If Ω is not a void, the solutions of these equations are uniquely determined.

When Ω is a void, the following matrix belongs to the first three groups:

$$\frac{1}{35(1-\nu)}\begin{pmatrix} 8-14\nu & 3-7\nu & 3-7\nu & -8 & -8 \\ 3-21\nu & 24-28\nu & 1-7\nu & -8 & 2-14\nu \\ 3-21\nu & 1-7\nu & 24-28\nu & 2-14\nu & -8 \\ -4+7\nu & -4+7\nu & 1 & 13-7\nu & -5+7\nu \\ -4+7\nu & 1 & -4+7\nu & -5+7\nu & 13-7\nu \end{pmatrix}. \tag{24.12}$$

Similarly, the matrix belonging to the last group is

$$\frac{1}{35(1-\nu)}\begin{pmatrix} 23-21\nu & -5+7\nu & -5+7\nu \\ -5+7\nu & 23-21\nu & -5+7\nu \\ -5+7\nu & -5+7\nu & 23-21\nu \end{pmatrix}. \tag{24.13}$$

The ranks of these two matrices are 4 and 3, respectively. Hence, we see that the rank of the matrix of coefficients in equation (24.10) is 15 when Ω is a void. Therefore, the homogeneous equations

$$A_{ijq} + \omega_{ijq}^{klp} A_{klp} = 0 \tag{24.14}$$

belonging to (24.10) have three independent non-trivial solutions $\overset{s}{A}_{klp}$

($s = 1, 2, 3$). Hence, if we denote by $\check{A}_{klp}$ an arbitrary particular solution of the system of equations (24.10), we can express the general solution as

$$A_{klp} = \sum_{s=1}^{3} C_s \overset{s}{A}_{klp} + \check{A}_{klp}, \tag{24.15}$$

where C_s are arbitrary constants. We can easily ascertain from (24.12) that $\overset{1}{A}_{klp}$ is the following vector:

$$\overset{1}{A}_{111} = 2 + \frac{2\nu}{1-2\nu}, \quad \overset{1}{A}_{221} = \overset{1}{A}_{331} = \frac{2\nu}{1-2\nu}, \quad \overset{1}{A}_{122} = \overset{1}{A}_{133} = 1. \tag{24.16}$$

Similarly, $\overset{2}{A}_{klp}$ and $\overset{3}{A}_{klp}$ are obtained by the cyclic permutation of the indices. The eigenstrain ϵ_{ij}^{*1} corresponding to this solution is obtained from (24.5′) and (24.9) as

$$\epsilon_{11}^{*1} = \frac{x_1}{\mu}, \quad \epsilon_{22}^{*1} = \epsilon_{33}^{*1} = 0, \quad \epsilon_{12}^{*1} = \frac{x_2}{2\mu}, \quad \epsilon_{13}^{*1} = \frac{x_3}{2\mu}, \quad \epsilon_{23}^{*1} = 0. \tag{24.17}$$

We obtain ϵ_{ij}^{*2} and ϵ_{ij}^{*3} by the cyclic permutation of the indices. It can be shown that the eigenstrain components in (24.17) do not introduce any stress field in D.

We conclude, therefore, that the eigenstrain (24.17) is impotent.

If the applied stress is the pure torsion around the x_3-axis, that is, $\sigma_{23}^0 = x_1$, $\sigma_{13}^0 = -x_2$, it is easily verified from (24.13) that the following vector is the particular solution of equation (24.10):

$$\check{A}_{132} = -\tfrac{5}{4}, \quad \check{A}_{231} = \tfrac{5}{4}, \tag{24.18}$$

with all other components equal to zero.

We have considered a rather special case by choosing a sphere and an isotropic material. We can generalize the results to cases where the shape of the inclusion and the anisotropy of the material are completely arbitrary. The preceding example of the impotent inclusion suggests the following generalization.

Suppose that eigenstrain components ϵ_{ij}^* in Ω are derived from functions u_i^* by

$$\epsilon_{ij}^* = \tfrac{1}{2}(u_{i,j}^* + u_{j,i}^*). \tag{24.19}$$

It is assumed that u_i^* are continuous in $\mathscr{D}$ and sufficiently smooth in the interior of Ω. Then, if $u_i^* = 0$ on the boundary $|\Omega|$ of Ω, the stress components σ_{ij} vanish everywhere in $\mathscr{D}$ and this eigenstrain is impotent (Ω is an impotent inclusion).

Before considering the analytical proof of this theorem, we make an intuitive observation. Since the boundary displacement is zero, the remainder $\mathscr{D}-\Omega$ is not disturbed by Ω; hence, the displacement and stress in $\mathscr{D}-\Omega$ vanish. On the other hand, the eigenstrain defined by (24.19) is compatible in Ω.

The analytical proof is as follows. When ϵ^*_{ij} is given by (24.19), the use of Gauss' theorem leads to

$$\int_\Omega G_{mk,l}(\mathbf{x}-\mathbf{x}')C_{klrp}u^*_{r,p}(\mathbf{x}')d\mathbf{x}' = \frac{\partial}{\partial x_l}\int_{|\Omega|} C_{klrp}G_{mk}(\mathbf{x}-\mathbf{x}')u^*_r(\mathbf{x}')n_p(\mathbf{x}')ds$$
$$+ C_{klrp}\frac{\partial^2}{\partial x_p \partial x_l}\int_\Omega G_{mk}(\mathbf{x}-\mathbf{x}')u^*_r(\mathbf{x}')d\mathbf{x}' \tag{24.20}$$

where $n_p(\mathbf{x}')$ stands for the outward normal at the point $\mathbf{x}'$ on $|\Omega|$. Because of $u^*_i=0$ on $|\Omega|$ and the definition (5.2) of Green's function G_{mk}, (24.20) becomes

$$\int_\Omega G_{mk,l}(\mathbf{x}-\mathbf{x}')C_{klrp}u^*_{r,p}(\mathbf{x}')d\mathbf{x}' = \begin{cases} -u^*_m(\mathbf{x}), & \mathbf{x}\in\Omega, \\ 0, & \mathbf{x}\in\mathscr{D}-\Omega. \end{cases} \tag{24.21}$$

Hence, (24.6) leads to $\sigma_{ij}=0$ everywhere.

When the boundary of Ω is expressed by an implicit function $f(\mathbf{x})=0$, we construct a function u^*_i explicitly. Assume that f is sufficiently smooth. Define the function $\hat{f}$ as

$$\hat{f}(\mathbf{x}) = \begin{cases} f(\mathbf{x}), & x\in\Omega, \\ 0, & x\in\mathscr{D}-\Omega. \end{cases} \tag{24.22}$$

Let g_i be a sufficiently smooth arbitrary function and set

$$u^*_i(\mathbf{x}) = g_i(\mathbf{x})\hat{f}(\mathbf{x}). \tag{24.23}$$

$u^*_i(\mathbf{x})$ is continuous in $\mathscr{D}$ but, in general, is not differentiable on the boundary of Ω. If Ω is an ellipsoid defined by equation (11.2), then the function u^*_i has the form

$$u^*_i(x) = \begin{cases} g_i(\mathbf{x})\left(\frac{x_1^2}{a_1^2}+\frac{x_2^2}{a_2^2}+\frac{x_3^2}{a_3^2}-1\right), & x\in\Omega \\ 0, & x\in\mathscr{D}-\Omega, \end{cases} \tag{24.24}$$

where g_i is an arbitrary smooth function. Since g_i is arbitrary, we can construct an infinite number of impotent eigenstrains from (24.23) and (24.19).

If ϵ^*_{ij} is a polynomial eigenstrain of degree one, the function u^*_i corresponding to this eigenstrain is a polynomial of degree two and can be expressed as

$$u^*_i(\mathbf{x}) = g_i(a_{jk}x_jx_k + a_jx_j + a_0), \tag{24.25}$$

where a_{jk}, a_j, a_0, and g_i are some constants. The eigenstrain derived from this function is impotent if, and only if, the domain Ω is bounded by the quadratic surface

$$a_{jk}x_jx_k + a_jx_j + a_0 = 0. \tag{24.26}$$

It is well known that there is no bounded quadratic surface except the ellipsoidal surface. Hence, if ϵ_{ij}^* is a polynomial impotent eigenstrain of degree one in a bounded domain Ω, then Ω is an ellipsoid.

Similarly, if ϵ_{ij}^* is a uniform impotent eigenstrain in a domain Ω, then Ω is a half-space.

There are not only smooth impotent eigenstrains, but also discontinuous ones. Let $\Omega_1, \ldots, \Omega_J$ be subsets of D. Suppose that u_i^{*p} is continuous in D, equal to zero outside Ω_p, and sufficiently smooth in the interior of Ω_p. Since the eigenstrains ϵ_{ij}^{*p} derived from these functions are impotent in Ω_p, the eigenstrain $\epsilon_{ij}^* = \epsilon_{ij}^{*1} + \cdots + \epsilon_{ij}^{*J}$ is also impotent in the union of $\Omega_1, \ldots, \Omega_J$. If the interior of intersection $\Omega_r \cap \Omega_q$ for integers $r \neq q$ is not empty, this eigenstrain is discontinuous on the boundary of $\Omega_r \cap \Omega_q$. For example, if we define the function u_i^{*p} as

$$u_i^{*p}(\mathbf{x}) = \begin{cases} \delta_{i1}(|p\mathbf{x}|^2 - 1), & |\mathbf{x}| \leq \dfrac{1}{p}, \\ 0, & |\mathbf{x}| > \dfrac{1}{p}, \end{cases} \tag{24.27}$$

then the impotent eigenstrain derived from these functions has $J - 1$ discontinuous surfaces in the unit sphere $|\mathbf{x}| < 1$.

25. Energies of inhomogeneities

The elastic strain energy, potential energy, interaction energy, and the Gibbs free energy will be discussed for a body which contains inhomogeneities.

Elastic strain energy

First we consider the case of a body D containing inhomogeneous inclusions and free from any external force or surface constraint. A given eigenstrain in Ω is denoted by ϵ_{ij}^p, where Ω is the subdomain of D occupied by the inhomogeneous inclusions. The (elastic) strain energy is the same as (13.1),

$$W^* = (1/2)\int_D \sigma_{ij}e_{ij}dD, \tag{25.1}$$

where $e_{ij} = \epsilon_{ij} - \epsilon^p_{ij}$ and $\epsilon_{ij} = (1/2)(u_{i,j} + u_{j,i})$. Integrating by parts, we obtain from (25.1)

$$W^* = -(1/2) \int_\Omega \sigma_{ij} \epsilon^p_{ij} dD. \tag{25.2}$$

If Ω is an ellipsoid and ϵ^p_{ij} is uniform, (25.2) becomes

$$W^* = -(1/2) V \sigma_{ij} \epsilon^p_{ij}, \tag{25.3}$$

where V is the volume of Ω.

If Ω consists of two ellipsoids, Ω_1 and Ω_2, W^* has the same form as (25.2) and $\Omega = \Omega_1 + \Omega_2$, $\sigma_{ij} = \sigma^{(1)}_{ij} + \sigma^{(2)}_{ij}$, where $\sigma^{(m)}_{ij}$ is the stress caused by ϵ^p_{ij} in Ω_m. The interaction energy between Ω_1 and Ω_2 is $-\int_{\Omega_1} \sigma^{(2)}_{ij} \epsilon^p_{ij} dD$ which is equal to $-\int_{\Omega_2} \sigma^{(1)}_{ij} \epsilon^p_{ij} dD$.

Next, consider the case when a body D, containing an inhomogeneous inclusion Ω with eigenstrain ϵ^p_{ij}, is subjected to an external surface force F_i. The displacement field is denoted by $u^0_i + u_i$ and the stress field by $\sigma^0_{ij} + \sigma_{ij}$. u^0_i and σ^0_{ij} are the displacement and stress when F_i acts in the absence of Ω. Then

$$W^* = (1/2) \int_D (\sigma^0_{ij} + \sigma_{ij})(u^0_{i,j} + u_{i,j} - \epsilon^p_{ij}) dD, \tag{25.4}$$

where

$$\sigma^0_{ij} = C_{ijkl} u^0_{k,l}, \quad \sigma^0_{ij} + \sigma_{ij} = C^*_{ijkl}(u^0_{k,l} + u_{k,l} - \epsilon^p_{kl}). \tag{25.5}$$

Here C_{ijkl} and C^*_{ijkl} are the respective elastic moduli of the matrix and inhomogeneous inclusion.

The expression for W^* given in (25.4) may be further simplified. An equation similar to (13.6) holds,

$$\int_D \sigma_{ij}(u^0_{i,j} + u_{i,j}) dD = 0. \tag{25.6}$$

Furthermore, we have

$$\begin{aligned} \sigma^0_{ij}(u_{i,j} - \epsilon^p_{ij}) &= C_{ijkl} u^0_{k,l}(u_{i,j} - \epsilon^p_{ij}) \\ &= C_{ijkl} u^0_{k,l}(u_{i,j} - \epsilon^p_{ij} - \epsilon^*_{ij} + \epsilon^*_{ij}) \\ &= u^0_{k,l} \sigma_{kl} + C_{ijkl} u^0_{k,l} \epsilon^*_{ij} \\ &= \sigma_{ij} u^0_{i,j} + \sigma^0_{ij} \epsilon^*_{ij}, \end{aligned} \tag{25.7}$$

where ϵ^*_{ij} is the fictitious eigenstrain introduced in the equivalent inclusion method, and

$$\sigma_{ij} = C_{ijkl}(u_{k,l} - \epsilon^p_{kl} - \epsilon^*_{kl}) \text{ in } \Omega. \tag{25.8}$$

Thus, we have

$$\int_D \sigma^0_{ij}(u_{i,j} - \epsilon^p_{ij})dD = \int_\Omega \sigma^0_{ij}\epsilon^*_{ij}dD, \tag{25.9}$$

since $\int_D \sigma_{ij}u^0_{i,j}dD = 0$. The elastic strain energy, therefore, becomes

$$W^* = (1/2)\int_D \sigma^0_{ij}u^0_{i,j}dD + (1/2)\int_\Omega \sigma^0_{ij}\epsilon^*_{ij}dD - (1/2)\int_\Omega \sigma_{ij}\epsilon^p_{ij}dD. \tag{25.10}$$

If Ω is an inhomogeneity ($\epsilon^p_{ij} = 0$), the elastic strain energy is

$$W^* = (1/2)\int_D \sigma^0_{ij}u^0_{i,j}dD + (1/2)\int_\Omega \sigma^0_{ij}\epsilon^*_{ij}dD. \tag{25.11}$$

Interaction energy

When a body D contains an inhomogeneity Ω and is subjected to F_i on S, the total potential energy of the system (the Gibbs free energy) is defined by the sum of the elastic strain energy and the potential energy of F_i,

$$\begin{aligned} W &= (1/2)\int_D (\sigma^0_{ij} + \sigma_{ij})(u^0_{i,j} + u_{i,j})dD - \int_S F_i(u^0_i + u_i)dS \\ &= (1/2)\int_D \sigma^0_{ij}u^0_{i,j}dD + (1/2)\int_\Omega \sigma^0_{ij}\epsilon^*_{ij}dD - \int_S F_i(u^0_i + u_i)dS, \end{aligned} \tag{25.12}$$

from (25.11).

If the material is homogeneous, W becomes

$$W_0 = (1/2)\int_D \sigma^0_{ij}u^0_{i,j}dD - \int_S F_iu^0_i dS. \tag{25.13}$$

The interaction energy between F_i and the inhomogeneity is defined by

$$\Delta W = W - W_0 = (1/2)\int_\Omega \sigma^0_{ij}\epsilon^*_{ij}dD - \int_S F_iu_idS. \tag{25.14}$$

It holds that

$$\begin{aligned} \int_S F_iu_idS = \int_S \sigma^0_{ij}n_ju_idS &= \int_D \sigma^0_{ij}u_{i,j}dD \\ &= \int_D \sigma^0_{ij}(u_{i,j} - \epsilon^*_{ij})dD + \int_D \sigma^0_{ij}\epsilon^*_{ij}dD \end{aligned} \tag{25.15}$$

and

$$\int_D \sigma^0_{ij}(u_{i,j} - \epsilon^*_{ij})dD = \int_D C_{ijkl}u^0_{k,l}(u_{i,j} - \epsilon^*_{ij})dD = \int_D u^0_{k,l}\sigma_{kl}dD = 0. \quad (25.16)$$

Therefore, we have

$$\Delta W = -(1/2)\int_\Omega \sigma^0_{ij}\epsilon^*_{ij}dD. \quad (25.17)$$

If σ^0_{ij} is uniform and Ω is an ellipsoidal inhomogeneity, ϵ^*_{ij} also becomes uniform and

$$\Delta W = -(1/2)V\sigma^0_{ij}\epsilon^*_{ij}, \quad (25.18)$$

where V is the volume of Ω.

When a body D contains inhomogeneous inclusion Ω with eigenstrain ϵ^p_{ij}, and it is subjected to F_i, the Gibbs free energy of the system is

$$W = (1/2)\int_D (\sigma^0_{ij} + \sigma_{ij})(u^0_{i,j} + u_{i,j} - \epsilon^p_{ij})dD - \int_S F_i(u^0_i + u_i)dS. \quad (25.19)$$

This total potential energy is compared with W_0 as defined by (25.13). The difference between W and W_0 is

$$\Delta W = W - W_0. \quad (25.20)$$

Since the first term on the right-hand side in (25.19) can be expressed by (25.10), we have

$$\Delta W = (1/2)\int_\Omega \sigma^0_{ij}\epsilon^*_{ij}dD - (1/2)\int_\Omega \sigma_{ij}\epsilon^p_{ij}dD - \int_S F_iu_idS. \quad (25.21)$$

It also holds that

$$\int_S F_iu_idS = \int_S \sigma^0_{ij}n_ju_idS = \int_D \sigma^0_{ij}u_{i,j}dD = \int_D \sigma^0_{ij}(u_{i,j} - \epsilon^p_{ij} - \epsilon^*_{ij})dD$$
$$+ \int_D \sigma^0_{ij}(\epsilon^p_{ij} + \epsilon^*_{ij})dD \quad (25.22)$$

and

$$\int_D \sigma^0_{ij}(u_{i,j} - \epsilon^p_{ij} - \epsilon^*_{ij})dD = \int_D C_{ijkl}u^0_{k,l}(u_{i,j} - \epsilon^p_{ij} - \epsilon^*_{ij})dD$$
$$= \int_D \sigma_{kl}u^0_{k,l}dD = 0. \quad (25.23)$$

Therefore, we have

$$\Delta W = -(1/2)\int_\Omega \sigma^0_{ij}\epsilon^*_{ij}dD - (1/2)\int_\Omega \sigma_{ij}\epsilon^p_{ij}dD - \int_\Omega \sigma^0_{ij}\epsilon^p_{ij}dD. \tag{25.24}$$

Colonnetti's theorem

The variation of Gibbs free energy associated with a change in a given physical system plays an important role in certain cases as shown in Chapter 7.

Suppose that a body D containing inhomogeneities Ω is subjected to an applied force F_i on the boundary S of D. The Gibbs free energy is expressed by (25.12), or

$$W = W^* + V^*, \tag{25.25}$$

with

$$W^* = (1/2)\int_D \sigma^A_{ij}e^A_{ij}dD$$
$$V^* = -\int_S F_i u^A_i dS, \tag{25.26}$$

where $\sigma^A_{ij} = \sigma^0_{ij} + \sigma_{ij}$, $e^A_{ij} = (1/2)(u^A_{i,j} + u^A_{j,i})$, and $u^A_i = u^0_i + u_i$. e^A_{ij} is the elastic strain, W^* the elastic strain energy, and V^* is the potential energy of F_i. If the inhomogeneities in D are further subjected to the eigenstrain e^p_{ij}, the Gibbs free energy of the system changes to (25.19). The new stress, elastic strain, and displacement become $\sigma^A_{ij} + \sigma^B_{ij}$, $e^A_{ij} + e^B_{ij}$, $u^A_i + u^B_i$, where σ^B_{ij}, e^B_{ij}, and u^B_i are increments (changes) in the fields due to ϵ^p_{ij}. The Gibbs free energy of the new system (25.19) can be written as

$$W = W^* + V^* \tag{25.27}$$

with

$$W^* = (1/2)\int_D \sigma^A_{ij}e^A_{ij}dD + (1/2)\int_D \sigma^B_{ij}e^B_{ij}dD,$$
$$V^* = -\int_S F_i u^A_i dS - \int_S F_i u^B_i dS. \tag{25.28}$$

This result has been obtained from Colonetti's theorem (1921) as shown below. The elastic strain energy is

$$W^* = (1/2)\int_D (\sigma_{ij}^A + \sigma_{ij}^B)(e_{ij}^A + e_{ij}^B)dD$$

$$= (1/2)\int_D \sigma_{ij}^A e_{ij}^A dD + (1/2)\int_D \sigma_{ij}^B e_{ij}^B dD + \int_D \sigma_{ij}^B e_{ij}^A dD, \qquad (25.29)$$

since $\sigma_{ij}^A e_{ij}^B = C_{ijkl}^{**} e_{kl}^A e_{ij}^B = \sigma_{kl}^B e_{kl}^A$, where C_{ijkl}^{**} is C_{ijkl} in $D-\Omega$ and C_{ijkl}^{*} in Ω. On the other hand,

$$\int_D \sigma_{ij}^B e_{ij}^A dD = \int_D \sigma_{ij}^B u_{i,j}^A dD = \int_S \sigma_{ij}^B n_j u_i^A dS - \int_D \sigma_{ij,j}^B u_i^A dD = 0, \qquad (25.30)$$

since $\sigma_{ij}^B n_j = 0$ on S and $\sigma_{ij,j}^B = 0$ in D. Equation (25.30) is Colonetti's theorem. The variations of W^* and V^* associated with the change from the system corresponding to (25.26) to the system corresponding to (25.28), are

$$\Delta W^* = (1/2)\int_D \sigma_{ij}^B e_{ij}^B dD = (1/2)\int_D \sigma_{ij}^B (u_{i,j}^B - \epsilon_{ij}^p) dD$$

$$= -(1/2)\int_D \sigma_{ij}^B \epsilon_{ij}^p dD \qquad (25.31)$$

$$\Delta V^* = -\int_S F_i u_i^B dS = -\int_S \sigma_{ij}^A u_i^B n_j dS$$

$$= -\int_D \sigma_{ij}^A u_{i,j}^B dS = -\int_D \sigma_{ij}^A (e_{ij}^B + \epsilon_{ij}^p) dD$$

$$= -\int_D \sigma_{ij}^A \epsilon_{ij}^p dD \qquad (25.32)$$

or

$$\Delta V^* = -\int_S \sigma_{ij}^0 u_i^B n_j dS = -\int_D \sigma_{ij}^0 u_{i,j}^B dD$$

$$= -\int_D \sigma_{ij}^0 (u_{i,j}^B - \epsilon_{ij}^{**}) dD - \int_D \sigma_{ij}^0 \epsilon_{ij}^{**} dD, \qquad (25.33)$$

where ϵ_{ij}^{**} has been defined in (22.15). Since

$$\int_D \sigma_{ij}^0 (u_{i,j}^B - \epsilon_{ij}^{**}) dD = \int_D u_{i,j}^0 \sigma_{ij}^B dD = 0, \qquad (25.34)$$

(25.33) leads to

$$\Delta V^* = -\int_\Omega \sigma^0_{ij}\epsilon^{**}_{ij} dD. \tag{25.35}$$

The terms $(1/2)\int_D \sigma^A_{ij} e^A_{ij} dD - \int_S F_i u^A_i dS$ have been calculated from (25.12). Thus, we have

$$\Delta W = W - W_0 = -(1/2)\int_\Omega \sigma^0_{ij}\epsilon^{*A}_{ij} dD - (1/2)\int_\Omega \sigma^B_{ij}\epsilon^p_{ij} dD - \int_\Omega \sigma^A_{ij}\epsilon^p_{ij} dD, \tag{25.35.1}$$

where W_0 is defined by (25.13) and ϵ^{*A}_{ij} is ϵ^*_{ij} in (25.17) or (22.5), which is the equivalent eigenstrain for the inhomogeneity under applied stress σ^0_{ij}. The above expression (25.35.1) is an alternative expression of (25.24).

The expressions of the interaction energies derived so far are valid also for the interaction between an inhomogeneous inclusion Ω and a different internal stress. σ^0_{ij} and u^0_i in the previous expressions are interpreted as the internal stress and displacement caused by the other internal stress source.

Uniform plastic deformation in a matrix

As an application of Colonnetti's theorem, let us consider a composite material under an applied force F_i. The material D consists of inhomogeneities Ω and matrix $D - \Omega$. The elastic stress is denoted by σ^A_{ij}. Assume the inhomogeneities to be stronger than the matrix and that a uniform plastic strain ϵ^p_{ij} is distributed in the matrix. Due to the plastic strain misfit with the inhomogeneities, the stress state changes from σ^A_{ij} to $\sigma^A_{ij} + \sigma^B_{ij}$. Then, the Gibbs free energy of the system becomes

$$W = W^* + V^* \tag{25.36}$$

with

$$W^* = (1/2)\int_D \sigma^A_{ij} e^A_{ij} dD + (1/2)\int_D \sigma^B_{ij} e^B_{ij} dD$$
$$V^* = -\int_S F_i u^A_i dS - \int_S F_i u^B_i dS, \tag{25.37}$$

where e^A_{ij} and e^B_{ij} are the elastic strains associated with σ^A_{ij} and σ^B_{ij}, and u^A_i and u^B_i are the corresponding displacements. For the discussion in Section 43 we will investigate the dependency of W on ϵ^p_{ij}. The first terms in (25.37) are independent of ϵ^p_{ij}. The second terms can be rewritten as follows

$$(1/2)\int_D \sigma_{ij}^B e_{ij}^B dD = (1/2)\int_D \sigma_{ij}^B(u_{i,j}^B - \epsilon_{ij}^p)dD = -(1/2)\int_{D-\Omega} \sigma_{ij}^B \epsilon_{ij}^p dD,$$

$$-\int_S F_i u_i^B dS = -\int_D \sigma_{ij}^A u_i^B n_j dS = -\int_D \sigma_{ij}^A(e_{ij}^B + \epsilon_{ij}^p)dD = -\int_{D-\Omega} \sigma_{ij}^A \epsilon_{ij}^p dD \tag{25.38}$$

Since $\epsilon_{ij}^p = 0$ in Ω and ϵ_{ij}^p is uniform in $D-\Omega$, the integrals in (25.38) are actually defined in $D-\Omega$. However it is convenient to extend the integral domain into D and subtract Ω from D. Then, since $\int_D \sigma_{ij}^B dD = 0$ (see (42.3)), we can write

$$\begin{aligned}-(1/2)\int_{D-\Omega} \sigma_{ij}^B \epsilon_{ij}^p dD &= -(1/2)\epsilon_{ij}^p \int_{D-\Omega} \sigma_{ij}^B dD \\ &= (-1/2)\epsilon_{ij}^p\left(\int_D \sigma_{ij}^B dD - \int_\Omega \sigma_{ij}^B dD\right) \\ &= (1/2)\epsilon_{ij}^p \int_\Omega \sigma_{ij}^B dD.\end{aligned} \tag{25.39}$$

Since $\sigma_{ij}^A = \sigma_{ij}^0 + \sigma_{ij}$ and $\int_D \sigma_{ij} dD = 0$,

$$\begin{aligned}-\int_{D-\Omega} \sigma_{ij}^A \epsilon_{ij}^p dD &= -\epsilon_{ij}^p \int_{D-\Omega} \sigma_{ij}^A dD = -\epsilon_{ij}^p\left(\int_D \sigma_{ij}^A dD - \int_\Omega \sigma_{ij}^A dD\right) \\ &= -\epsilon_{ij}^p \sigma_{ij}^0 \int_D dD + \epsilon_{ij}^p \int_\Omega \sigma_{ij}^A dD.\end{aligned} \tag{25.40}$$

Thus, the part of W which depends on ϵ_{ij}^p becomes

$$\Delta W = (1/2)\epsilon_{ij}^p \int_\Omega \sigma_{ij}^B dD - \epsilon_{ij}^p \sigma_{ij}^0 \int_D dD + \epsilon_{ij}^p \int_\Omega \sigma_{ij}^A dD. \tag{25.41}$$

σ_{ij}^B can be calculated as the eigenstress by assuming that Ω has eigenstrain $-\epsilon_{ij}^p$ and that the matrix has no plastic strain, since the misfit strain is a relative quantity between the matrix and the inclusions.

Energy balance

In order to understand the physical meaning of the Gibbs free energy change ΔW, we consider the energy dissipated when ϵ_{ij}^p in (25.27) or (25.36) is changed by $\delta\epsilon_{ij}^p$. The work done by applied force F_i during the change is

$$\int_S F_i \delta u_i^B dS, \tag{25.42}$$

where F_i is kept constant during the variation. The change of elastic strain energy is

$$\int_D (\sigma_{ij}^A + \sigma_{ij}^B)\delta e_{ij}^B dD, \tag{25.43}$$

since e_{ij}^A does not change during the variation, where σ_{ij}^A is the stress due to σ_{ij}^0 and the inhomogeneity, and σ_{ij}^B the stress caused by ϵ_{ij}^p and the inhomogeneity. The energy δQ dissipated during the variation is the difference between (25.42) and (25.43),

$$\delta Q = \int_S F_i \delta u_i^B dS - \int_D (\sigma_{ij}^A + \sigma_{ij}^B)\delta e_{ij}^B dD, \tag{25.44}$$

where

$$(1/2)(\delta u_{i,j}^B + \delta u_{j,i}^B) = \delta e_{ij}^B + \delta \epsilon_{ij}^p. \tag{25.45}$$

Thus, we have

$$\delta Q = -\delta(\Delta W) \tag{25.46}$$

which means that the change of the Gibbs free energy is minus the energy dissipated during the variation $\delta\epsilon_{ij}^p$.

Expression (25.44) can be rewritten as

$$\delta Q = \int_D (\sigma_{ij}^A + \sigma_{ij}^B)\delta \epsilon_{ij}^p dD, \tag{25.47}$$

where (25.45), $(\sigma_{ij}^A + \sigma_{ij}^B)n_j = F_i$ on S, and $\sigma_{ij,j}^A = 0$, $\sigma_{ij,j}^B = 0$ in D are used. Expression (25.47) can also be obtained from (25.46) and (25.35.1) since σ_{ij}^B in (25.35.1) is a linear function of ϵ_{ij}^p as seen from (22.13a).

When ϵ_{ij}^p is defined uniformly only in the matrix, (25.47) becomes

$$\begin{aligned}\delta Q &= \int_{D-\Omega} (\sigma_{ij}^A + \sigma_{ij}^B)\delta \epsilon_{ij}^p dD \\ &= \delta\epsilon_{ij}^p \left\{\int_D (\sigma_{ij}^A + \sigma_{ij}^B) dD - \int_\Omega (\sigma_{ij}^A + \sigma_{ij}^B) dD\right\}.\end{aligned} \tag{25.48}$$

From (42.3) and the definitions of σ_{ij}^A and σ_{ij}^B, we have

$$\int_D \sigma_{ij}^B dD = 0, \quad \int_D \sigma_{ij}^A dD = \int_D \sigma_{ij}^0 dD. \tag{25.49}$$

Therefore,

$$\delta Q = \delta\epsilon_{ij}^p \left\{\int_D \sigma_{ij}^0 dD - \int_\Omega (\sigma_{ij}^A + \sigma_{ij}^B) dD\right\}. \tag{25.50}$$

which can also be obtained from (25.46) and (25.41). $\sigma_{ij}^A + \sigma_{ij}^B = \sigma_{ij}^0 + \sigma_{ij}$ can be calculated from (22.13), where ϵ_{ij}^p is taken $-\epsilon_{ij}^p$ in Ω, since σ_{ij} is equivalent to the eigenstress (stress disturbance) caused by the misfit $-\epsilon_{ij}^p$ in Ω and the applied stress σ_{ij}^0.

Let us consider an example of the simple tension σ_{33}^0. It is assumed that the matrix is a perfectly plastic solid with the yield stress σ_y and that the inhomogeneities $\Omega = \Sigma_p \Omega_p$ are perfectly elastic. The plastic strain has no dilatation and $\epsilon_{11}^p = \epsilon_{22}^p = -\epsilon_{33}^p/2$. The energy dissipated during $\delta\epsilon_{ij}^p$ is

$$\delta Q = \int_{D-\Omega} \sigma_y \delta\epsilon_{33}^p dD. \tag{25.51}$$

Then, (25.50) becomes

$$\sigma_y \delta\epsilon_{13}^p \int_{D-\Omega} dD = \delta\epsilon_{33}^p \left\{ \sigma_{33}^0 \int_{D-\Omega} dD - \int_\Omega (\sigma_{33} - \sigma_{11}/2 - \sigma_{22}/2) dD \right\}. \tag{25.52}$$

The last terms are calculated from (22.13), resulting in

$$\sigma_{33} - (\sigma_{11} + \sigma_{22})/2 = A\epsilon_{33}^p + B\sigma_{33}^0, \tag{25.53}$$

where

$$A = E^*/2(1-\nu^*), \quad B = (E^*\nu - E\nu^*)/E(1-\nu^*), \tag{25.54}$$

for penny-shaped particles (the disk plane is perpendicular to the x_3 axis);

$$A = \frac{(3/2)E^*E(7-5\nu)}{E^*(1+\nu)(8-10\nu) + E(7-5\nu)(1+\nu^*)},$$
$$B = \frac{(7-5\nu)\{E^*(1+\nu) - E(1+\nu^*)\}}{E^*(1+\nu)(8-10\nu) + E(7-5\nu)(1+\nu^*)}, \tag{25.55}$$

for spherical particles; and

$$A = \frac{(1/2)E^*\{3E(1-2\nu^*) + 2E^*(1+\nu)\}}{E^*(1+\nu) + E(1+\nu^*)(1-2\nu^*)},$$
$$B = \frac{(E^*)^2(1+\nu)/E - 2E^*\nu^*(1+\nu) - E(1+\nu^*)(1-2\nu^*)}{E^*(1+\nu) + E(1+\nu^*)(1-2\nu^*)}, \tag{25.56}$$

for rod-shaped particles (the axis of the rod is parallel to the x_3 axis). The values of A and B for anisotropic media have been calculated by Lin et al. (1973). When (25.53) is substituted into (25.52), we have

$$(1-f)\sigma_y = (1-f)\sigma_{33}^0 - f(A\epsilon_{33}^p + B\sigma_{33}^0) \tag{25.57}$$

or

$$\sigma_{33}^{0}=\frac{fA\epsilon_{33}^{p}}{1-f(1+B)}+\frac{(1-f)}{1-f(1+B)}\sigma_{y}, \tag{25.58}$$

where f is the volume fraction of particles. Equation (25.58) has been obtained by Tanaka and Mori (1970). The stress-strain of the composite material displays linear work-hardening as shown in Fig. 43.1. The details are presented in Section 43. The yield point also increases since$(1-f)/\{1-(1+B)f\}>1$. It decreases, however, when $E^*<E$ and $\nu^*\approx\nu$.

Let us consider a creep problem under a constant applied stress σ_{33}^{0}. If we assume that the matrix behaves as a Newtonian viscous solid (the stress is $\tilde{\mu}d\epsilon_{33}^{p}/dt$), δQ in (25.50) can be written as

$$\delta Q=(1-f)\tilde{\mu}\dot{\epsilon}_{33}^{p}\delta\epsilon_{33}^{p}, \tag{25.59}$$

then (25.50) becomes

$$(1-f)\tilde{\mu}\dot{\epsilon}_{33}^{p}=(1-f)\sigma_{33}^{0}-f(A\epsilon_{33}^{p}+B\sigma_{33}^{0}). \tag{25.60}$$

The above differential equation has the solution

$$\epsilon_{33}^{p}=\frac{1-(1+B)f}{Af}\sigma_{33}^{0}[1-\exp\{-fAt/(1-f)\tilde{\mu}\}] \tag{25.61}$$

when the initial condition $\epsilon_{33}^{p}=0$ at $t=0$ is used. The composite material, as a whole, behaves like a Kelvin's material (Mura 1972). Cyclic creep problems have been investigated by Mura et al. (1975) and Shetty et al. (1975) by using equation (25.50).

26. Precipitates and martensites

Precipitates and martensites in alloys are typical examples of inhomogeneous inclusions. The elastic strain energy caused by these particular inclusions will be discussed in this section. If interfacial energy can be neglected, the morphology of precipitates and martensites can be determined by the condition that the elastic strain energy (25.2) take a minimum value. σ_{ij} in (25.2) or (25.3) can be evaluated by the equivalent inclusion method described by (22.13), where $\epsilon_{ij}^{0}=0$ and $\sigma_{ij}^{0}=0$, since no applied load is considered here.

Isotropic precipitates

Consider an infinite isotropic matrix with shear modulus μ and Poisson's ratio ν, containing an ellipsoidal precipitate whose elastic constants are μ^* and ν^*. Barnett et al. (1974) calculated the elastic strain energy per unit volume of the precipitate when

$$\begin{aligned} &\epsilon_{ij}^p = \delta_{ij}\epsilon \\ &a_1 = a_2 = a, \quad 0 \le \beta = a_3/a \le \infty, \end{aligned} \tag{26.1}$$

where ϵ is a constant misfit strain. Their results are

$$W^*/V = 3\epsilon^2\gamma^*\left[1 - \left\{1 + \frac{\gamma}{\gamma^*}\left(\frac{2K+1}{K+M-KM}\frac{1-\nu}{1+\nu} - 1\right)\right\}^{-1}\right], \tag{26.2}$$

where

$$\begin{aligned} \gamma &= \mu(1+\nu)/(1-2\nu), \quad \gamma^* = \mu^*(1+\nu^*)/(1-2\nu^*), \\ M &= 1/(1-\beta^2) - \beta(1-\beta^2)^{-3/2}\cos^{-1}\beta; \ \beta < 1, \\ &= 1/3; \quad \beta = 1, \\ &= 1/(1-\beta^2) + \beta(\beta^2-1)^{-3/2}\cosh^{-1}\beta; \quad \beta > 1, \\ K &= \frac{1 + (\mu^*/\mu - 1)\{M(5-4\nu) - 3(3M-1)/2(1-\beta^2)\}/2(1-\nu)}{1 + (\mu^*/\mu - 1)\{(1-2\nu) + 2M(1+\nu) - 3(3M-1)/2(1-\beta^2)\}/2(1-\nu)}. \end{aligned} \tag{26.3}$$

When $\mu^* = \mu$ and $\nu^* = \nu$, (26.2) reduces to (13.21),

$$W_0/V = 2\mu\epsilon^2(1+\nu)/(1-\nu). \tag{26.4}$$

The normalized strain energy W^*/W_0 is plotted in Fig. 26.1 against β for several values of μ^*/μ. Barnett et al. (1974) concluded that the minimum strain energy is achieved by a flat ellipsoid when $\mu^* < \mu$, and by a sphere when $\mu^* > \mu$. The same conclusion has been stated by Lászlo (1943).

Anisotropic precipitates

Here we calculate the elastic strain energy when an infinite anisotropic matrix contains an ellipsoidal anisotropic precipitate. The precipitate is an inhomogeneous inclusion with eigenstrain ϵ_{ij}^p. The dependency of the results on the orientation and shape of the precipitates is discussed.

Let us denote the elastic moduli of the matrix by C_{ijkl} and of the precipitate by C_{ijkl}^*. The x_i coordinate system is taken along crystallographic directions of the matrix ([100], [010], [001]). The orientation of the precipitate is characterized by θ and ϕ (see Fig. 26.2).

Mori et al. (1978) have obtained contour lines of the elastic strain

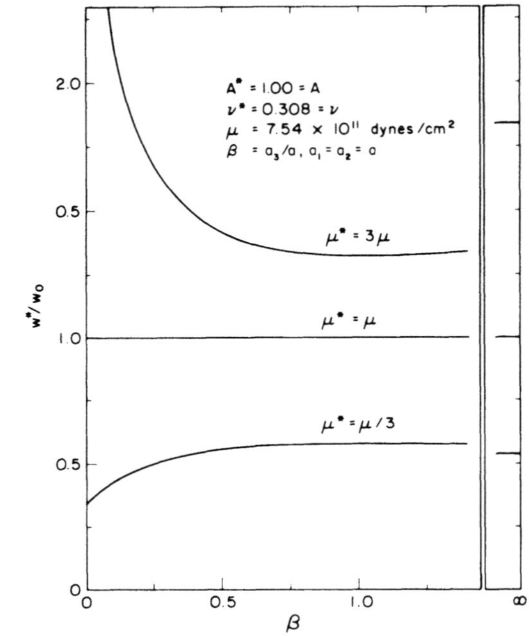

Fig. 26.1. Normalized strain energy of an isotropic coherent ellipsoidal precipitate in an isotropic matrix vs. aspect ratio. A, ν, and μ are anisotropy ratio, Poisson's ratio, and shear modulus, respectively.

energy for a needle-shaped precipitate (Fig. 26.3) and those for a disk-shaped precipitate (Fig. 26.4), when the matrix is copper ($C_{11} = 16.84 \times 10^{10}\ \text{N/m}^2$, $C_{12} = 12.14 \times 10^{10}\ \text{N/m}^2$, $C_{44} = 7.54 \times 10^{10}\ \text{N/m}^2$) and

$$\epsilon^p_{ij} = \delta_{ij}\epsilon, \quad C^*_{ij} = fC_{ij}. \tag{26.5}$$

These conditions assume that the phase transformation strain is dilatational and the precipitate crystalline directions coincide with those of the matrix.

It can be seen that the minimum energy is achieved when the disk plane is perpendicular to [001]. This is due to the fact that the Zener anisotropy factor (Zener 1948)

$$A = 2C_{44}/(C_{11} - C_{12}) \tag{26.6}$$

is greater than 1 and C^*_{ij} is assumed to equal fC_{ij} in the energy calculation. The stress components in an ellipsoidal inhomogeneity containing eigenstrain ϵ^p_{ij} are expressed by (22.13a). The right hand side in

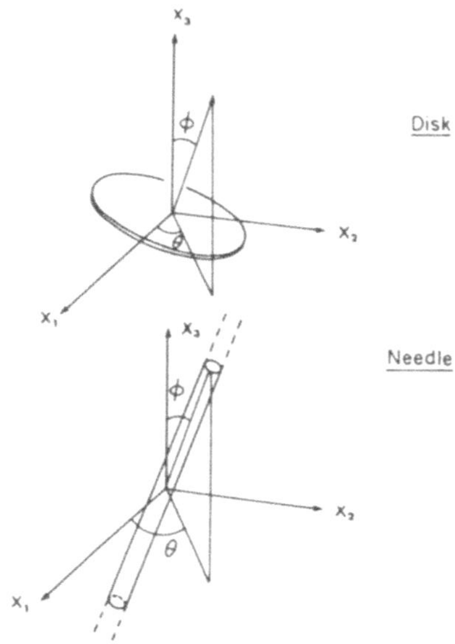

Fig. 26.2. Orientations of precipitates with respect to crystallographic directions x_i.

(22.13a) becomes (17.47) for a flat ellipsoidal inclusion, where ϵ_{ij}^{**} is used for ϵ_{ij}^*. As explained by (17.37′) and (17.42), the displacement and the traction at the flat interface are zero. Using the principle of equivalent inclusion, we can find that the displacement and the traction at the flat interface of the inhomogeneous inclusion are also zero. This means that the inclusion is stressed but the matrix is not. The stress

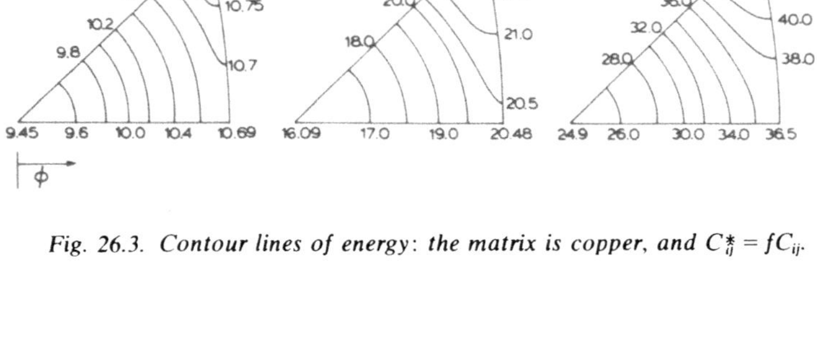

Fig. 26.3. Contour lines of energy: the matrix is copper, and $C_{ij}^* = fC_{ij}$.

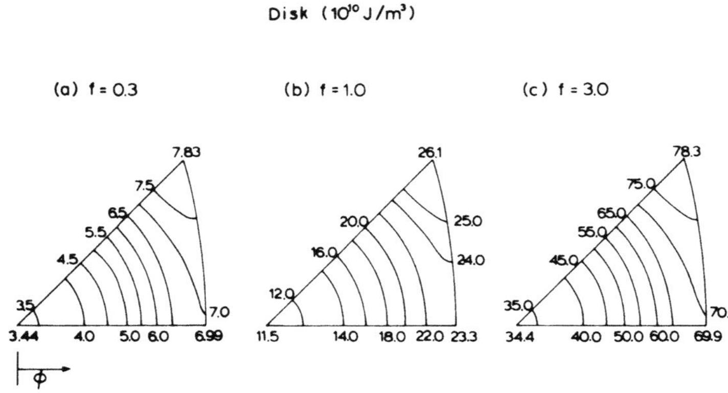

Fig. 26.4. Contour lines of energy: the matrix is copper and $C^*_{ij} = fC_{ij}$.

state is independent of the matrix moduli and remains the same even if the moduli of the matrix are equal to C^*_{ijkl}. Therefore, this inhomogeneity inclusion becomes the homogeneous inclusion with ϵ^p_{ij}. The stress is

$$\sigma_{pq} = C^*_{pqik}\{C^*_{jlmn}\epsilon^p_{mn}N_{ij}(\bar{\xi})D^{-1}(\bar{\xi})\bar{\xi}_k\bar{\xi}_l - \epsilon^p_{ki}\}, \tag{26.7}$$

from (17.47), where $N_{ij}(\bar{\xi})$ and $D(\bar{\xi})$ are expressed in terms of C^*_{ijkl}, and $\bar{\xi}$ is the unit vector normal to the flat surface. The fact that the elastic stress state is independent of the matrix elastic moduli has been established by Mura and Lin (1974) and Lee et al. (1977) who have pointed out that the elastic strain energy of a disk-shaped precipitate is independent of the elastic moduli of the matrix. When C^*_{ijkl} has cubic symmetry and $\epsilon^p_{ij} = \delta_{ij}\epsilon^*$, the elastic strain energy W^* is calculated from (26.7) and (25.3)

$$W^* = (V/2)(C^*_{11} + 2C^*_{12})(\epsilon^*)^2\{3 - (C^*_{11} + 2C^*_{12})F\}, \tag{26.8}$$

where

$$\begin{aligned} F = {} & \{C^{*2}_{44} + 2C^*_{44}(C^*_{11} - C^*_{12} - 2C^*_{44})\Gamma + 3(C^*_{11} - C^*_{12} - 2C^*_{44})^2\Delta\} \\ & /\{C^*_{11}C^{*2}_{44} + C^*_{44}(C^*_{11} + C^*_{12})(C^*_{11} - C^*_{12} - 2C^*_{44})\Gamma + (C^*_{11} - C^*_{12} - 2C^*_{44})^2 \\ & \times (C^*_{11} + 2C^*_{12} + C^*_{44})\Delta\}, \\ \Gamma = {} & \bar{\xi}_1^2\bar{\xi}_2^2 + \bar{\xi}_2^2\bar{\xi}_3^2 + \bar{\xi}_3^2\bar{\xi}_1^2, \\ \Delta = {} & \bar{\xi}_1^2\bar{\xi}_2^2\bar{\xi}_3^2. \end{aligned} \tag{26.8.1}$$

It can be seen that (26.8) has an extreme value when either one of $|\bar{\xi}_1|$, $|\bar{\xi}_2|$, $|\bar{\xi}_3|$ is equal to 1 or $|\bar{\xi}_1| = |\bar{\xi}_2| = |\bar{\xi}_3| = 1/\sqrt{3}$. When the Zener factor

$A^* = 2C^*_{44}/(C^*_{11} - C^*_{12}) > 1$, the minimum of W^* is found at $\bar{\xi} = (0, 0, 1)$, which is

$$W^* = V(\epsilon^*)^2(C^*_{11} + 2C^*_{12})(C^*_{11} - C^*_{12})/C^*_{11}, \tag{26.8.2}$$

in agreement with the result of Mura and Lin (1974). When $A^* < 1$, W^* takes a minimum at $\bar{\xi} = (1/\sqrt{3}, 1/\sqrt{3}, 1/\sqrt{3})$.

From Fig. 26.3 it is found that the elastic strain energy of a needle-shaped precipitate is a minimum when the needle is aligned parallel to one of the crystallographic directions ⟨001⟩. However, the minimum of the elastic strain energy of the needle can not be determined by A and A^* alone. In fact Mori et al. (1978) have shown that the orientation of the needle (the axes direction) for minimum energy (for $A > 1$, $A^* < 1$) is [111] or [001] depending on the combination of C_{ij} and C^*_{ij}.

It is important to note from (26.7) that conclusions (17.42) and (17.46) with condition (17.44) are obtained even for an inhomogeneous flat ellipsoidal (disk shape) inclusion, where ϵ^*_{ij} is replaced by ϵ^p_{ij}. The strain energy calculation of the spinodal alloy is essentially identical to that of a disk-shaped precipitate. The elastic strain energy of the spinodally decomposed cubic alloy has the same orientation dependence as (26.8) (e.g., see Hilliard 1970). For example, when a solute concentration, c, fluctuates along the direction $\bar{\xi}$, c and the corresponding eigenstrain are expressed by

$$\begin{aligned} c &= \sum_{n=1}^{\infty} c(n)\cos(2\pi nx/\lambda), \\ \epsilon^p_{ij} &= \epsilon^0_{ij}\sum_{n=1}^{\infty} c(n)\cos(2\pi nx/\lambda), \end{aligned} \tag{26.8.3}$$

where λ is the wave length of the fluctuation and x is the distance along $\bar{\xi}$. Applying (3.14) and (25.2) with $\epsilon^0_{ij} = \epsilon^*\delta_{ij}$, we obtain the elastic strain energy W^* as

$$W^* = \tfrac{1}{2}\langle c^2\rangle(C^*_{11} + 2C^*_{12})(\epsilon^*)^2\{3 - (C^*_{11} + 2C^*_{12})F\} \tag{26.8.4}$$

per unit volume of a cubic crystal. Here, $\langle c^2\rangle$ is defined as

$$\langle c^2\rangle = (1/\lambda)\int_0^{\lambda} c^2 dx \tag{26.8.5}$$

and F has been defined in (26.8.1). It has often been said that the coefficient $G^* = (C^*_{11} + 2C^*_{12})\{3 - (C^*_{11} + 2C^*_{12})F\}$ in the above equation is the elastic modulus in the direction along which the concentration of solute atoms fluctuates. However, such a concept leads to a misunderstanding, since the stress components normal to the disk surface (σ_{3p}) are zero as is shown by (17.42), while G^* is simply evaluated from (26.8). This elastic strain energy is caused by the stress components parallel to the disk surface ($\sigma_{11}, \sigma_{22}, \sigma_{12}$). This energy is somewhat related to the elastic modulus perpendicular to the direction of solute atom fluctuation.

One can extend (26.8.3) to the situation where ϵ_{ij}^0 is not pure dilatation. Such a case might occur when a carbon atom concentration fluctuates in martensitic steel that has a high carbon concentration. In this case, because of the interaction of carbon atoms, almost all the carbon atoms are accommodated in a particular type of sublattice (Zener 1946, Mori, et al. 1976). For example, ϵ_{ij}^0 in (26.8.3) has components of $\epsilon_{33}^0 \neq \epsilon_{11}^0 = \epsilon_{22}^0$. It is known that $\epsilon_{33}^0 = 0.93$ and $\epsilon_{11}^0 = \epsilon_{22}^0 = -0.08$. If the concentration of the carbon atoms fluctuates as expressed by (26.8.3), we obtain the elastic energy per unit volume as

$$W^* = \tfrac{1}{2}\langle c^2\rangle\{H - G\} \tag{26.8.6}$$

from (3.14) and (25.2), where

$$\begin{aligned} H = \{&2C_{11}(\sigma_{11}^0)^2 - 4C_{12}\sigma_{11}^0\sigma_{33}^0 \\ &+ (C_{11} + C_{12})(\sigma_{33}^0)^2\}/\{(C_{11} + 2C_{12})(C_{11} - C_{12})\}, \end{aligned} \tag{26.8.7}$$

$$\begin{aligned} G = [&(\sigma_{11}^0)^2\{(C_{44})^2(\bar\xi_1^2 + \bar\xi_2^2) + 2C_{44}(C_{11} - C_{12} - 2C_{44})\bar\xi_1^2\bar\xi_2^2 \\ &+ 2(C_{11} - C_{44})(C_{11} - C_{12} - 2C_{44})\bar\xi_1^2\bar\xi_2^2\bar\xi_3^2 + C_{44}(C_{11} - C_{44})(\bar\xi_1^2 + \bar\xi_2^2)\bar\xi_3^2\} \\ &- 2\sigma_{11}^0\sigma_{33}^0(C_{12} + C_{44})\{C_{44}(\bar\xi_1^2 + \bar\xi_2^2)\bar\xi_3^2 + 2(C_{11} - C_{12} - 2C_{44})\bar\xi_1^2\bar\xi_2^2\bar\xi_3^2\} \\ &+ (\sigma_{33}^0)^2\{(C_{44})^2\bar\xi_3^2 + C_{44}(C_{11} - C_{44})(\bar\xi_1^2 + \bar\xi_2^2)\bar\xi_3^2 \\ &+ (C_{11} - C_{12} - 2C_{44})(C_{11} + C_{12})\bar\xi_1^2\bar\xi_2^2\bar\xi_3^2\}] \\ &/\{(C_{44})^2C_{11} + C_{44}(C_{11} - C_{12} - 2C_{44})(C_{11} + C_{12})(\bar\xi_1^2\bar\xi_2^2 + \bar\xi_2^2\bar\xi_3^2 + \bar\xi_3^2\bar\xi_1^2) \\ &+ (C_{11} - C_{12} - 2C_{44})^2(C_{11} + 2C_{12} + C_{44})\bar\xi_1^2\bar\xi_2^2\bar\xi_3^2\}, \end{aligned}$$

with

$$\begin{aligned} \sigma_{11}^0 &= (C_{11} + C_{12})\epsilon_{11}^0 + C_{12}\epsilon_{33}^0, \\ \sigma_{33}^0 &= 2C_{12}\epsilon_{11}^0 + C_{11}\epsilon_{33}^0. \end{aligned} \tag{26.8.8}$$

It can be shown that W^* has maxima at $\bar\xi_1^2 = 1$, $\bar\xi_2^2 = 1$ or $\bar\xi_3^2 = 1$ and minima at

$$\bar\xi_1^2 = \sin^2\theta,\ \bar\xi_2^2 = 0 \text{ and } \bar\xi_3^2 = \cos^2\theta \tag{26.8.9}$$

or

$$\bar\xi_1^2 = 0,\ \bar\xi_2^2 = \sin^2\theta \text{ and } \bar\xi_3^2 = \cos^2\theta \tag{26.8.10}$$

with θ given by

$$\theta = \cos^{-1}(\sqrt{A}), \tag{26.8.11}$$

where A is the root of

$$\begin{aligned}
&(C_{11} - C_{12} - 2C_{44})(C_{11} + C_{12})C_{44}\{(\sigma_{11}^0)^2 - (\sigma_{33}^0)^2\}A^2 \\
&- 2[C_{44}C_{11}\{(C_{11} - C_{44})(\sigma_{11}^0)^2 + (C_{11} - C_{44})(\sigma_{33}^0)^2 - 2(C_{12} + C_{44})\sigma_{11}^0\sigma_{33}^0\} \\
&- (C_{11} - C_{12} - 2C_{44})(C_{11} + C_{12})C_{44}(\sigma_{11}^0)^2]A + C_{44}^2C_{11}\{(\sigma_{11}^0)^2 + (\sigma_{22}^0)^2\} \\
&- C_{44}C_{11}\{(C_{11} - C_{44})(\sigma_{11}^0)^2 + (C_{11} - C_{44})(\sigma_{33}^0)^2 - 2(C_{12} + C_{44})\sigma_{11}^0\sigma_{33}^0\} \\
&+ (C_{11} - C_{12} - 2C_{44})(C_{11} + C_{12})C_{44}(\sigma_{11}^0)^2 = 0 \qquad (26.8.12) \\
&(0 \leq A \leq 1).
\end{aligned}$$

By putting $C_{11} = 2.37 \times 10^{11}$ N/m^2, $C_{12} = 1.41 \times 10^{11}$ N/m^2 and $C_{44} = 1.16 \times 10^{11}$ N/m^2, one obtains $\theta = 19.3°$. This example has been treated by Michiko Toyoshima (1980) for the discussion of the modulated structure observed in iron-carbon martensites (Toyoshima and Nagakura, 1979). Toyoshima has found that the direction of the fluctuation of carbon atom concentration is expressed either by [hol] or [ohl], and the angle θ depends on the amount of carbon in the martensite; as the carbon concentration goes to zero, θ approaches 22°.

Incoherent precipitates

The precipitates discussed in the last two subsections are coherent precipitates. When a second phase is nucleated, the interface between the precipitate and the matrix is classified as either coherent or incoherent. By coherent it is meant that there is a one-to-one correspondence between atoms at the interface; an incoherent interface refers to a lack of conservation of lattice sites across the interface. For an incoherent precipitate the stress is purely in hydrostatic fashion along the interface. The associated elastic strain energy has been obtained by Nabarro (1940) for the isotropic case, and by Kröner (1954) and Lee and Johnson (1978) for the anisotropic case.

Let us consider an ellipsoidal incoherent precipitate in an infinite anisotropic medium. The elastic moduli of the precipitate and the matrix are denoted by C^*_{ijkl} and C_{ijkl}, respectively. The equivalent inclusion method is employed. From (22.11) we have

$$C^*_{ijkl}(\epsilon_{kl} - \epsilon^p_{kl}) = C_{ijkl}(\epsilon_{kl} - \epsilon^p_{kl} - \epsilon^*_{kl}), \tag{26.9}$$

where no applied stress is considered. The stress components given by

(26.9) must be hydrostatic,

$$C_{ijkl}(\epsilon_{kl} - \epsilon^{p}_{kl} - \epsilon^{*}_{kl}) = \delta_{ij}\sigma. \tag{26.10}$$

Furthermore, the incoherent precipitate is characterized by giving

$$\epsilon^{p}_{kk} = \epsilon. \tag{26.11}$$

This means that for the incoherent precipitate the six components of ϵ^{p}_{ij} are not given, but only ϵ^{p}_{kk} is prescribed to be a constant, ϵ. By assuming the relation (22.12), the equivalent eigenstrain ϵ^{*}_{ij} can be expressed in terms of ϵ^{p}_{kl} from (26.9), as in the coherent case. When the result, ϵ^{*}_{ij}, is substituted into (26.10), the six components of ϵ^{p}_{ij} can be found to be linear functions of σ. The unknown σ is then determined from condition (26.11), while the associated elastic strain energy can be determined from (25.3).

Lee and Johnson (1978) have calculated W^*/V for the spheroidal precipitate ($a_1 = a_2 = a$, $a_3/a = \beta$), where the crystallographic directions of the precipitate and the matrix are parallel to the principal axes of the spheroid. Their result is

$$W^*/V = (\epsilon^2/2)K^*c^0/(K^* + c^0), \tag{26.12}$$

where $K^* = (C^{*}_{11} + 2C^{*}_{12})/3$ is the bulk modulus of the precipitate, and

$$c^0 = \frac{K\{(S_{1111} + S_{1122} - 1)(1 - S_{3333}) + 2S_{1133}S_{3311}\}}{(S_{1111} + S_{1122} - 1)(S_{3333} - 2/3) + \frac{2}{3}(S_{3333} - S_{1133} - S_{3311} - 1) - 2S_{1133}S_{3311}} \tag{26.13}$$

with $K = (C_{11} + 2C_{12})/3$. C^{*}_{ij} and C_{ij} are the elastic moduli of cubic crystals. The Eshelby tensor S_{ijkl} has been given by (17.27) and (17.28), where $\beta = 1/\rho$. Kröner (1954) has denoted by c^0 the compression modulus. Figure 26.5 shows the numerical results of $W^*/VC_{44}\epsilon^2$ calculated by Lee and Johnson (1978). The elastic strain energy reaches its maximum when the precipitate is a sphere, vanishing as the shape becomes a thin, flat ellipsoid.

The incoherent state of the precipitates characterized by (26.10) and (26.11) can be interpreted as a free energy minimum state of coherent precipitates with given ϵ^{p}_{ij}, when additional inelastic deformation $\Delta\epsilon^{p}_{ij}$ is allowable. Consider a coherent precipitate with eigenstrain ϵ^{p}_{ij}. The corresponding stress field is denoted by σ_{ij} and is calculated from (26.9). Let us consider a relaxation which occurs inside or at the surface of the coherent precipitate. Dislocations punching out of a precipitate represent the realization process occurring outside, which may be caused by actual plastic deformation, or by mass diffusion along the

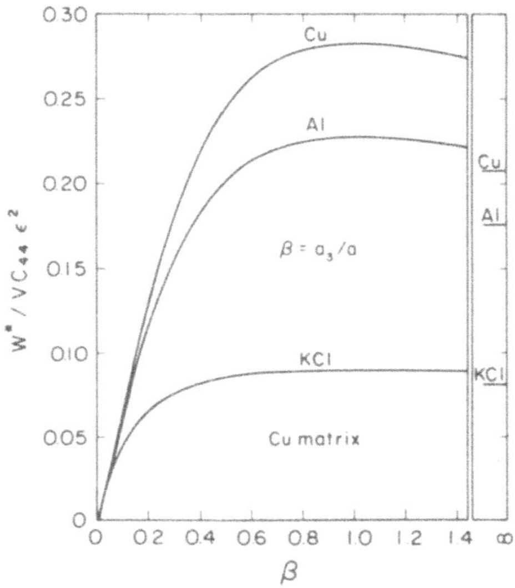

Fig. 26.5. Normalized elastic strain energy, per unit volume of the incoherent precipitate vs. aspect ratio for the Cu precipitate, Al precipitate and KCl precipitate in a Cu matrix.

matrix-precipitate interface, or through the bulk of the precipitate as suggested by Christian (1965). This inelastic deformation due to relaxation is denoted by $\Delta\epsilon_{ij}^{p}$. Either case of diffusion or plastic deformation does not cause volume changes,

$$\Delta\epsilon_{ii}^{p} = 0. \tag{26.13.1}$$

The total stress in the precipitate is $\sigma_{ij} + \Delta\sigma_{ij}$ where $\Delta\sigma_{ij}$ is the eigenstress due to $\Delta\epsilon_{ij}^{p}$. The associate elastic strain energy W^* is

$$W^* = -(1/2)V(\sigma_{ij} + \Delta\sigma_{ij})(\epsilon_{ij}^{p} + \Delta\epsilon_{ij}^{p}) \tag{26.13.2}$$

from (25.3).

The minimum principle of free energy W^* is introduced. $\Delta\epsilon_{ij}^{p}$ is created so that W^* becomes minimum for a given ϵ_{ij}^{p}. This extreme condition leads to

$$(\sigma_{ij} + \Delta\sigma_{ij})\delta(\Delta\epsilon_{ij}^{p}) = 0, \tag{26.13.3}$$

since $\sigma_{ij} + \Delta\sigma_{ij}$ is a linear function of $\epsilon_{ij}^{p} + \Delta\epsilon_{ij}^{p}$, which is the solution of (26.9) when ϵ_{ij}^{p} is replaced by $\epsilon_{ij}^{p} + \Delta\epsilon_{ij}^{p}$. Condition (26.13.3) is satisfied when

$$\sigma_{ij} + \Delta\sigma_{ij} = \delta_{ij}\sigma, \tag{26.13.4}$$

because of (26.13.1), and W^* becomes

$$W^* = -(1/2)V\sigma\epsilon^p_{ii}. \tag{26.13.5}$$

Condition (26.13.4) is no more than (26.10); that is, the situation considered above is identical to that of an incoherent precipitate.

Martensitic transformation

One of the important applications of inclusion theory is in the area of martensitic transformation. Although various aspects of martensitic transformation have been investigated (e.g., Bilby and Christian 1956, Christian 1965, Wayman 1964, etc.), not until quite recently have the details of the application of the inclusion theory appeared in the literature.

The problem to be discussed here is the elastic field and elastic strain energy associated with the martensitic transformation. We assume that the orientation of martensite blades relative to the crystalline directions is determined by the principle of minimum elastic strain energy.

Mura, Mori, and Kato (1976) have considered a thin disk-shaped martensite blade ($a_1 = a_2 = a$, $\beta = a_3/a \ll 1$) as shown in Fig. 26.6. The **x** coordinate system is chosen parallel to certain crystallographic directions of the matrix, while the **x**′ coordinate system coincides with the principal axes of the ellipsoid. In many ferrous alloys, martensites take the form of a thin flat ellipsoid which consists of alternative twins, as shown schematically in Fig. 26.6. Each twin can be considered to be formed by the respective transformation strain (eigenstrains). In the case of martensitic transformation from a face-centered cubic to a body

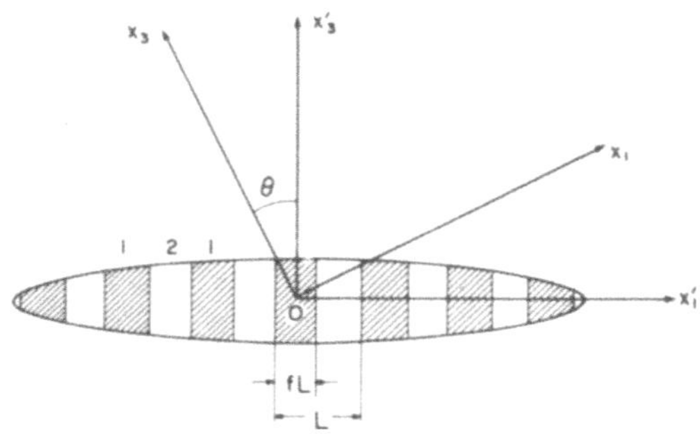

Fig. 26.6. Disk-shaped martensite blade, $a_1 = a_2 = a$, $\beta = a_3/a \ll 1$.

centered tetragonal, the eigenstrain components in the **x** coordinate system are given by

$$\epsilon_{ij}^{p} = \begin{pmatrix} \epsilon_1 & 0 & 0 \\ 0 & \epsilon_2 & 0 \\ 0 & 0 & \epsilon_1 \end{pmatrix} \quad \text{in region 1,}$$

$$= \begin{pmatrix} \epsilon_2 & 0 & 0 \\ 0 & \epsilon_1 & 0 \\ 0 & 0 & \epsilon_1 \end{pmatrix} \quad \text{in region 2,} \tag{26.14}$$

and (see e.g. Wechsler, Lieberman and Read 1953)

$$\epsilon_1 = \sqrt{2}b/b_0 - 1, \quad \epsilon_2 = c/b_0 - 1, \tag{26.15}$$

where b_0 is the lattice constant for f.c.c. phase and b and c are those for the b.c.t. phase. Region 1 is denoted by hatching in Fig. 26.6.

The Fourier series expression of (26.14) is

$$\epsilon_{ij}^{p} = \bar{\epsilon}_{ij}^{p}(0) + \sum_{n=1}^{\infty} \bar{\epsilon}_{ij}^{p}(n) \cos(2\pi n x_1'/L), \tag{26.16}$$

where

$$\bar{\epsilon}_{ij}^{p}(0) = (1/L) \int_{-L/2}^{L/2} \epsilon_{ij}^{p} dx_1',$$
$$\bar{\epsilon}_{ij}^{p}(n) = (2/L) \int_{-L/2}^{L/2} \epsilon_{ij}^{p} \cos(2\pi n x_1'/L) dx_1'. \tag{26.17}$$

L is the wave-length of the alternating twins. $\bar{\epsilon}_{ij}^{p}(0)$ is the mean value of ϵ_{ij}^{p}. If the volume ratio of regions 1 and 2 is denoted by f, it holds that

$$\bar{\epsilon}_{11}^{p}(0) = f\epsilon_1 + (1-f)\epsilon_2,$$
$$\bar{\epsilon}_{22}^{p}(0) = f\epsilon_2 + (1-f)\epsilon_1, \tag{26.18}$$
$$\bar{\epsilon}_{33}^{p}(0) = f\epsilon_1 + (1-f)\epsilon_1 = \epsilon_1.$$

The eigenstrain components in the **x**′ coordinate system are given by

$$\epsilon_{i'j'}^{p} = a_{ki'} a_{lj'} \epsilon_{kl}^{p}, \tag{26.19}$$

where $a_{ki'}$ is the direction cosine between the x_k- and the x_i'-axes.

As a first approximation, let us start with the case where $\epsilon_{ij}^{p} = \bar{\epsilon}_{ij}^{p}(0)$. This is similar to the approximation adopted by Shibata and Ono (1975, 1977) and Easterling and Thölén (1976). Equation (26.7) can be used for the elastic field. When

$$\epsilon^{p}_{1'1'} = \epsilon^{p}_{2'2'} = \epsilon^{p}_{1'2'} = 0, \tag{26.20}$$

the stress components vanish everywhere. Therefore, the associated elastic strain energy becomes zero. Solving (26.18) with (26.19), it is found that the condition (26.20) can be satisfied if

$$a_{33'} = \cos\theta = (-\epsilon_1/\epsilon_2)^{1/2}, \quad a_{13'} = (1+\epsilon_1/\epsilon_2)^{1/2}, \quad a_{23'} = 0, \tag{26.21}$$

and

$$f = \epsilon_1/(\epsilon_1 - \epsilon_2), \tag{26.21'}$$

where θ is denoted in Fig. 26.6.

In the phenomenological crystallographic theory of martensitic transformation, the deformation in the martensite is assumed to be uniform, corresponding to the consideration of only the first term in (26.16). Furthermore, the phenomenological theory assumes that the magnitude and the angle between any two vectors lying parallel to the flat interface do not change (Wayman 1968). This is equivalent to (26.20). The postulates in the phenomenological crystallographic theory are interpreted in such a manner that the elastic strain energy vanishes when the uniform transformation is assumed. Thus the prediction of f and the oreintation of the martensite plate based on the phenomenological theory should agree with (26.21) and (26.21'). In fact, this is the case when $|\epsilon_1|, |\epsilon_2| \ll 1$ (Wechsler et al. 1953). The last restriction is that of the small deformation theory, while the calculations in the phenomenological theory are based on the finite deformation theory.

When we consider the effect of higher order terms in (26.16), the equivalent inclusion condition (22.11) is written as

$$\sigma_{ij} = C^{*}_{ijkl}(\epsilon_{kl} - \epsilon^{p}_{kl}) = C_{ijkl}(\epsilon_{kl} - \epsilon^{p}_{kl} - \epsilon^{*}_{kl}), \tag{26.22}$$

where ϵ^{*}_{ij} is the equivalent eigenstrain. The Fourier series expression is assumed for ϵ^{**}_{ij},

$$\epsilon^{**}_{ij} = \bar{\epsilon}^{**}_{ij}(0) + \sum_{n=1}^{\infty} \bar{\epsilon}^{**}_{ij}(n)\cos(2\pi n x_1/L), \tag{26.23}$$

where

$$\epsilon^{**}_{ij} = \epsilon^{p}_{ij} + \epsilon^{*}_{ij}. \tag{26.24}$$

From (17.67), condition (26.22) is written as

$$C^{*}_{ijkl}\Big\{C_{pqst}\bar{\epsilon}^{**}_{st}(0)\bar{\xi}_l\bar{\xi}_q N_{kp}(\bar{\xi})D^{-1}(\bar{\xi}) - \bar{\epsilon}^{p}_{kl}(0)$$

$$+\sum_{n=1}^{\infty} C_{pqst}\bar{\epsilon}^{**}_{st}(n)\bar{\xi}_l\bar{\xi}_q J_0(2\pi an/L)N_{kp}(\bar{\boldsymbol{\xi}})D^{-1}(\bar{\boldsymbol{\xi}})$$

$$\left.-\sum_{n=1}^{\infty} \bar{\epsilon}^{p}_{kl}(n)\cos(2\pi n x'_1/L)\right\}$$

$$= C_{ijkl}\left\{C_{pqst}\bar{\epsilon}^{**}_{st}(0)\bar{\xi}_l\bar{\xi}_q N_{kp}(\bar{\boldsymbol{\xi}})D^{-1}(\bar{\boldsymbol{\xi}}) - \bar{\epsilon}^{**}_{kl}(0)\right.$$

$$+\sum_{n=1}^{\infty} C_{pqst}\bar{\epsilon}^{**}_{st}(n)\bar{\xi}_l\bar{\xi}_q J_0(2\pi an/L)N_{kp}(\bar{\boldsymbol{\xi}})D^{-1}(\bar{\boldsymbol{\xi}})$$

$$\left.-\sum_{n=1}^{\infty} \bar{\epsilon}^{**}_{kl}(n)\cos(2\pi n x'_1/L)\right\}, \tag{26.25}$$

where $\bar{\boldsymbol{\xi}}$ is the unit vector along the x'_3-direction. When the orientation of the flat ellipsoid and f are chosen as defined by (26.21) and (26.21′), the first two terms on both sides of (26.25) disappear by taking

$$\bar{\epsilon}^{**}_{ij}(0) = \bar{\epsilon}^{p}_{ij}(0). \tag{26.26}$$

Further simplification of (26.25) can be made when the terms containing J_0 are neglected. Observation by electron microscope has shown that $a \approx 10\ \mu$m and $L \approx 100$ Å (e.g., Tamura et al. 1946). Thus, $J_0(2\pi an/L) \approx J_0(\infty) \approx 0$. Finally, equation (26.25) becomes

$$-C^{*}_{ijkl}\sum_{n=1}^{\infty} \bar{\epsilon}^{p}_{kl}(n)\cos(2\pi n x'_1/L) = -C_{ijkl}\sum_{n=1}^{\infty} \bar{\epsilon}^{**}_{kl}(n)\cos(2\pi n x'_1/L) \tag{26.27}$$

which is also equal to σ_{ij}. Therefore, from (26.16),

$$\sigma_{ij} = -C^{*}_{ijkl}\{\epsilon^{p}_{kl} - \bar{\epsilon}^{p}_{kl}(0)\}. \tag{26.28}$$

The elastic strain energy is

$$W^* = -(1/2)\int_V \sigma_{ij}\epsilon^{p}_{ij}\,dD, \tag{26.29}$$

where V is the volume of the flat ellipsoid. When (26.14) and (26.18) are used in the above two formulae, we have

$$W^*/V = f(1-f)(C^{*}_{11} - C^{*}_{12})(\epsilon_1 - \epsilon_2)^2, \tag{26.30}$$

where f is given by (26.21′). This is surprisingly large, presumably because the very thin flat ellipsoid is assumed in the calculation.

Suezawa (1978) has used the method of Khachaturyan (1967) and evaluated the strain energy of martensitic transformation in terms of L (see also a review paper by Christian 1980).

Shibata and Ono (1975, 1977) have considered the martensitic transformation of titanium. They assume elastic isotropy for both the matrix and the martensite. A homogeneous spheroidal inclusion ($a_1 = a_3 = a$, $k = a_2/a$) is subjected to a uniform transformation strain ϵ^P_{ij} (eigenstrain). The elastic strain energy is expressed by (13.20′), where ϵ^*_{ij} is replaced by $\epsilon^P_{i'j'}$ (see Fig. 26.7).

The transformation of the b.c.c. phase of Ti and Zr alloys during quenching is martensitic and results in at least four structures: h.c.p., f.c.c., orthorhombic and f.c.c. orthorhombic (Williams 1973). The h.c.p. martensite is most common and occurs in Ti, Zr, Ti–Mo alloys, Zr–Nb alloys and others. In pure metals and other dilute alloys, the lath martensite forms, whereas the plate martensite occurs with increasing solute content. The orientation relationship of the plate martensite to the matrix has been established by Burgers (1934),

$$(011)_B \| (0001)_H,\qquad [1\bar{1}1]_B \| [11\bar{2}0]_H. \tag{26.31}$$

By taking the x_1, x_2, x_3-directions as $[\bar{1}1\bar{1}]_B$, $[\bar{2}\bar{1}1]_B$ and $[011]_B$, respectively, Shibata and Ono (1977) have used

$$\epsilon^P_{ij} = \begin{pmatrix} 0.0381 & -0.0884 & 0 \\ -0.0884 & -0.0434 & 0 \\ 0 & 0 & 0.0088 \end{pmatrix}, \tag{26.32}$$

where the ratios of lattice parameters in b.c.c. and h.c.p. phases are taken as $a_H/a_B = 0.899$ and $c_H/a_H = 1.587$. Denoting the principal axes of the spheroid by x'_1, x'_2 and x'_3, we have the relation (26.19). The rotated coordinate system $\mathbf{x}'$ is obtained by rotating first about the x_3-axis by θ, followed by the rotation about the x'_1-axis by ψ (see Fig. 26.7). For a given ϵ^P_{ij}, the elastic strain energy per unit volume of martensite becomes a function of θ, ψ and $k = a_2/a$. Shibata and Ono have calculated θ, ψ

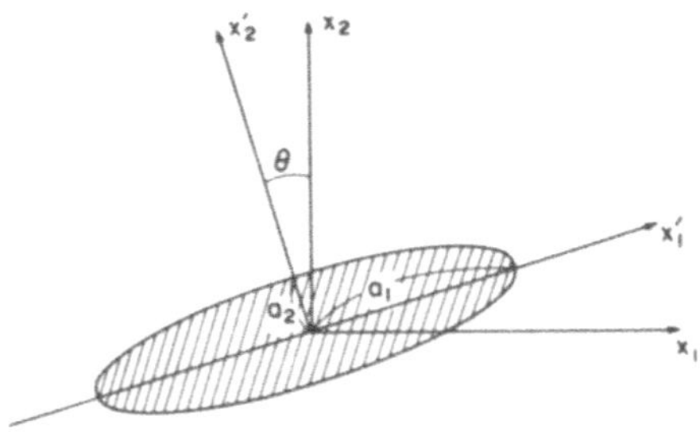

Fig. 26.7. Spheroidal martensite $a_1 = a_3 = a$, $\beta = a_2/a$.

and k, under which the elastic strain energy attains its minimum value. Their results indicate that the elastic strain energy is minimized when the morphology of h.c.p. Ti martensite is a thin disk shape lying on a plane close to $(11X)_B$, where X is equal to $1{\cdot}2 \sim 1{\cdot}3$. This habit plane agrees with experimental observations.

A spherical Fe precipitate, which is formed in a Cu matrix and is f.c.c. before martensitic transformation, becomes b.c.c. with the banded structure of alternating twins after transformation (Easterling and Miekk-oja, 1967). The elastic strain energy of an Fe particle with the banded structure has recently been evaluated by Kinsman et al. (1975) using an unrealistic approximation based on a simplified model. The physical situation they have envisaged is a spherical region undergoing martensitic transformation, with the upper and the lower halves shearing in anti-parallel directions, as shown schematically in Fig. 26.8(a). Thus, the surface and the eigenstrain of the transformed particle Ω are

$$(x_1^2 + x_2^2 + x_3^2)/a^2 = 1,$$
$$\epsilon^p_{31}(\mathbf{x}) = \pm(1/2)S, \quad x_3 \gtrless 0 \text{ in } \Omega, \tag{26.33}$$

where a is the radius of the spherical particle and S is a constant to be determined by the crystallographic information. However, Kinsman et al. (1975) approximated (26.33) with the eigenstrain

$$\epsilon^p_{31}(\mathbf{x}) = tx_3/a \text{ in } \Omega, \tag{26.34}$$

with $t = S$, in order to utilize Asaro and Barnett's work (1975). Mura, Mori and Kato (1976) have calculated the elastic strain energy associated

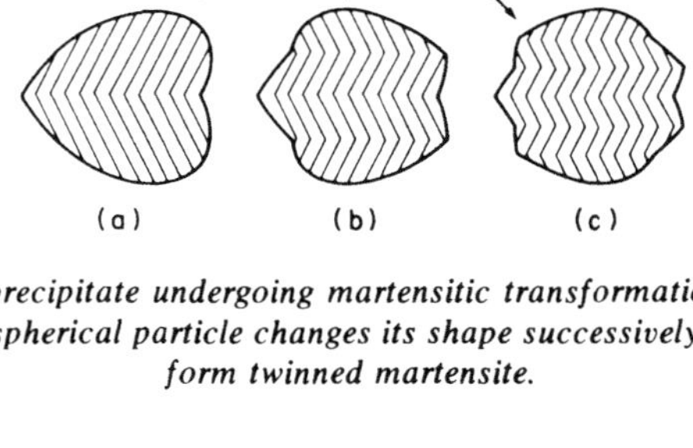

Fig. 26.8. Spherical precipitate undergoing martensitic transformation. When matrix constraint is absent, the spherical particle changes its shape successively to (a), (b) and (c) to form twinned martensite.

with situation (26.33) as follows. The Fourier series expression for the eigenstrain (26.33) is

$$\epsilon^p_{31}(\mathbf{x}) = (2S/\pi)\sum_{n=1}^{\infty}(2n-1)^{-1}\sin\{(2n-1)\pi x_3/a\}. \tag{26.35}$$

For simplicity, assume that the spherical precipitate and the matrix have the same isotropic elastic constants. From (17.56), the only non-vanishing stress component σ_{31} pertinent to the strain energy evaluation, is

$$\sigma_{31}(\mathbf{x}) = \sum_{n=1}^{\infty}\bar{\sigma}_{31}(\mathbf{x}, n), \tag{26.36}$$

with

$$\begin{aligned}\bar{\sigma}_{31}(\mathbf{x}, n) = (\mu S a^3/\pi C_3)\int_{S^2}&[J_0(C_0R)\sin\{(\mathbf{x}\cdot\bar{\boldsymbol{\xi}})C_3\bar{\xi}_3/a\}\\&+2J_0(C_0R)C_3\bar{\xi}_3(\mathbf{x}\cdot\bar{\boldsymbol{\xi}})\cos\{(\mathbf{x}\cdot\bar{\boldsymbol{\xi}})C_3\bar{\xi}_3/a\}/a\\&+J_1(C_0R)\{(\mathbf{x}\cdot\bar{\boldsymbol{\xi}})^2C_0/(a^2R)+C_3^2\bar{\xi}_3^2R/C_0\}\sin\{(\mathbf{x}\cdot\bar{\boldsymbol{\xi}})C_3\bar{\xi}_3/a\}]\\&\times(\bar{\xi}_1^2+\bar{\xi}_3^2-3\bar{\xi}_1^2\bar{\xi}_3^2)dS(\bar{\boldsymbol{\xi}})-(4\mu S/C_3)\sin(C_3x_3/a),\end{aligned} \tag{26.37}$$

where $C_3 = (2n-1)\pi$, μ is the shear modulus, and Poisson's ratio is taken to be 1/3. Since the analytical integration involved in (26.37) is difficult, the stress and the elastic strain energy to be determined from (26.29) have been evaluated numerically on a computer. The result is

$$W^*/V = 0.182\,\mu S^2. \tag{26.38}$$

The above value is a little larger than that of Kinsman et al. (1975).

Similar calculations have been performed for the cases where

$$\epsilon^p_{31}(\mathbf{x}) = \left\{\begin{array}{ll}\tfrac{1}{2}S, & 2Na/M < x_3 < (2N+1)a/M\\ -\tfrac{1}{2}S, & (2N+1)a/M < x_3 < (2N+2)a/M\end{array}\right\} \text{ in } \Omega, \tag{26.39}$$

with $M = 2, 3$ (see Figs. 26.8(b), (c)), (the case $M = 1$ corresponds to (26.33)). In (26.39) N takes on integer values. The calculated elastic strain energies are plotted against M in Fig. 26.9, which also includes results for the case $M = 0$, corresponding to uniform shear and the value of Kinsman et al. (1975) for the case $M = 1$. It can be seen that as M increases, the elastic strain energy decreases. Thus, as far as the elastic strain energy is concerned, the alternating twin structure with the shorter interval is favored. This conclusion agrees with intuition, based on an examination of Fig. 26.8; as the interval between the alternating twins decreases, the shape change of the particle caused by the transformation becomes smaller as a whole.

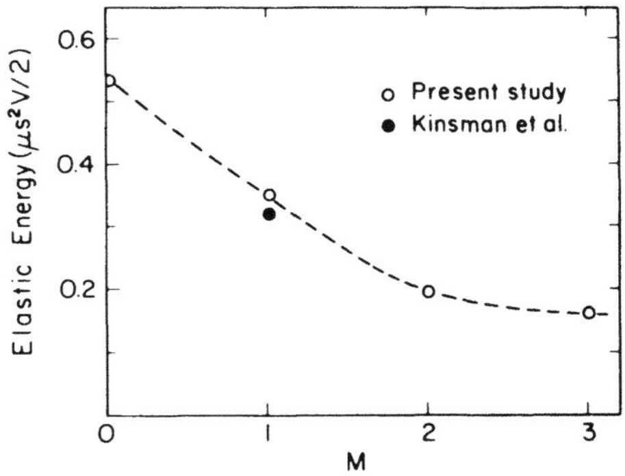

Fig. 26.9. Elastic strain energy of a martensite particle with M pairs of twins in Fig. 26.8.

Stress orienting precipitation

As a typical example, we discuss the precipitation of α'' phases, $Fe_{16}N_2$, from a supersaturated Fe–N alloy. The eigenstrain (misfit) of this precipitate has the components

$$\epsilon_{11}^{p} = \epsilon_{22}^{p} = -0.0023, \ \epsilon_{33}^{p} = 0.0971 \tag{26.40}$$

(Jack, 1951). There are two more types of the crystallographically equivalent $Fe_{16}N_2$, eigenstrains of which are obtained by the cyclic permutation of the index in (26.40). The type characterized by (26.40) is called the (001) variant, since the precipitates having (26.40) are formed in disk shapes, approximately parallel to the (001) plane. This can be understood by noting that $|\epsilon_{11}^{p}| = |\epsilon_{22}^{p}| \ll |\epsilon_{33}^{p}|$ and that the elastic energy of the disk-shaped precipitate is determined by the eigenstrains within the disk plane (see (17.42), (26.7) and the related discussions).

Statistically, the three types of the variants, (100), (010), and (001), are formed in an equal amount, when there is no external or internal stress, except the stress due to the precipitates. However, when an external stress is applied, a particular type of the variant is preferentially found; e.g., the presence of the tensile stress along [001] favors the formation of the (001) variants against others (Nakada, Leslie and Chray, 1967). This phenomenon can be interpreted by the interaction between the precipitates and the applied stress; the interaction energy (25.24) is far smaller for the (001) variant than for the (100) or (010) variant, when the tensile stress is applied along the [001]. This effect, known as stress orienting effect, is remarkable, and it has been shown that as the tensile stress increases, not only does the relative number but

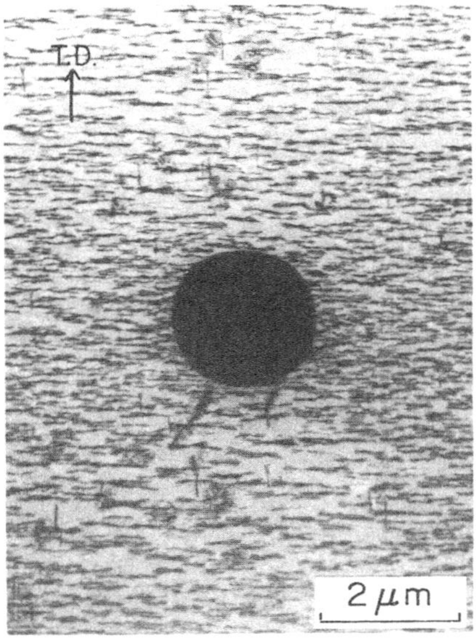

Fig. 26.10. Precipitates around a non-metalic inclusion subjected to a tensile stress (after Sato et al. 1979).

also the absolute number of the (001) variants become larger (Tanaka, Sato and Mori, 1978).

Sato, Mori and Tanaka (1979) examined the precipitation orientations around a non-metallic inclusion subjected to a tensile stress as shown in Fig. 26.10. In the figure, the tensile stress $\sigma_{33}^0 = 98\ \mathrm{MN/m^2}$ has been applied along the [001] (vertical direction) during the precipitation of $Fe_{16}N_2$ in a Fe–N alloy single crystal. It is noted that most of the precipitates are parallel to the (001) plane. With a close examination it can be seen that the density of the precipitates is larger in the close neighborhood of the inclusion (black contrasted spheroid) along the [010] (horizontal direction) and is small at the top or the bottom of the inclusion, the [001] direction. Sato et al. have calculated the stress outside the inclusion from (18.6) together with the method of the equivalent inclusion in Section 22. The elastic constants are $C_{1111} = 2.37 \times 10^{11}\ \mathrm{N/m^2}$, $C_{1122} = 1.41 \times 10^{11}\ \mathrm{N/m^2}$ and $C_{1212} = 1.16 \times 10^{11}\ \mathrm{N/m^2}$ for Fe. Since the non-metallic inclusion is identified as a slug particle, the isotropic elastic constants are assigned to the inclusion; the shear modulus $2.9 \times 10^{10}\ \mathrm{N/m^2}$ and the Poisson ration 0.22, for typical glass. Figure 26.11 shows the equi-contour lines of the stress disturbance, the

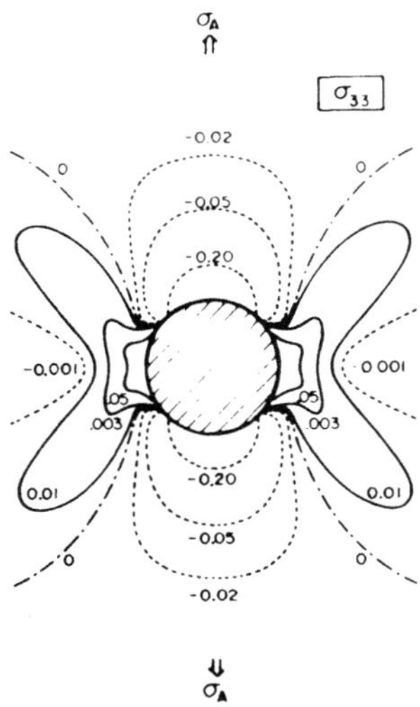

Fig. 26.11. Equi-contour lines of the stress disturbance around an inclusion (after Sato et al. 1979).

disturbance is expressed in unit of σ_A. In agreement with Fig. 26.10, the stress at the [010] edges is significantly large while that at the [001] edges is smaller than σ_A.

5

Cracks

An elliptical crack is a special case of an ellipsoidal void where one of the principal axes of the ellipsoid becomes vanishingly small. The Griffith fracture criterion and stress intensity factors of cracks are the primary topics discussed in this chapter.

27. Critical stresses of cracks in isotropic media

Penny-shaped cracks

Cracks can be considered as a special case of voids when one of the principal axes, say a_3, is infinitesimally small compared to the others. The simplest example, a penny-shaped crack ($a_1 = a_2 = a$, $a_3 = 0$), is considered first. Assume that an infinitely extended isotropic material containing a crack Ω is subjected to a uniform applied stress σ^0_{33} at infinity. The equivalent inclusion method is employed for the analysis.

The crack is simulated by a penny-shaped inclusion with an eigenstrain ϵ^*_{ij}. The corresponding eigenstress σ_{ij} is given by (11.26). The necessary conditions for the application of the equivalent inclusion method are

$$\left.\begin{aligned} &\sigma^0_{33} + \sigma_{33} = 0 \\ &\sigma_{31} = \sigma_{32} = 0 \end{aligned}\right\} \quad \text{in } \Omega \tag{27.1}$$

which correspond to traction-free conditions on the crack surface. It can be seen in (11.26) that no finite values of ϵ^*_{ij} satisfy the first condition in (27.1) for $a_3 \to 0$. Therefore, we allow ϵ^*_{33} to be unbounded in such a manner that

$$\lim_{a_3 \to 0} a_3 \epsilon^*_{33} = \text{finite} = \epsilon^*. \tag{27.2}$$

All other components of ϵ_{ij}^* are taken to be zero. Thus,

$$\sigma_{33} = -\frac{\mu\pi}{2(1-\nu)}\frac{\epsilon^*}{a},$$
$$\sigma_{31} = \sigma_{32} = 0. \tag{27.3}$$

Conditions (27.1) are satisfied if ϵ^* is chosen as

$$\epsilon^* = \frac{2(1-\nu)a}{\mu\pi}\sigma_{33}^0. \tag{27.4}$$

The interaction energy (the decrease in the total potential energy of the system due to the crack) is given by (25.18),

$$\Delta W = -(1/2)(4\pi/3)a_1a_2a_3\sigma_{33}^0\epsilon_{33}^*$$
$$= -(2\pi/3)a^2\sigma_{33}^0\epsilon^* = -\frac{4(1-\nu)a^3}{3\mu}(\sigma_{33}^0)^2. \tag{27.5}$$

The Griffith (1921) fracture criterion is

$$\frac{\partial}{\partial a}(\Delta W + 2\pi a^2\gamma) = 0, \tag{27.6}$$

where $2\pi a^2$ is the total surface of the crack and γ is the surface energy per unit area. Condition (27.6) leads to the critical stress

$$\sigma_{33}^0 = \left\{\frac{\pi\mu\gamma}{(1-\nu)a}\right\}^{1/2}. \tag{27.7}$$

A given penny-shaped crack with radius a becomes unstable (starts to grow) when the applied stress σ_{33}^0 reaches the critical value given by (27.7).

A penny-shaped crack under a uniform shear σ_{31}^0 at infinity can be treated in a similar manner. All components of ϵ_{ij}^* except ϵ_{31}^* are taken to be zero in (11.27), where the unbounded ϵ_{31}^* is such that

$$\lim_{a_3\to 0} a_3\epsilon_{31}^* = \text{finite} = \epsilon^*. \tag{27.8}$$

We then, have,

$$\sigma_{31} = -\frac{\mu\pi(2-\nu)}{2(1-\nu)}\frac{\epsilon^*}{a}. \tag{27.9}$$

The traction-free conditions on the crack surface are

$$\begin{aligned} &\sigma^0_{31} + \sigma_{31} = 0, \\ &\sigma_{33} = \sigma_{32} = 0, \end{aligned} \tag{27.10}$$

which are satisfied by

$$\epsilon^* = \frac{2(1-\nu)a}{\mu\pi(2-\nu)}\sigma^0_{31}. \tag{27.11}$$

The interaction energy thus becomes

$$\Delta W = -(4\pi/3)a^2\sigma^0_{31}\epsilon^* = -\frac{8(1-\nu)a^3}{3\mu(2-\nu)}(\sigma^0_{31})^2, \tag{27.12}$$

and condition (27.6) leads to the critical stress

$$\sigma^0_{31} = \left\{\frac{\pi\mu(2-\nu)\gamma}{2(1-\nu)a}\right\}^{1/2}. \tag{27.13}$$

Results (27.7) and (27.13) have been obtained by Sack (1946) and Sneddon (1946). The interaction energy and the crack opening displacement for a pair of coplaner penny-shaped cracks are given by Fu and Keer (1969) for the shear case and by Fu (1973) for thermal cracks.

Slit-like cracks

When one of the principal axes of an ellipsoidal void, say a_3, becomes infinite and the minor axis a_2 becomes infinitely small, the ellipsoid becomes a slit-like crack. When the unbounded solid which contains this crack is subjected to a uniform tension σ^0_{22} at infinity, the stress disturbance due to the crack can be simulated by (11.22a). In (11.22a) we choose all $\epsilon^*_{ij} = 0$ except the unbounded ϵ^*_{22} which must satisfy

$$a_2\epsilon^*_{22} = \text{finite} = \epsilon^* \quad \text{for } a_2 \to 0. \tag{27.14}$$

Then, we have

$$\sigma_{22} = -\frac{\mu}{1-\nu}\frac{\epsilon^*}{a_1}. \tag{27.15}$$

The traction-free conditions on the crack surface are

$$\begin{aligned} &\sigma^0_{22} + \sigma_{22} = 0, \\ &\sigma_{21} = \sigma_{23} = 0, \end{aligned} \tag{27.16}$$

which are satisfied when

$$\epsilon^* = (1-\nu)a_1\sigma_{22}^0/\mu. \tag{27.17}$$

The interaction energy per unit length of the crack in the x_3-direction becomes

$$\begin{aligned}\Delta W &= -(1/2)\pi a_1 a_2 \sigma_{22}^0 \epsilon_{22}^* = -(1/2)\pi a_1 \sigma_{22}^0 \epsilon^* \\ &= -\pi(1-\nu)a_1^2(\sigma_{22}^0)^2/2\mu,\end{aligned} \tag{27.18}$$

and the Griffith fracture criterion

$$\frac{\partial}{\partial a_1}(\Delta W + 4a_1\gamma) = 0 \tag{27.19}$$

leads to

$$\sigma_{22}^0 = \left\{\frac{4\mu\gamma}{\pi(1-\nu)a_1}\right\}^{1/2}. \tag{27.20}$$

When the body containing this slit crack is subjected to a uniform applied stress σ_{12}^0 at infinity, a similar analysis leads to

$$\sigma_{12}^0 = \left\{\frac{4\mu\gamma}{\pi(1-\nu)a_1}\right\}^{1/2}. \tag{27.21}$$

The above result, (27.20) and (27.21), is for the plane strain case. The corresponding result for the plane stress is obtained by replacing $E/(1-\nu^2)$ by E and $\nu/(1-\nu)$ by ν. Then, we have the critical stresses

$$\sigma_{22}^0 = \left(\frac{2E\gamma}{\pi a_1}\right)^{1/2}, \quad \sigma_{12}^0 = \left(\frac{2E\gamma}{\pi a_1}\right)^{1/2}. \tag{27.22}$$

The above results have been obtained by Griffith (1921).

Flat ellipsoidal cracks

When a crack is a flat ellipsoid ($a_1 > a_2$, $a_3 = 0$), the stress disturbance can be simulated by (11.26) or (11.27) depending on the applied stress.

For simple tension σ_{33}^0, we have

$$\begin{aligned}&\lim_{a_3\to 0} a_3\epsilon_{33}^* = \sigma_{33}^0 a_2(1-\nu)/\mu E(k), \\ &\qquad \Delta W = -(2/3)\pi(\sigma_{33}^0)^2 a_1 a_2^2(1-\nu)/\mu E(k),\end{aligned} \tag{27.23}$$

where $k^2 = 1 - a_2^2/a_1^2$, and $E(k)$, and $F(k)$ in (27.24) are the complete elliptic integrals defined by (11.25).

For simple shear σ_{31}^0, we have

$$\lim_{a_3\to 0} a_3\epsilon_{13}^* = \sigma_{31}^0/2\mu\left[\frac{\nu}{1-\nu}\,\frac{a_2\{F(k)-E(k)\}}{a_1^2-a_2^2}+\frac{E(k)}{a_2}\right],$$
$$\Delta W = -(2/3)\pi(\sigma_{31}^0)^2 a_1 a_2/\mu\left[\frac{\nu}{1-\nu}\,\frac{a_2\{F(k)-E(k)\}}{a_1^2-a_2^2}+\frac{E(k)}{a_2}\right]. \tag{27.24}$$

The expressions for ΔW in (27.23) and (27.24) have been obtained by Kassir and Sih (1967).

The critical stress can be obtained from

$$\frac{\partial}{\partial a_1}(\Delta W + 2\pi a_1 a_2 \gamma) = 0,$$
$$\frac{\partial}{\partial a_2}(\Delta W + 2\pi a_1 a_2 \gamma) = 0, \tag{27.25}$$

if a_1 and a_2 are regarded as independent variables. The following formulae are useful for calculation of (27.25):

$$dE(k)/dk = \{E(k) - F(k)\}/k,$$
$$dF(k)/dk = \{E(k)/(1-k^2) - F(k)\}/k, \tag{27.26}$$
$$\partial k/\partial a_1 = a_2^2/a_1^3 k,\ \partial k/\partial a_2 = -a_2/a_1^2 k.$$

For simple tension, the first equation in (27.25) yields

$$\sigma_{33}^0 = \left[\frac{3\mu\gamma k^2 E^2(k)}{(1-\nu)a_2\{(1-2(k')^2)E(k)+(k')^2F(k)\}}\right]^{1/2}, \tag{27.27}$$

and the second equation gives

$$\sigma_{33}^0 = \left[\frac{3\mu\gamma k^2 E^2(k)}{(1-\nu)a_2\{(2-(k')^2)E(k)-(k')^2F(k)\}}\right]^{1/2}, \tag{27.27.1}$$

where

$$k^2 = 1 - a_2^2/a_1^2,\ (k')^2 = a_2^2/a_1^2. \tag{27.27.2}$$

The solution of the simultaneous equations (27.25) leads to $a_1 = a_2$ and σ_{33}^0 given by (27.7). For a given set of a_1 and a_2, where $a_1 \neq a_2$, the critical value of σ_{33}^0 is given as the smaller of (27.27) and (27.27.1). When $a_1 > a_2$, the smaller value is (27.27.1). The crack grows from the minor axis of the ellipse. This is consistent with the fact that the stress intensity factor at the minor axis is greater than that at the major axis.

For simple shear, the first equation in (27.25) yields

$$\sigma_{31}^{0} = \left[\frac{3\mu\gamma}{(1-\nu)a_2} \times \frac{\{(k^2-\nu)E(k)+\nu(k')^2F(k)\}^2}{\{(1-\nu)-(3+\nu)(k')^2+2(k')^4\}E(k)+\{(2\nu+1)(k')^2-(k')^4\}F(k)}\right]^{1/2} \tag{27.28}$$

and the second equation gives

$$\sigma_{31}^{0} = \left[\frac{3\mu\gamma}{(1-\nu)a_2} \times \frac{\{(k^2-\nu)E(k)+\nu(k')^2F(k)\}^2}{\{2-2\nu+(4\nu-3)(k')^2+(k')^4\}E(k)+\{(\nu-1)(k')^2+(1-3\nu)(k')^4\}F(k)}\right]^{1/2}. \tag{27.28.1}$$

When $1 > a_2/a_1 > 0.7$, ($\nu = 0.3$), σ_{31}^{0} given by (27.28) is smaller than σ_{31}^{0} given by (27.28.1). Therefore, the crack starts to grow from the major axis of the ellipse. If, however, $a_2/a_1 < 0.7$, σ_{31}^{0} given by (27.28.1) becomes smaller than σ_{31}^{0} given by (27.28). Then, the crack starts to grow from the minor axis. When $a_2/a_1 > 1$, expressions (27.28) and (27.28.1) must be rewritten by using the following formulae:

$$\begin{aligned}
&E(k) = (a_2/a_1)E(\bar{k}),\\
&F(k) = (a_1/a_2)F(\bar{k}),\\
&\bar{k}^2 = 1 - a_1^2/a_2^2,\\
&k^2 = \bar{k}^2/(\bar{k}^2-1),\\
&(k')^2 = 1/(1-\bar{k}^2).
\end{aligned} \tag{27.28.2}$$

Smith (1971) has investigated the case when δa_1 and δa_2 satisfy the condition $m = \delta a_2/\delta a_1$. He considered a penny-shaped crack which in its initial state is subjected to an applied stress field σ_{33}^{0} and σ_{31}^{0}. This penny-shaped crack may grow into an elliptical crack. The Griffith fracture criterion may be obtained from (27.23) and (27.24) as

$$\delta\Big\{-(2/3)\pi(\sigma_{33}^{0})^2a_1a_2^2(1-\nu)/\mu E(k) - (2/3)\pi(\sigma_{31}^{0})^2a_1a_2/\mu\left[\frac{\nu}{1-\nu}\frac{a_2\{F(k)-E(k)\}}{a_1^2-a_2^2}+\frac{E(k)}{a_2}\right]+2\pi a_1a_2\gamma\Big\}=0, \tag{27.29}$$

where

$$\delta\{\ \} = \left[\frac{\partial}{\partial a_1}\{\ \} + m\frac{\partial}{\partial a_2}\{\ \}\right]\delta a_1. \tag{27.30}$$

The limits $a_1 \to a$ and $a_2 \to a$ are taken in (27.29), since the crack is initially penny-shaped. Then, (27.29) and (27.30) lead to

$$\frac{\pi\mu\gamma}{(1-\nu)a} = (\sigma_{33}^0)^2 + \frac{(\sigma_{31}^0)^2\{(4-\nu) + m(4-3\nu)\}}{(2-\nu)^2(1+m)}. \tag{27.31}$$

The right-hand side of (27.31) is a maximum at $m = 0$. It implies that for a penny-shaped crack with a given size the minimum σ_{33}^0 is obtained when $m = 0$ under a constant σ_{31}^0, or the minimum σ_{31}^0 is obtained when $m = 0$ under a constant σ_{33}^0. Therefore, it can be seen that a penny-shaped crack under σ_{31}^0 and σ_{33}^0 grows into an ellipse, and, further, that there is no growth perpendicular to the direction of the applied shear stress. Generally it is believed that a penny-shaped crack extends to another penny-shaped crack under the action of σ_{33}^0 alone. According to Smith's theory, a penny-shaped crack extends to an elliptic crack under σ_{33}^0 when a non-zero σ_{31}^0 is also acting.

Crack opening displacement

Since crack is regarded as an ellipsoidal void by letting $a_3 \to 0$, the crack opening displacement is defined by

$$u_i = h\epsilon_{3i}^*, \tag{27.32}$$

where h is the half thickness of the ellipsoidal void in the x_3-direction. Since

$$h = a_3(1 - x_1^2/a_1^2 - x_2^2/a_2^2)^{1/2}, \tag{27.33}$$

the crack opening displacement is

$$u_i = a_3\epsilon_{3i}^*(1 - x_1^2/a_1^2 - x_2^2/a_2^2)^{1/2}, \tag{27.34}$$

where $a_3\epsilon_{3i}^*$ is finite as $a_3 \to 0$. The above result can also be applied to the anisotropic case.

28. Critical stresses of cracks in anisotropic media

Uniform applied stress

Consider an elliptic crack ($x_1^2/a_1^2 + x_2^2/a_2^2 = 1$) in an anisotropic medium. The x coordinate axes are chosen parallel to the principal axes of the ellipsoid, $x_1^2/a_1^2+x_2^2/a_2^2+x_3^2/a_3^2=1$. The elliptic crack is degenerated from the ellipsoid by taking $a_3\to 0$. Let σ_{ij}^0 be the applied stress at infinity.

The boundary conditions on the crack surface are

$$\begin{aligned}&\sigma_{33}^0+\sigma_{33}=0,\\&\sigma_{31}^0+\sigma_{31}=0,\\&\sigma_{32}^0+\sigma_{32}=0,\end{aligned}\tag{28.1}$$

where σ_{ij} is the stress disturbance due to the crack. This stress disturbance is simulated by the eigenstress derived from an equivalent ellipsoidal inclusion with eigenstrain ϵ_{ij}^*. From (17.12) we have

$$\sigma_{ij}=C_{ijkl}\left\{\frac{a_1a_2a_3}{4\pi}C_{pqmn}\epsilon_{mn}^*\int_{S^2}G_{kplq}(\bar{\boldsymbol{\xi}})\zeta^{-3}dS(\bar{\boldsymbol{\xi}})-\epsilon_{kl}^*\right\},\tag{28.2}$$

where

$$\begin{aligned}&\zeta=(a_1^2\bar{\xi}_1^2+a_2^2\bar{\xi}_2^2+a_3^2\bar{\xi}_3^2)^{1/2},\\&G_{kplq}(\bar{\boldsymbol{\xi}})=N_{kp}(\bar{\boldsymbol{\xi}})\bar{\xi}_l\bar{\xi}_q/D(\bar{\boldsymbol{\xi}}).\end{aligned}\tag{28.3}$$

Since we are interested in the limiting case $a_3\to 0$, the integral in (28.2) is modified. As shown in (17.42), σ_{3j} becomes zero for this case when ϵ_{ij}^* is finite. However, σ_{3j} becomes nonzero if ϵ_{ij}^* is infinite and $a_3\epsilon_{ij}^*$ is finite for $a_3\to 0$. Nonzero values of σ_{ij} are necessary in order to satisfy the boundary conditions (28.1).

It holds that

$$a_1a_2a_3\int_{S^2}G_{kplq}(\bar{\boldsymbol{\xi}})\zeta^{-3}dS(\bar{\boldsymbol{\xi}})=\int_{S^2}\frac{G_{kplq}(\bar{\boldsymbol{\xi}})dS(\bar{\boldsymbol{\xi}})}{a_1^2a_2^2a_3^2(\bar{\xi}_1^2/a_2^2a_3^2+\bar{\xi}_2^2/a_3^2a_1^2+\bar{\xi}_3^2/a_1^2a_2^2)^{3/2}}.\tag{28.4}$$

Introducing the following transformation:

$$\begin{aligned}&\bar{\xi}_1=(1-\bar{\xi}_3^2)^{1/2}\cos\theta,\\&\bar{\xi}_2=(1-\bar{\xi}_3^2)^{1/2}\sin\theta,\\&t=\bar{\xi}_3/(1-\bar{\xi}_3^2)^{1/2},\\&dS(\bar{\boldsymbol{\xi}})=d\bar{\xi}_3d\theta=(1-\bar{\xi}_3^2)^{3/2}dtd\theta,\end{aligned}\tag{28.5}$$

and since $G_{kplq}(\bar{\boldsymbol{\xi}})$ is a polynomial of degree zero, we express (28.4) as

$$\int_0^{2\pi}\frac{d\theta}{a_1^2a_2^2a_3^2}\int_{-\infty}^{\infty}\frac{G_{kplq}(\cos\theta,\sin\theta,t)dt}{(\cos^2\theta/a_2^2a_3^2+\sin^2\theta/a_3^2a_1^2+t^2/a_1^2a_2^2)^{3/2}}.\tag{28.6}$$

When integration by parts with respect to t is performed, using

$$\frac{\partial}{\partial t}\{t/(\cos^2\theta/a_2^2a_3^2+\sin^2\theta/a_3^2a_1^2+t^2/a_1^2a_2^2)^{1/2}\}$$
$$=(\cos^2\theta/a_2^2a_3^2+\sin^2\theta/a_3^2a_1^2)/(\cos^2\theta/a_2^2a_3^2+\sin^2\theta/a_3^2a_1^2+t^2/a_1^2a_2^2)^{3/2} \tag{28.7}$$

and

$$(G_{kplq})_{t=\infty}=(G_{kplq})_{t=-\infty}=G_{kplq}(0,0,1), \tag{28.8}$$

(28.6) becomes

$$2G_{kplq}(0,0,1)\int_0^{2\pi}\frac{d\theta}{a_1a_2(\cos^2\theta/a_2^2+\sin^2\theta/a_1^2)}-a_3\int_0^{2\pi}\frac{d\theta}{a_1^2a_2^2}$$
$$\times\int_{-\infty}^{\infty}\frac{t\partial G_{kplq}(\cos\theta,\sin\theta,t)/\partial t dt}{(\cos^2\theta/a_2^2+\sin^2\theta/a_1^2)(\cos^2\theta/a_2^2+\sin^2\theta/a_1^2+a_3^2t^2/a_1^2a_2^2)^{1/2}}. \tag{28.9}$$

Noting that

$$\int_0^{2\pi}\frac{d\theta}{\cos^2\theta/a_2^2+\sin^2\theta/a_1^2}=2\pi a_1a_2 \tag{28.10}$$

and that $a_3 \ll a_1, a_2$, we finally have

$$\sigma_{ij}=(1/4\pi)C_{ijkl}C_{pqmn}\epsilon^*_{mn}\{4\pi G_{kplq}(0,0,1)-a_3\Pi_{kplq}\}-C_{ijkl}\epsilon^*_{kl}, \tag{28.11}$$

where

$$\Pi_{kplq}=\int_0^{2\pi}\frac{d\theta}{a_1^2a_2^2}\int_{-\infty}^{\infty}\frac{t\partial G_{kplq}(\cos\theta,\sin\theta,t)/\partial t dt}{(\cos^2\theta/a_2^2+\sin^2\theta/a_1^2)^{3/2}}$$
$$=-\int_0^{2\pi}\frac{d\theta}{a_1^2a_2^2}\int_{-\infty}^{\infty}\frac{G^*_{kplq}(\cos\theta,\sin\theta,t)}{(a_2^{-2}\cos^2\theta+a_1^{-2}\sin^2\theta)^{3/2}}dt \tag{28.12}$$

after integrating by parts, where G^*_{kplq} is chosen so that $[tG^*_{kplq}]_{t=-\infty}^{t=\infty}=0$ and $G_{kplq}-G^*_{kplq}=$ constant, namely, the quotient of two polynomials of degree 6 divided by a polynomial of degree 6 is rewritten in a constant plus a polynomial of degree 4 divided by a polynomial of degree 6. The constant is ignored when the integration by parts is performed. $G_{kplq}=G^*_{kplq}$ when l and q are not 3 at the same time.

As has been shown already by (17.42), the first and third terms in (28.11) cancel each other out on the crack surface. The boundary conditions (28.1) therefore become

$$\sigma^0_{3j}-(1/4\pi)C_{3jkl}C_{pqmn}a_3\epsilon^*_{mn}\Pi_{kplq}=0 \qquad (j=1,2,3) \tag{28.13}$$

or

$$\sigma^0_{3j} + L_{3jmn} a_3 \epsilon^*_{mn} = 0 \quad (j = 1, 2, 3), \tag{28.14}$$

where

$$L_{ijmn} = -(1/4\pi) C_{ijkl} C_{pqmn} \Pi_{kplq}. \tag{28.15}$$

The set of three equations (28.14) can be solved for the remaining eigenstrains.

Mura and Lin (1974) and Cheng (1977) have considered (28.14) in some detail when the crack is oriented as shown in Fig. 28.1, where the x_2-axis coincides with one of the crystallographic directions, say the x_2'-axis. The other two crystallographic directions are taken as the x_1'-and x_3'-axes.

For a simple tension case ($\sigma^0_{3j} = \sigma^0_{33}$, normal to the crack surface), they have found that

$$a_3 \epsilon^*_{33} = -\sigma^0_{33} L_{3131}/\Delta, \qquad a_3 \epsilon^*_{31} = \sigma^0_{33} L_{3133}/2\Delta, \tag{28.16}$$

and for a simple shear case ($\sigma^0_{3j} = \sigma^0_{31}$, tangential to the crack surface),

$$a_3 \epsilon^*_{33} = \sigma^0_{31} L_{3331}/\Delta, \qquad a_3 \epsilon^*_{31} = -\sigma^0_{31} L_{3333}/2\Delta, \tag{28.17}$$

where

$$\Delta = L_{3333} L_{3131} - L_{3133} L_{3331}. \tag{28.18}$$

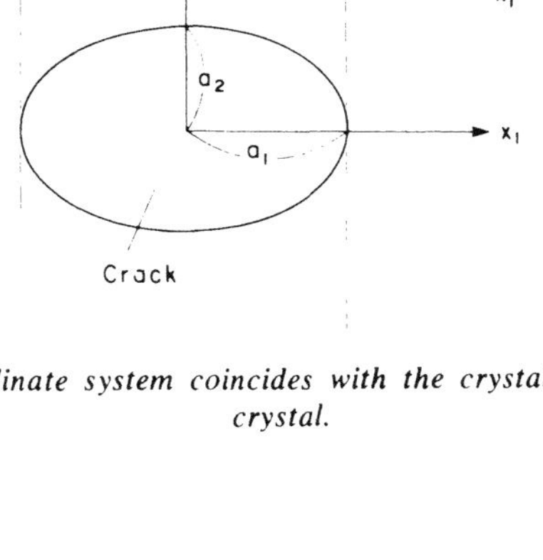

Fig. 28.1. The **X**′ *coordinate system coincides with the crystallographic directions of a crystal.*

All other components of ϵ^*_{ij} can be taken as zero. It is found that $L_{3133} = 0$, $L_{3331} = 0$ for $\alpha = 0$.

The interaction energy is

$$\Delta W = -(1/2)(4\pi/3)a_1a_2\sigma^0_{3j}a_3\epsilon^*_{3j}. \tag{28.19}$$

For a penny-shaped crack ($a_1 = a_2 = a$), the critical stress is obtained from

$$\partial(\Delta W + 2\pi a^2\gamma)/\partial a = 0, \tag{28.20}$$

where γ is the surface energy. Numerical results obtained by Mura and Lin (1974) and Cheng (1977) are shown in Fig. 28.2 and Fig. 28.3 for simple tension and shear, respectively. Two materials, iron and zinc, are examined. In the case of zinc, the x'_1, x'_2-plane is taken as the basal plane. The physical constants are taken from the American Institute of Physic's Handbook (1972). For iron, $\lambda = 14.1$, $\mu = 11.6$, and $\mu' = -13.6$ with units of 10^{11} dyne/cm^2, and $\gamma = 2.0 \times 10^3$ dyne/cm. For zinc, $C_{11} = 16.1$, $C_{33} = 6.1$, $C_{44} = 3.83$, $C_{12} = 3.42$, and $C_{13} = 5.01$ with unit 10^{11} dyne/cm^2, and $\gamma = 708$ dyne/cm. It is interesting to observe that the critical σ^0_{33} of iron takes a minimum value at $\alpha = 90°$, while the cor-

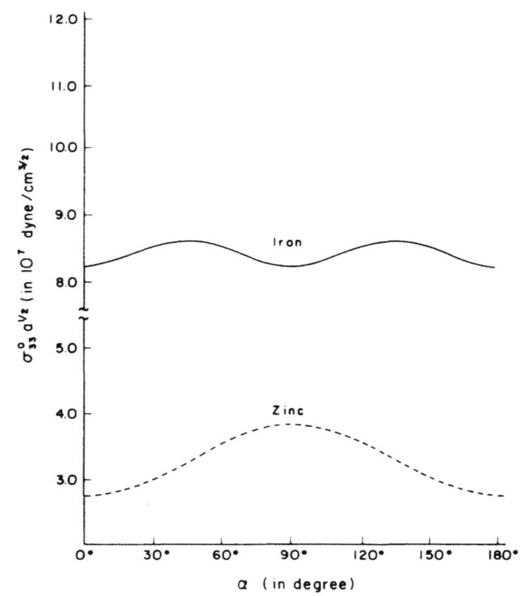

Fig. 28.2. The critical stress in simple tension.

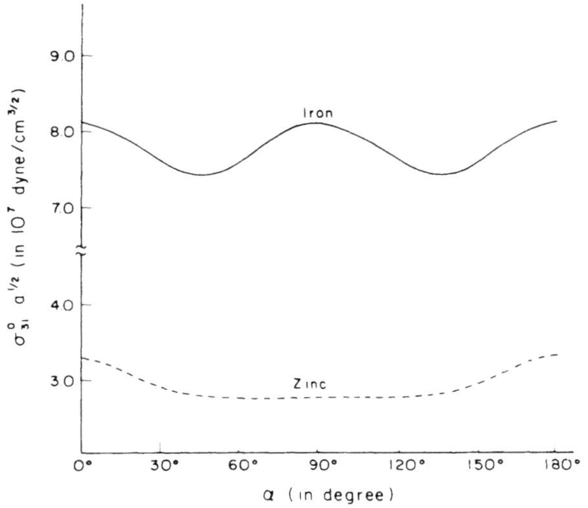

Fig. 28.3. The critical stress in simple shear.

responding quantity for zinc obtains its maximum at $\alpha = 90°$. The critical value of σ^0_{31} for iron reaches a maximum value at $\alpha = 90°$, but that of zinc takes a minimum value at $\alpha = 90°$.

Non-uniform applied stress

If the applied stress is given in the form of

$$\sigma^0_{3j} = \sigma^p_{3j} x_p / a_p, \tag{28.21}$$

the eigenstrain used in the equivalent inclusion method becomes

$$\epsilon^*_{ij} = \epsilon^p_{ij} x_p / a_p. \tag{28.22}$$

The eigenstress inside the equivalent inclusion Ω can be obtained from (19.12), where $S^* = S^2$. When $a_3 \ll a_1, a_2$, we have, as seen in (19.17),

$$\sigma_{ij} = C_{ijkl}\{C_{pqmn}\epsilon^s_{mn}(x_s/a_s)G_{kplq}(0, 0, 1) - \epsilon^p_{kl}x_p/a_p\} + a_3 L^{st}_{ijmn}\epsilon^s_{mn}x_t/a_t, \tag{28.23}$$

where

$$L^{st}_{ijmn} = -(3/4\pi)C_{ijkl}C_{pqmn}\int_{S^2}\frac{a_1a_2\zeta_s\zeta_t\bar{\xi}_3\partial G_{kplq}(\bar{\xi})/\partial\bar{\xi}_3}{(a_1^2\cos^2\theta + a_2^2\sin^2\theta)^{5/2}(1-\bar{\xi}_3^2)^{1/2}}\,dS(\bar{\xi}), \tag{28.24}$$

and

$$\begin{aligned}
&\zeta_1 = a_1\xi_1, \zeta_2 = a_2\xi_2, \zeta_3 = 0,\\
&\boldsymbol{\xi} = (\cos\theta, \sin\theta, 0),\\
&\bar{\xi}_1 = (1-\bar{\xi}_3^2)^{1/2}\cos\theta,\\
&\bar{\xi}_2 = (1-\bar{\xi}_3^2)^{1/2}\sin\theta,\\
&dS(\bar{\boldsymbol{\xi}}) = d\bar{\xi}_3 d\theta.
\end{aligned} \tag{28.25}$$

The boundary conditions are

$$\sigma_{3j}^0 + \sigma_{3j} = 0\,(j = 1, 2, 3). \tag{28.26}$$

It can also be shown that the first two terms in (28.23) cancel each other for σ_{3j}. However, the above boundary conditions can only be satisfied by taking $a_3\epsilon_{mn}^s$ = finite for $a_3 \to 0$. The applied stress (28.21) cannot be given arbitrarily. Since $\sigma_{3j,j}^0 = 0$ must be satisfied, we restrict attention to the case where

$$\sigma_{31}^1 = \sigma_{32}^2 = \sigma_{33}^3 = 0. \tag{28.27}$$

For simplicity, the crack plane is assumed to be one of the crystallographic planes ($\alpha = 0$ in Fig. 28.1). Then, for simple tension, $\sigma_{33}^0 = \sigma_{33}^1 x_1/a_1 + \sigma_{33}^2 x_2/a_2$,

$$\begin{aligned}
a_3\epsilon_{33}^1 &= -\sigma_{33}^1/L_{3333}^{11},\\
a_3\epsilon_{33}^2 &= -\sigma_{33}^2/L_{3333}^{22},
\end{aligned} \tag{28.28}$$

and for simple shear, $\sigma_{31}^0 = \sigma_{31}^2 x_2/a_2$,

$$a_3\epsilon_{31}^2 = -\sigma_{31}^2/L_{3131}^{22}. \tag{28.29}$$

If the material is isotropic,

$$\begin{aligned}
L_{3333}^{11} &= -\mu\{(k')^2F(k) + (2k^2-1)E(k)\}/(1-\nu)a_2k^2,\\
L_{3333}^{22} &= \mu\{(k')^2F(k) - (k^2+1)E(k)\}/(1-\nu)a_2k^2,\\
L_{3131}^{22} &= \mu\{(k')^2F(k) - (k^2+1)E(k)\}/a_2k^2,
\end{aligned} \tag{28.30}$$

where $k^2 = 1 - a_2^2/a_1^2$, $(k')^2 = a_2^2/a_1^2$, and $E(k)$ and $F(k)$ are defined by (11.25).

*Π *integrals for a penny-shaped crack* ($a_1 = a_2$)

The integrals defined by (28.12) are given below for cubic and hexagonal crystals when the crystallographic directions coincide with the x coordinate system ($\alpha = 0$ in Fig. 28.1). The crack plane is the basal plane for hexagonal crystals.

$*\Pi$ *integrals for cubic crystals*

$$\Pi_{1111} = \Pi_{2222} = -\frac{\pi}{aa_1}\int_0^1 \frac{(1-x^2)^{1/2}(\kappa+\gamma x^2)}{p(p+q)}\,dx$$

$$-\frac{2\pi}{aa_1}\int_0^1 \left[\frac{\mu^2+\kappa x^2}{pq(1-x^2)^{1/2}} - \frac{a}{\mu\sqrt{2}(1-x)^{1/2}}\right]dx - \frac{2\pi\sqrt{2}}{\mu a_1},$$

$$\Pi_{3333} = -\frac{\pi\gamma}{aa_1}\int_0^1 \frac{x^2(1-x^2)^{1/2}}{p(p+q)}\,dx$$

$$-\frac{\pi}{aa_1}\int_0^1 \frac{\mu^2 x^4(1-x^2)^{1/2}(c-2b-3cx^2)}{p^3 q}\,dx$$

$$-\frac{4\pi}{aa_1}\int_0^1 \left[\frac{x^2(\kappa-2\mu^2)}{pq(1-x^2)^{1/2}} - \frac{\kappa-2\mu^2}{\sqrt{2}(1-x)^{1/2}}\right]dx - \frac{4\pi\sqrt{2}}{aa_1}(\kappa-2\mu^2)$$

$$-\frac{4\pi}{aa_1}\int_0^1 \left[\frac{(p^2+q^2)\mu^2 b x^4(1-2x^2)}{(pq)^3(1-x^2)^{1/2}} + \frac{\mu^2 b}{\sqrt{2}(1-x)^{1/2}}\right]dx$$

$$+\frac{4\pi\sqrt{2}}{aa_1}\mu^2 b,$$

$$\Pi_{3311} = \Pi_{3322} = -\frac{\pi\gamma}{2aa_1}\int_0^1 \frac{(1-x^2)^{3/2}}{p(p+q)}\,dx$$

$$-\frac{2\pi}{aa_1}\int_0^1 \left[\frac{\mu^2+\kappa-\kappa x^2}{pq(1-x^2)^{1/2}} - \frac{\mu^2}{\sqrt{2}(1-x)^{1/2}}\right]dx - \frac{2\pi\sqrt{2}\mu^2}{aa_1},$$

$$\Pi_{1122} = \Pi_{2211} = \frac{\pi}{aa_1}\int_0^1 \frac{(\kappa+\gamma x^2)(1-x^2)^{1/2}}{p(p+q)}\,dx$$

$$-\frac{2\pi}{aa_1}\int_0^1 \frac{(\kappa+\gamma x^2)(1-x^2)^{1/2}}{pq}\,dx$$

$$-\frac{2\pi}{aa_1}\int_0^1 \left[\frac{\mu^2+\kappa x^2}{pq(1-x^2)^{1/2}} - \frac{a}{\mu\sqrt{2}(1-x)^{1/2}}\right]dx - \frac{2\pi\sqrt{2}}{a_1\mu},$$

$$\Pi_{1133} = \Pi_{2233} = -\frac{\pi}{aa_1}\int_0^1 \frac{x^4(1-x^2)^{1/2}(\mu^2+\kappa x^2)(c-2b-3cx^2)}{p^3 q}\,dx$$

$$
-\frac{4\pi}{aa_1}\int_0^1\left[\frac{(p^2+q^2)bx^4(1-2x^2)(\mu^2+\kappa x^2)}{(pq)^3(1-x^2)^{1/2}}+\frac{2ba}{\mu\sqrt{2}(1-x)^{1/2}}\right]dx
$$

$$
+\frac{8\pi\sqrt{2}b}{a_1\mu}+\frac{2\pi}{aa_1}\int_0^1\left[\frac{(8\kappa-\gamma)x^4+(4\mu^2-\kappa)x^2}{pq(1-x^2)^{1/2}}\right.
$$

$$
\left.-\frac{7\kappa-\gamma+4\mu^2}{\sqrt{2}(1-x)^{1/2}}\right]dx+\frac{2\pi\sqrt{2}}{aa_1}(7\kappa-\gamma+4\mu^2),
$$

$$
\Pi_{1212}=\frac{\pi(\lambda+\mu)}{aa_1}\int_0^1\frac{(1-x^2)^{1/2}(\mu+\mu' x^2)}{p(p+q)}\,dx,
$$

$$
\Pi_{1313}=\Pi_{2323}=\frac{\pi\mu'(\lambda+\mu)}{aa_1}\int_0^1\frac{x^2(1-x^2)^{1/2}}{p(p+q)}\,dx
$$

$$
+\frac{2\pi\mu(\lambda+\mu)}{aa_1}\int_0^1\left[\frac{x^2}{pq(1-x^2)^{1/2}}-\frac{1}{\sqrt{2}(1-x)^{1/2}}\right]dx
$$

$$
+\frac{2\pi\sqrt{2}\mu(\lambda+\mu)}{aa_1},
$$

where

$$
\begin{aligned}
a &= \mu^2(\lambda+2\mu+\mu'),\\
b &= a^{-1}\mu\mu'(2\lambda+2\mu+\mu'),\\
c &= a^{-1}\mu'^2(3\lambda+3\mu+\mu'),\\
\kappa &= \mu(\lambda+\mu+\mu'),\\
\gamma &= \mu'(2\lambda+2\mu+\mu'),\\
p &= [1+bx^2(1-x^2)+(1-x^2)^2(b+cx^2)/4]^{1/2},\\
q &= [1+bx^2(1-x^2)]^{1/2}.
\end{aligned}
$$

$*\Pi$ integrals for transversely isotropic materials

$$
\Pi_{1111}=\Pi_{2222}
$$

$$
=-\frac{\pi}{2a_1}\int_0^1\left\{\frac{[f(1-x^2)+hx^2][(d+3e)(1-x^2)+4fx^2]-g^2x^2(1-x^2)}{\Delta^{-1}(1-x^2)^{1/2}}\right.
$$

$$
\left.-\frac{4}{\sqrt{2}f(1-x)^{1/2}}\right\}dx-\frac{2\sqrt{2}\pi}{a_1 f},
$$

$$\Pi_{3333} = -\frac{4\pi}{a_1}\int_0^1 \Delta e x^2(1-x^2)^{1/2}(d-4fx^2)dx$$

$$-\frac{4\pi}{a_1}\int_0^1\left[\frac{\Delta f x^4(d-4fx^2)}{(1-x^2)^{1/2}} - \frac{f^3h(d-4f)}{\sqrt{2}(1-x)^{1/2}}\right]dx - \frac{4\pi\sqrt{2}}{a_1}f^3h(d-4f)$$

$$+\frac{8\pi}{a_1}\int_0^1\Big\{(1-x^2)^{-1/2}\Delta^2[e(1-x^2)+fx^2]^2 fx^6\{g^2(1-2x^2)+(f-h)$$

$$[d(1-x^2)+fx^2]+(d-f)[f(1-x^2)+hx^2]\}$$

$$-\frac{(f^2-g^2-2fh+dh)}{\sqrt{2}(1-x)^{1/2}fh^2}\Big\}dx + \frac{8\pi\sqrt{2}}{a_1}\,\frac{f^2-g^2-2fh+dh}{fh^2},$$

$$\Pi_{1122} = \Pi_{2211}$$

$$= -\frac{\pi}{2a_1}\int_0^1\Big\{\frac{[f(1-x^2)+hx^2][(d+3e)(1-x^2)+4fx^2]-3g^2x^2(1-x^2)}{\Delta^{-1}(1-x^2)^{1/2}}$$

$$-\frac{4}{f\sqrt{2}(1-x)^{1/2}}\Big\}dx - \frac{2\pi\sqrt{2}}{a_1 f},$$

$$\Pi_{3311} = \Pi_{3322} = -\frac{\pi}{2a_1}\int_0^1\Big\{\Delta(1-x^2)^{-1/2}[d(1-x^2)+fx^2][e(1-x^2)+fx^2]$$

$$-\frac{1}{\sqrt{2}h(1-x)^{1/2}}\Big\}dx - \frac{\sqrt{2}\pi}{2a_1 h},$$

$$\Pi_{1133} = \Pi_{2233} = -\frac{2\pi}{a_1}\int_0^1 \Delta x^2 f(d+e)(1-x^2)^{1/2}dx$$

$$-\frac{2\pi}{a_1}(2f^2+hd+he-g^2)\int_0^1\left[\frac{\Delta x^4}{(1-x^2)^{1/2}} - \frac{1}{\sqrt{2}(1-x)^{1/2}f^2h}\right]dx$$

$$-\frac{2\pi\sqrt{2}}{a_1}\,\frac{(2f^2+hd+he-g^2)}{f^2h}$$

$$+\frac{24\pi}{a_1}fh\int_0^1\left[\frac{\Delta x^6}{(1-x^2)^{1/2}} - \frac{1}{f^2h\sqrt{2}(1-x)^{1/2}}\right]dx + \frac{24\pi\sqrt{2}}{a_1 f}$$

$$-\frac{2\pi}{a_1}\int_0^1 4\Delta^2 x^8 fh(1-x^2)^{-1/2}\{(f-e)[df(1-x^2)^2$$

$$+(f^2+dh-g^2)x^2(1-x^2)+fhx^4]+[e(1-x^2)+fx^2]$$

$$\times[(f^2-2fd+dh-g^2)(1-x^2)+(2fh-dh-f^2+g^2)x^2]\}dx$$

$$+\frac{8\pi}{a_1}\int_0^1 \frac{3fh-dh-eh-f^2+g^2}{f^2h\sqrt{2}(1-x)^{1/2}}\,dx$$

$$-\frac{8\pi\sqrt{2}}{a_1 f^2 h}(3fh-dh-eh-f^2+g^2),$$

$$\Pi_{1212} = -\frac{\pi}{2a_1}\int_0^1 \Delta(1-x^2)^{1/2}\{g^2x^2-(d-e)[f(1-x^2)+hx^2]\}dx,$$

$$\Pi_{1313} = \Pi_{2323} = \frac{2\pi}{a_1}\int_0^1 \left\{\Delta g x^2(1-x^2)^{-1/2}[e(1-x^2)+fx^2]\right.$$

$$\left. -\frac{g}{fh\sqrt{2}(1-x)^{1/2}}\right\}dx + \frac{2\pi\sqrt{2}g}{a_1 fh},$$

where $\Delta = 1/\Delta^{-1}$,

$$\Delta^{-1} = [e(1-x^2)+fx^2]\{[d(1-x^2)+fx^2][f(1-x^2)+hx^2]-g^2x^2(1-x^2)\},$$

and d, h, f, g, and e are defined in (17.30).

29. Stress intensity factors for a flat ellipsoidal crack

Stress intensity factors of cracks as well as critical stresses are important quantities in fracture mechanics, as has been pointed out by Irwin (1948). The stress intensity factors are also important in the study of crack growth rates of fatigue cracks in connection with the Paris law (1964). Stress fields around cracks and their stress intensity factors are well documented in the literature, e.g., Sneddon and Lowengrub (1969), Sneddon (1961), Irwin (1958), Sih and Liebowitz (1968), Paris and Sih (1964), Kassir and Sih (1975), among others. However, little is known about the stress intensity factors of materials which are anisotropic or when crack shapes are arbitrary.

In this section we follow the work of Willis (1968) on the stress field around a flat ellipsoidal crack in general anisotropic elastic media. The applied stress, σ^0_{ij}, is assumed to be a polynomial function of the coordinates. The **x** coordinate system is chosen as shown in Fig. 28.1. The stress disturbance is denoted by σ_{ij}. The total stress field is $\sigma^0_{ij}+\sigma_{ij}$. The displacement disturbance u_i is related to σ_{ij} through

$$\sigma_{ij} = C_{ijkl}u_{k,l}. \qquad (29.1)$$

The boundary conditions are

$$\begin{aligned} &\sigma^0_{3i} + \sigma_{3i} = 0 \qquad \text{when } x_3 = 0, \qquad x_1^2/a_1^2 + x_2^2/a_2^2 \leq 1, \\ &u_i(\mathbf{x}) \to 0 \qquad \text{when } |\mathbf{x}| \to \infty \end{aligned} \tag{29.2}$$

where σ^0_{3i} is a given polynomial function of x_1 and x_2.

It can be expected that the displacement field u_i and the corresponding stress σ_{ij} have properties similar to those derived from eigenstrains ϵ^*_{ij}. From (16.2) we have

$$u_i(\mathbf{x}) = i(2\pi)^{-3} \iint_{-\infty}^{\infty} C_{jlmn}\epsilon^*_{nm}(\mathbf{x}')\xi_l N_{ij}(\boldsymbol{\xi})D^{-1}(\boldsymbol{\xi}) \exp\{-i\boldsymbol{\xi}\cdot(\mathbf{x}-\mathbf{x}')\}d\boldsymbol{\xi}d\mathbf{x}'. \tag{29.3}$$

When

$$\epsilon^*_{3m}(\mathbf{x}')d\mathbf{x}' = -b_m(\mathbf{x}')dS_3(\mathbf{x}') = b_m(\mathbf{x}')dx_1'dx_2' \tag{29.4}$$

is defined on the crack surface, (29.3) becomes

$$\begin{aligned} u_i(\mathbf{x}) = i(2\pi)^{-3} \int_{-\infty}^{\infty} \int_S C_{jlm3} b_m(\mathbf{x}')\xi_l N_{ij}(\boldsymbol{\xi})D^{-1}(\boldsymbol{\xi}) \\ \times \exp\{-i\boldsymbol{\xi}\cdot(\mathbf{x}-\mathbf{x}')\}d\boldsymbol{\xi}dx_1'dx_2', \end{aligned} \tag{29.5}$$

where $S = S_3$ is the crack surface and $\mathbf{b}$ is called the Burgers vector of the Somigliana dislocation. Eshelby (1957) has found that $|\mathbf{b}|$ is proportional to $(1 - x_1^2/a_1^2 - x_2^2/a_2^2)^{1/2}$ when the applied stress at infinity is uniform. In the present case where the applied stress is polynomial, $\mathbf{b}(\mathbf{x})$ is chosen as

$$\mathbf{b}(\mathbf{x}) = \bar{\mathbf{b}}(x_1/a_1 + ix_2/a_2)^p(x_1/a_1 - ix_2/a_2)^q(1 - x_1^2/a_1^2 - x_2^2/a_2^2)^{1/2}, \tag{29.6}$$

where $\bar{\mathbf{b}}$ is a constant vector of a complex number which will be determined later together with p and q, and $p \geq q$. The ξ_3-integration can be performed by using Cauchy's theorem. Then, (29.5) becomes

$$\begin{aligned} u_i(\mathbf{x}) = (2\pi)^{-2} \sum_{N=1}^{3} \iint_{-\infty}^{\infty} d\xi_1 d\xi_2 \frac{C_{jlm3}\xi_l^N N_{ij}(\boldsymbol{\xi}^N)}{\partial D(\boldsymbol{\xi}^N)/\partial \xi_3} \iint_S b_m(\mathbf{x}') \\ \times \exp\{-i\boldsymbol{\xi}^N\cdot(\mathbf{x}-\mathbf{x}')\}dx_1'dx_2', \end{aligned} \tag{29.7}$$

where

$$\boldsymbol{\xi}^N = (\xi_1, \xi_2, \xi_3^N(\xi_1, \xi_2)) \tag{29.8}$$

and $\xi_3 = \xi_3^N(\xi_1, \xi_2)$ are the roots of the equation

$$D(\boldsymbol{\xi}) = 0 \tag{29.9}$$

with negative imaginary parts for $x_3 > 0$. For isotropic materials there is only one double root with a negative imaginary part ($N = 1$). For a double root the residual is obtained from $\partial\{(\xi_q\xi_l N_{kp}(\boldsymbol{\xi})D^{-1}(\boldsymbol{\xi}))(\xi_3 - \xi_3^N)^2\}/\partial\xi_3$.

The stress components are

$$\sigma_{ij}(\mathbf{x}) = \frac{-i}{4\pi^2}\sum_{N=1}^{3}\iint_{-\infty}^{\infty} d\xi_1 d\xi_2 \frac{C_{ijkl}C_{pqm3}\xi_q^N\xi_l^N N_{kp}(\boldsymbol{\xi}^N)}{\partial D(\boldsymbol{\xi}^N)/\partial\xi_3}\iint_S b_m(\mathbf{x}')$$
$$\times \exp\{-i\boldsymbol{\xi}^N\cdot(\mathbf{x}-\mathbf{x}')\}dx_1'dx_2'. \tag{29.10}$$

The functions $\xi_3^N(\xi_1, \xi_2)$ are homogeneous of degree 1. It should be noted that $N_{kp}(\boldsymbol{\xi})$ and $D(\boldsymbol{\xi})$ are homogeneous of degrees 4 and 6, respectively.

The complicated integrations in (29.7) and (29.10), when $\mathbf{b}(\mathbf{x})$ is given by (29.6), are carried out by Willis. The results are

$$u_i(\mathbf{x}) = \frac{(-1)^{q+1}}{2\pi}\bar{b}_m\int_0^{2\pi} d\phi\sum_{N=1}^{3} F_{im}(\eta_1/a_1, \eta_2/a_2, \xi_3^N(\eta_1/a_1, \eta_2/a_2))e^{i(p-q)\phi}$$
$$\times\frac{\partial}{\partial g}\int_1^{\{(g+1)/(g-1)\}^{1/2}}\left\{\left[-g+\frac{\sqrt{g^2-1}}{2}\left(w+\frac{1}{w}\right)\right]^p\right.$$
$$\times\left[g+\frac{\sqrt{g^2-1}}{2}\left(w+\frac{1}{w}\right)\right]^q \tag{29.11}$$
$$\left.+\left[-g+\frac{\sqrt{g^2-1}}{2}\left(w+\frac{1}{w}\right)\right]^{p+1}\left[g+\frac{\sqrt{g^2-1}}{2}\left(w+\frac{1}{w}\right)\right]^{q+1}\right\}\frac{dw}{w},$$

$$\sigma_{ij}(\mathbf{x}) = \frac{(-1)^{q+1}}{2\pi}\bar{b}_m\int_0^{2\pi} d\phi\sum_{N=1}^{3} F_{ijm}(\eta_1/a_1, \eta_2/a_2, \xi_3^N(\eta_1/a_1, \eta_2/a_2))e^{i(p-q)\phi}$$
$$\times\frac{\partial^2}{\partial g^2}\int_1^{\{(g+1)/(g-1)\}^{1/2}}\left\{\left[-g+\frac{\sqrt{g^2-1}}{2}\left(w+\frac{1}{w}\right)\right]^p\right.$$
$$\times\left[g+\frac{\sqrt{g^2-1}}{2}\left(w+\frac{1}{w}\right)\right]^q$$
$$\left.+\left[-g+\frac{\sqrt{g^2-1}}{2}\left(w+\frac{1}{w}\right)\right]^{p+1}\left[g+\frac{\sqrt{g^2-1}}{2}\left(w+\frac{1}{w}\right)\right]^{q+1}\right\}\frac{dw}{w}, \tag{29.12}$$

where

$$F_{im}(\xi_1, \xi_2, \xi_3^N) = C_{jlm3}\xi_l^N N_{ij}(\boldsymbol{\xi}^N)/\partial D(\boldsymbol{\xi}^N)/\partial \xi_3,$$
$$F_{ijm}(\xi_1, \xi_2, \xi_3^N) = C_{ijkl}C_{pqm3}\xi_q^N\xi_l^N N_{kp}(\boldsymbol{\xi}^N)/\partial D(\boldsymbol{\xi}^N)/\partial \xi_3; \tag{29.13}$$

$$\eta_1 = \cos\phi, \qquad \eta_2 = \sin\phi, \qquad \eta_1 + i\eta_2 = \exp(i\phi),$$
$$y_1 = x_1/a_1, \qquad y_2 = x_2/a_2, \qquad y = (y_1^2 + y_2^2)^{1/2},$$
$$g = -x_3\xi_3^N(\eta_1/a_1, \eta_2/a_2) - \boldsymbol{\eta}\cdot\mathbf{y}, \tag{29.14}$$
$$D(\xi_1, \xi_2, \xi_3^N) = 0.$$

When $x_3 = 0$,

$$\sigma_{ij}(x_1, x_2, 0) = \frac{i}{4}\bar{b}_m\int_0^{2\pi}\sum_{N=1}^{3} F_{ijm}(\eta_1/a_1, \eta_2/a_2, \xi_3^N(\eta_1/a_1, \eta_2/a_2))e^{i(p-q)\phi}$$
$$\frac{\partial^2}{\partial(\boldsymbol{\eta}\cdot\mathbf{y})^2}[K_{pq}(\boldsymbol{\eta}\cdot\mathbf{y}) - K_{p+1,q+1}(\boldsymbol{\eta}\cdot\mathbf{y})]d\phi, \tag{29.15}$$

where

$$K_{pq}(\boldsymbol{\eta}\cdot\mathbf{y}) = \sum_{\substack{m=0\\ m+n:\text{even}}}^{p}\sum_{n=0}^{q}(-1)^{(m-n)/2}\binom{p}{m}\binom{q}{n}\binom{m+n}{\frac{m+n}{2}}\left(\frac{1-(\boldsymbol{\eta}\cdot\mathbf{y})^2}{4}\right)^{(m+n)/2}$$
$$\times(\boldsymbol{\eta}\cdot\mathbf{y})^{p+q-m-n}. \tag{29.16}$$

It is seen that $\sigma_{ij}(x_1, x_2, 0)$ is a polynomial of degree $(p+q)$ in y_1, y_2 when $|\mathbf{y}| < 1$. The boundary conditions (29.2) are satisfied by choosing suitable constants $\bar{b}_1, \bar{b}_2, \bar{b}_3$ when σ_{3i}^0 is a polynomial of degree $(p+q)$ in y_1, y_2.

The relative displacement of the two crack surfaces is

$$u_i(x_1, x_2, +0) - u_i(x_1, x_2, -0) = b_i(x_1, x_2). \tag{29.17}$$

This is based upon the fact that the Burgers vector $\mathbf{b}$ (or dislocation density on a plane) is defined by the multiple-valuedness of the displacement across the crack plane. Since the displacement is symmetric about the crack plane, $u_i(x_1, x_2, +0) = \frac{1}{2}b_i(x_1, x_2)$. When $\bar{\mathbf{b}}$ is determined by the boundary conditions, the elastic field is completely determined by (29.11) and (29.12).

The interaction energy (25.17) becomes

$$\Delta W = -\frac{1}{2}\int_v \sigma_{m3}^0\epsilon_{3m}^* d\mathbf{x}' = \frac{1}{2}\int_S \sigma_{3m}^0 b_m(\mathbf{x}')dx_1'dx_2' \tag{29.18}$$

which is used for the Griffith fracture criterion calculation.

In (29.16) we have $K_{00} = 1$, $K_{11} = \frac{1}{2}(\boldsymbol{\eta}\cdot\mathbf{y})^2 + \frac{1}{2}$, $K_{10} = (\boldsymbol{\eta}\cdot\mathbf{y})$, and $K_{21} =$

$\frac{1}{2}(\boldsymbol{\eta}\cdot\mathbf{y})^3+\frac{1}{2}(\boldsymbol{\eta}\cdot\mathbf{y})$. These values are used for a constant applied stress $p=q=0$ and a linear applied stress $p=1, q=0$.

Uniform applied stresses

The integral with respect to w in (29.12) is denoted by I. Then, for $p=q=0$,

$$\frac{\partial^2 I}{\partial g^2}=-\frac{1}{2}\log\frac{g+1}{g-1}+\frac{g}{g^2-1}. \tag{29.19}$$

Since the imaginary part of g is positive, $-\pi\leq\arg(g+1)/(g-1)\leq 0$. When $x_3=0$, and $y<1$, we have $g=-(\boldsymbol{\eta}\cdot\mathbf{y})$ and $\log(g+1)/(g-1)=\log|(1+g)/(1-g)|-i\pi$. Furthermore, $\log|(1+g)/(1-g)|$ and $g/(g^2-1)$ are odd functions of $\boldsymbol{\eta}$ for $x_3=0$. On the other hand, $\Sigma_{N=1}^{3}F_{ijm}$ is an even function of $\boldsymbol{\eta}$. Therefore, for $x_3=0$ and $y<1$, $-\frac{1}{2}\log|(1+g)/(1-g)|$ and $g/(g^2-1)$ do not contribute to (29.12) and we have

$$\sigma_{ij}=\frac{-i}{4}\int_0^{2\pi}\sum_{N=1}^{3}F_{ijm}(\eta_1/a_1,\ \eta_2/a_2,\ \xi_3^N(\eta_1/a_1,\ \eta_2/a_2))\bar{b}_m d\phi. \tag{29.20}$$

The boundary conditions (29.2) determine the unknown constants $\bar{b}_1$, $\bar{b}_2$, and $\bar{b}_3$.

The stress intensity factor at crack tip is defined by

$$k_{ij}=\lim_{r\to 0}(2\pi r)^{1/2}\sigma_{ij}, \tag{29.21}$$

where r is the normal distance from the edge of the crack. For a flat ellipsoidal crack,

$$r=(y-1)/(x_1^2/a_1^4+x_2^2/a_2^4)^{1/2}. \tag{29.22}$$

The stress components σ_{ij} in (29.21) can be evaluated from (29.12) in the neighborhood of the crack edge. When $\mathbf{x}$ is close to the crack edge, the second term in (29.19) is dominant. Accordingly, we consider the integral

$$J(\mathbf{x})=\frac{1}{2\pi}\int_0^{2\pi}\frac{f(\boldsymbol{\eta})}{g^2-1}\,d\phi, \tag{29.23}$$

where $g=-(\boldsymbol{\eta}\cdot\mathbf{y})$. We have

$$J(\mathbf{x})=\frac{1}{2\pi}\int_0^{2\pi}\frac{f(\boldsymbol{\eta})-f(\bar{\mathbf{y}})}{g^2-1}\,d\phi+\frac{f(\bar{\mathbf{y}})}{2\pi}\int_0^{2\pi}\frac{d\phi}{g^2-1}, \tag{29.24}$$

where $\bar{\mathbf{y}} = \mathbf{y}/y$. The first term in (29.24) remains bounded as $y \to 1$, so that

$$J(\mathbf{x}) = \frac{f(\bar{\mathbf{y}})}{2\pi} \int_0^{2\pi} \frac{d\phi}{g^2 - 1} + 0(1) = if(\bar{\mathbf{y}})/(y^2 - 1)^{1/2} + 0(1). \tag{29.25}$$

Thus,

$$\sigma_{ij} = \frac{-1}{2\pi} \int_0^{2\pi} \sum_{N=1}^{3} F_{ijm} \bar{b}_m \frac{g}{g^2 - 1} \, d\phi + 0(1) \tag{29.26}$$

$$= \left[\frac{i(\boldsymbol{\eta} \cdot \mathbf{y})}{(y^2 - 1)^{1/2}} \sum_{N=1}^{3} F_{ijm}(\eta_1/a_1, \eta_2/a_2, \xi_3^N(\eta_1/a_1, \eta_2/a_2)) \bar{b}_m \right]_{\substack{y \to 1 \\ \eta \to y}} + 0(1).$$

The stress intensity factor (29.21) becomes

$$k_{ij} = i\sqrt{\pi}(x_1^2/a_1^4 + x_2^2/a_2^4)^{-1/4} \sum_{N=1}^{3} F_{ijm}(x_1/a_1^2, x_2/a_2^2, \xi_3^N(x_1/a_1^2, x_2/a_2^2)) \bar{b}_m. \tag{29.27}$$

The stress intensity factor of a flat ellipsoidal crack in transversely isotropic media has been investigated by Shield (1951), Kassir and Sih (1968) and Hoenig (1978). Their methods differ from the Willis' approach (1968). Explicit expressions for k_I, k_{II}, and k_{III} are presented by Hoenig.

For isotropic materials, (29.9) has two roots. The root with a negative imaginary part is

$$\xi_3(\xi_1, \xi_2) = -i(\xi_1^2 + \xi_2^2)^{1/2}. \tag{29.28}$$

Using (3.34), we can write (29.13) as

$$F_{311}(\xi_1, \xi_2, \xi_3) = \frac{-i\mu(\lambda + \mu)}{\lambda + 2\mu} \left(\xi_1^2 + \frac{1}{2} \frac{\lambda + 2\mu}{\lambda + \mu} \xi_2^2 \right) / (\xi_1^2 + \xi_2^2)^{1/2},$$

$$F_{312} = F_{321}(\xi_1, \xi_2, \xi_3) = \frac{-i\mu(\lambda + \mu)}{\lambda + 2\mu} \frac{\lambda}{2(\lambda + \mu)} \xi_1 \xi_2 / (\xi_1^2 + \xi_2^2)^{1/2},$$

$$F_{313} = F_{323} = F_{331} = F_{332} = 0, \tag{29.29}$$

$$F_{322}(\xi_1, \xi_2, \xi_3) = \frac{-i\mu(\lambda + \mu)}{\lambda + 2\mu} \left(\frac{1}{2} \frac{\lambda + 2\mu}{\lambda + \mu} \xi_1^2 + \xi_2^2 \right) / (\xi_1^2 + \xi_2^2)^{1/2},$$

$$F_{333}(\xi_1, \xi_2, \xi_3) = \frac{-i\mu(\lambda + \mu)}{\lambda + 2\mu} (\xi_1^2 + \xi_2^2)^{1/2};$$

and (29.20) becomes

$$\sigma_{31} = -\frac{\mu}{4} \frac{\lambda + \mu}{\lambda + 2\mu} \int_0^{2\pi} \left(\frac{\eta_1^2}{a_1^2} + \frac{1}{2} \frac{\lambda + 2\mu}{\lambda + \mu} \frac{\eta_2^2}{a_2^2} \right) \bar{b}_1 \frac{d\phi}{(\eta_1^2/a_1^2 + \eta_2^2/a_2^2)^{1/2}},$$

$$\sigma_{32} = -\frac{\mu}{4}\frac{\lambda+\mu}{\lambda+2\mu}\int_0^{2\pi}\left(\frac{1}{2}\frac{\lambda+2\mu}{\lambda+\mu}\frac{\eta_1^2}{a_1^2}+\frac{\eta_2^2}{a_2^2}\right)\bar{b}_2\frac{d\phi}{(\eta_1^2/a_1^2+\eta_2^2/a_2^2)^{1/2}}, \tag{29.30}$$

$$\sigma_{33} = -\frac{\mu}{4}\frac{\lambda+\mu}{\lambda+2\mu}\int_0^{2\pi}\bar{b}_3(\eta_1^2/a_1^2+\eta_2^2/a_2^2)^{1/2}d\phi.$$

If $\sigma_{ij}^0 = \sigma_{33}^0$, boundary conditions (29.2) yield $\bar{b}_1 = \bar{b}_2 = 0$, and

$$\bar{b}_3 = \frac{2(1-\nu)}{\mu E(k)}a_2\sigma_{33}^0. \tag{29.31}$$

$E(k)$ is defined by (11.25). Result (29.31) is consistent with (27.23), since $2a_3\epsilon_{33}^* = \bar{b}_3$. The stress intensity factor k_{33} in (29.27) is denoted by k_I for $\sigma_{ij}^0 = \sigma_{33}^0$, i.e.

$$k_I = \sigma_{33}^0 a_2\sqrt{\pi}(x_1^2/a_1^4+x_2^2/a_2^4)^{1/4}/E(k). \tag{29.32}$$

For a penny-shaped crack where $a_1 = a_2 = a$ and $E(k) = \pi/2$,

$$k_I = \sigma_{33}^0(\pi a)^{1/2}/(\pi/2). \tag{29.32'}$$

The result (29.32) has been obtained by Sadowsky and Sternberg (1949) while the special case of a penny-shaped crack has been considered by Sneddon (1946).

In a similar manner, we obtain the stress intensity factor for shear stresses. When $\sigma_{ij}^0 = \sigma_{31}^0$, boundary conditions (29.2) and (29.30) lead to

$$\bar{b}_1 = \sigma_{31}^0 a_2 k^2 / \frac{\mu}{2(1-\nu)}\{(k^2-\nu)E(k)+\nu(1-k^2)F(k)\} \tag{29.33}$$

and $\bar{b}_2 = \bar{b}_3 = 0$, where $F(k)$ and $E(k)$ are defined in (11.25). The stress intensity factors (29.27) become

$$k_{31} = \sigma_{31}^0 a_2\sqrt{\pi}k^2\frac{(x_1^2/a_1^4+x_2^2/a_2^4)^{-3/4}\{x_1^2/a_1^4+(1-\nu)x_2^2/a_2^4\}}{(k^2-\nu)E(k)+\nu(1-k^2)F(k)}, \tag{29.34}$$

$$k_{32} = \sigma_{31}^0 a_2\sqrt{\pi}k^2\frac{(x_1^2/a_1^4+x_2^2/a_2^4)^{-3/4}\nu(x_1/a_1^2)(x_2/a_2^2)}{(k^2-\nu)E(k)+\nu(1-k^2)F(k)}. \tag{29.34'}$$

Denoting the normal vector along the ellipse by $\boldsymbol{\nu}$, we define the stress intensity factor of the second kind k_{II} by

$$k_{II} = k_{31}\nu_1 + k_{32}\nu_2, \tag{29.35}$$

where

$$\begin{aligned} \nu_1 &= (x_1/a_1^2)/(x_1^2/a_1^4+x_2^2/a_2^4)^{1/2}, \\ \nu_2 &= (x_2/a_2^2)/(x_1^2/a_1^4+x_2^2/a_2^4)^{1/2}. \end{aligned} \tag{29.36}$$

From this, we have

$$k_{II} = \sigma_{31}^0 a_2 \sqrt{\pi} k^2 \frac{(x_1^2/a_1^4 + x_2^2/a_2^4)^{-1/4}(x_1/a_1^2)}{(k^2-\nu)E(k) + \nu(1-k^2)F(k)}. \tag{29.37}$$

In the derivation, the following integrals are used:

$$\begin{aligned} &\int_0^{\pi/2} \left(\frac{\cos^2\phi}{a_1^2} + \frac{\sin^2\phi}{a_2^2}\right)^{1/2} d\phi = E(k)/a_2, \\ &\int_0^{\pi/2} \frac{\sin^2\phi}{a_2}\left(\frac{\cos^2\phi}{a_1^2} + \frac{\sin^2\phi}{a_2^2}\right)^{-1/2} d\phi = (1-1/k^2)F(k) + E(k)/k^2. \end{aligned} \tag{29.38}$$

Expression (29.37) is consistent with Kassir and Sih's result (1966).

Non-uniform applied stresses

Let us consider the case where the applied stress before the disturbance due to the crack, is a linear function of x_1 and x_2. Then, $p = 1$ and $q = 0$ are chosen in (29.6). When the integral with respect to w in (29.12) is denoted by I, we have

$$\frac{\partial^2 I}{\partial g^2} = \frac{3}{2} g \log\frac{g+1}{g-1} + \frac{2-3g^2}{g^2-1}. \tag{29.39}$$

For $x_3 = 0$ and $y < 1$, $\log(g+1)/(g-1) = \log|(1+g)/(1-g)| - i\pi$, $g\log|(1+g)/(1-g)|$ and $(2-3g^2)/(g^2-1)$ are even functions of $\boldsymbol{\eta}$. Since $\Sigma F_{ijm}e^{i\phi}$ is an odd function of $\boldsymbol{\eta}$, (29.12) becomes

$$\begin{aligned} \sigma_{ij} = &-\frac{3i}{4}\int_0^{2\pi} \sum_{N=1}^{3} F_{ijm}(\eta_1/a_1, \eta_2/a_2, \xi_3^N(\eta_1/a_1, \eta_2/a_2)) \\ &\times(\eta_1 + i\eta_2)\bar{b}_m(\boldsymbol{\eta}\cdot\mathbf{y})d\phi. \end{aligned} \tag{29.40}$$

The boundary conditions (29.2) determine the unknown constants.

For a stress near the crack edge, the second term in (29.39) is dominant. Thus, for $x_3 = 0$ and $y \to 1$,

$$\begin{aligned} \sigma_{ij} &= \frac{-1}{2\pi}\int_0^{2\pi} \sum_{N=1}^{3} F_{ijm}\bar{b}_m e^{i\phi}\frac{2-3g^2}{g^2-1}d\phi + 0(1) \\ &= \left[\frac{-i\{2-3(\boldsymbol{\eta}\cdot\mathbf{y})^2\}}{(y^2-1)^{1/2}} \sum_{N=1}^{3} F_{ijm}(\eta_1/a_1, \eta_2/a_2, \xi_3^N(\eta_1/a_1, \eta_2/a_2)) \right. \\ &\quad \left. \times(\eta_1 + i\eta_2)\bar{b}_m\right]_{\substack{y\to 1\\ \boldsymbol{\eta}\to\mathbf{y}}} + 0(1). \end{aligned} \tag{29.41}$$

The stress intensity factor becomes

$$k_{ij} = i\sqrt{\pi}(x_1^2/a_1^4 + x_2^2/a_2^4)^{-1/4} \sum_{N=1}^{3} F_{ijm}(x_1/a_1^2, x_2/a_2^2, \xi_3^N(x_1/a_1^2, x_2/a_2^2)) \times (x_1/a_1 + ix_2/a_2)\bar{b}_m. \tag{29.42}$$

If the material is isotropic and the applied stress is given by

$$\sigma_{33}^0 = \frac{3\mu}{4} \frac{\lambda + \mu}{\lambda + 2\mu}(A_1 y_1 + A_2 y_2), \tag{29.43}$$

we choose

$$\bar{b}_3 = b + ib', \tag{29.44}$$

where b and b' are real numbers. Then, condition $\sigma_{33} = -\sigma_{33}^0$ applied to (29.40) leads to

$$\begin{aligned} b &= A_1 / \int_0^{2\pi} \eta_1^2(\eta_1^2/a_1^2 + \eta_2^2/a_2^2)^{1/2} d\phi, \\ b' &= -A_2 / \int_0^{2\pi} \eta_2^2(\eta_1^2/a_1^2 + \eta_2^2/a_2^2)^{1/2} d\phi. \end{aligned} \tag{29.45}$$

The stress intensity factor k_{33} is denoted by k_I,

$$k_I = a_2(\sqrt{\pi}/3)(x_1^2/a_1^4 + x_2^2/a_2^4)^{1/4}\left(x_1 \frac{\partial \sigma_{33}^0}{\partial x_1}/E_1(k) + x_2 \frac{\partial \sigma_{33}^0}{\partial x_2}/E_2(k)\right), \tag{29.46}$$

where

$$\begin{aligned} E_1(k) &= \int_0^{\pi/2} \sin^2\phi(1 - k^2\sin^2\phi)^{1/2} d\phi, \\ E_2(k) &= \int_0^{\pi/2} \cos^2\phi(1 - k^2\sin^2\phi)^{1/2} d\phi, \\ k^2 &= (a_1^2 - a_2^2)/a_1^2. \end{aligned} \tag{29.47}$$

The stress intensity factor k_{II} defined by (29.35) is derived in a similar manner, while the stress intensity factor k_{III} of the third kind is defined by

$$k_{III} = k_{31}t_1 + k_{32}t_2, \tag{29.48}$$

where vector $\mathbf{t}$ is the tangential unit vector along the ellipse.

Kassir and Sih (1968) have found that k_I under a uniform σ_{33}^0 is independent of the elastic moduli of a transversely isotropic solid. Barnett and Asaro (1972) have found that for a slit-like crack the stress intensity factor K_{ij} is independent of the elastic moduli for any arbitrarily anisotropic material.

Sekine and Mura (1979) have modified Willis' method (1968) and proved that if the displacement discontinuity of the elliptical Somigliana's dislocation has the form $P_N(x_1, x_2)(1 - x_1^2/a_1^2 - x_2^2/a_2^2)^{1/2}$, where P_N is a homogeneous polynomial of degree N, the stresses on the plane of the Somigliana's dislocation are inhomogeneous polynomials in the coordinates, whose terms are of degree N, $(N-2)$, $(N-4)$, ... The same subject has been discussed by Kassir and Sih (1967) under the more restrictive condition $P_N = P_N(x_1^2, x_2^2)$.

Certain special cases relating to elliptical disk-shaped cracks have been examined in the literature. Chen (1966) has solved the problem of an elliptical crack subjected to linearly varying loads in a transversely isotropic body. Stress intensity factors for an embedded elliptical crack or a semi-elliptical crack approaching the free surface of the semi-infinite solid under uniform tension perpendicular to the plane of crack, have been investigated by Miyamoto and Miyoshi (1971), Shah and Kobayashi (1973), and Nisitani and Murakami (1974). No analytical expressions are available for the stress intensity factors.

30. Stress intensity factor for a slit-like crack

The stress intensity factors of a slit-like crack in anisotropic media can be obtained from (29.27) and (29.42) as a special case when $a_1 \to \infty$ and $x_2 \to a_2 = a$. However, more explicit expressions can be obtained by the method of continuously distributed dislocations (Bilby and Eshelby 1968). We follow the work of Barnett and Asaro (1972) in this section.

Consider a slit-like crack in an infinite anisotropic elastic medium (Fig. 30.1). σ_{ij}^0 denotes the applied stress at infinity. The stress dis-

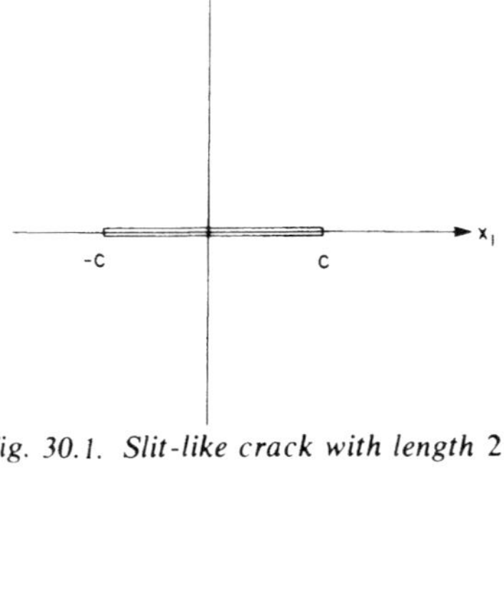

Fig. 30.1. Slit-like crack with length 2C.

turbance induced by the crack is denoted by σ_{ij}, so that the total stress field is $\sigma^0_{ij} + \sigma_{ij}$. The boundary conditions are

$$\sigma^0_{2i} + \sigma_{2i} = 0, \qquad \text{for } x_2 = 0, \qquad |x_1| \leq c. \tag{30.1}$$

σ_{ij} is simulated by the stress field due to continuously distributed dislocations on $x_2 = 0, |x_1| \leq c$, where the dislocation line direction is parallel to the x_3-axis. From the elastic theory of dislocations (Willis 1970 and Barnett and Swanger 1971), the stress at $x_1, x_2 = 0$ due to a single dislocation located at $x_1 = t, x_2 = 0$ with the Burgers vector b_j, is expressed as

$$\sigma_{2i} = K_{ij}b_j/(x_1 - t). \tag{30.2}$$

The boundary conditions (30.1) may be satisfied if the dislocations are continuously distributed with strength $b_j(t)$ (Burgers vector times distribution density). Then, (30.1) becomes

$$K_{ij} \int_{-c}^{c} b_j(t)(x_1 - t)^{-1}dt = -\sigma^0_{2i} \tag{30.3}$$

or

$$\int_{-c}^{c} b_j(t)(x_1 - t)^{-1}dt = -K^{-1}_{ji}\sigma^0_{2i}. \tag{30.4}$$

The inverse matrix components K^{-1}_{ji} are given by

$$K^{-1}_{ji} = K^{-1}_{ij} = \epsilon_{jmn}\epsilon_{irs}K_{mr}K_{ns}/2\epsilon_{abc}K_{1a}K_{2b}K_{3c}, \tag{30.5}$$

so that

$$K^{-1}_{ji}K_{is} = K_{ji}K^{-1}_{is} = \delta_{js}. \tag{30.6}$$

Equation (30.4) is the Hilbert integral equation for $b_j(t)$.

Uniform applied stresses

For constant σ^0_{2i}, as seen in Appendix 4, the solution of (30.4) is easily obtained as

$$b_j(t) = K^{-1}_{ji}\sigma^0_{2i}t/\pi(c^2 - t^2)^{1/2}, \tag{30.7}$$

where Q_0 in (A.4.5) of Appendix 4 is taken to be zero in view of condition

$$\int_{-c}^{c} b_j(t)dt = 0. \tag{30.8}$$

This condition is satisfied, since there is no relative displacement of the crack surface at $x_1 = \pm c$. The relative displacement (crack opening displacement) Δu_j at x_1 is given by

$$\Delta u_j = \int_{x_1}^{c} b_j(t)dt \tag{30.9}$$

from dislocation theory.

The stress traction acting on the plane of the crack ($|x_1| > c, x_2 = 0$) becomes

$$\begin{aligned} &\sigma_{2i}^0 + K_{ij} \int_{-c}^{c} K_{jm}^{-1} \sigma_{2m}^0 (t/\pi)(c^2 - t^2)^{-1/2}(x_1 - t)^{-1} dt \\ &= \sigma_{2i}^0 + \int_{-c}^{c} \sigma_{2i}^0 (t/\pi)(c^2 - t^2)^{-1/2}(x_1 - t)^{-1} dt \\ &= \sigma_{2i}^0 |x_1|/(x_1^2 - c^2)^{1/2} \qquad (|x_1| > c). \end{aligned} \tag{30.10}$$

The stress intensity factor defined by (29.21) leads to

$$k_{2i} = \sigma_{2i}^0 \sqrt{\pi c} \tag{30.11}$$

where $r = (x_1 - c)$. The last equation is Barnett and Asaro's remarkable result. The stress intensity factor k_{2i} due to σ_{2i}^0 is independent of elastic moduli. Result (30.11) is a special case of (29.32) and (29.34) for $a_1 \to \infty$, $a_2 \to c$.

From (30.7) and (30.9), the crack opening displacement is

$$\Delta u_j = K_{ji}^{-1} \sigma_{2i}^0 (c^2 - x_1^2)^{1/2}/\pi. \tag{30.12}$$

The interaction energy ΔW is expressed as

$$\Delta W = -(1/2) \int_{-c}^{c} \sigma_{2j}^0 \Delta u_j dx_1 \tag{30.13}$$

which becomes

$$\Delta W = -(1/4) K_{ji}^{-1} \sigma_{2i}^0 \sigma_{2j}^0 c^2. \tag{30.14}$$

The crack extension force is then given by

$$-\partial(\Delta W)/\partial c = (1/2) K_{ji}^{-1} \sigma_{2i}^0 \sigma_{2j}^0 c. \tag{30.15}$$

The critical applied stress can be obtained by the use of (27.19) where $a_1 = c$.

When the material is isotropic, $K_{11} = K_{22} = \mu/2\pi(1-\nu)$, $K_{33} = \mu/2\pi$ and $K_{ij} = 0$ $(i \neq j)$ (see Read, 1953). The critical stresses obtained from (30.15) and (27.19) agree with (27.20) and (27.21).

The tensor components appearing in (30.14) and (30.15) are referred to the x coordinate system shown in Fig. 30.1. Since the form of (30.14) and (30.15) is invariant for coordinate systems, we can choose a new coordinate system in the crystallographic directions in which the elastic moduli appear in their simplest form.

Non-uniform applied stresses

When an applied stress σ_{2i}^0 is a function of x_1 on $x_2 = 0, |x_1| \leq c$, the integral equation (30.4) still holds. The solution has been obtained by Barnett and Asaro (1972). Since the involved analysis is essentially the same as that in the following section for the Dugdale crack, the calculation is not reported here.

Isotropic materials

The stress intensity factors of a slit-like crack in an infinitely extended isotropic material are the same as (30.11). These factors, $k_{22} = k_I$, $k_{21} = k_{II}$, $k_{23} = k_{III}$, have been obtained by Westergaard (1939). Irwin (1948) pointed out the relation between the critical stress and the stress intensity factor (see also Irwin 1957, 1958). The associated stress intensity factors for cases where an infinitesimally thin notch (crack) is introduced into a semi-infinite slab have been investigated by many researchers (see Wigglesworth 1957, Lachenbruch 1961, Koiter 1965, Nishitani and Murakami 1972, and others). For a strip with a center notch (crack), numerical calculations for the stress intensity factors have been performed by Isida (1970, 1971). The stress intensity factor for an inner or edge crack in arbitrarily shaped plates has been investigated by Murakami (1978) by using the Green function for the traction-free crack in an infinite plate (see also Erdogan 1962). Yamamoto and Sumi (1978) recently reported that a slit-like crack is not a two-dimensional problem (neither plane strain nor plane stress), since for moderate values of the ratio of crack length to the plate thickness and because of the effect of Poisson's ratio, the necessary conditions for neither plane stress nor plane strain can be satisfied. They evaluate the variation of the stress intensity factor along the crack front.

Stress intensity factors of notches and cracks in various standard test specimens and machine elements under various applied loads have been investigated by a number of researchers, and many of the results are listed in a handbook by Tada (1973), and in a special publication by Kamei and Yokobori (1974).

A recent development is the use of the finite element method in the evaluation of the stress intensity factors (see Zienkiewicz, 1971). Early original work in this area are found in papers by Watwood (1969), Chan, Tuba and Wilson (1970), and Dixson and Pook (1969). Since the stress field in the neighborhood of the crack tip is proportional to $1/\sqrt{r}$ (r is the distance from the crack tip), the proportionality constant, i.e., the stress intensity factor, can be obtained by plotting the finite element solution as a function r. The stress intensity factor can also be found by plotting the crack opening displacement versus r. These two methods are called the stress and displacement methods. A third method, an energy method, has been proposed, based on the fact that the energy release rate due to the crack extension is proportional to the square of the stress intensity factor. The first two methods do not give satisfactory results unless very fine meshes are employed, while moderate accuracy can be obtained by the energy method. The energy method, however, cannot be applied to mixed mode or three-dimensional problems. Recently, the stress method has been reconsidered by Miyata and Kusumoto (1975) and Murakami (1976, 1978). They have found empirical relations between stress intensity factors and stress solutions obtained by the finite element method, and they have obtained stress intensity factors for several three-dimensional cracks with relatively coarse meshes.

Byskov(1970), Rao et al. (1971) and Pian et al. (1971) proposed special elements by using the hybrid variational method or the collocation method. These methods introduce additional freedom in conjunction with Williams' coefficients (1957) for stress singularities. Another interesting approach, proposed by Yamamoto (1971), is illustrated in Fig. 30.2. The exact stress field σ is $\sigma \approx \sigma^a - \sigma^r + \sigma^e$, where σ^a is an analytical solution for an infinite body, σ^r is a solution by the finite element method in the finite body subjected to a boundary force with the same traction as at the corresponding imaginary boundary in an infinite body, and σ^e is a solution of the finite body with the originally given load by the finite element method. Yamamoto postulated that, in the neighborhood of the crack tip, $\sigma = \alpha(\sigma^a - \sigma^r) + \sigma^e = \alpha\sigma^a$, where α is introduced as a Lagrangian multiplier which coincides with the stress intensity factor. Thus, we have $\alpha = \sigma^e/\sigma^r$. This method has been further extended by Yamamoto and Sumi (1978, 1979) to include three-dimensional crack problems.

The theory of stability of interacting tension cracks in brittle solids has been recently developed by Nemat-Nasser (1978), Nemat-Nasser et al. (1978), Keer et al. (1979), and Sumi et al. (1980), in connection with thermally induced edge cracks in plane strain conditions. These authors

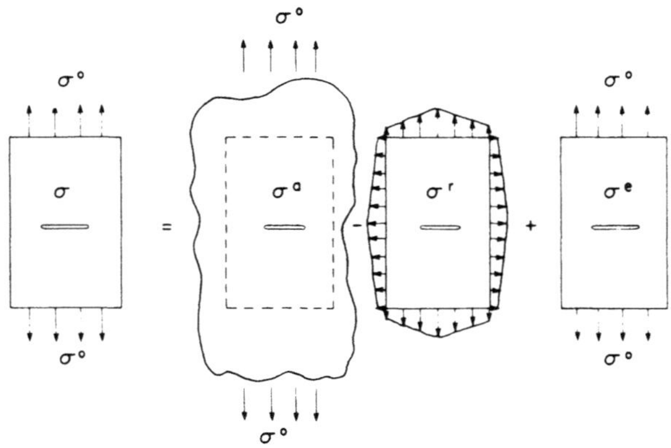

Fig. 30.2. Superposition method.

show that as a half-space is cooled on its free surface, tension cracks develop and penetrate into the half-space. However, a critical state is then reached at which some cracks stop, as others grow at a faster rate. Moreover, depending on the cooling mechanism, another critical state may be attained at which the cracks which had stopped may actually snap closed as the remaining ones jump into a finitely longer length. The growth process is highly imperfection sensitive, as shown by Nemat-Nasser et al. (1980). In a series of experiments on glass plates, Geyer and Nemat-Nasser (1981) have obtained good qualitative and quantitative verification of the theory.

31. Stress concentration factors

Let us consider an ellipsoidal inhomogeneity in an infinite medium (see Fig. 31.1). σ_{ij}^0 denotes the applied stress, and σ_{ij} the stress disturbance due to the inhomogeneity. σ_{ij} can be simulated by the eigenstress associated with eigenstrain ϵ_{ij}^* of an equivalent inclusion in an homogeneous medium. The stress immediately outside the inclusion can be written as

$$\sigma_{ij}^T = [\sigma_{ij}] + \sigma_{ij}(\text{in}) + \sigma_{ij}^0, \tag{31.1}$$

where

$$[\sigma_{ij}] = \sigma_{ij}(\text{out}) - \sigma_{ij}(\text{in}); \tag{31.2}$$

see (6.11). The equivalent eigenstrain ϵ_{ij}^* in (6.11) must be determined from the equivalency condition, (22.7).

The stress concentration factor κ of σ_{ij}^T for a given $\sigma_{kl}^0 = S$ is defined by

$$\kappa = \sigma_{ij}^T / S. \tag{31.3}$$

If the inhomogeneity is a void, $\sigma_{ij}(\text{in}) + \sigma_{ij}^0 = 0$, and

$$\kappa = [\sigma_{ij}]/S. \tag{31.4}$$

Mura and Cheng (1977) have calculated κ for a three-dimensional lens-shaped crack, where a_3 is much smaller than a_1 or a_2. Since $a_3 \ll a_1, a_2$, the results (28.16) and (28.17) can be used. For simplicity the case $\alpha = 0$ in Fig. 28.1 is considered. The definition (31.3) does not mean that κ is same for all stress components.

Simple tension

Let σ_{33}^0 be prescribed with all other components of σ_{ij}^0 being zero. Combining (31.4), (6.11), and (28.16), we have the stress concentration factor,

$$\kappa = (4\pi/a_3)\{C_{3333} - C_{33kl}C_{pq33}G_{kplq}(\mathbf{n})\}/C_{33mn}C_{st33}\Pi_{msnt}, \tag{31.5}$$

where $G_{kplq}(\mathbf{n}) = N_{kp}(\mathbf{n})n_l n_q / D(\mathbf{n})$ and Π_{msnt} is defined by (28.12). The interesting points to be investigated are on the crack boundary with $x_3 = 0$. The radius ρ at a point corresponding to angle β in Fig. 31.1 is

$$\rho = (a_3^2/a_2)\{\sin^2\beta + (a_2/a_1)^2\cos^2\beta\}^{1/2}. \tag{31.6}$$

Equation (31.5) is then written as

$$\kappa = \frac{4\pi}{(\rho a_2)^{1/2}} \frac{\{C_{3333} - C_{33kl}C_{pq33}G_{kplq}(\mathbf{n})\}}{C_{33mn}C_{st33}\Pi_{msnt}} \left(\sin^2\beta + \frac{a_2^2}{a_1^2}\cos^2\beta\right)^{1/4}. \tag{31.7}$$

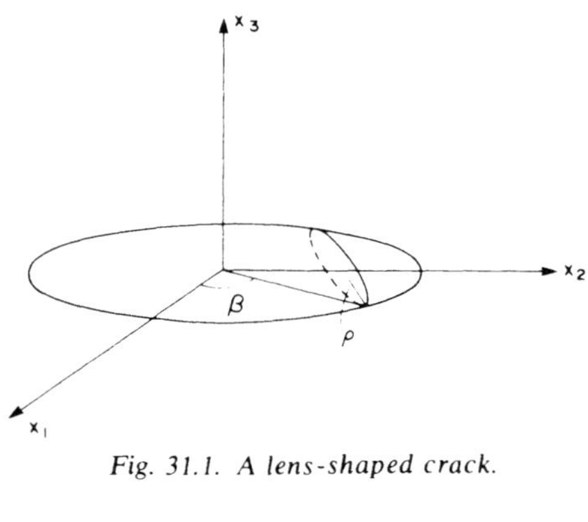

Fig. 31.1. A lens-shaped crack.

For isotropic materials it becomes

$$\kappa = 2(a_2/\rho)^{1/2}(\sin^2 \beta + \frac{a_2^2}{a_1^2}\cos^2 \beta)^{1/4}/E(k), \tag{31.8}$$

where $E(k)$ is defined in (11.25). When κ is considered at the point $\beta = 90°$ with $a_1 \to \infty$, it agrees with the well-known solution given by Inglis (1913).

Numerical values of $\kappa/(a_2/\rho)^{1/2}$ at $\beta = 90°$, where the stress concentration factor becomes maximum, are plotted in Fig. 31.2.

Pure shear

Let σ_{31}^0 be prescribed with all other components of σ_{ij}^0 being zero. From (31.4), (6.11), (28.17), and (31.6) the stress concentration factor of σ_{31}^T is

$$\kappa = \frac{4\pi}{(\rho a_2)^{1/2}} \frac{\{C_{3131} - C_{31kl}C_{pq31}G_{kplq}(\mathbf{n})\}}{C_{31mn}C_{st31}\Pi_{msnt}} \left(\sin^2 \beta + \frac{a_2^2}{a_1^2}\cos^2 \beta\right)^{1/4}. \tag{31.9}$$

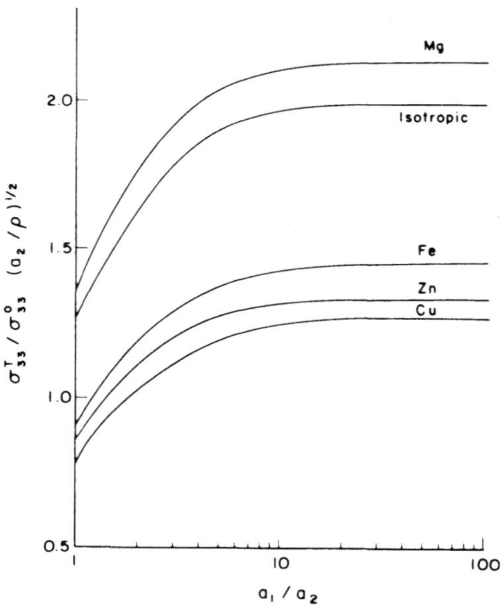

Fig. 31.2. Maximum tensile stress concentration factor divided by $(a_2/\rho)^{1/2}$ for various ratio of a_1/a_2.

For isotropic materials and $\beta = 90°$, it becomes

$$\kappa = (a_2/\rho)^{1/2}k^2 / \left[\frac{k^2}{1-\nu}\{F(k) - E(k)\} + \{E(k) - (k')^2F(k)\}\right], \tag{31.10}$$

where $F(k)$ is defined in (11.25) and $k'^2 = 1 - k^2$.

Numerical values of $\kappa/(a_2/\rho)^{1/2}$ at $\beta = 90°$ are plotted in Fig. 31.3. The elastic moduli in Figs. 31.2 and 31.3 are taken from the American Institute of Physics Handbook. It is observed that when a_1/a_2 approaches infinity, the values of $\kappa(a_2/\rho)^{1/2}$ approach constants which agree with the values obtained from Lekhnitskii's method (1968). The resulting displacement field agrees with Chen's solution (1966) for an elliptical disk-shaped crack in transversely isotropic materials.

The state of stress due to an ellipsoidal cavity in an infinite isotropic medium has been given by Sadowsky and Sternberg (1949) and Miyamoto (1953) for the case of three constant normal stresses, applied at infinity, having their directions aligned with the principal axes of the ellipsoid. Earlier solutions obtained by Neuber (1937) consider loading in applied shear as well as normal stresses for the special case of the spheroidal cavity. Mirandy and Paul (1976) also employ the equivalent inclusion method. They investigate the effect of Poisson's ratio on the stress concentration factors of an ellipsoidal cavity subjected to a completely arbitrary constant state of stress at infinity.

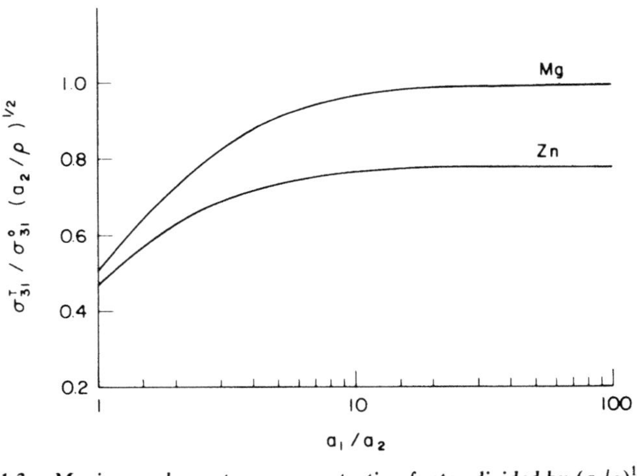

Fig. 31.3. Maximum shear stress concentration factor divided by $(a_2/\rho)^{1/2}$ for various ratio of a_1/a_2.

It should be pointed out that the stress intensity factors generally cannot be derived by the limiting process

$$k = \lim \sqrt{2\pi\rho}\,\kappa\sigma^0_{ij}. \tag{31.11}$$

32. Dugdale–Barenblatt cracks

In the absence of plastic deformations around crack tips (or notches), some stress components become infinite there. In reality, however, the stress field cannot exceed a certain limit because of the plastic deformation in the neighborhood of crack tips. Dugdale (1960) has simplified the situation and has assumed that the plastic region is one-dimensional, being confined on (c, a) and $(-a, -c)$, as shown in Fig. 32.1. He has related the size of the plastic domain to the applied stress and crack length $2c$ by assuming the material to be elastic–perfectly plastic (Fig. 32.2a). One method of coping with this problem of stress singularity is that of Barenblatt (1960). He postulates a system of cohesive stresses in the corresponding region (a, c) and $(-a, -c)$ in Fig. 32.1. The magnitude of these stresses is just sufficient to cause the usual stress singularity to vanish.

The boundary conditions for a Dugdale crack under a simple tension, $\sigma^0_{22} = S$, are

$$\begin{aligned}
&\sigma_{2j} = 0 \quad (j = 1, 2) && \text{at } x_2 = 0 \quad |x_1| < c,\\
&\sigma_{22} = k_0 && \text{at } x_2 = 0 \quad c < |x_1| < a,\\
&\sigma_{22} = S && \text{at } x_2 = \infty,\\
&u_2 = 0, \quad \sigma_{21} = 0 && \text{at } x_2 = 0 \quad a < |x_1|,
\end{aligned} \tag{32.1}$$

where σ^0_{22} is the applied tensile stress and k_0 is the tensile yield stress.

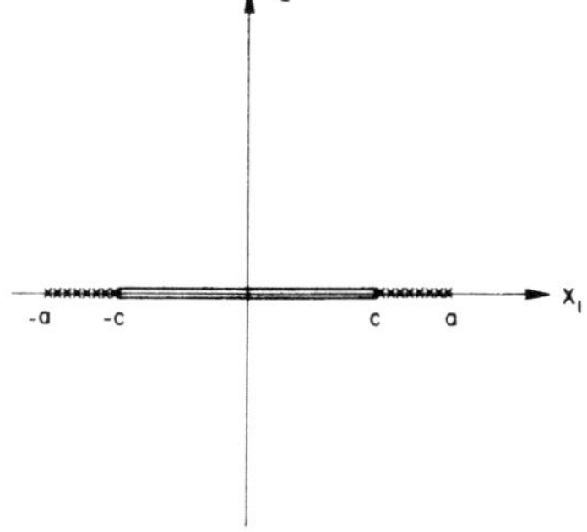

Fig. 32.1. Dugdale crack.

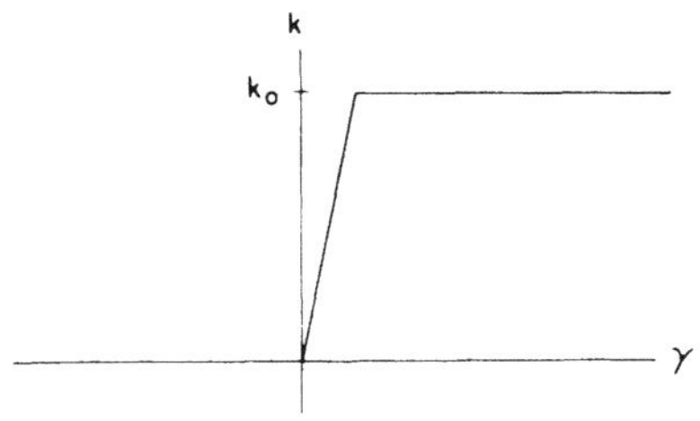

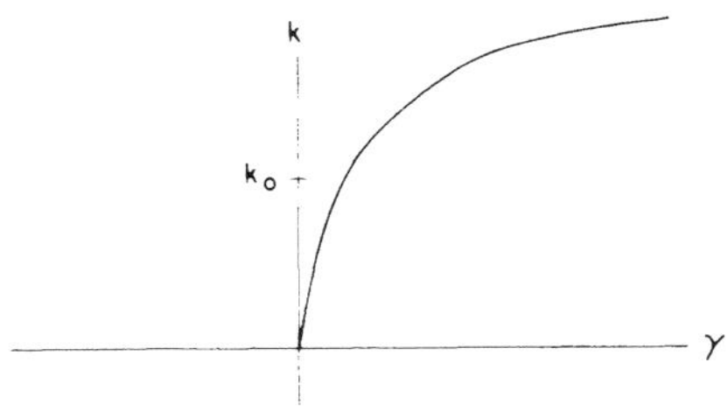

Fig. 32.2. Work hardening material.

The following stress analysis shows that $\sigma_{11} = \sigma_{22} = \sigma_{33}$ for the plane strain and $\sigma_{11} = \sigma_{22}$, $\sigma_{33} = 0$ for the plane stress at the crack tips. The boundary condition $\sigma_{22} = k_0$ is therefore valid only for the plane stress assumption, since $\sigma_{22} = k_0$ comes from the maximum shear theory $(\sigma_{22} - \sigma_{33})/2 = k_0/2$.

The differential equations to be solved are, for the isotropic materials,

$$\mu u_{i,jj} + (\mu + \lambda) u_{k,ki} = 0. \tag{32.2}$$

When the Papkovich–Neuber potential Ψ is used,

$$\begin{aligned} 2\mu u_1 &= \frac{1-\kappa}{2}\Psi_{,1} - x_2\Psi_{,12}, \\ 2\mu u_2 &= \frac{1+\kappa}{2}\Psi_{,2} - x_2\Psi_{,22}, \end{aligned} \tag{32.3}$$

where

$$\Psi_{,11} + \Psi_{,22} = 0, \tag{32.4}$$

$\kappa = (3-\nu)/(1+\nu)$ for the plane stress, and $\kappa = 3 - 4\nu$ for the plane strain, where ν is Poisson's ratio. The stress components derived from (32.3) are

$$\begin{aligned}\sigma_{22} &= \Psi_{,22} - x_2\Psi_{,222}\\ \sigma_{12} &= -x_2\Psi_{,122} \qquad\qquad (32.5)\\ \sigma_{11} &= -\Psi_{,11} - x_2\Psi_{,112}.\end{aligned}$$

Before we apply (32.5) to (32.1), it is convenient to rewrite (32.1) as

$$\begin{aligned}&\sigma_{22} = 0 && \text{at } x_2 = \infty,\\ &\sigma_{12} = 0 && \text{at } x_2 = \infty,\\ &\sigma_{12} = 0 && \text{at } x_2 = 0,\\ &\sigma_{22} = -S && \text{at } x_2 = 0, \quad |x_1| < c \qquad\qquad (32.6)\\ &\sigma_{22} = k_0 - S && \text{at } x_2 = 0, \quad c < |x_1| < a\\ &u_2 = 0 && \text{at } x_2 = 0, \quad a < |x_1|.\end{aligned}$$

The solution for (32.1) is obtained from the solution for (32.6) by superimposing S onto σ_{22}.

From the symmetry of the problem with respect to x_1, Ψ is chosen as

$$\Psi = -\int_0^\infty \xi^{-1}A(\xi)e^{-\xi x_2}\cos\xi x_1 d\xi, \qquad (32.7)$$

which satisfies (32.4). The first three conditions in (32.6) are automatically satisfied. With (32.5), the fourth and fifth conditions yield

$$\int_0^\infty \xi A(\xi)\cos\xi x_1 d\xi = \sigma(x_1), \qquad (32.8)$$

where

$$\begin{aligned}\sigma(x_1) &= S && \text{for } |x_1| < c\\ &= S - k_0 && \text{for } c < |x_1| < a. \qquad\qquad (32.9)\end{aligned}$$

Assuming

$$A(\xi) = \int_0^a q(t)J_0(\xi t)dt, \qquad (32.10)$$

where J_0 is the Bessel function of zero order, from (32.3) and (32.7) we have, at $x_2 = 0$,

$$\begin{aligned}\frac{4\mu}{1+\kappa}u_2 &= \int_0^a q(t)dt \int_0^\infty J_0(\xi t)\cos \xi x_1 d\xi \\ &= \int_0^a q(t)\frac{H(t-x_1)}{(t^2-x_1^2)^{1/2}}dt \\ &= \int_{x_1}^a \frac{q(t)}{(t^2-x_1^2)^{1/2}}dt \quad \text{for } |x_1|<a \\ &= 0 \qquad\qquad\qquad \text{for } a<|x_1|. \end{aligned} \tag{32.11}$$

The last condition in (32.6) is thereby satisfied. The unknown function q is determined from (32.8). Substituting (32.10) into (32.8), we obtain

$$\begin{aligned}\sigma(x_1) &= \frac{d}{dx_1}\int_0^a q(t)dt\int_0^\infty J_0(\xi t)\sin \xi x_1 d\xi = \frac{d}{dx_1}\int_0^a q(t)\frac{H(x_1-t)}{(x_1^2-t^2)^{1/2}}dt \\ &= \frac{d}{dx_1}\int_0^{x_1}\frac{q(t)}{(x_1^2-t^2)^{1/2}}dt \quad \text{for } |x_1|<a. \end{aligned} \tag{32.12}$$

Integrating with respect to x_1 we have Abel's integral equation

$$\int_0^{x_1}\frac{q(t)}{(x_1^2-t^2)^{1/2}}dt = P(x_1), \tag{32.13}$$

where

$$P(x_1) = \int_0^{x_1}\sigma(s)ds.$$

The solution is

$$q(t) = \frac{2}{\pi}\frac{d}{dt}\int_0^t \frac{x_1 P(x_1)dx_1}{(t^2-x_1^2)^{1/2}}$$

or integrating by parts,

$$q(t) = \frac{2}{\pi}t\int_0^t \frac{\sigma(x_1)}{(t^2-x_1^2)^{1/2}}dx_1. \tag{32.14}$$

When $\sigma(x_1)$ is given as (32.9), (32.14) becomes

$$\begin{aligned} q(t) &= St \quad \text{for } t<c \\ q(t) &= \frac{2}{\pi}t\left\{(S-k_0)\cos^{-1}\frac{c}{t} + S\sin^{-1}\frac{c}{t}\right\} \qquad \text{for } c<t<a. \end{aligned} \tag{32.15}$$

All quantities of the elastic field can be obtained from (32.15) by using (32.7) and (32.10). In order to have a finite value of stress, however, we must have

$$q(a) = 0, \tag{32.16}$$

because, at $x_2 = 0$, $a < |x_1|$, we have

$$\begin{aligned}\sigma_{22} &= -\frac{d}{dx_1}\int_0^a q(t)\frac{H(x_1 - t)}{(x_1^2 - t^2)^{1/2}}\,dt = -\frac{d}{dx_1}\int_0^a q(t)d\,\sin^{-1}\frac{t}{|x_1|}\\ &= \frac{aq(a)}{x_1(x_1^2 - a^2)^{1/2}} + \frac{1}{x_1}[q'(t)(x_1^2 - t^2)^{1/2}]_0^a - \int_0^a q''(t)(x_1^2 - t^2)^{1/2}dt.\end{aligned}$$

From (32.15) and (32.16) we have

$$\cos^{-1}\frac{c}{a} = \frac{\pi}{2}\frac{S}{k_0}. \tag{32.17}$$

The above equation gives the relationship between the applied stress and the plastic zone size. The relation has been found by Dugdale (1960).

Of particular interest is the displacement component u_2 in the plastic domain. It is obtained, from (32.11) and (32.15) for $x_2 = 0^+$,

$$\begin{aligned}u_2(x_1, c) &= \frac{(1+\kappa)k_0}{4\pi\mu}\left[(x_1 + c)\cosh^{-1}\left|\frac{m}{c + x_1} + n\right|\right.\\ &\quad\left. - (x_1 - c)\cosh^{-1}\left|\frac{m}{c - x_1} + n\right|\right]\\ &= \frac{(1+\kappa)k_0}{4\pi\mu}\left[x_1 \log\frac{x_1(a^2 - c^2)^{1/2} - c(a^2 - x_1^2)^{1/2}}{x_1(a^2 - c^2)^{1/2} + c(a^2 - x_1^2)^{1/2}}\right.\\ &\quad\left. + c\log\frac{(a^2 - c^2)^{1/2} + (a^2 - x_1^2)^{1/2}}{(a^2 - c^2)^{1/2} - (a^2 - x_1^2)^{1/2}}\right],\end{aligned} \tag{32.18}$$

where $m = (a^2 - c^2)/a$, $n = c/a$. The value at $x^2 = 0^-$ is the negative of (32.18). The displacement at the upper tip of the crack is obtained from (32.18) by setting $x_1 = c$,

$$u_2 = \frac{(1+\kappa)k_0 c}{2\pi\mu}\cosh^{-1}\left\{\frac{1}{2}\left(\frac{c}{a} + \frac{a}{c}\right)\right\}. \tag{32.19}$$

When $a/c \approx 1$ and $S/k_0 \ll 1$, (32.17) and (32.19) are approximated as

$$\begin{aligned}1 - c/a &= \frac{\pi^2}{8}(S/k_0)^2,\\ u_2 &= \frac{(1+\kappa)k_0}{2\pi\mu}c(1 - c/a) = \frac{(1+\kappa)}{16\mu}\pi c\,\frac{S^2}{k_0},\end{aligned} \tag{32.20}$$

and we have

$$u_2(x_1, c)/u_2(c, c) = (1-\xi)^{1/2} - \frac{\xi}{2}\log\frac{1+(1-\xi)^{1/2}}{1-(1-\xi)^{1/2}}, \tag{32.20.1}$$

where $\xi = (x_1 - c)/(a - c)$.

Orowan (1945, 1949) and Irwin (1948, 1957) have proposed that the applied critical stress S required to propagate a crack of length $2c$ in a semi-brittle solid of elastic constant μ, should be written in the form of a Griffith equation, i.e.,

$$S \simeq (\mu\gamma/c)^{1/2}, \tag{32.21}$$

in which γ represents the sum of the true surface energy and the plastic work required to increase the area of the crack by unit amount. This Orowan-Irwin formula can be obtained from (32.20). We postulate that the crack becomes unstable when the crack opening displacement $2u_2$ given in (32.20) reaches a critical value Φ. Then,

$$S = \left(\frac{8\mu k_0 \Phi}{\pi(1+\kappa)c}\right)^{1/2}. \tag{32.22}$$

The strain hardening effect can be taken into account in this result if an approximation is employed as done by Theocaris and Gdoutos (1974). Alternative interpretations of (32.21) are given by Weertman (1978) and Vilmann and Mura (1979).

A similar result can be obtained for the anti-plane problem. A shear stress $\sigma_{32}^0 = S$ is applied at infinity. The boundary conditions are

$$\begin{aligned}
&\sigma_{32} = S && \text{at } x_2 = \infty, \\
&\sigma_{32} = 0 && \text{at } x_2 = 0, \quad |x_1| < c, \\
&\sigma_{32} = k_0 && \text{at } x_2 = 0, \quad c < |x_1| < a \\
&w = 0 && \text{at } x_2 = 0, \quad a < |x_1|,
\end{aligned} \tag{32.23}$$

where w is the displacement in the x_3-direction. Again we rewrite (32.23) as

$$\begin{aligned}
&\sigma_{32} = 0 && \text{at } x_2 = \infty \\
&\sigma_{32} = -S && \text{at } x_2 = 0, \quad |x_1| < c \\
&\sigma_{32} = k_0 - S && \text{at } x_2 = 0, \quad c < |x_1| < a \\
&w = 0 && \text{at } x_2 = 0, \quad a < |x_1|.
\end{aligned} \tag{32.24}$$

The solution for (32.23) is obtained from the solution for (32.24) by superimposing S onto 0. Hooke's law yields $\sigma_{31} = \mu w_{,1}$, $\sigma_{32} = \mu w_{,2}$ and the equation of equilibrium $\sigma_{31,1} + \sigma_{32,2} = 0$ leads to

$$w_{,11} + w_{,22} = 0. \tag{32.25}$$

We assume w has the form

$$w = (1/\mu) \int_0^\infty A(\xi) e^{-\xi x_2} \cos \xi x_1 d\xi. \tag{32.26}$$

Then,

$$\sigma_{32} = -\int_0^\infty \xi A(\xi) e^{-\xi x_2} \cos \xi x_1 d\xi \tag{32.27}$$

which satisfies the first condition in (32.24). The second and third conditions in (32.24) yield

$$\int_0^\infty \xi A(\xi) \cos \xi x_1 d\xi = \sigma(x_1), \tag{32.28}$$

where

$$\begin{aligned} \sigma(x_1) &= S \quad \text{for } |x_1| < c \\ &= S - k_0 \quad \text{for } c < |x_1| < a. \end{aligned} \tag{32.29}$$

The above result is the same as (32.8) with (32.9). Taking

$$A(\xi) = \int_0^a q(t) J_0(\xi t) dt, \tag{32.30}$$

we have, from (32.15),

$$\begin{aligned} q(t) &= St \quad \text{for } t < c, \\ q(t) &= \frac{2}{\pi} t \left\{ (S - k_0) \cos^{-1} \frac{c}{t} + S \sin^{-1} \frac{c}{t} \right\} \quad \text{for } c < t < a. \end{aligned} \tag{32.31}$$

w at $x_2 = 0$ is calculated from (32.26) and (32.11) as

$$\begin{aligned} \mu w &= \int_x^a \frac{q(t)}{(t^2 - x_1^2)^{1/2}} dt \quad \text{for } |x_1| < a \\ &= 0 \quad \text{for } a < |x_1|. \end{aligned} \tag{32.32}$$

Substituting (32.31), we have, similar to (32.18),

$$w = \frac{k_0}{\pi\mu}\left[(x_1+c)\cosh^{-1}\left|\frac{m}{c+x_1}+n\right| - (x_1-c)\cosh^{-1}\left|\frac{m}{c-x_1}+n\right|\right], \tag{32.33}$$

where $m = (a^2 - c^2)/a$ and $n = c/a$.

The condition for the finite value of σ_{32} at $x_1 = a$ is (32.16); therefore,

$$\cos^{-1}\frac{c}{a} = \frac{\pi}{2}\frac{S}{k_0}. \tag{32.34}$$

When $a/c \approx 1$ and $S/k_0 \ll 1$, (32.34) and w at $x_1 = c$ are written approximately as

$$1 - c/a = \frac{\pi^2}{8}(S/k_0)^2,$$

$$w = \frac{2}{\pi\mu}k_0(a - c). \tag{32.35}$$

We can see from this result that the critical stress for the anti-plane problem also has the form given by (32.21).

BCS model

Bilby, Cottrell and Swinden (1963) have solved the anti-plane problem given by (32.23) or (32.24) using a completely different method. The stress disturbance caused by the crack and plastic region is simulated by

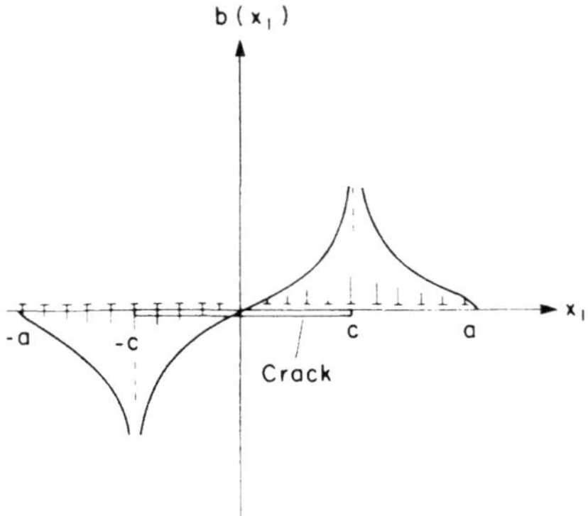

Fig. 32.3. Distribution of dislocations.

the dislocation stress. The dislocations are distributed continuously in the domain $(-a, a)$. The idea that cracks can be simulated by arrayed dislocations has originated in the work of Eshelby (1957).

Consider a crack as shown in Fig. 32.3. $\sigma_{23}^0 = S$ is an applied stress. The dislocation stress σ_{23} at $x_2 = 0$ due to a screw dislocation at $x_1 = t$, $x_2 = 0$ is $(\mu b/2\pi)/(x_1 - t)$. The boundary conditions corresponding to (32.23) can be reduced to the Hilbert integral equation

$$S + \int_{-a}^{a} (\mu/2\pi) b(t)(x_1 - t)^{-1} dt = \begin{cases} 0 & \text{for } |x_1| < c, \\ k_0 & \text{for } c < |x_1| < a, \end{cases} \tag{32.36}$$

where $b(t)$ is the strength of the Burgers vector (Burgers vector times the density of dislocation distribution). As seen in Appendix 4, the work of Head and Louat (1955) leads to the solution for (32.36),

$$\begin{aligned} b(x_1) &= \frac{2k_0}{\pi\mu} \log \frac{x_1(a^2 - c^2)^{1/2} + c(a^2 - x_1^2)^{1/2}}{x_1(a^2 - c^2)^{1/2} - c(a^2 - x_1^2)^{1/2}} \quad \text{for } c < |x_1| < a, \\ b(x_1) &= \frac{2k_0}{\pi\mu} \log \frac{x_1(a^2 - c^2)^{1/2} + c(a^2 - x_1^2)^{1/2}}{c(a^2 - x_1^2)^{1/2} - x_1(a^2 - c^2)^{1/2}} \quad \text{for } |x_1| < c, \end{aligned} \tag{32.37}$$

provided that condition (A.4.7) is satisfied. This condition gives (A.4.9) or

$$c/a = \cos(\pi S/2k_0) \tag{32.38}$$

which coincides with (32.34).

The relative displacement Δu_3 at $x_2 = 0$ (crack opening displacement in $|x_1| < c$, and plastic displacement in $c < |x_1| < a$) is obtained from

$$\Delta u_3 = \int_{x_1}^{a} b(t) dt. \tag{32.39}$$

It becomes

$$\begin{aligned} \Delta u_3 = \frac{2k_0}{\pi\mu} \Bigg\{ & x_1 \log \frac{x_1(a^2 - c^2)^{1/2} - c(a^2 - x_1^2)^{1/2}}{x_1(a^2 - c^2)^{1/2} + c(a^2 - x_1^2)^{1/2}} \\ & + c \log \frac{(a^2 - c^2)^{1/2} + (a^2 - x_1^2)^{1/2}}{(a^2 - c^2)^{1/2} - (a^2 - x_1^2)^{1/2}} \Bigg\}. \end{aligned} \tag{32.40}$$

The relative displacement at $x_1 = c$ is obtained by the limiting process,

$$\frac{(x_1/c)(a^2 - c^2)^{1/2} - (a^2 - x_1^2)^{1/2}}{(a^2 - c^2)^{1/2} - (a^2 - x_1^2)^{1/2}} \to \frac{a^2}{c^2}, \text{ becoming}$$

$$\Delta u_3 = (4k_0/\pi\mu) c \log(a/c). \tag{32.41}$$

For small plastic zones, $(a-c)/c \ll 1$, (32.38) and (32.41) become, approximately,

$$c/a = 1 - \pi^2 S^2/8k_0^2,$$
$$\Delta u_3 = (4k_0/\pi\mu)(a-c) = \pi c S^2/2\mu k_0, \tag{32.42}$$

which agree with (32.35) where $\Delta u_3 = 2w$. The critical stress is similar to (32.22). The distribution of dislocations (32.37) is sketched in Fig. 32.3.

The BCS model has been extended by many researchers. In order to investigate plastic yielding from sharp notches, Bilby, Cottrell, Smith, and Swinden (1964), Bilby and Swinden (1965) have considered an infinite array of relaxed cracks. Fracture characteristics of solids containing doubly-periodic arrays of cracks are investigated by Karihaloo (1978, 1979).

Barnett and Asaro (1972) have considered the BCS model in an arbitrarily anisotropic elastic medium. As shown by (30.2), the dislocation stress at $x_1 = x_1$, $x_2 = 0$ due to a single dislocation located at $x_1 = t$, $x_2 = 0$ with Burgers vector $\mathbf{b}$, is

$$\sigma_{2i} = K_{ij}b_j/(x_1 - t). \tag{32.43}$$

Assuming a continuous distribution of dislocations with the strength $\mathbf{b}(t)$, we write the boundary conditions given in the domain $x_2 = 0, |x_1| < a$, as

$$\sigma_{2i}^0 + \int_{-a}^{a} K_{ij}b_j(t)(x_1 - t)^{-1}dt = 0 \quad \text{for } |x_1| < c,$$
$$\sigma_{2i}^0 + \int_{-a}^{a} K_{ij}b_j(t)(x_1 - t)^{-1}dt = k_i(x_1) \quad \text{for } c < |x_1| < a, \tag{32.44}$$

where σ_{2i}^0 $(i = 1, 2, 3)$ are components of the applied stress at infinity, and k_i is the component of yielding shear stress. It is convenient to rewrite (32.44) as

$$\int_{-a}^{a} b_j(t)(x_1 - t)^{-1}dt = K_{ij}^{-1}(k_i(x_1) - \sigma_{2i}^0), \tag{32.45}$$

where $k_i(x_1) = 0$ for $|x_1| < c$. When k_i is constant, the solution of (32.45) has the same form as (A.4.8), where k is replaced by $K_{ij}^{-1}k_i$.

The BCS model has been further extended to cracks with inclined slip planes (Fig. 32.4) by Atkinson and Kay (1971), Vitek (1975, 1976), Riedel (1976), Atkinson and Kanninen (1977), Karihaloo (1979), Takeuchi (1979), and Hayashi and Nemat-Nasser (1981). A further extension to the two-dimensional dislocation distribution is presented in Section 41.

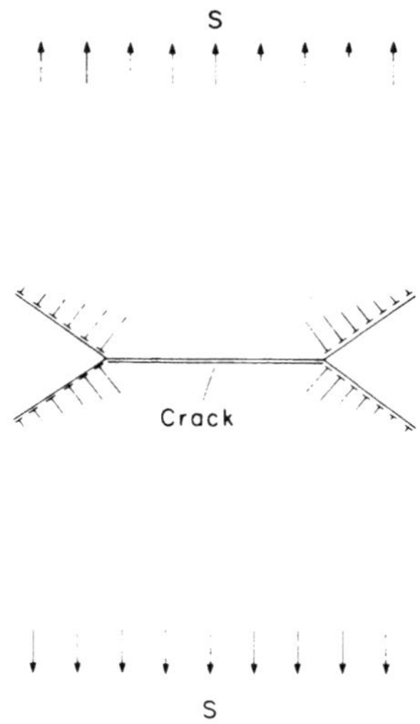

Fig. 32.4. Crack with inclined slip plane.

The Dugdale model has been used to obtain a dynamic solution by Goodier and Field (1963), and Kanninen et al. (1968). The dislocation distribution preidcted by the BCS model is observed by an electron microscope by Kobayashi and Ohr (1979) and Ohr and Narayan (1980).

Penny-shaped crack

Let us consider a penny-shaped crack with radius c as shown in Fig. 32.5. The material is assumed to be isotropic and infinitely extended. When a uniform tension S is applied at infinity, the tip of the crack is

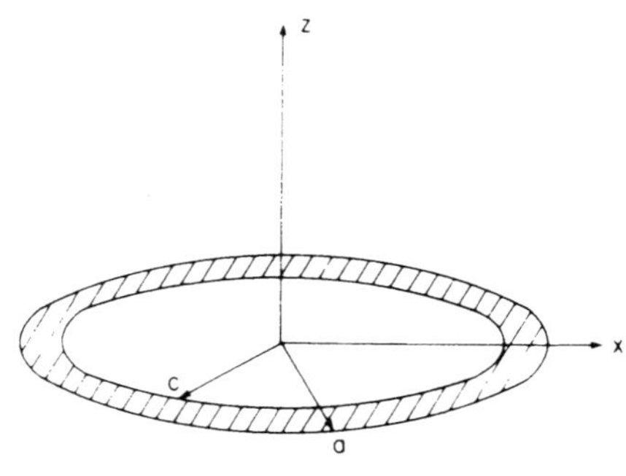

Fig. 32.5. Penny-shaped Dugdale crack.

subjected to the infinitely large stress even for a small S. The stress, however, is relaxed if a plastic zone at the neighborhood of the crack tip is developed, as indicated by the hatched area $c < r < a$ in Fig. 32.5. When the crack surface is normal to the tensile direction and the maximum shear theory of yielding is employed for the plastic domain, Dugdale's hypothesis requires the following boundary conditions for an axially symmetrical problem:

$$\begin{aligned}
&\sigma_z = S && \text{at } z = \infty, \\
&\sigma_{zr} = \sigma_{z\theta} = 0 && \text{at } z = 0, \\
&\sigma_z = 0 && \text{at } z = 0, \quad 0 < r < c, \\
&\tfrac{1}{2}(\sigma_z - \sigma_\theta) = k_0 && \text{at } z = 0, \quad c < r < a, \\
&w = 0 && \text{at } z = 0, \quad a < r,
\end{aligned} \tag{32.46}$$

where w is the displacement in the z direction and k_0 is a point on the stress-strain curve as shown in Fig. 32.2. Since the material in the plastic zone is work-hardening, the stress state in this zone is not uniform. However, for mathematical simplicity we assume that the stress state is uniform in the plastic zone. It is convenient to consider, instead of (32.46), the following boundary conditions:

$$\begin{aligned}
&\sigma_z = 0 && \text{at } z = \infty, \\
&\sigma_{zr} = \sigma_{z\theta} = 0 && \text{at } z = 0, \\
&\sigma_z = -S && \text{at } z = 0, \quad 0 < r < c, \\
&\tfrac{1}{2}(\sigma_z - \sigma_\theta) = k_0 - S/2 && \text{at } z = 0, \quad c < r < a, \\
&w = 0 && \text{at } z = 0 \quad 0 < r.
\end{aligned} \tag{32.47}$$

Proof that $\frac{1}{2}(\sigma_z - \sigma_\theta)$ gives the maximum stress in Tresca's yield condition will be given after the solution has been obtained. Keer and Mura (1966) have solved this problem as described below.

According to Inglis (1913), Collins (1962), and England (1963), the solution for the axially symmetric problems can be expressed as

$$\sigma_z = -\frac{\partial^2 \phi}{\partial z^2} + z\frac{\partial^3 \phi}{\partial z^3},$$

$$\sigma_\theta = -\frac{\partial^2 \phi}{\partial z^2} - (1-2\nu)\frac{\partial^2 \phi}{\partial r^2} + \frac{z}{r}\frac{\partial^2 \phi}{\partial r \partial z},$$

$$\sigma_{zr} = z\frac{\partial^3 \phi}{\partial r \partial z^2},$$

$$\sigma_{z\theta} = \frac{z}{r}\frac{\partial^3 \phi}{\partial\theta\partial z^2}, \tag{32.48}$$

$$\sigma_r = \frac{\partial^2\phi}{\partial r^2} + \frac{2\nu}{r}\frac{\partial\phi}{\partial r} + z\frac{\partial^3\phi}{\partial r^2 \partial z},$$

$$2\mu w = -2(1-\nu)\frac{\partial\phi}{\partial z} + z\frac{\partial^2\phi}{\partial z^2},$$

where

$$\frac{\partial\phi}{\partial z} = \frac{1}{2i}\int_{-a}^{a} f(t)[r^2 + (z+it)^2]^{-1/2}dt,$$

$$\frac{\partial\phi}{\partial r} = \frac{r}{2i}\int_{-a}^{a} f(t)\{z + it + [r^2 + (z+it)^2]^{1/2}\}^{-1}[r^2 + (z+it)^2]^{-1/2}dt, \tag{32.49}$$

$$f(t) = -f(-t).$$

The unknown function $f(t)$ is determined from the third and fourth condition in (32.47), using (32.48). Since we have

$$\lim_{z\to 0}[r^2 + (z+it)^2]^{1/2} = (r^2 - t^2)^{1/2} \qquad (r > t),$$

and

$$\lim_{z\to 0}[r^2 + (z+it)^2]^{1/2} = i(t^2 - r^2)^{1/2} \qquad (t > r),$$

the derivatives of ϕ as $z \to 0$ become (when $f(0) = 0$)

$$\left(\frac{\partial\phi}{\partial z}\right)_{z=0} = -\int_r^a \frac{f(t)dt}{(t^2 - r^2)^{1/2}} \qquad (0 \le r \le a),$$

$$= 0 \qquad (a < r < \infty),$$

$$\left(\frac{\partial^2\phi}{\partial z^2}\right)_{z=0} = \frac{1}{r}\frac{d}{dr}\int_0^r \frac{tf(t)dt}{(r^2 - t^2)^{1/2}} = \int_0^r \frac{f'(t)dt}{(r^2 - t^2)^{1/2}} \qquad (0 \le r \le a), \tag{32.50}$$

$$\left(\frac{\partial\phi}{\partial r}\right)_{z=0} = -\frac{1}{r}\int_0^r \frac{tf(t)dt}{(r^2 - t^2)^{1/2}} \qquad (0 \le r \le a).$$

Conditions (32.47), therefore, lead to

$$-\frac{1}{r}\frac{d}{dr}\int_0^r \frac{tf(t)dt}{(r^2 - t^2)^{1/2}} = -S \qquad (r < c),$$

$$-\frac{(1-2\nu)}{2}\frac{d}{dr}\left\{\frac{1}{r}\int_0^r \frac{tf(t)dt}{(r^2 - t^2)^{1/2}}\right\} = k - \frac{S}{2} \qquad (c < r < a). \tag{32.51}$$

The function $f(t)$ can be determined from (32.51) by the method described by Green and Zerna (1954),

$$
\begin{aligned}
&\frac{\pi}{2} f(t) = St \qquad (0 < t < c), \\
&\frac{\pi}{2} f(t) = St - \frac{4k_0 - (1+2\nu)S}{1-2\nu}\left\{(t^2 - c^2)^{1/2} - \frac{c}{2}\sin^{-1}\frac{c}{t}\right\} \qquad (c < t < a).
\end{aligned} \tag{32.52}
$$

It is seen that $(1/2)(\sigma_z - \sigma_\theta)$ is the maximum shear stress.

Using (32.50) and (32.48), we have

$$
\begin{aligned}
&\sigma_z - \sigma_\theta = 2k_0 - S, \\
&\sigma_z - \sigma_r = 2k_0 - S - \left\{2k_0 - (1+2\nu)\frac{S}{2}\right\}\frac{c}{r}, \\
&\sigma_r - \sigma_\theta = \left\{2k_0 - (1+2\nu)\frac{S}{2}\right\}\frac{c}{r},
\end{aligned} \tag{32.53}
$$

at $z = 0$, $c < r < a$. Furthermore, we have

$$
(\sigma_z)_{z=0,\, r=a} = -\lim_{t\to a}\frac{f(t)}{(a^2 - t^2)^{1/2}} + \int_0^a \frac{tf(t)dt}{(a^2 - t^2)^{3/2}} \tag{32.54}
$$

by applying integration by parts to the second formula in (32.50). The requirement that the stress singularity vanish at $r = a$ leads to

$$
f(a) = 0 \tag{32.55}
$$

which becomes, from (32.52),

$$
\frac{(1-2\nu)\left(\dfrac{S}{2k_0}\right)}{2 - (1+2\nu)\left(\dfrac{S}{2k_0}\right)} = \left\{1 - \left(\frac{c}{a}\right)^2\right\}^{1/2} - \frac{c}{2a}\cos^{-1}\frac{c}{a}. \tag{32.56}
$$

The displacement w can be easily obtained from (32.48) as

$$
\begin{aligned}
\frac{\pi}{4}\frac{2\mu(w)_{z=0^+}}{(1-\nu)} &= -\frac{\pi}{4}\frac{2\mu(w)_{z=0^-}}{(1-\nu)} \\
&= S(a^2 - r^2)^{1/2} - \frac{4k_0 - (1+2\nu)S}{(1-2\nu)}\left[\int_r^a \frac{(t^2 - c^2)^{1/2}}{(t^2 - r^2)^{1/2}}dt\right. \\
&\qquad \left. - \frac{c}{2}\int_r^a \cos^{-1} c/t \, \frac{dt}{(t^2 - r^2)^{1/2}}\right] \qquad (c \le r \le a).
\end{aligned} \tag{32.57}
$$

The value at the tip of crack is

$$\frac{2\mu(w)_{z=0,\,r=c}}{(1-\nu)ak_0}=\frac{16}{\pi}$$

$$\frac{(1-c^2/a^2)^{1/2}\left\{(1-c^2/a^2)^{1/2}-\dfrac{c}{2a}\cos^{-1}\dfrac{c}{a}\right\}-1+\dfrac{c}{a}+\dfrac{c}{2a}\displaystyle\int_{c/a}^{1}\frac{\cos^{-1}(c/at)\,dt}{(t^2-c^2/a^2)^{1/2}}}{(1+2\nu)\left\{(1-c^2/a^2)^{1/2}-\dfrac{c}{2a}\cos^{-1}\dfrac{c}{a}\right\}+(1-2\nu)} \tag{32.58}$$

Equation (32.56) can be approximated as

$$1-c/a=\frac{1}{8}(1-2\nu)^2(S/k_0)^2 \tag{32.59}$$

and equation (32.58) as

$$2w\approx\frac{24(1-\nu)k_0}{\pi(1-2\nu)\mu}c(1-c/a), \tag{32.60}$$

when $c/a\approx 1$ and $S/k_0\ll 1$.
Eliminating $(1-c/a)$ from (32.59) and (32.60), we have

$$S\approx\left\{\frac{\pi k_0\mu 2w}{3(1-\nu)(1-2\nu)c}\right\}^{1/2}. \tag{32.61}$$

If it is postulated that the crack becomes unstable when the crack opening displacement $2w$ reaches a critical value Φ, then the critical applied stress can be expressed as

$$S=\left\{\frac{\pi\mu k_0\Phi}{3(1-\nu)(1-2\nu)c}\right\}^{1/2}. \tag{32.62}$$

The ratio of (32.22) to (32.62) is about 0.42 when $\nu=1/3$ and the κ for the plane stress case is used. This means that the slit-like crack is 0.42 times weaker than the penny-shaped crack for the same crack length c.

It will be shown that the approach using the dislocation distribution and that using the Dugdale plasticity hypothesis are equivalent for the case of the penny-shaped crack.

According to Kroupa (1960), the stress σ_z due to a circular dislocation located on the plane $z=0$, is

$$\sigma_z=\frac{b\mu}{2(1-\nu)}\int_0^\infty s\xi J_0(\xi r)J_1(\xi s)d\xi,\qquad z=0, \tag{32.63}$$

where b is the Burgers vector in the z-direction and s is the radius of the dislocation. Keer and Mura (1966) have shown that the equivalent stress and displacement fields can be obtained when the distribution of dislocations is chosen as

$$\alpha_{\theta z} = -\frac{2(1-\nu)}{\mu}\frac{d}{dr}\int_r^a \frac{f(t)dt}{(t^2-r^2)^{1/2}}, \tag{32.64}$$

where $f(t)$ is defined by (32.52).

*33. Stress intensity factor for an arbitrarily shaped plane crack

The three-dimensional elasticity problem of finding the stress distribution near a flat crack embedded in an infinitely extended homogeneous, isotropic solid, and opened up under prescribed internal normal pressure, has been solved analytically by a number of investigators for cracks having circular and elliptical contours. The problem becomes much more difficult, and analytical solutions are not available when the crack contour is other than circular or elliptical. For such cases, attempts have been made to formulate the problem in terms of integral equations which might be solved by numerical techniques.

An integral equation formulation of the general three-dimensional elasticity problem is given by Cruse (1969), who obtains the limiting form of Somigliana's identity for displacements inside the body due to known surface tractions and displacements, when the point at which the displacement is evaluated approaches the surface of the body. This results in a system of coupled integral equations for the unknown surface tractions and displacements. A numerical technique is discussed for the solution of these integral equations. Further refinement of the numerical method is given in a paper by Lacht and Watson (1976). These papers, however, do not address the crack problems. In another paper Cruse (1973) has reported an application of the numerical method to the solution of the penny-shaped crack problem. By the use of single layer and double layer potentials, proposed by Kinoshita and Mura (1956) (see Kupradze 1965, Mikhlin 1965), Bui (1975, 1977) obtains a singular integral equation for the crack displacement discontinuity. The shape of the crack can be rather general and the domain of integration is the surface of the crack. Bui has numerically solved the integral equation for the penny-shaped, elliptical and square crack under constant normal pressure. Weaver (1977) has derived from Somigliana's equation a singular integral equation for the unknown dislocations which are defined over the crack area only. He has numerically solved the problem of a crack in the form of a rectangle under constant normal pressure.

Considerable work in this area has been done by Soviet researchers.

One formulation is given by Panasyuk (1968), who has reduced the crack problem to the solution of a singular integral equation for the crack displacement, but has not elaborated on the development of a solution technique for an arbitrarily shaped crack. He also discusses in some detail an approximate solution for an almost circular crack. Martynenko (1970), using Green's functions of a Riemann space, gives the exact solution for a circular crack. He also obtains an approximate solution for a crack in the form of a strip which becomes exact when the strip becomes a half-plane. Andreykiv and Stadnik (1974) consider a plane crack whose contour is a convex, closed piece-wise smooth line, consisting of n ($n = 1,2,\ldots$) circular arcs forming sections of the circumferences of circles, each of which encloses the entire crack domain, and m ($m = 1, 2, \ldots$) straight-line segments. They reduce the problem to the solution of a system of two dimensional Fredholm integral equations with the stresses unknown outside the crack domain. They numerically solve this system for semi-circular and square cracks under constant pressure.

What follows is a formulation of the plane crack problem in the work of Mastrojannis, Keer, and Mura (1978). To test the formulation, a numerical analysis is given involving plane cracks of various contour shapes embedded in an infinite isotropic solid and subjected to uniform and linearly varying internal normal pressure.

Consider an unbounded homogeneous elastic solid weakened by a plane crack S having a smooth contour ∂S. Let x_i ($i = 1, 2, 3$) be the Cartesian coordinates of the point $\mathbf{x}$, and let the plane $x_3 = 0$ be the plane of the crack (Fig. 33.1). The faces of the crack are loaded symmetrically

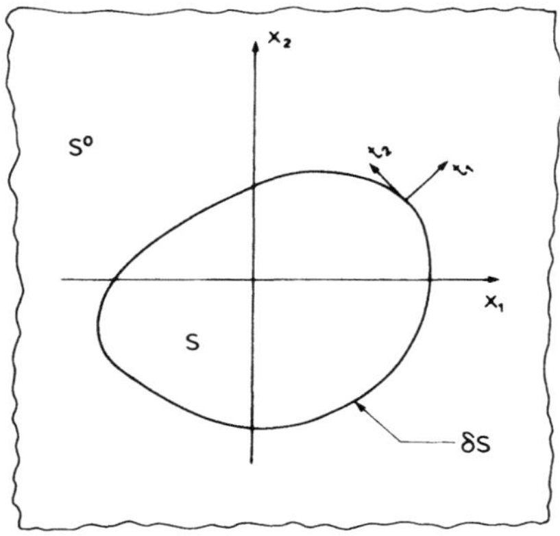

Fig. 33.1. An arbitrary plane crack S on $x_3 = 0$ plane.

with respect to the $x_3 = 0$ plane with an arbitrary normal pressure of intensity $P(x_1, x_2)$; at infinity, the solid is stress-free. It is necessary to determine the stress $\sigma_{33}(x_1, x_2, x_3 = 0)$ In the region S^0 outside the crack domain, the normal displacement $u_3(x_1, x_2, x_3 = 0)$ of points inside the crack, and the stress intensity factor k_I (Mode I) along the crack edge.

The determination of the stress $\sigma_{33}(x_1, x_2, 0)$ and the displacement $u_3(x_1, x_2, 0)$ reduces to a problem for the elastic half-space $x_3 \geq 0$ with the following boundary conditions (Sneddon 1966):

$$\begin{aligned} &u_3(x_1, x_2, 0) = 0 \qquad \text{for } (x_1, x_2) \in S^0, \\ &\sigma_{33}(x_1, x_2, 0) = -P(x_1, x_2) \qquad \text{for } (x_1, x_2) \in S, \\ &\sigma_{31}(x_1, x_2, 0) = \sigma_{32}(x_1, x_2, 0) = 0 \qquad \text{for } (x_1, x_2) \in (S + S^0). \end{aligned} \tag{33.1}$$

The equations of equilibrium

$$u_{i,jj} + \frac{1}{1-2\nu} u_{j,ji} = 0 \qquad (i, j = 1, 2, 3) \tag{33.2}$$

and the conditions of zero tangential stresses on the plane $x_3 = 0$ are satisfied by taking the displacement vector (e.g., Panasyuk, 1968) in the form

$$u_i = \Phi_i - \frac{x_3}{2(1-\nu)} \Phi_{3,i}, \tag{33.3}$$

where ν is the Poisson's ratio and Φ_1, Φ_2, and Φ_3 are harmonic functions of $\mathbf{x}$ such that

$$\Phi_{i,3} = -\frac{1-2\nu}{2(1-\nu)} \Phi_{3,i} \qquad (i = 1, 2). \tag{33.4}$$

The stresses σ_{33}, σ_{31}, and σ_{32} can be expressed in terms of Φ_3 as follows:

$$\begin{aligned} \sigma_{33}(\mathbf{x}) &= \frac{E}{2(1-\nu^2)} [\Phi_{3,3} - x_3 \Phi_{3,33}], \\ \sigma_{31}(\mathbf{x}) &= -\frac{E}{2(1-\nu^2)} x_3 \Phi_{3,31}, \\ \sigma_{32}(\mathbf{x}) &= -\frac{E}{2(1-\nu^2)} x_3 \Phi_{3,32}, \end{aligned} \tag{33.5}$$

where E is the Young's modulus. Thus, the elastic problem posed by equations (33.1) reduces to that of finding a harmonic function $\Phi_3(\mathbf{x})$

satisfying the mixed boundary conditions

$$\begin{aligned} &\Phi_3(x_1, x_2, 0) = 0 \qquad \text{for } (x_1, x_2) \in S^0, \\ &\left.\frac{\partial \Phi_3(x_1, x_2, x_3)}{\partial x_3}\right|_{x_3=0} = -\frac{2(1-\nu^2)}{E} P(x_1, x_2) \qquad \text{for } (x_1, x_2) \in S. \end{aligned} \tag{33.6}$$

To solve this problem, the harmonic function $\Phi_3(\mathbf{x})$ is expressed as a Fourier integral (Andreykiv and Stadnik 1974),

$$\Phi_3(\mathbf{x}) = -\iint_{-\infty}^{+\infty} \frac{\exp[-x_3(\xi_1^2+\xi_2^2)^{1/2} + i(x_1\xi_1 + x_2\xi_2)]}{(\xi_1^2+\xi_2^2)^{1/2}} A(\xi_1, \xi_2)\,d\xi_1 d\xi_2, \tag{33.7}$$

where $A(\xi_1, \xi_2)$ is an unknown function. Letting

$$q(x_1, x_2) = \begin{cases} -\dfrac{2(1-\nu^2)}{2\pi E} P(x_1, x_2) & \text{for } (x_1, x_2) \in S \\[2ex] \dfrac{2(1-\nu^2)}{2\pi E} \sigma_{33}(x_1, x_2, 0) & \text{for } (x_1, x_2) \in S^0, \end{cases} \tag{33.8}$$

we obtain from (33.6) and (33.7),

$$q(x_1, x_2) = \frac{1}{2\pi} \iint_{-\infty}^{\infty} \exp[+i(x_1\xi_1 + x_2\xi_2)] A(\xi_1, \xi_2)\,d\xi_1 d\xi_2 \tag{33.9}$$

which, upon application of the inverse Fourier transform theorem, gives

$$A(\xi_1, \xi_2) = \frac{1}{2\pi} \iint_{-\infty}^{\infty} \exp[-i(\xi_1\eta_1 + \xi_2\eta_2)] q(\eta_1, \eta_2)\,d\eta_1 d\eta_2. \tag{33.10}$$

Using (33.10) in (33.7) and noting that

$$\frac{1}{2\pi} \iint_{-\infty}^{\infty} \frac{\exp[i(x_1-\eta_1)\xi_1 + i(x_2-\eta_2)\xi_2]}{(\xi_1^2+\xi_2^2)^{1/2}} d\xi_1 d\xi_2 = \frac{L1}{[(x_1-\eta_1)^2 + (x_2-\eta_2)^2]^{1/2}}, \tag{33.11}$$

we have

$$\begin{aligned} u_3(x_1, x_2, 0) = \frac{(1-\nu^2)}{\pi E} \Bigg\{ &\iint_S \frac{P(\xi_1, \xi_2)\,d\xi_1 d\xi_2}{[(x_1-\xi_1)^2 + (x_2-\xi_2)^2]^{1/2}} \\ &- \iint_{S^0} \frac{\sigma_{33}(\xi_1, \xi_2, 0)\,d\xi_1 d\xi_2}{[(x_1-\xi_1)^2 + (x_2-\xi_2)^2]^{1/2}} \Bigg\}. \end{aligned} \tag{33.12}$$

Finally, from the first condition of (33.1), one obtains the following integral equation for the stress $\sigma_{33}(x_1, x_2, 0)$ in S^0:

$$\iint_S \frac{P(\xi_1, \xi_2)d\xi_1 d\xi_2}{[(x_1-\xi_1)^2+(x_2-\xi_2)^2]^{1/2}} = \iint_{S^0} \frac{\sigma_{33}(\xi_1, \xi_2, 0)d\xi_1 d\xi_2}{[(x_1-\xi_1)^2+(x_2-\xi_2)^2]^{1/2}}$$

$$\text{for } (x_1, x_2) \in S^0. \qquad (33.13)$$

Once the stress $\sigma_{33}(x_1, x_2, 0)$ acting on the plane of the crack is known, the stress intensity factor k_I (Mode I) for a plane crack can be determined from the equation

$$k_I = \lim_{t_1 \to 0}(2\pi t_1)^{1/2}\sigma_{33}(t_1, t_2 = 0, t_3 = 0), \qquad (33.14)$$

where t_1, t_2, and t_3 constitute a local orthogonal coordinate system on the crack contour, as shown in Fig. 33.1.

The analytical solution of (33.13) for an arbitrarily shaped crack appears to be beyond the present mathematical techniques. For this reason, numerical methods must be used in order to obtain specific results. It is well known that the stress $\sigma_{33}(\xi_1, \xi_2, 0)$ vanishes at infinity and has a square-root singularity on the smooth contour ∂S of the crack. Therefore, we assume that the stress distribution $\sigma_{33}(\xi_1, \xi_2, 0)$ in (33.13) has the form

$$\sigma_{33}(\xi_1, \xi_2, 0) = B(\xi_1, \xi_2)/[2\pi w(\xi_1, \xi_2)]^{1/2} \qquad \text{for } (\xi_1, \xi_2) \in S^0, \qquad (33.15)$$

where $B(\xi_1, \xi_2)$ is an unknown bounded function, and the function $w(\xi_1, \xi_2)$ represents the shortest distance from the point (ξ_1, ξ_2) to the crack contour. Thus, $w(\xi_1, \xi_2)$ behaves like t_1 in (33.14) near the crack neighborhood, and, at a point at infinity, $w(\xi_1, \xi_2) \to \infty$. Rvachev (1975) and Goncharyuk (1972) have given a method for the construction of $w(\xi_1, \xi_2)$.

Let the smooth crack contour ∂S be approximated by a set of n straight line segments with end-points $M_i(x_1, y_1)$, $N_i(x_2, y_2)$ ($i = 1, 2, 3, \ldots, n$), where x_1, y_1 are the coordinates of point M_i. When the crack contour is convex, the shortest distance of point (ξ_1, ξ_2) from the ith straight line can be written as

$$f_i(\xi_1, \xi_2) = d_i - \xi_1 \cos\theta_i - \xi_2 \sin\theta_i, \qquad (33.16)$$

where d_i is the normal distance from the origin of the ξ_1, ξ_2-coordinate system to the straight line $f_i(\xi_1, \xi_2) = 0$, and $(\cos\theta_i, \sin\theta_i)$ is a unit vector along d_i. Then, we write

$$w(\xi_1, \xi_2) = |f_i(\xi_1, \xi_2)\Lambda_1 f_2(\xi_1, \xi_2) \cdots \Lambda_1 f_n(\xi_1, \xi_2)|, \tag{33.17}$$

where

$$f_i(\xi_1, \xi_2)\Lambda_1 f_j(\xi_1, \xi_2) = \tfrac{1}{2}[f_i(\xi_1, \xi_2) + f_j(\xi_1, \xi_2) - |f_i(\xi_1, \xi_2) - f_j(\xi_1, \xi_2)|]. \tag{33.18}$$

The right-hand side in (33.17) has been called the R_1-conjunction by Rvachev (1975). It defines analytically the minimum value amongst $|f_1(\xi_1, \xi_2)|, |f_2(\xi_1, \xi_2)|, \ldots, |f_n(\xi_1, \xi_2)|$. Furthermore, we assume that the function $B(\xi_1, \xi_2)$ has the following form:

$$B(\xi_1, \xi_2) = G(\xi_1, \xi_2) \iint_S \frac{[w(\eta_1, \eta_2)]^{1/2} P(\eta_1, \eta_2) d\eta_1 d\eta_2}{(\xi_1 - \eta_1)^2 + (\xi_2 - \eta_2)^2} \quad \text{for } (\xi_1, \xi_2) \in S^0, \tag{33.19}$$

where $G(\xi_1, \xi_2)$ is a new unknown function to be determined.

With the above assumption, (33.15) becomes

$$\sigma_{33}(\xi_1, \xi_2, 0) = \frac{G(\xi_1, \xi_2)}{[2\pi w(\xi_1, \xi_2)]^{1/2}} \iint_S \frac{[w(\eta_1, \eta_2)]^{1/2} P(\eta_1, \eta_2) d\eta_1 d\eta_2}{(\xi_1 - \eta_1)^2 + (\xi_2 - \eta_2)^2} \quad \text{for } (\xi_1, \xi_2) \in S^0 \tag{33.20}$$

and (33.14) leads to

$$k_I(\xi_1, \xi_2) = G(\xi_1, \xi_2) \iint_S \frac{[w(\eta_1, \eta_2)]^{1/2} P(\eta_1, \eta_2) d\eta_1 d\eta_2}{(\xi_1 - \eta_1)^2 + (\xi_2 - \eta_2)^2} \quad \text{for } (\xi_1, \xi_2) \in \partial S. \tag{33.21}$$

The integral equation (33.13) is reduced to the following one which defines $G(\xi_1, \xi_2)$ in S^0:

$$\iint_S \frac{P(\xi_1, \xi_2) d\xi_1 d\xi_2}{[(x_1 - \xi_1)^2 + (x_2 - \xi_2)^2]^{1/2}}$$

$$= \iint_{S^0 - \Delta T} \frac{G(\xi_1, \xi_2) Q(\xi_1, \xi_2) d\xi_1 d\xi_2}{[2\pi w(\xi_1, \xi_2)]^{1/2} [(x_1 - \xi_1)^2 + (x_2 - \xi_2)^2)^{1/2}}$$

$$+ \iint_{\Delta T} \Big[\frac{G(\xi_1, \xi_2) Q(\xi_1, \xi_2)}{[2\pi w(\xi_1, \xi_2)]^{1/2}}$$

$$-\left.\frac{G(x_1, x_2)Q(x_1, x_2)}{[2\pi w(x_1, x_2)]^{1/2}}\right] \frac{d\xi_1 d\xi_2}{[(x_1-\xi_1)^2+(x_2-\xi_2)^2]^{1/2}}$$

$$+\frac{G(x_1, x_2)Q(x_1, x_2)}{[2\pi w(x_1, x_2)]^{1/2}} \iint_{\Delta T} \frac{d\xi_1 d\xi_2}{[(x_1-\xi_1)^2+(x_2-\xi_2)^2]^{1/2}} \quad \text{for } (x_1, x_2) \in S^0, \tag{33.22}$$

where ΔT is a small region of the domain S^0 inside which point (x_1, x_2) is ocated, and

$$Q(\xi_1, \xi_2) = \iint_S \frac{[w(\eta_1, \eta_2)]^{1/2} P(\eta_1, \eta_2) d\eta_1 d\eta_2}{(\xi_1-\eta_1)^2+(\xi_2-\eta_2)^2} \quad \text{for } (\xi_1, \xi_2) \in S^0. \tag{33.23}$$

The sum of the third and fourth terms on the right-hand side in (33.22) are zero, but these terms are artificially added to avoid numerical integration at the singularity $\boldsymbol{\xi} = \mathbf{x}$. The fourth integral is analytically performed as shown below.

When ΔT is taken as a quadrilateral with the centroidal (x_1, x_2), we have

$$\iint_{\Delta T} \frac{d\xi_1 d\xi_2}{[(x_1-\xi_1)^2+(x_2-\xi_2)^2]^{1/2}}$$
$$= R_1 \log(C_1) + R_2 \log(C_2) + R_3 \log(C_3) + R_4 \log(C_4), \tag{33.24}$$

where

$$\begin{aligned} R_1 &= [(Y_2-Y_1)x_1+(X_1-X_2)x_2+(X_2Y_1-Y_2X_1)]/ \\ &\quad [(X_1-X_2)^2+(Y_1-Y_2)^2]^{1/2}, \\ C_1 &= \{[(X_1-x_1)^2+(Y_1-x_2)^2]^{1/2}+[(X_1-x_1)^2+(Y_1-x_2)^2-R_1^2]^{1/2}\}/ \\ &\quad \{(X_2-x_1)^2+(Y_2-x_2)^2]^{1/2}-[(X_2-x_1)^2+(Y_2-x_2)^2-R_1^2]^{1/2}\}, \end{aligned} \tag{33.25}$$

and where (X_1, Y_1), (X_2, Y_2), (X_3, Y_3), and (X_4, Y_4) are the coordinates of corner points of the quadrilateral. The expressions for R_2, C_2, etc., can be obtained from (33.25) by the cyclic permutation of (1, 2, 3).

To treat the integral equation (33.22) numerically, the infinite domain of integration, S^0, is replaced by a finite domain, S_F^0, enclosed between the crack contour ∂S and a similarly oriented and shaped contour ∂S_F^0 at some distance from ∂S. The finite domain S_F^0 is then divided into N quadrilaterals over each of which the value of the unknown function $G(\xi_1, \xi_2)$ is assumed to be constant, where (ξ_1, ξ_2)

defines the centroid of the corresponding quadrilateral. Also, the crack domain S is divided into M triangles and M quadrilaterals, where M is equal to the number of the line segments that compose the crack contour. Replacing the integrations over S and S_F^0 by the sum of integrals over quadilaterals and triangles, the solution of (33.22) reduces to the solution of a system of N algebraic equations with N unknowns (the values of $G(\xi_1, \xi_2)$ on each quadilateral in S_F^0).

**Numerical examples*

The integral equation (33.22) has been solved and the stress intensity factor k_I calculated for cracks subjected to uniform and linearly varying internal pressure, i.e.,

$$\begin{aligned} P(x_1, x_2) &= P_0, \\ P(x_1, x_2) &= P_0 + P_1 x_1, \end{aligned} \tag{33.26}$$

where P_0 and P_1 are given constants. The cracks have contours in the shape of a circle, an ellipse, an egg, a square, and a triangle. The contour of each crack is approximated by 48 straight line segments, and the corners of the square and triangular cracks are rounded. The outer contour ∂S_F^0 is taken at a distance from the crack edge 3.5 times the radius of the inscribed circle. Choosing the center and the orientation of the coordinate system properly and utilizing the resulting symmetry about the ξ_1-axis, we reduce the solution of integral equation (33.22) to the solution of a system of algebraic equations with 144 unknowns. The running time for the solution of each crack problem with the two loading cases was 143 seconds on the CDC 6600 Computer, the only input data being the coordinates of the 48 points on the crack contour, the values of P_0 and P_1, the normal distance between ∂S and ∂S_F^0, and the number of unknowns.

Figures 33.2 and 33.3 are plots of $k_I/2P_0(R/\pi)^{1/2}$ against the polar angle θ, where R is the radius of the inscribed circle. The ratio P_1/P_0 is taken to be 0.2. The results for the square crack under constant normal pressure seem to be identical (excluding corners) to the results obtained numerically by Weaver (1977).

For polygonal plane-cracks, the stress singularity at points of abrupt changes in the slope of the crack contour is less severe than the square-root singularity along the smooth portions of the contour. For this reason, the stress intensity factor, as defined by (29.21), at those points assumes a zero value.

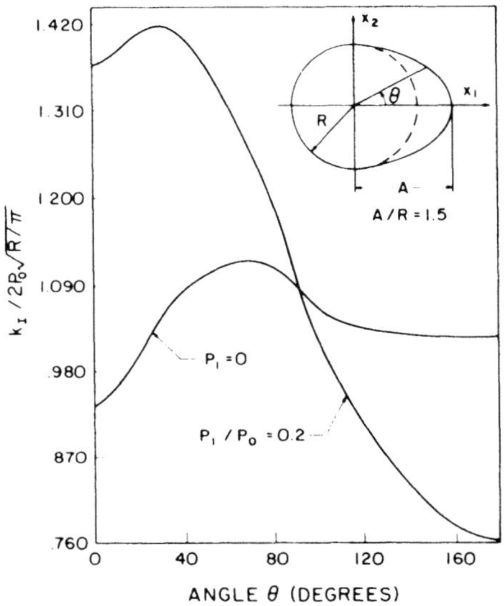

Fig. 33.2. The stress intensity factor along an egg shaped plane crack.

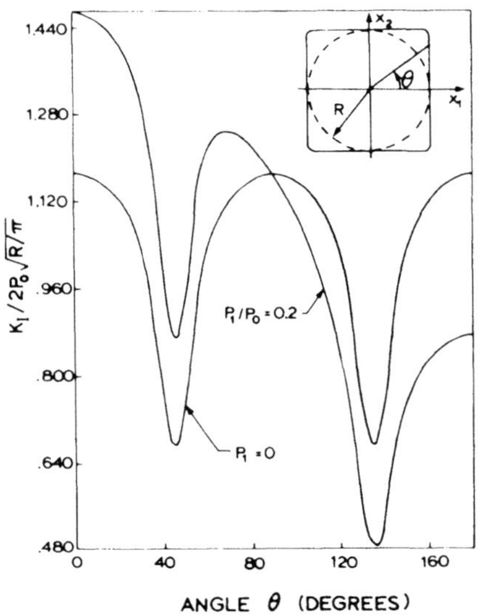

Fig. 33.3. The stress intensity factor along a rectangular plane crack.

The method discussed here differs from those employed by Cruse (1969, 1973), Bui (1975, 1977), Weaver (1977), Andreykiv and Stadnik (1974), and Hayashi and Abe (1980). The present method results in a single integral equation for the function $G(\xi_1, \xi_2)$ characterizing the normal stresses outside the crack domain.

Although the support of the integral equation (33.22) is infinite and truncation of the domain of integration introduces some error, the numerical results obtained by the present method show only 0.5 per cent error for the circular crack and 1.85 per cent error for the elliptical crack compared with the known analytical results. This is extremely satisfactory compared with the results reported by Cruse (1973), 16 per cent in error, and by Bui (1977), 20 per cent in error, for the circular crack under constant pressure. In Bui's method, the support of the integral equation is finite (integration over the crack domain only), but the treatment of the singularity seems to be the main difficulty. The satisfactory results of the method offered here are attributed mainly to the choice of the stress distribution $\sigma_{33}(\xi_1, \xi_2, 0)$ in the form shown in (33.20) with the unknown function $G(\xi_1, \xi_2)$.

34. Crack growth

A crack tends to grow when an applied stress reaches a critical value. The crack propagation may be dynamical for brittle materials. If the material is ductile, the propagation may be quasi-static and the crack length may increase slowly with load, maintaining at each instant a state of equilibrium. In the case of fatigue, cracks usually grow in the static fashion for any cyclic loading unless the tips of cracks are extremely slanted or residual stresses are built up to oppose the motion of the cracks.

Energy release rate

Let us consider a static growth of a crack A in a body D subjected to the applied force F_i on its surface S. The displacement and stress fields are denoted by u_i and σ_{ij}, when the crack surface is A. We are interested in evaluating the change of energy during a crack growth δA (Fig. 34.1) in order to define the generalized force at the crack tip necessary for the crack extension. Changes of the displacement and stress fields due to the crack extension δA are denoted by δu_i and $\delta\sigma_{ij}$, yielding

$$\int_S F_i \delta u_i dS = \int_D \sigma_{ij} \delta u_{i,j} dD - \int_B \sigma_{ij} n_j [\delta u_i] dS, \tag{34.1}$$

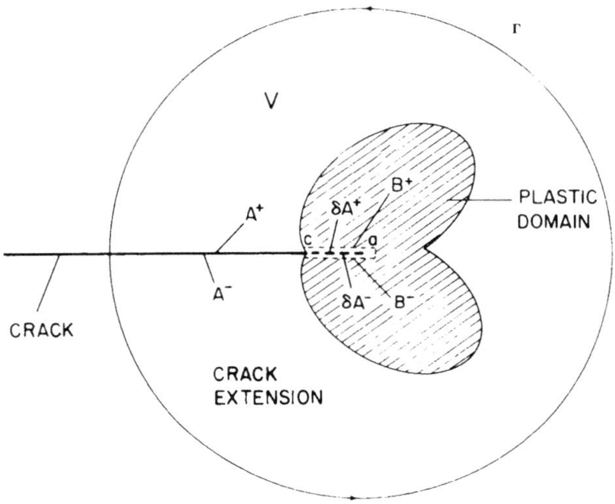

Fig. 34.1. Crack A grows by δA.

where B is a surface of discontinuity of δu_i, and $\mathbf{n}$ is the normal vector on $B^+ \equiv B$. For most cases, B is δA. For the BCS model, B includes the plastic zone and $[\delta u_i] = \delta u_i$ (on B^+) $- \delta u_i$ (on B^-). Equation (34.1) has been derived from $\int_S F_i \delta u_i dS = \int_S \sigma_{ij} n_j \delta u_i dS$ by applying Gauss' theorem, where the boundaries of D are S, A^+, A^-, B^+, and B^- (see Fig. 34.1) and where $\sigma_{ij,j} = 0$ in D, $\sigma_{ij} n_j = 0$ on $A^+ \equiv A$. The strain $(1/2)(u_{i,j} + u_{j,i})$ is the sum of elastic strain e_{ij} and plastic strain ϵ_{ij}^p,

$$(1/2)(u_{i,j} + u_{j,i}) = e_{ij} + \epsilon_{ij}^p. \tag{34.2}$$

Therefore, (34.1) can be written

$$\int_S F_i \delta u_i dS - \int_D \sigma_{ij} \delta e_{ij} dD = \int_D \sigma_{ij} \delta \epsilon_{ij}^p dD - \int_B \sigma_{ij} n_j [\delta u_i] dS. \tag{34.3}$$

$\int_S F_i \delta u_i dS$ is the work done by applied force F_i during the change of crack surface δA and can be written as $\delta \int_S F_i u_i dS$ under the assumption that $\mathbf{F}$ does not change during the infinitesimal crack growth. $\int_D \sigma_{ij} \delta e_{ij} dD = \delta[(1/2)\int_D \sigma_{ij} e_{ij} dD]$ is the change of elastic strain energy. $\int_D \sigma_{ij} \delta \epsilon_{ij}^p dD$ is the plastic work done during the crack growth. Finally, $-\int_B \sigma_{ij} n_j [\delta u_i] dS$ is a positive work done at the crack tip during the crack opening $[\delta u_i]$.

The left-hand side in (34.3) is identical to $-\delta(\Delta W)$ when $\epsilon_{ij}^p = 0$, where ΔW has been defined by (25.14).

The expression

$$\Pi = (1/2)\int_D \sigma_{ij}e_{ij}dD - \int_S F_iu_idS \tag{34.4}$$

is the mechanical energy portion of the Gibbs free energy. The Griffith fracture criterion can be written as

$$\delta G = \delta(\Pi + 2A\gamma) = 0, \tag{34.5}$$

where G is the Gibbs free energy and γ is the surface energy per unit area of the crack surface. Since the right-hand side in (34.3) expresses irreversible energies, $-\delta\Pi$ can be interpreted as the energy release during the crack growth δA. If a slit-like crack is changed by δA, the energy release rate, $\mathcal{G}_i$, is defined from

$$\mathcal{G}_i\delta\xi_i = -\delta\Pi = \int_D \sigma_{ij}\delta\epsilon^p_{ij}dD - \int_B \sigma_{ij}n_j[\delta u_i]dS, \tag{34.6}$$

where $\delta\xi_i$ is the displacement (change of position) of the crack tip due to the crack extension δA. For a three-dimensional crack, the left hand side in (34.6) must be modified (see e.g. Palaniswamy and Knauss 1978).

For a slit-like crack, the generalized force acting on the tip of the crack is $\mathcal{G}_i$ (crack extension force). For the slowly growing crack, the second term in (34.6) can be neglected since it is of order $(\delta A)^2$. Then, (34.6) means that the plastic work during the crack growth $\delta\xi_1 = \delta A$ is proportional to δA. This result agrees with the experiment reported by Lee and Liebowitz (1978).

Irwin (1958) has calculated $-\int_{\delta A}\sigma_{ij}n_j[\delta u_i]dS$ for a slit-like crack with length $2c$ under tension σ^0_{22} in an elastic medium. According to the elasticity analysis, $\sigma_{22} = k_I/\{2\pi(x_1 - X)\}^{1/2}$ and $u_2 = (1-\nu)k_I\{2(X - x_1)\}^{1/2}/\mu\sqrt{\pi}$, where X is the x_1 coordinate of the crack tip and $k_I = \sigma^0_{22}(\pi c)^{1/2}$. Irwin assumes that $[\delta u_2] = 2u_2 = 2(1-\nu)k_I\{2(X + \delta\xi_1 - x_1)\}^{1/2}/\mu\sqrt{\pi}$. Then, for a unit slice,

$$\begin{aligned}\mathcal{G}_1\delta\xi_1 &= 2\int_X^{X+\delta\xi_1}(1-\nu)k_I^2\left(\frac{X+\delta\xi_1 - x_1}{x_1 - X}\right)^{1/2}(\mu\pi)^{-1}dx_1\\ &= \frac{1-\nu}{\mu}k_I^2\delta\xi_1.\end{aligned} \tag{34.7}$$

The same integral $-\int_{\delta A}\sigma_{ij}n_j[\delta u_i]dS$ can be calculated for the Dugdale crack. From (32.20), $[\delta u_2] \approx (1+\kappa)\pi S^2\delta\xi_1/8\mu k_0$ and $\sigma_{22} \approx k_0(\delta c = \delta\xi_1)$. Then, the integral becomes $(1+\kappa)\pi S^2(\delta\xi_1)^2/8\mu$, which becomes zero at the limit $\delta\xi_1 \to 0$. Yokobori et al. (1966, 1968), and Rice (1966) have pointed out this paradoxical result. This inconsistency with the Orowan-Irwin theory has also been discussed later by Kfouri and Rice (1977). Wnuk (1972, 1974) has postulated that the integral domain in the second term in (34.6) is a material constant (process zone size) and has derived an expression for a quasi-static extension of a tensile crack. Denoting the process zone by Δ, he has defined the crack separation energy by

$$\mathcal{G}_1^\Delta \delta\xi_1 = \int_c^{c+\Delta} k_0[\delta u_2]dx_1, \tag{34.7.1}$$

where k_0 is defined in (32.1).

From (34.13′),

$$\begin{aligned}\mathcal{G}_1^\Delta &= -k_0\int_c^{c+\Delta}[u_{2,1}]dx_1\\ &= -2k_0\{u_2(c+\Delta, c) - u_2(c, c)\}\\ &\approx -2k_0\{u_2(c+\Delta, c) - u_2(c+\Delta, c+\Delta)\}.\end{aligned} \tag{34.7.2}$$

For small scale yielding, (32.18) can be approximated as

$$u_2(x_1, c) = \frac{4(1-\nu^2)k_0}{\pi E}\{R - \xi/2 - (\xi/2)\log(4R/\xi)\}, \tag{34.7.3}$$

where $a - c = R$ and $x_1 - c = \xi$. Then, (34.7.2) yields

$$\mathcal{G}_1^\Delta = \frac{8(1-\nu^2)k_0^2\Delta}{\pi E}\left(\frac{dR}{dc} + \frac{1}{2} + \frac{1}{2}\log\frac{4R}{\Delta}\right). \tag{34.7.4}$$

This is Wnuk's (1972) equation. A similar result is obtained by Rice and Sorensen (1978), Kfouri and Rice (1977), and Kfouri (1979).

From (32.20) we have, for small scale yielding,

$$\begin{aligned}R &= (c/2)Q^2,\\ Q &\equiv \pi S/2k_0,\end{aligned} \tag{34.7.5}$$

where Q is a loading parameter. If Δ and $\mathcal{G}_1^\Delta$ are constant, from (34.7.4) and (34.7.5) we obtain a relation, $Q = Q(c)$, which represents a stable crack growth. The catastrophic propagation is determined by the condition $dQ/dc = 0$. This idea has been developed by Cherepanov (1968), McClintock (1958), Rice (1968), and Wnuk (1979) among others.

The J-integral

The generalized force acting at the tip of a crack can be expressed in terms of a path-independent integral (J-integral) of a Maxwell tensor of elasticity taken over a surface. This fact has been developed by Eshelby (1951) and later by Rice (1968) and Eshelby (1970).

The three-dimensional expression of the J-integral is defined by

$$J_l = \int_\Gamma W n_l dS - \int_\Gamma \sigma_{ij} n_j u_{i,l} dS = \int_\Gamma P_{lj} n_j dS \tag{34.8}$$

with

$$P_{lj} = W\delta_{lj} - \sigma_{ij} u_{i,l}, \tag{34.8.1}$$

where Γ is an arbitrary closed surface (a closed curve for the two-dimensional case) including the crack tip, $\mathbf{n}$ is the outward unit normal on dS, P_{lj} is the energy-momentum tensor, and W is defined by

$$W_{,l} = \sigma_{ij} \epsilon_{ij,l}, \tag{34.9}$$

where $\epsilon_{ij} = (1/2)(u_{i,j} + u_{j,i})$. It is easy to show that J_l is a path-independent integral. If another path (surface) Γ^* is taken, application of Gauss' theorem leads to $\int_{\Gamma-\Gamma^*} W n_l dS - \int_{\Gamma-\Gamma^*} \sigma_{ij} n_j u_{i,l} dS = 0$.

For non-linear elastic materials, (34.9) can be integrated as

$$W = \int_0^{\epsilon_{ij}} \sigma_{ij} d\epsilon_{ij} \tag{34.9.1}$$

If the plastic unloading is involved, we cannot have (34.9.1) from (34.9).

Denoting by V the domain bounded by Γ, A^+, B^+, A^-, B^-, we write (34.8), by the Gauss' theorem, as (Fig. 34.1)

$$J_l = \int_V W_{,l} dD - \int_V \sigma_{ij} u_{i,lj} dD + \int_{\pm B} \sigma_{ij} n_j u_{i,l} dS, \tag{34.9.2}$$

where $W_{,l} = \sigma_{ij} u_{i,jl}$, $\sigma_{ij,j} = 0$, and $\sigma_{ij} n_j = 0$ on A. It has been assumed in the derivation of (34.9.2) that W on $A^+ + B^+$ is equal to W on $A^- + B^-$. The first and second integrals in (34.9.2) cancel each other and $\sigma_{ij} n_j$ on B^+ is $-\sigma_{ij} n_j$ on B^-. Then, (34.9.2) becomes

$$J_l = \int_B \sigma_{ij} n_j [u_{i,l}] dS, \tag{34.10}$$

where $[u_{i,l}] = (u_{i,l} \text{ on } B^+ = B) - (u_{i,l} \text{ on } B^-)$.

On the crack surface, u_i is multiple-valued. After the crack tip has moved by $\delta\boldsymbol{\xi}$, the change of displacement can be written as

$$\delta u_i = u_i(\mathbf{x} - \delta\boldsymbol{\xi}) - u_i(\mathbf{x}) + \Delta u_i, \tag{34.11}$$

where Δu_i is a single-valued function. Equation (34.11) is Eshelby's basic assumption. Taylor's expansion of $u_i(\mathbf{x}-\delta\boldsymbol{\xi})$ about point $\mathbf{x}$ is

$$u_i(\mathbf{x}-\delta\boldsymbol{\xi}) = u_i(\mathbf{x}) - u_{i,l}\delta\xi_l. \tag{34.12}$$

Then, (34.11) becomes

$$\delta u_i = -u_{i,l}\delta\xi_l + \Delta u_i \tag{34.13}$$

or

$$[\delta u_i] = -[u_{i,l}]\delta\xi_l, \tag{34.13'}$$

since Δu_i is a single-valued function. Therefore, we can write (34.10) in the form

$$J_l\delta\xi_l = -\int_B \sigma_{ij}n_j[\delta u_i]dS. \tag{34.14}$$

From (34.6) and (34.14) we have, for the slit-like crack,

$$\mathscr{G}_l\delta\xi_l = \int_D \sigma_{ij}\delta\epsilon^p_{ij}dD + J_l\delta\xi_l. \tag{34.15}$$

When $\delta\xi_l = \delta\xi_1 = \delta A$ and $\delta\epsilon^p_{ij} = 0$, we have

$$\mathscr{G}_1 = J_1. \tag{34.16}$$

The meaning of the J-integral in plane problems and in the absence of plastic deformations, is clearly shown by (34.16); however, the same is not the case when plastic deformations are involved. Miyamoto and Kageyama (1978) modified the J-integral for plastic deformations. It should be noted that J_l becomes zero when the material is perfectly plastic and $[\delta u_i]$ is of the order of $\delta\xi_1$, since $B = \delta\xi_1$, and $\sigma_{ij}n_j = k_0$ in (34.14). J_1 becomes $k_0\delta_t$ for the BCS model, where δ_t is the crack opening displacement at the crack tip.

If dislocations in the neighborhood of the crack tip are displaced by the same amount as the displacement of the crack tip, we can write

$$\delta\epsilon^p_{ji} = -\epsilon_{jlh}\delta\xi_l\alpha_{hi}, \tag{34.17}$$

where α_{ni} is the dislocation density tensor defined by (7.33) [see (38.18)]. Therefore, Mura (1979) has written (34.15) as

$$\mathscr{G}_l = -\int_D \sigma_{ij}\epsilon_{jlh}\alpha_{hi}dD + J_l. \tag{34.18}$$

The integrand in the integral represents the Peach-Koehler (1950) force on dislocations.

Other path-independent integrals related to energy-release rates have been investigated by Knowles and Sternberg (1972) and Budiansky and Rice (1973). The J-integral is also used for investigating singular behavior at the end of a tensile crack in a hardening material (Hutchinson 1968, Rice and Rosengren 1968).

Eshelby (1970) has suggested a broader application of the energy-momentum tensor to phase transformation problems. A martensitic transformation region B undergoes a change of natural shape and size, that is, a change of inhomogeneous inclusion. He writes the change of the mechanical energy, (34.4), caused by a migration $\delta\xi_i$ of the inclusion surface S as

$$\delta\Pi = -\int_S [P_{lj}]\delta\xi_l n_j dS, \tag{34.18.1}$$

where $[P_{lj}] = P_{lj}$ (out) $- P_{lj}$ (in), as defined in (6.4). Since

$$[u_{i,l}] = \left[\frac{\partial u_i}{\partial n}\right] n_l, \tag{34.18.2}$$

where $\partial/\partial n$ denotes differentiation along the normal, (34.18.1) is equivalent to the statement that there is an effective normal force

$$F_l = \left([W] - \sigma_{ij} n_j \left[\frac{\partial u_i}{\partial n}\right]\right) n_l \tag{34.18.3}$$

per unit area of the interface. Equation (34.18.3) with $F_l = 0$ can be used to find the equilibrium position of phase and twin boundaries in the presence of stresses produced by the transformation itself, or applied externally, or both.

Fatigue

It is well known that materials fail under repeated (cyclic) loading and unloading at stresses smaller than the static fracture stresses. The magnitude of the stress required to produce failure decreases as the number of stress cycles increases. Micro-cracks are created in early cycles and they grow with a velocity according to a certain law; e.g. the Paris law (1964). Weertman (1966, 1973) has derived the Paris law by using the BCS model. A similar extension of the BCS model to fatigue is given by Lardner (1968), Bilby and Heald (1968), McCartney and Gale (1971), and Mura and Lin (1974). Recently, Budiansky and Hutchinson (1978) have used the Dugdale method for the analysis of fatigue crack closure.

Let us consider a slit-like crack ($|x_1| < c, x_2 = 0$) as shown in Fig.

32.1, which may stand for an initial crack or a sharp notch. When $\sigma_{22}^0 = S$ is first attained at infinity, a plastic zone develops in the domain $(c < |x_1| < a, x_2 = 0)$. The displacement is given by (32.20), where k_0 is the flow stress at this stage. The plastic zone is assumed to be small compared with c, and k_0 can be defined uniformly throughout the plastic zone. If the first unloading is $-\Delta S$ (the applied stress is reduced to $S - \Delta S$), the change of stress and displacement is defined by the boundary conditions similar to (32.1),

$$
\begin{aligned}
&\sigma_{2j} = 0 \quad (j = 1, 2) \quad \text{at } x_2 = 0 \quad |x_1| < c,\\
&\sigma_{22} = -k_0 - (k_0 + \Delta k_1) \quad \text{at } x_2 = 0 \quad c < |x_1| < a_1,\\
&\sigma_{22} = -\Delta S \quad \text{at } x_2 = \infty,\\
&u_2 = 0, \quad \sigma_{21} = 0 \quad \text{at } x_2 = 0, \quad a < |x_1|,
\end{aligned}
\tag{34.19}
$$

where $c < |x_1| < a_1$ is a new plastic zone caused by the unloading. The stress and displacement in (34.19) are increments of the state described by (32.1). The real stress and displacement at the unloaded state are the sum of those corresponding to (32.1) and those corresponding to (34.19). The flow stress in the new plastic domain is $-(k_0 + \Delta k_1)$, under the assumption that there is no Bauschinger effect (see Fig. 34.2). The elasticity solution for the boundary conditions (34.19) is similar to (32.20). It is obtained from (32.20) by changing S to $-\Delta S$ and also changing k_0 to $-2k_0 - \Delta k_1$. It hence follows that,

$$
\begin{aligned}
&1 - c/a_1 = \frac{\pi^2}{8}\{\Delta S/(2k_0 + \Delta k_1)\}^2,\\
&u_2 = -\frac{(1+\kappa)(2k_0 + \Delta k_1)}{2\pi\mu} c(1 - c/a_1).
\end{aligned}
\tag{34.20}
$$

The quantities corresponding to the subsequent loading by ΔS (the

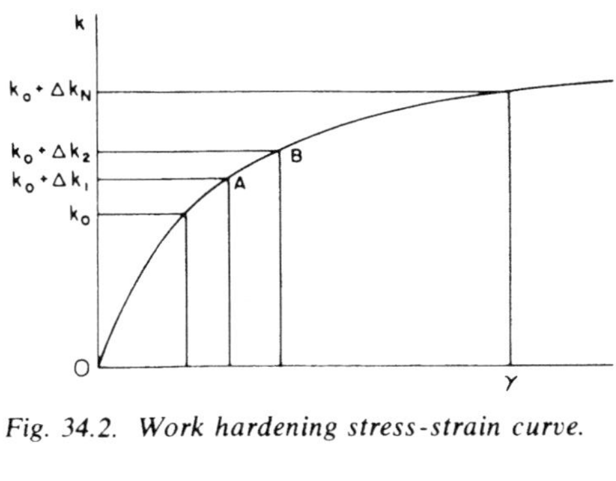

Fig. 34.2. Work hardening stress-strain curve.

applied stress becomes S again) are obtained in a similar: we use ΔS and $k_0+\Delta k_1+k_0+\Delta k_2$ in place of $-\Delta S$ and $-2k_0-\Delta k_1$ in (34.20). Then,

$$1-c/a_2=\frac{\pi^2}{8}\{\Delta S/(2k_0+\Delta k_1+\Delta k_2)\}^2,$$
$$u_2=\frac{(1+\kappa)(2k_0+\Delta k_1+\Delta k_2)}{2\pi\mu}c(1-c/a_2). \tag{34.21}$$

After n cycles of loading and unloading, we have

$$1-c/a_n=\frac{\pi^2}{8}\{\Delta S/(2k_0+\Delta k_{n-1}+\Delta k_n)\}^2,$$
$$u_2=\frac{(1+\kappa)(2k_0+\Delta k_{n-1}+\Delta k_n)}{2\pi\mu}c(1-c/a_n). \tag{34.22}$$

It is noted that

$$a>a_1>a_2>\cdots>a_n \tag{34.23}$$

for work-hardening materials, and

$$a>a_1=a_2=\cdots=a_n \tag{34.24}$$

for perfectly plastic materials.

Suppose that the initial slit starts to grow after the critical number of cycles, N. We seek to establish some relation between the applied stress, the slit length, and N. For this purpose we eliminate $(1-c/a_n)$ from the two equations in (34.22) to obtain

$$u_2=\frac{(1+\kappa)\pi}{16\mu}c(\Delta S)^2/(2k_0+\Delta k_{n-1}+\Delta k_n)$$
$$\approx\frac{(1+\kappa)}{16\mu}\pi c(\Delta S)^2/2(k_0+\Delta k_n). \tag{34.25}$$

We assume that the crack starts to grow when the total plastic work done at the tip of the crack attains a critical value, U. The total plastic work is $2k_0u_2+\Sigma_{n=1}^{N}2(k_0+\Delta k_n)2u_2$, where u_2 in the first term is (32.20) and the second u_2 is (34.25), and the factor 2 comes from the work done at the upper and lower surfaces at the crack tip. The other factor 2 in the second term comes from the fact that N refers to both N loadings and N unloadings. Then, our energy criterion leads to

$$U=\frac{1+\kappa}{8\mu}\pi cS^2+N\frac{1+\kappa}{8\mu}\pi c(\Delta S)^2 \tag{34.26}$$

or

$$N = \left(U - \frac{1+\kappa}{8\mu} K^2\right) \frac{8\mu}{1+\kappa} (\Delta K)^{-2}, \tag{34.27}$$

where $\Delta K = (\pi c)^{1/2}\,\Delta S$, and $K = (\pi c)^{1/2} S$. K is called the stress intensity factor and ΔK is called the stress amplitude intensity factor. If the slit (length $2c$) is located in the center of a strip (width $2L$), $\Delta K = (\pi c)^{1/2}\Delta S[(2L/\pi c)\tan(\pi c/2L)]^{1/2}$. This ΔK is also used approximately for a double-edge-notched strip (notch length a and strip width $2L$). For a single edge-notched strip (notch length c and strip width $2L$), the correction factor, to be multiplied to $(\pi c)^{1/2}\Delta S$, varies from 1.15 (for $c/L = 0.1$) to 2.86 (for $c/L = 1$).

From the first equation in (34.22), we have, approximately,

$$a_N - c = \frac{\pi}{32} \frac{(\Delta K)^2}{(k_0 + \Delta k_N)^2}. \tag{34.28}$$

The rate of crack growth is,

$$dc/dN = (a_N - c)/N \tag{34.29}$$

which leads to

$$\frac{dc}{dN} = \frac{\pi(1+\kappa)}{256\mu k_N^2} (\Delta K)^4 \Big/ \left(U - \frac{1+\kappa}{8\mu} K^2\right), \tag{34.30}$$

where $k_N = k_0 + \Delta k_N$. This is the fourth power law of Paris (1964). It has been assumed in the derivation that the crack grows in the same fashion after N cycles where the initial crack in the second step is the final crack in the first step. Experiments (e.g., Izumi and Fine, 1979) seem to support the tendency shown in (34.30).

A similar result can be obtained for the anti-plane problem where a shear stress $\sigma_{32}^0 = S$ is applied at infinity and σ_{32}^0 is cyclically changed with amplitude ΔS. Then, we have

$$\frac{dc}{dN} = \frac{\pi(\Delta K)^4}{64\mu k_N^2} / (U - K^2/2\mu). \tag{34.31}$$

If a penny-shaped crack is considered under cyclic tension with amplitude ΔS, we have, from (32.59) and (32.60),

$$\begin{aligned} a_N - c &= (1-2\nu)^2 c(\Delta S)^2/32(k_0 + \Delta k_N)^2, \\ w &= 3(1-\nu)(1-2\nu)c(\Delta S)^2/4\pi\mu(k_0 + \Delta k_n), \end{aligned} \tag{34.32}$$

assuming $\Delta k_n \approx \Delta k_{n-1}$. Our crack growth assumption leads to

$$U = 2k_0 w + \sum_{n=1}^{N} 2(k_0 + \Delta k_n)2w, \tag{34.33}$$

where w in the first term is (32.60) and w in the second term is given by (34.32). Then,

$$U = 3(1-\nu)(1-2\nu)\{cS^2 + Nc(\Delta S)^2\}/\pi\mu \tag{34.34}$$

and, therefore, (34.29) gives

$$\frac{dc}{dN} = \frac{3(1-\nu)(1-2\nu)^3\pi(\Delta K)^4}{512\mu k_N^2\{U - 3(1-\nu)(1-2\nu)K^2/4\mu\}}, \tag{34.35}$$

where $k_N = k_0 + \Delta k_N$, $K = 2(c/\pi)^{1/2}S$, $\Delta K = 2(c/\pi)^{1/2}\Delta S$.

Comparing (34.31) and (34.35), it can be said that the crack growth rate of a slit-like crack is about 100 times that of a penny-shaped crack.

Dynamic crack growth

When a crack extends at speeds high enough to necessitate the inclusion of inertia effects, the associated kinetic energy must enter the energy argument of the previous subsections. In recent years a rather extensive body of literature has become available for the analysis of dynamic problems of crack propagation in homogeneous, isotropic, linearly elastic solids. Erdogan (1968), Achenbach (1972), Freund (1975) and Kanninen (1978) have presented review articles on this subject.

Let us consider a dynamic growth of a crack A in a body D subjected to an applied force F_i on the surface S of D. The rate of work done by F_i is

$$\int_S F_i\dot{u}_i dS = \int_D \sigma_{ij}\dot{u}_{i,j} dD + \int_D \rho\ddot{u}_i\dot{u}_i dD - \int_B \sigma_{ij}n_j[\dot{u}_i]dS, \tag{34.36}$$

where B is the domain where $\dot{u}_i$ has a multiple value. The right-hand side in (34.36) has been derived from $\int_S \sigma_{ij}n_j\dot{u}_i dS$ by applying Gauss' theorem, where $\sigma_{ij,j} = \rho\ddot{u}_i$ in D, $\sigma_{ij}n_j = 0$ on A, $[\dot{u}_i] = \dot{u}_i$ (on B^+) $- \dot{u}_i$ (on B^-), and $\mathbf{n}$ is the unit normal on $B^+ \equiv B$. Since $(1/2)(\dot{u}_{i,j} + \dot{u}_{j,i}) = \dot{e}_{ij} + \dot{\epsilon}_{ij}^p$, (34.36) can be written as

$$\int_S F_i\dot{u}_i dS - (d/dt)\left[(1/2)\int_D \sigma_{ij}e_{ij} dD + (1/2)\int_D \rho\dot{u}_i\dot{u}_i dD\right]$$

$$= \int_D \sigma_{ij}\dot{\epsilon}_{ij}^p dD - \int_B \sigma_{ij}n_j[\dot{u}_i]dS = \mathcal{G}_i\dot{\xi}_i \tag{34.37}$$

for a slit-like crack, where e_{ij} and ϵ_{ij}^p are elastic and plastic strains, respectively, $\mathscr{G}_i$ is the energy release rate, and $\dot{\xi}_i$ is the velocity of crack. Achenbach (1972) further investigated $\mathscr{G}_i$ for a slit-like crack in an elastic medium as Irwin (1958) did for a static crack [see (34.7)]. Suppose at time $t = t_0$ a slit-like crack of width $2c$ starts to grow and at time $t > t_0$ the position of the crack tip is $x_1 = X(t - t_0)$ (see Fig. 34.3), where $X(0) = 0$. The singular parts of the stresses in the plane of the crack are of the general form

$$\sigma_{2i}(x_1, 0, t) = T_{2i}[x_1 - X(t - t_0)]^{-1/2}, \tag{34.38}$$

and the singular parts of the particle velocities assume the forms

$$\begin{aligned} \dot{u}_i(x_1, 0^+, t) &= \dot{U}_i[X(t - t_0) - x_1]^{1/2}, \\ \dot{u}_i(x_1, 0^-, t) &= -\dot{U}_i[X(t - t_0) - x_1]^{1/2}. \end{aligned} \tag{34.39}$$

The functions T_{2i} and $\dot{U}_i$, which generally depend on time and on the geometrical and material parameters, are termed the stress-intensity function and the velocity-intensity function, respectively. Achenbach has evaluated

$$-\int_B \sigma_{ij} n_j [\dot{u}_i] dS = \int_{X(t-t_0)}^{X(t-t_0)+\epsilon} \sigma_{2i}[\dot{u}_i] dx_1 = \pi T_{2i} \dot{U}_i, \tag{34.40}$$

where ϵ is an arbitrarily small positive number. More specific expressions for T_{2i} and $\dot{U}_i$ were obtained by Achenbach (1972) for the case of anti-plane shear. The case of plane strain is examined by Nuismer and Achenbach (1972), and Freund (1972). For the symmetric crack opening (Mode I), the instantaneous hoop stress near the crack tip may be expressed as

$$\sigma_\theta \approx (2\pi r)^{-1/2} k_I(t, v) T_\theta^I(\theta, v), \tag{34.40.1}$$

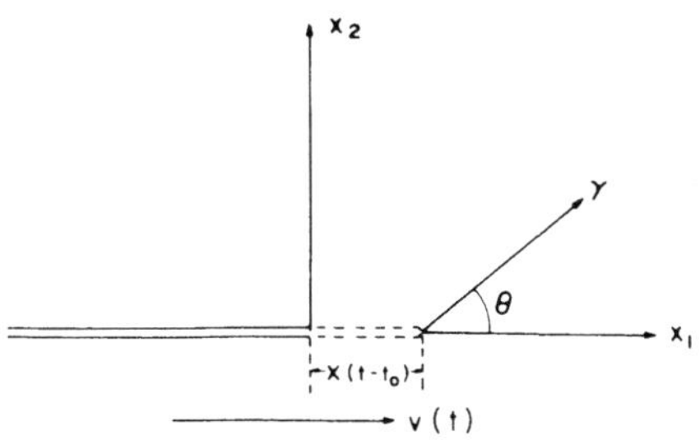

Fig. 34.3. Crack growth.

where $T_\theta^I(0, v) = 1$, $k_I(t, v)$ is the Mode I elastodynamic stress intensity factor, and $v = \dot{\xi}$ is the uniform velocity of crack. The function $T_\theta^I(\theta, v)$ is universal in that it is independent of the overall geometry and the loading.

The conditions governing crack motion can be expressed in terms of $k_I(t, v)$ and an experimentally determined critical value that is assumed to be a material property. According to a brief review by Achenbach and Kanninen (1978), the initiation of crack growth is determined by

$$k_I(t, 0) = K_d(\dot{\sigma}), \tag{34.40.2}$$

where $\dot{\sigma}$ represents the loading rate. For perfectly brittle fracture, $K_c = K_d$. Similarly, for a propagating crack,

$$k_I(t, v) = K_D(v), \tag{34.40.3}$$

where K_D is known as the dynamic fracture toughness. A crack is arrested after time t_a when $k_I < K_D$ for all $t > t_a$.

The J-integral (34.8) can also be modified for the dynamic case, where we add the kinetic energy density K to obtain

$$J_l = \int_\Gamma (W + K) n_l dS - \int_\Gamma \sigma_{ij} n_j u_{i,l} dS, \tag{34.41}$$

where

$$K = (1/2)\rho \dot{u}_i \dot{u}_i \tag{34.42}$$

and W is defined by (34.9). Multiplying (34.41) by $\dot{\xi}_l$ and applying the Gauss theorem, we have, for a slit-like crack,

$$\begin{aligned} J_l \dot{\xi}_l = &\int_V (\sigma_{ij} u_{i,jl} + \rho \dot{u}_i \dot{u}_{i,l}) \dot{\xi}_l dD - \int_V \sigma_{ij} u_{i,lj} \dot{\xi}_l dD \\ &- \int_V \rho \ddot{u}_i u_{i,l} \dot{\xi}_l dD + \int_{B^\pm} \sigma_{ij} n_j u_{i,l} \dot{\xi}_l dS, \end{aligned} \tag{34.43}$$

where V is interpreted to be bounded by Γ, A^+, B^+, A^- and B^-, and W and K on $A^+ + B^+$ are assumed equal to W and K on $A^- + B^-$. $u_{i,l}$ on B^+ is not equal to $u_{i,l}$ on B^- as seen in (34.13′) or

$$[\dot{u}_i] = -[u_{i,l}]\dot{\xi}_l \tag{34.44}$$

by dividing (34.13′) by dt. The first and third terms on the right-hand side in (34.43) cancel each other. Furthermore, the second and fourth terms also cancel each other if V is sufficiently small, since from (34.13)

$$\dot{u}_i \approx -u_{i,l}\dot{\xi}_l$$
$$\ddot{u}_i \approx -\dot{u}_{i,l}\dot{\xi}_l. \tag{34.45}$$

Therefore, (34.43) becomes

$$J_l\dot{\xi}_l = -\int_B \sigma_{ij}n_j[\dot{u}_i]dS. \tag{34.46}$$

Then, the flux of energy release (34.37) can be written as

$$\mathscr{G}_i\dot{\xi}_i = \int_D \sigma_{ij}\dot{\epsilon}^p_{ij}dD + J_i\dot{\xi}_i. \tag{34.47}$$

For the elastic crack of mode I we have

$$\mathscr{G}_1 = J_1 = \frac{1-\nu^2}{E}A(v)k_I^2, \tag{34.47.1}$$

where E and ν are Young's modulus and Poisson's ratio, respectively, and A is a geometry independent function of the crack speed given by

$$A(v) = \frac{(v/c_T)^2(1-v^2/c_L^2)^{1/2}}{(1-\nu)[4(1-v^2/c_L^2)^{1/2}(1-v^2/c_T^2)^{1/2}-(2-v^2/c_T^2)^2]}. \tag{34.47.2}$$

Here c_L and c_T are the longitudinal and shear wave speeds, respectively. The function $A(v)$ is unity at zero crack-tip speed, and increases monotonically to become unbounded as $v \to c_R$, where c_R is the speed of Rayleigh waves in the material. As $v \to c_R$ the elastodynamic stress intensity factor vanishes, and we find $\mathscr{G}_1 \to 0$ as $v \to c_R$.

Contrary to the static case, the J-integral (34.41) is path-dependent. However, it can be path-independent if Γ is chosen as a small loop around the crack tip, or a rectangle made up of segments $|x_1 - X(t-t_0)| = \epsilon$ and $|x_2| = \delta$. An alternative expression for the energy flux,

$$J_l\dot{\xi}_l = \int_\Gamma (W+K)n_l\dot{\xi}_l dS + \int_\Gamma \sigma_{ij}n_j\dot{u}_i dS, \tag{34.48}$$

can be derived directly from (34.41) and (34.45). The above expression has been obtained by Atkinson and Eshelby (1968), and Freund (1972).

Several solutions of the equations of linear elasticity have been found for moving cracks. Yoffé (1951) has treated a crack of constant length being opened at one end and closed at the other end with a constant speed. Craggs (1960) has considered a semi-infinite crack extending at constant speed. Broberg (1960, 1967) has treated crack growth, starting from zero crack length. Baker (1962), on the other hand considered a semi-infinite crack which suddenly appears in an elastic solid and then extends at constant speed under the action of a spatially uniform and time-independent normal pressure on the crack surfaces. In contrast to the work of Broberg and Baker, Freund (1972) has treated the extension of an existing equilibrium crack in a body subjected to general time-independent loads.

When a Dugdale crack with a constant length is propagating with a constant speed v, the crack tip opening displacement has been calculated by Kanninen et al. (1968). Their result can be written as

$$\delta(v) = A(v)\delta(0), \tag{34.49}$$

where $A(v)$ is the function given by (34.47.2) and $\delta(0)$ denotes the value of the crack tip opening displacement in the static case given by (32.20).

A more realistic solution is obtained by Atkinson (1967) by considering a crack expanding at a uniform speed with collinear strip yield zones.

The order of stress singularity at the elastic crack is 0.5 as seen in (30.10). The order of stress singularity decreases when plastic deformation is allowed at the crack tip. It decreases with decreasing of workhardening rate as shown by Hutchinson (1968) and Rice and Rosengren (1968). It becomes zero for the perfectly plastic solid. Recently, Achenbach and Kanninen (1978) showed that for a moving elasto-plastic crack the order of the singularity decreases with increasing crack velocity.

6

Dislocations

Certain fundamental formulae for the elastic field caused by a dislocation loop have been presented in Section 7. In this chapter, the integrations appearing in these formulae are completed, and applications to the continuum plasticity are discussed.

An extensive literature is available on solutions of elastic fields caused by dislocations in various states of motion in bodies with various materials, and geometries. The books by Hirth and Lothe (1968), Steeds (1973), and Lardner (1974) contain explicit expressions for the elastic fields. On the physics of dislocations, the following books are mentioned: Read (1953), Cottrell (1953), Weertman and Weertman (1964), Hull (1965), Friedel (1964), Suzuki (1967), and Nabarro (1967). Reviews and proceedings on dislocation theory include Seerger (1955), deWit (1960), Mura (1968), Kröner (1968), Simmons, deWit and Bullough (1970), Bacon, Barnett and Scattergood (1978), and Nabarro (1979). Continuum theories of continuous distribution of dislocations and plasticity are found in Kondo (1955), Bilby (1960), Kröner (1958), and Kunin (1975).

35. Displacement fields

The displacement field of a dislocation loop L is expressed by (7.6),

$$u_i(\mathbf{x}) = \int_S C_{jlmn} b_m G_{ij,l}(\mathbf{x}-\mathbf{x}') n_n dS(\mathbf{x}') \tag{35.1}$$

or, from (3.19),

$$u_i(\mathbf{x}) = i(2\pi)^{-3} \int_{-\infty}^{\infty} \int_S C_{jlmn} b_m \xi_l N_{ij}(\boldsymbol{\xi}) D^{-1}(\boldsymbol{\xi}) \exp\{i\boldsymbol{\xi}\cdot(\mathbf{x}-\mathbf{x}')\} d\boldsymbol{\xi} n_n dS(\mathbf{x}'), \tag{35.2}$$

where S is the slip plane, L is the boundary of S, and $\mathbf{n}$ is the unit vector normal to S.

Parallel dislocations

Suppose that the slip plane $S(\mathbf{x})$ is defined by

$$-\infty < x < \infty, \quad -R < y < 0, \quad z = 0. \tag{35.3}$$

The plane S is a strip shown in Fig. 35.1. It should be noted that the parallel dislocations (bounding S) are not necessarily parallel to the crystalline directions x_1, x_2, x_3. Vectors $\boldsymbol{\xi}$ and $\mathbf{x}$ have the respective components (ξ_1, ξ_2, ξ_3) and (x_1, x_2, x_3), in the x_i-coordinate system, but have the components (ξ, η, ζ) and (x, y, z) in the x, y, z-coordinate system. That is,

$$\begin{aligned}\boldsymbol{\xi} \cdot (\mathbf{x} - \mathbf{x}') &= \xi_1(x_1 - x_1') + \xi_2(x_2 - x_2') + \xi_3(x_3 - x_3') \\ &= \xi(x - x') + \eta(y - y') + \zeta(z - z').\end{aligned} \tag{35.4}$$

The unit vectors in the x, y, z-directions are denoted by $\boldsymbol{\nu}$, $\mathbf{m}$ and $\mathbf{n}$, respectively.

The following calculations have been performed by Willis (1970). Equation (35.2) becomes

$$\begin{aligned}u_i(\mathbf{x}) = i(2\pi)^{-3}C_{jlmn}b_m n_n \int_{-R}^{0} dy' \int_{-\infty}^{\infty} dx' \iiint_{-\infty}^{\infty} \frac{\xi_l N_{ij}(\boldsymbol{\xi})}{D(\boldsymbol{\xi})} \\ \times \exp[i\{\xi(x - x') + \eta(y - y') + \zeta z\}]d\xi d\eta d\zeta\end{aligned} \tag{35.5}$$

Integration with respect to x' produces a factor $2\pi\delta(\xi)$, so that after integration with respect to ξ, (35.5) becomes

$$\begin{aligned}u_i(\mathbf{x}) = i(2\pi)^{-2}C_{jlmn}b_m n_n \int_{-R}^{0} dy' \iint_{-\infty}^{\infty} \frac{\xi_l N_{ij}(\boldsymbol{\xi})}{D(\boldsymbol{\xi})} \\ \times \exp[i\{\eta(y - y') + \zeta z\}]d\eta d\zeta.\end{aligned} \tag{35.6}$$

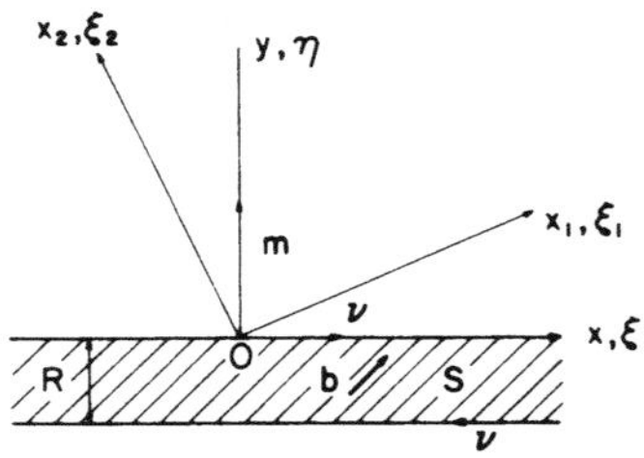

Fig. 35.1. Parallel dislocations bounding the slip plane S.

Next, we integrate with respect to ζ, regarding it as a complex variable. If $z > 0$, we close the contour in the upper half of the complex ζ-plane (the lower half plane for $z < 0$) and then use Cauchy's theorem to obtain

$$u_i(\mathbf{x}) = -(2\pi)^{-1} C_{jlmn} b_m n_n \sum_{N=1}^{3} \int_{-R}^{0} dy' \int_{-\infty}^{\infty} \frac{\xi_l^N N_{ij}(\boldsymbol{\xi}^N)}{n_k \dfrac{\partial D}{\partial \xi_k}(\boldsymbol{\xi}^N)} \times \exp[i\{\eta(y-y') + \zeta^N z\}] d\eta, \tag{35.7}$$

where

$$\boldsymbol{\xi}^N = \mathbf{m}\eta + \mathbf{n}\zeta^N(\eta), \quad (\xi_l^N = m_l\eta + n_l\zeta^N(\eta)), \tag{35.8}$$

and ζ^N are the three roots, with positive imaginary parts, of

$$D(\mathbf{m}\eta + \mathbf{n}\zeta) = 0. \tag{35.9}$$

$\boldsymbol{\nu}$, $\mathbf{m}$, and $\mathbf{n}$ are orthogonal unit vectors shown in Fig. 35.1. Since $\xi_l N_{ij}(\boldsymbol{\xi})/n_k \partial D/\partial \xi_k$ is a homogeneous algebraic equation of degree zero, we divide the numerator and the denominator by η^5. Then (35.7) becomes, by dividing the domain of integration,

$$\begin{aligned} u_i(\mathbf{x}) = -(2\pi)^{-1} C_{jlmn} b_m n_n \sum_{N=1}^{3} \Bigg[& \frac{(m_l + n_l\zeta^N) N_{ij}(\mathbf{m} + \mathbf{n}\zeta^N)}{n_k \dfrac{\partial D}{\partial \xi_k}(\mathbf{m} + \mathbf{n}\zeta^N)} \times \int_{-R}^{0} dy' \int_{0}^{\infty} \exp\{i\eta(y - y' + \zeta^N z)\} d\eta \\ & + \frac{(m_l + n_l\bar{\zeta}^N) N_{ij}(\mathbf{m} + \mathbf{n}\bar{\zeta}^N)}{n_k \dfrac{\partial D}{\partial \xi_k}(\mathbf{m} + \mathbf{n}\bar{\zeta}^N)} \times \int_{-R}^{0} dy' \int_{-\infty}^{0} \exp\{i\eta(y - y' + \bar{\zeta}^N z)\} d\eta \Bigg], \end{aligned} \tag{35.10}$$

where ζ^N are the three roots of

$$D(\mathbf{m} + \mathbf{n}\zeta^N) = 0 \tag{35.11}$$

with positive imaginary parts and $\bar{\zeta}^N$ denotes the complex conjugate of ζ^N. The integrals can be easily evaluated, since ζ^N and $\bar{\zeta}^N$ are independent of η, and the domain of integrals is divided to have convergency. The result is,

$$\begin{aligned} u_i(\mathbf{x}) = \frac{i}{2\pi} C_{jlmn} b_m n_n \sum_{N=1}^{3} \Bigg[& \frac{(m_l + n_l\zeta^N) N_{ij}(\mathbf{m} + \mathbf{n}\zeta^N)}{n_k \dfrac{\partial D}{\partial \xi_k}(\mathbf{m} + \mathbf{n}\zeta^N)} \log \frac{y + \zeta^N z}{y + \zeta^N z + R} \\ & - \frac{(m_l + n_l\bar{\zeta}^N) N_{ij}(\mathbf{m} + \mathbf{n}\bar{\zeta}^N)}{n_k \dfrac{\partial D}{\partial \xi_k}(\mathbf{m} + \mathbf{n}\bar{\zeta}^N)} \log \frac{y + \bar{\zeta}^N z}{y + \bar{\zeta}^N z + R} \Bigg]. \end{aligned} \tag{35.12}$$

A straight dislocation

The solution for a single dislocation along the x-axis is obtained by $R\to\infty$. The terms of $\log R$ are neglected, assuming that they are infinitely large constants. Then, we have

$$u_i(\mathbf{x})=\frac{i}{2\pi}C_{jlmn}b_m n_n\sum_{N=1}^{3}\left[\frac{(m_l+n_l\zeta^N)N_{ij}(\mathbf{m}+\mathbf{n}\zeta^N)}{n_k\dfrac{\partial D}{\partial\xi_k}(\mathbf{m}+\mathbf{n}\zeta^N)}\log(y+\zeta^N z)\right.$$
$$\left.-\frac{(m_l+n_l\bar{\zeta}^N)N_{ij}(\mathbf{m}+\mathbf{n}\bar{\zeta}^N)}{n_k\dfrac{\partial D}{\partial\xi_k}(\mathbf{m}+\mathbf{n}\bar{\zeta}^N)}\log(y+\bar{\zeta}^N z)\right]. \tag{35.13}$$

36. Stress fields

The elastic distortion of a dislocation loop L is expressed by (7.16),

$$\beta_{ji}(\mathbf{x})=i(2\pi)^{-3}\int_L dx_h'\int_{-\infty}^{\infty}\epsilon_{jnh}C_{klmn}\xi_l N_{ik}(\boldsymbol{\xi})D^{-1}(\boldsymbol{\xi})b_m\exp\{i\boldsymbol{\xi}\cdot(\mathbf{x}-\mathbf{x}')\}d\boldsymbol{\xi}. \tag{36.1}$$

The stress field can be obtained from

$$\sigma_{ij}=C_{ijkl}\beta_{lk}. \tag{36.2}$$

Dislocation segments

Since an arbitrary dislocation loop can be approximated by a sum of a finite number of dislocation segments, the solution for a dislocation segment is fundamental for any shape of dislocation loop. The following calculation is based upon the work of Willis (1970), and Mura and Mori (1976).

Let us consider a dislocation segment $\mathbf{yz}$ shown in Fig. 36.1. An arbitrary point $\mathbf{x}'$ on the segment is expressed by

$$\mathbf{x}'=\mathbf{y}+l(\mathbf{z}-\mathbf{y}),\quad 0\le l\le 1,$$
$$dx_h'=(z_h-y_h)dl, \tag{36.3}$$

and the volume element in the $\boldsymbol{\xi}$-space is given by

$$d\boldsymbol{\xi}=\xi^2 d\xi dS(\bar{\boldsymbol{\xi}}), \tag{36.4}$$

where $dS(\bar{\boldsymbol{\xi}})$ is the surface element of the unit sphere S^2 in the $\boldsymbol{\xi}$-space and $\bar{\boldsymbol{\xi}}=\boldsymbol{\xi}/\xi$, $\xi=|\boldsymbol{\xi}|$. Willis has integrated (36.1) first with respect to l and

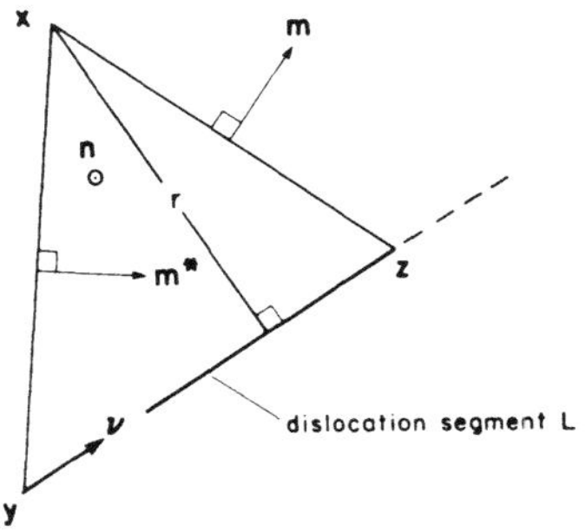

Fig. 36.1. Dislocation segment **yz**. *Unit vector* **n** *is normal to the plane* **xyz**.

then with respect to ξ, obtaining

$$\beta_{ji}(\mathbf{x})=\frac{-i}{(2\pi)^3}\epsilon_{jnh}C_{klmn}b_m(z_h-y_h)\int_{S^2}\frac{\bar{\xi}_l N_{ik}(\bar{\xi})dS(\bar{\xi})}{D(\bar{\xi})\{\bar{\boldsymbol{\xi}}\cdot(\mathbf{y}-\mathbf{x})\}\{\bar{\boldsymbol{\xi}}\cdot(\mathbf{z}-\mathbf{x})\}}. \tag{36.5}$$

In order to have a convergent integral, the exponential term in (36.1) has been written as $\exp[\xi\{-\epsilon+i\bar{\boldsymbol{\xi}}\cdot(\mathbf{x}-\mathbf{x}')\}]$, where $\epsilon>0$. The limit $\epsilon\to 0$ has been taken after the integration.

$N_{ik}(\bar{\boldsymbol{\xi}})$ and $D(\bar{\boldsymbol{\xi}})$ are homogeneous polynomials of degree 4 and 6 respectively and the integrand in (36.5) is an odd function of $\bar{\boldsymbol{\xi}}$. Thus, the integral on the unit sphere S^2 vanishes almost everywhere except on the two unit circles satisfying $\bar{\boldsymbol{\xi}}\cdot(\mathbf{y}-\mathbf{x})=0$ and $\bar{\boldsymbol{\xi}}\cdot(\mathbf{z}-\mathbf{x})=0$; see Fig. 36.2. It is sufficient to reduce the integral domain in (36.5) into the narrow strips of width 2ϵ along these two unit circles.

Let us introduce three unit vectors **n**, **m***, and **m**, as seen in Fig. 36.1. **n** is normal to the plane containing points **x**, **y**, **z**; **m*** is orthogonal to **n** and $\mathbf{y}-\mathbf{x}$; and **m** is orthogonal to **n** and $\mathbf{z}-\mathbf{x}$. The vector products

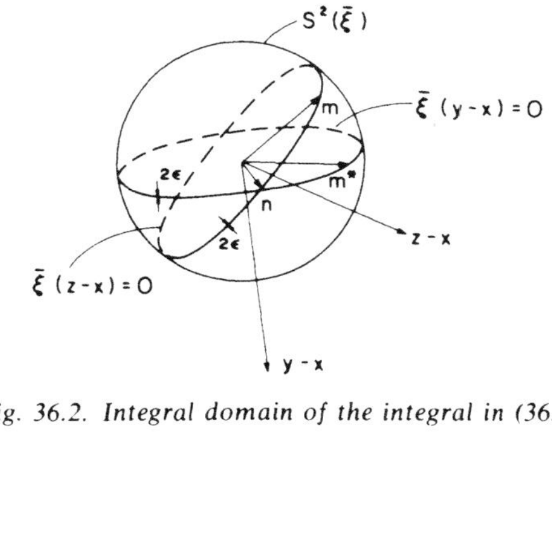

Fig. 36.2. Integral domain of the integral in (36.5).

$\mathbf{m}^* \times \mathbf{n}$ and $\mathbf{m} \times \mathbf{n}$ are directed parallel to the $\mathbf{y}-\mathbf{x}$ and $\mathbf{z}-\mathbf{x}$ directions, respectively.

The principal part of the integrand in (36.5) can be written as

$$\frac{1}{\{\bar{\xi}\cdot(\mathbf{y}-\mathbf{x})\}\{\bar{\xi}\cdot(\mathbf{z}-\mathbf{x})\}} = \frac{1}{\{\bar{\xi}\cdot(\mathbf{z}-\mathbf{y})\}\{\bar{\xi}\cdot(\mathbf{y}-\mathbf{x})\}} - \frac{1}{\{\bar{\xi}\cdot(\mathbf{z}-\mathbf{y})\}\{\bar{\xi}\cdot(\mathbf{z}-\mathbf{x})\}}. \tag{36.6}$$

When the integral on the narrow strip along the circle $\bar{\xi}\cdot(\mathbf{y}-\mathbf{x})=0$ is considered, the first term on the right-hand side in (36.6) is used; on the other hand, when the integral on the narrow strip along the circle $\bar{\xi}\cdot(\mathbf{z}-\mathbf{x})=0$ is considered, the second term on the right-hand side in (36.6) is used. The plane of circle $\bar{\xi}\cdot(\mathbf{y}-\mathbf{x})=0$ is perpendicular to $(\mathbf{y}-\mathbf{x})$ and, therefore, contains vectors $\mathbf{n}$ and $\mathbf{m}^*$. The plane of circle $\bar{\xi}\cdot(\mathbf{z}-\mathbf{x})=0$ contains vectors $\mathbf{n}$ and $\mathbf{m}$. On the narrow strip along the circle $\bar{\xi}\cdot(\mathbf{y}-\mathbf{x})=0$,

$$\bar{\xi} = -\mathbf{m}^*\sin\phi + \mathbf{n}\cos\phi + h(\mathbf{y}-\mathbf{x})/|\mathbf{y}-\mathbf{x}|, \quad |h|\le\epsilon\ll 1, \tag{36.7}$$

with parameters ϕ and h, where h is infinitesimally small. Then the integral in (36.5) defined on this narrow strip becomes

$$\int_{S^2} = -\int_0^{2\pi} d\phi \int_{-\epsilon}^{\epsilon} dh\, \frac{(-m_l^*\sin\phi + n_l\cos\phi)N_{ik}(-\mathbf{m}^*\sin\phi+\mathbf{n}\cos\phi)}{D(-\mathbf{m}^*\sin\phi+\mathbf{n}\cos\phi)\sin\phi\{\mathbf{m}^*\cdot(\mathbf{z}-\mathbf{y})\}h|\mathbf{y}-\mathbf{x}|}, \tag{36.8}$$

where

$$\begin{aligned} \bar{\xi}\cdot(\mathbf{z}-\mathbf{y}) &= -\sin\phi\{\mathbf{m}^*\cdot(\mathbf{z}-\mathbf{y})\}, \\ \bar{\xi}\cdot(\mathbf{y}-\mathbf{x}) &= h|\mathbf{y}-\mathbf{x}|, \end{aligned} \tag{36.9}$$

are used. Furthermore, we have

$$\begin{aligned} \{\mathbf{m}^*\cdot(\mathbf{z}-\mathbf{y})\}|\mathbf{y}-\mathbf{x}| &= r|\mathbf{z}-\mathbf{y}| = 2\times \text{area of triangle } \mathbf{xyz}, \\ (z_h-y_h)/|\mathbf{z}-\mathbf{y}| &= \nu_h, \end{aligned} \tag{36.10}$$

from Fig. 36.1, and

$$\int_{-\epsilon}^{\epsilon} \frac{dh}{h} = i\pi, \tag{36.10a}$$

where r is the distance from $\mathbf{x}$ to the dislocation line, and Cauchy's principal value is taken for the last integral with respect to h.

A similar calculation is performed for the integral on the narrow strip along the circle $\bar{\boldsymbol{\xi}} \cdot (\mathbf{z}-\mathbf{x}) = 0$ by putting

$$\bar{\boldsymbol{\xi}} = -\mathbf{m}\sin\phi + \mathbf{n}\cos\phi + h(\mathbf{z}-\mathbf{x})/|\mathbf{z}-\mathbf{x}|, \quad |h| \le \epsilon \ll 1. \tag{36.11}$$

Then, (36.5) becomes

$$\beta_{ji}(\mathbf{x}) = \frac{1}{(2\pi)^2}\epsilon_{jnh}C_{klmn}b_m\nu_h r^{-1}\Bigg[-\frac{1}{2}\int_0^{2\pi} \frac{(-m_l^*\sin\phi + n_l\cos\phi)N_{ik}(-\mathbf{m}^*\sin\phi + \mathbf{n}\cos\phi)}{\sin\phi\, D(-\mathbf{m}^*\sin\phi + \mathbf{n}\cos\phi)}\,d\phi + \frac{1}{2}\int_0^{2\pi}\frac{(-m_l\sin\phi + n_l\cos\phi)N_{ik}(-\mathbf{m}\sin\phi + \mathbf{n}\cos\phi)}{\sin\phi D(-\mathbf{m}\sin\phi + \mathbf{n}\cos\phi)}\,d\phi\Bigg] \tag{36.12}$$

or

$$\begin{aligned}\beta_{ji}(\mathbf{x}) = \frac{1}{(2\pi)^2}\epsilon_{jnh}C_{klmn}b_m\nu_h r^{-1}\Bigg(&\frac{1}{2}\int_0^{2\pi}\Bigg[\frac{m_l^*N_{ik}(-\mathbf{m}^*\sin\phi + \mathbf{n}\cos\phi)}{D(-\mathbf{m}^*\sin\phi + \mathbf{n}\cos\phi)} \\ &-\frac{\cos\phi}{\sin\phi}n_l\left\{\frac{N_{ik}(-\mathbf{m}^*\sin\phi + \mathbf{n}\cos\phi)}{D(-\mathbf{m}^*\sin\phi + \mathbf{n}\cos\phi)} - \frac{N_{ik}(\mathbf{n})}{D(\mathbf{n})}\right\}\Bigg]d\phi \\ &-\frac{1}{2}\int_0^{2\pi}\Bigg[\frac{m_lN_{ik}(-\mathbf{m}\sin\phi + \mathbf{n}\cos\phi)}{D(-\mathbf{m}\sin\phi + \mathbf{n}\cos\phi)} \\ &-\frac{\cos\phi}{\sin\phi}n_l\left\{\frac{N_{ik}(-\mathbf{m}\sin\phi + \mathbf{n}\cos\phi)}{D(-\mathbf{m}\sin\phi + \mathbf{n}\cos\phi)} - \frac{N_{ik}(\mathbf{n})}{D(\mathbf{n})}\right\}\Bigg]d\phi\Bigg),\end{aligned} \tag{36.13}$$

where $\int_0^{2\pi}(\cos\phi/\sin\phi)n_l(N_{ik}(\mathbf{n})/D(\mathbf{n}))d\phi$ is added to the first integral in (36.12) and subtracted from the second integral in order to eliminate the singularities at $\phi = 0$, π, and 2π.

By introducing the definition

$$\begin{aligned}I_{lik}(\mathbf{m}^*, \mathbf{n}) = \int_0^{2\pi}\Bigg[&\frac{m_l^*N_{ik}(-\mathbf{m}^*\sin\phi + \mathbf{n}\cos\phi)}{D(-\mathbf{m}^*\sin\phi + \mathbf{n}\cos\phi)} - \frac{\cos\phi}{\sin\phi}n_l \\ &\times\left\{\frac{N_{ik}(-\mathbf{m}^*\sin\phi + \mathbf{n}\cos\phi)}{D(-\mathbf{m}^*\sin\phi + \mathbf{n}\cos\phi)} - \frac{N_{ik}(\mathbf{n})}{D(\mathbf{n})}\right\}\Bigg]d\phi,\end{aligned} \tag{36.14}$$

(36.13) can be written as

$$\beta_{ji}(\mathbf{x}) = \frac{1}{(2\pi)^2}\epsilon_{jnh}C_{lkmn}b_m\nu_h r^{-1}\left\{\frac{1}{2}I_{lik}(\mathbf{m}^*, \mathbf{n}) - \frac{1}{2}I_{lik}(\mathbf{m}, \mathbf{n})\right\} \tag{36.15}$$

For an infinitely long straight dislocation, $\mathbf{m}^* = -\mathbf{m}$, and $\mathbf{m}^*$ and $\mathbf{m}$

are perpendicular to $\boldsymbol{\nu}$. Then, $I_{lik}(\mathbf{m}^*, \mathbf{n}) = -I_{lik}(\mathbf{m}, \mathbf{n})$ and, therefore

$$\beta_{ji}(\mathbf{x}) = -(2\pi)^{-2}\epsilon_{jnh}C_{klmn}b_m\nu_h r^{-1}I_{lik}(\mathbf{m}, \mathbf{n}) \equiv \nu_h r^{-1} \textstyle\sum_{jih}(\mathbf{m}, \mathbf{n}), \tag{36.16}$$

where $\mathbf{n} = \boldsymbol{\nu} \times \mathbf{m}$.

Figure 36.3 shows the atomic arrangement around the center of a straight edge dislocation in BCC Fe. This arrangement of atoms has been determined from (36.16) with (36.14) or (36.24) when $\boldsymbol{\nu} = [\bar{1}10]/\sqrt{2}$, $\mathbf{b} = a[111]/2$, where a is the lattice constant of BCC Fe. Looking through the BCC structure along the $[\bar{1}10]$ direction, we notice two types of $(1\bar{1}0)$ planes, alternatingly stacked perpendicular to this direction. The atoms on these planes are denoted by the filled and open circles. In this calculation, the elastic moduli are taken as $C_{1111} = 2.37 \times 10^{11}\ N/m^2$, $C_{1122} = 1.41 \times 10^{11}\ N/m^2$, and $C_{1212} = 1.16 \times 10^{11}\ N/m^2$.

Willis' formula

There are several alternative expressions of (36.15) which lead to Willis' result (1970) and the result of Asaro et al. (1973). Willis obtained the solution in terms of residues of a complex plane. Employing the transformation

$$p = -\cos\phi/\sin\phi,\ dp = d\phi/\sin^2\phi, \tag{36.17}$$

neglecting the terms which have been added and subtracted artificially,

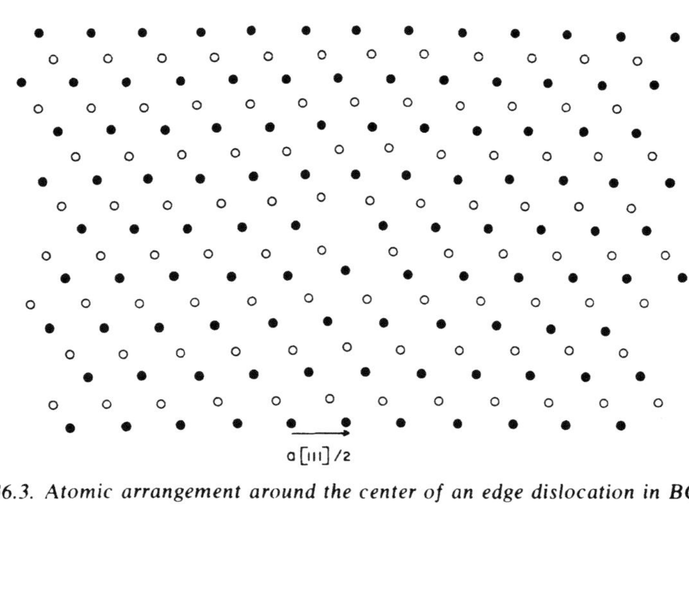

Fig. 36.3. Atomic arrangement around the center of an edge dislocation in BCC Fe.

and considering the fact that $N_{ik}(\boldsymbol{\xi})/D(\boldsymbol{\xi})$ is a homogeneous function of $\boldsymbol{\xi}$ of degree -2, we write (36.14) as

$$I_{lik}(\mathbf{m}^*, \mathbf{n}) = 2\int_{-\infty}^{\infty} \frac{(m_l^* + pn_l)N_{ik}(\mathbf{m}^* + p\mathbf{n})}{D(\mathbf{m}^* + p\mathbf{n})}\,dp. \tag{36.18}$$

$I_{lik}(\mathbf{m}, \mathbf{n})$ is also transformed to a similar form. The integral in (36.18) can be expressed in terms of residues in the complex p-plane by Cauchy's theorem. Then, (36.15) becomes

$$\beta_{ji}(\mathbf{x}) = i(2\pi)^{-1}\epsilon_{jnh}b_m C_{klmn}\nu_h r^{-1}\sum_{N=1}^{3}\left[\frac{(m_l^* + n_l\zeta^{*N})N_{ik}(\mathbf{m}^* + \mathbf{n}\zeta^{*N})}{n_s\dfrac{\partial D}{\partial \xi_s}(\mathbf{m}^* + \mathbf{n}\zeta^{*N})} - \frac{(m_l + n_l\zeta^{N})N_{ik}(\mathbf{m} + \mathbf{n}\zeta^{N})}{n_s\dfrac{\partial D}{\partial \xi_s}(\mathbf{m} + \mathbf{n}\zeta^{N})}\right], \tag{36.19}$$

where ζ^N and ζ^{*N} are complex numbers with positive imaginary parts and are the roots of the following equations:

$$D(\mathbf{m} + \mathbf{n}\zeta^N) = 0,$$
$$D(\mathbf{m}^* + \mathbf{n}\zeta^{*N}) = 0. \tag{36.20}$$

Since $D(\mathbf{m} + \mathbf{n}\zeta^N)$ is the sextic equation in ζ^N, considerable computation is required to find the roots of (36.20). Expression (36.13) is usually more convenient for numerical calculation than (36.19).

The Asaro et al. formula

The result of Asaro, Hirth, Barnett, and Lothe (1973) can also be obtained from (36.15). The following notation is introduced:

$$\mathbf{n}' = -\mathbf{m}^*\sin\phi + \mathbf{n}\cos\phi,$$
$$\mathbf{m}' = \mathbf{m}^*\cos\phi + \mathbf{n}\sin\phi,$$
$$(m', n')_{st} = C_{sptq}m'_p n'_q,$$
$$(n', n')^{-1}_{ik} = N_{ik}(\mathbf{n}')/D(\mathbf{n}'). \tag{36.21}$$

Then, we have

$$(n', n')(n', n')^{-1} = I,$$
$$(n', n') = \cos^2\phi(n, n) - \cos\phi\sin\phi\{(m^*, n) + (n, m^*)\} + \sin^2\phi(m^*, m^*),$$

$$(m', n') = \cos^2 \phi\{(m^*, n) + (n, m^*)\} - \cos \phi \sin \phi\{(m^*, m^*) - (n, n)\} - (n, m^*), \tag{36.22}$$

where I is the unit matrix. By substituting the second equation into the first and multiplying by $(n, n)^{-1} \cos \phi/\sin \phi$, we have

$$\{I \cos \phi/\sin \phi - (n, n)^{-1}(n, m^*) - (n, n)^{-1}(m', n')\}(n', n')^{-1} = (n, n)^{-1} \cos \phi/\sin \phi, \tag{36.23}$$

where the third equation in (36.22) has been used. The second integrand in (36.14) can be written as the ik component of $n_l\{(n', n')^{-1} - (n, n)^{-1}\} \cos \phi/\sin \phi$, which becomes from (36.23) $n_l\{(n, n)^{-1}(n, m^*) + (n, n)^{-1}(m', n')\}(n', n')^{-1}$. Therefore, we have

$$I_{lik}(\mathbf{m}^*, \mathbf{n}) = \int_0^{2\pi} [m_l^*(n', n')^{-1}_{ik} - n_l(n, n)^{-1}_{is}\{(n, m^*)_{st} + (m', n')_{st}\}(n', n')^{-1}_{tk}]\, d\phi. \tag{36.24}$$

The expression (36.15) with (36.24) is the result obtained by Asaro et al. These expressions can be represented in alternative ways, as given by Barnett (1972), Barnett and Swanger (1971), Barnett et al. (1972), Malén (1970, 1971), and Barnett and Lothe (1973). These results illustrate further development of the works of Eshelby, Read and Shockley (1953), Stroh (1958, 1962), and Bullough and Bilby (1954). On the other hand, Lothe (1967), Brown (1967), and Indenbom and Orlov (1968) have shown how the in-plane strain field of the general planar dislocation loop is determined directly from the strain field of a straight dislocation and the derivatives of these strains with respect to variables describing the direction of the line in the plane as shown by (7.17). The details of the stress analysis of dislocations, along with the work of Barnett et al. and Lothe and Brown, can be found in the review article by Bacon, Barnett and Scattergood (1978). The stress analysis of an infinite straight dislocation based upon the work of Eshelby et al. and Stroh can be found in the book by Steeds (1973).

Dislocation loops

The elastic distortion of a dislocation loop L is calculated by the integration of (36.15) along the dislocation line (see Fig. 36.4),

$$\beta_{ji}(\mathbf{x}) = \frac{-1}{8\pi^2} \epsilon_{jnh} C_{klmn} b_m \oint_L \frac{\nu_h}{r} \frac{d}{dl} I_{lik}(\mathbf{m}, \mathbf{n})\, dl, \tag{36.25}$$

where the difference of I_{lik} at two neighboring points in (36.15) has been

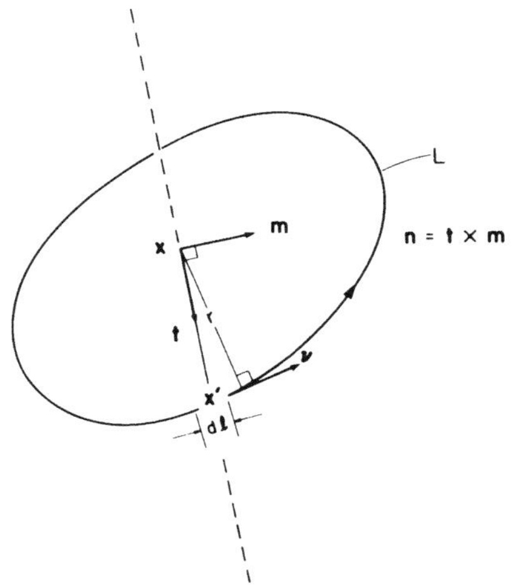

Fig. 36.4. Dislocation loop L.

expressed by the differential. The unit vectors **m**, **n**, and $\boldsymbol{\nu}$ are shown in Fig. 36.4. Vector **n** is normal to the plane containing dl and **x**, and vector **m** is lying on the plane.

If the dislocation loop is a smooth curve, (36.25) is written,

$$\beta_{ji}(\mathbf{x}) = \frac{1}{8\pi^2} \epsilon_{jnh} C_{klmn} b_m \oint_L \frac{d}{dl}\left(\frac{\nu_h}{r}\right) I_{lik}(\mathbf{m}, \mathbf{n})\, dl \tag{36.26}$$

after integration by parts, or

$$\beta_{ji}(\mathbf{x}) = -\frac{1}{2} \oint_L \frac{d}{dl}\left(\frac{\nu_h}{r}\right) \sum\nolimits_{jih} (\mathbf{m}, \mathbf{n})\, dl \tag{36.27}$$

with the use of notation in (36.16). If a vector **t** is introduced along the line **xx**′, $t_h \Sigma_{jih}$ (**m**, **n**) becomes the *ji* component of the elastic distortion caused by a fictitious infinite straight dislocation containing points **x** and **x**′ with direction **t**. This distortion is measured at the end points of vector **m**. The expression (36.27) is more convenient to use than the Indenbom and Orlov expression (7.17) or the Lothe (1967) and Brown (1967) expression, since they contain higher derivatives of the straight dislocation solution.

If the dislocation loop L is a smooth convex planar loop and point **x** is a field point inside L, then (36.27) can be reduced to a more convenient form. Let us denote by α the angle between the tangential direction at a point **x**′ on the dislocation loop and some datum, and by θ

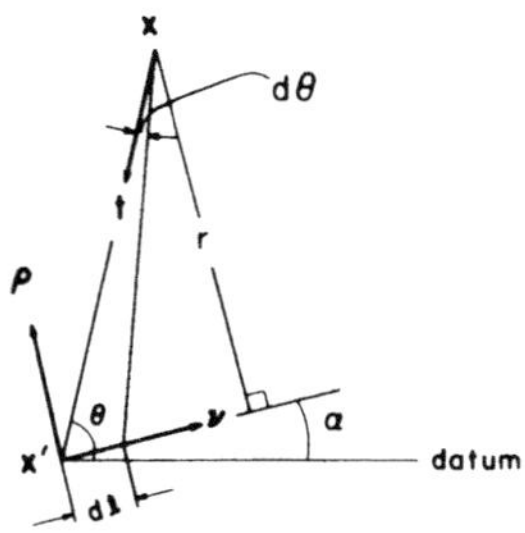

Fig. 36.5. Geometry at dislocation line element dl.

the angle between $\mathbf{x}-\mathbf{x}'$ and the same datum (Fig. 36.5). Then, we have

$$\begin{aligned} dl\,\sin(\theta-\alpha) &= |\mathbf{x}-\mathbf{x}'|d\theta, \\ dl\,\cos(\theta-\alpha) &= -d|\mathbf{x}-\mathbf{x}'|, \\ r &= |\mathbf{x}-\mathbf{x}'|\sin(\theta-\alpha), \\ d\alpha &= \kappa dl, \\ d\nu_h &= \rho_h d\alpha, \end{aligned} \tag{36.28}$$

where κ is the curvature of L at $\mathbf{x}'$ and $\boldsymbol{\rho}$ is the unit vector directed toward the center of curvature. Elementary calculus yields

$$\begin{aligned} \frac{d}{d\theta}\left(\frac{\nu_h}{r}\right) &= \frac{\kappa}{\sin^3(\theta-\alpha)}\{\rho_n \sin(\theta-\alpha)+\nu_h\cos(\theta-\alpha)\} \\ &= -\kappa t_h \csc^3(\theta-\alpha), \end{aligned} \tag{36.29}$$

where (36.28) is used. Expression (36.27) is written, therefore, as

$$\beta_{ji}(\mathbf{x}) = (1/2)\oint_L \kappa \sum\nolimits_{ji}(\theta)\csc^3(\theta-\alpha)d\theta, \tag{36.30}$$

where

$$\sum\nolimits_{ji}(\theta) = t_h \sum\nolimits_{jih}(\mathbf{m},\mathbf{n}) \tag{36.31}$$

is the ji component of the elastic distortion at the unit distance caused by the fictitious infinite straight dislocation containing $\mathbf{x}$ and $\mathbf{x}'$. The last result (36.30) has been obtained by Asaro and Barnett (1976). When point x is located outside L, we can have a result similar to (36.30).

For a planar loop, the Indenbom and Orlov expression (7.17)

reduces to Brown's formula (1967). Simple calculus gives for two dimensions (see Fig. 36.5),

$$\nu_p \frac{\partial}{\partial x_p} = \cos(\theta - \alpha) \frac{\partial}{\partial |\mathbf{x} - \mathbf{x}'|} - \frac{\sin(\theta - \alpha)}{|\mathbf{x} - \mathbf{x}'|} \frac{\partial}{\partial \theta}. \tag{36.32}$$

The elastic distortion of an infinite and straight dislocation is inversely proportional to the distance from the dislocation line to the field point. Therefore, (7.18) is written as

$$\beta_{ji}(\boldsymbol{\nu}; \mathbf{t}) = \{1/\sin(\theta - \alpha)\} \sum\nolimits_{ji}(\theta). \tag{36.33}$$

Applying the operator of (36.32) twice to (36.33), we transform (7.17) to

$$\begin{aligned}\beta_{ji}(\mathbf{x}) &= \frac{1}{2} \oint_L \frac{\sin(\theta - \alpha)}{|\mathbf{x} - \mathbf{x}'|^2} \Big(\sum\nolimits_{ji}(\theta) + \sum\nolimits_{ji}''(\theta)\Big) dl(\mathbf{x}') \\ &= \frac{1}{2} \oint_L \frac{1}{|\mathbf{x} - \mathbf{x}'|} \Big(\sum\nolimits_{ji}(\theta) + \sum\nolimits_{ji}''(\theta)\Big) d\theta. \end{aligned} \tag{36.34}$$

The above expression has been derived by Brown (1967) as a refinement of a similar expression by Lothe (1967).

Complicated configurations of dislocations can often be obtained by superposition of angular dislocations. Yoffe (1960) has calculated the elastic fields for angular dislocations in the unbounded material. Yoffe (1961), Comninou and Dundurs (1975), and Comninou (1977) have obtained similar results for a half-space.

Elastic solutions for a circular dislocation loop are obtained by Keller (1957), and Kroupa (1960) for the unbounded medium and by Salamon and Dundurs (1971) for a two-phase material. Elastic behaviors of disclination loops near a free surface and a two-phase material are investigated by T.W. Chou (1971), Chou and Lu (1972), Chou and Pan (1973) and Chou, and Lu (1973). Recently, Sekine and Mura (1979) have obtained the elastic field of a dislocation dipole segment. The elastic field of planar periodic dislocation networks has been obtained by Mura (1964). Owen and Mura (1967), Owen (1971), and Saada (1976). The solution for a helical dislocation has been obtained by Owen and Mura (1967).

37. Dislocation density tensor

The dislocation density tensor of Nye (1953) has already been introduced in (7.33). In this section, the tensor will be introduced by a different method.

First, consider a dislocation loop L which is the boundary of slip plane S, Fig. 7.1. The slip $\mathbf{b}$ on S introduces a plastic distortion β^P_{ji},

$$\beta^P_{ji} d\mathbf{x} = -b_i n_j dS = -b_i dS_j, \tag{37.1}$$

where $d\mathbf{x} = dx_1 dx_2 dx_3$, dS is the surface element of S, and $\mathbf{n}$ is the unit normal vector of S. Kroupa (1962) has called β^P_{ji} the dislocation loop density tensor. The Fourier integral representation of β^P_{ji} is

$$\beta^P_{ji}(\mathbf{x}) = \int_{-\infty}^{\infty} \bar{\beta}^P_{ji}(\boldsymbol{\xi}) \exp(i\boldsymbol{\xi} \cdot \mathbf{x}) d\boldsymbol{\xi}, \tag{37.2}$$

where

$$\bar{\beta}^P_{ji}(\boldsymbol{\xi}) = (2\pi)^{-3} \int_{-\infty}^{\infty} \beta^P_{ji}(\mathbf{x}) \exp(-i\boldsymbol{\xi} \cdot \mathbf{x}) d\mathbf{x}. \tag{37.3}$$

Since b_i is constant on S, (37.3) becomes

$$\bar{\beta}^P_{ji}(\boldsymbol{\xi}) = -(2\pi)^{-3} b_i \int_S n_j \exp(-i\boldsymbol{\xi} \cdot \mathbf{x}) dS, \tag{37.4}$$

where (37.1) is used.

Multiplying (37.4) by $-i\epsilon_{hlj}\xi_l$ and applying the Stokes theorem, we have

$$-i\epsilon_{hlj}\xi_l \bar{\beta}^P_{ji}(\boldsymbol{\xi}) = (2\pi)^{-3} b_i \int_L \exp(-i\boldsymbol{\xi} \cdot \mathbf{x}) \nu_h dl, \tag{37.5}$$

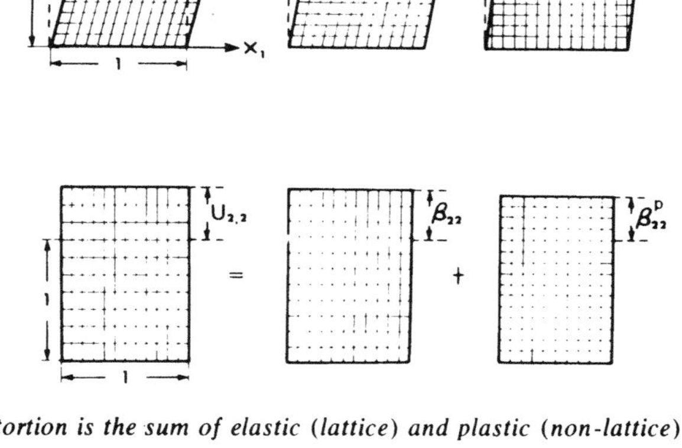

Fig. 37.1. Distortion is the sum of elastic (lattice) and plastic (non-lattice) distortions.

where dl is the line element of L, and ν_h is its unit tangent vector. The equality $-i\xi_l \exp(-i\boldsymbol{\xi}\cdot\mathbf{x}) = (\partial/\partial x_l)\exp(-i\boldsymbol{\xi}\cdot\mathbf{x})$ has been used in the derivation.

The dislocation density tensor is defined by

$$\alpha_{hi}d\mathbf{x} = b_i\nu_h dl = b_i dl_h. \tag{37.6}$$

Its Fourier integral expression is

$$\alpha_{hi}(\mathbf{x}) = \int_{-\infty}^{\infty} \bar{\alpha}_{hi}(\boldsymbol{\xi})\exp(i\boldsymbol{\xi}\cdot\mathbf{x})d\boldsymbol{\xi}, \tag{37.7}$$

where

$$\bar{\alpha}_{hi}(\boldsymbol{\xi}) = (2\pi)^{-3}\int_{-\infty}^{\infty} \alpha_{hi}(\mathbf{x})\exp(-i\boldsymbol{\xi}\cdot\mathbf{x})d\mathbf{x} \tag{37.8}$$

or

$$\bar{\alpha}_{hi}(\boldsymbol{\xi}) = (2\pi)^{-3}b_i\int_L \exp(-i\boldsymbol{\xi}\cdot\mathbf{x})\nu_h dl. \tag{37.9}$$

Comparing (37.5) and (37.9), we have

$$\bar{\alpha}_{hi} = -i\epsilon_{hlj}\xi_l\bar{\beta}^p_{ji}. \tag{37.10}$$

When this is substituted into (37.7), we obtain

$$\alpha_{hi} = -\epsilon_{hlj}\beta^p_{ji,l}, \tag{37.11}$$

which is Kröner's expression (1955, 1956).

Although the result (37.11) has been obtained for a single dislocation loop, it also holds for the case of continuously distributed dislocations, where α_{hi} and β^p_{ji} are spatial functions. In this case we can interpret

$$\alpha_{hi}d\mathbf{x} = \sum b_i\nu_h dl, \tag{37.12}$$

where the summation is taken on all dislocation segments contained in the infinitesimal cube $d\mathbf{x}$.

The dislocation density tensor α_{hi} expresses the x_i-component of the total Burgers vector of dislocations threading the unit surface perpendicular to the x_h-direction. Denoting the Burgers circuit by c (see Fig. 10.2), the total Burgers vector can be expressed as $\int_S \alpha_{hi}\nu_h dS$ which must be equal to the multiple value of displacement expressed by (10.11). It may also be expressed by $-\int_c \beta^p_{ji}dx_j$ or $-\int_S \epsilon_{hlj}\beta^p_{ji,l}\nu_h dS$ after the use of Stokes' theorem (7.10). Equating the two expressions, we have (37.11).

The stress field due to a continuous distribution of dislocations can be obtained from (37.12) and (7.15),

$$\sigma_{ij}(\mathbf{x}) = C_{ijkl} \int \epsilon_{lnh} C_{pqmn} G_{kp,q}(\mathbf{x}-\mathbf{x}')\alpha_{hm}(\mathbf{x}')d\mathbf{x}'. \tag{37.13}$$

An isolated dislocation line is a special case when the dislocation density tensor takes the form of Dirac's delta function. For instance, the dislocation line shown in Fig. 4.1 is expressed by $\alpha_{33} = b_3\delta(x_1)\delta(x_2)$, $\beta^p_{23} = b_3\delta(x_2)H(-x_1)$, and the dislocation line in Fig. 4.2 by $\alpha_{31} = b_1\delta(x_1)\delta(x_2)$, $\beta^p_{21} = b_1\delta(x_2)H(-x_1)$.

For the continuous distribution of dislocations, it hold that

$$u_{i,j} = \beta_{ji} + \beta^p_{ji}, \tag{37.14}$$

where β_{ji} is the elastic distortion: the total distortion is the sum of elastic and plastic distortions. The plastic distortion does not produce any distortion among lattice points since it is caused by gliding (slip), while the elastic distortion is originated in an elastic deformation of the lattice as shown in Fig. 37.1.

Surface dislocation density

When β^p_{ji} jumps at a surface Σ, the surface dislocation density tensor is defined by

$$\begin{aligned} \alpha_{hi} &= -\epsilon_{hlj}\{\beta^p_{ji}(\mathrm{II}) - \beta^p_{ji}(\mathrm{I})\}n_l \\ &= -\epsilon_{hlj}[\beta^p_{ji}]n_l, \end{aligned} \tag{37.15}$$

where $\beta^p_{ji}(\mathrm{I})$ and $\beta^p_{ji}(\mathrm{II})$ are values of β^p_{ji} in domain I and II on Σ, respectively, and **n** is the normal vector on Σ directed toward domain II from domain I. The surface dislocation density tensor has been defined by Bullough and Bilby (1956). The formulation of (37.15) is a natural consequence of (37.11) for discontinuous quantities on an interface.

Figure 37.2 shows a uniform plastic distortion β^p_{31} in the matrix II around an inclusion. Equation (37.15) gives the surface dislocation density tensor,

$$\alpha_{11} = -\beta^p_{31}n_2, \quad \alpha_{21} = \beta^p_{31}n_1, \tag{37.16}$$

which is defined on the surface of the inclusion. These dislocations are Orowan's loops (1948, 1959), discussed in Section 43.

Another interesting example of the surface dislocations can be found in the crack problems discussed in Sections 27 and 32. A slit-like

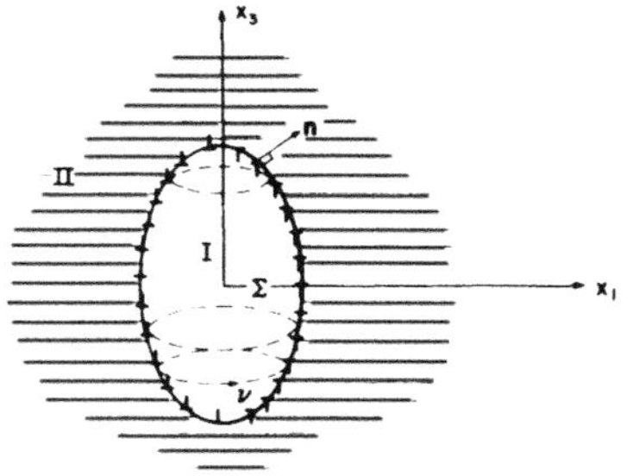

Fig. 37.2. Uniform plastic distortion around an inclusion. Orowan's dislocation loops $\alpha_{\theta 1}$ are created by the misfit.

crack with length $2a_1$ along the x_1 axis subjected to σ_{22}^0 at infinity was simulated by the infinitesimally thin elliptical inclusion ($a_2 \to 0$),

$$x_1^2/a_1^2 + x_2^2/a_2^2 = 1. \tag{37.17}$$

The equivalent eigenstrain ϵ_{22}^* is given by (27.14) and (27.17). This inclusion can be replaced by the surface dislocations (37.15) where $\beta_{ji}^p(\mathrm{II}) = 0$ and $\beta_{ji}^p(\mathrm{I}) = \epsilon_{22}^*$. Then, we have

$$\alpha_{32} = \epsilon_{22}^* n_1, \tag{37.18}$$

where n_1 is

$$n_1 = x_1/(x_1^2 + x_2^2 a_1^4/a_2^4)^{1/2} \tag{37.19}$$

from (37.17). From (27.14) and (27.17),

$$\alpha_{32} = (1-\nu)\sigma_{22}^0(a_1/\mu a_2)n_1 \tag{37.20}$$

and, therefore,

$$2\alpha_{32} = 2(1-\nu)\sigma_{22}^0(x_1/a_1)/\mu(1 - x_1^2/a_1^2)^{1/2} \tag{37.21}$$

after using (37.17) and letting $a_2 \to 0$. The above dislocation distribution (37.21) defined in $|x_1| < a_1$ simulates the stress disturbance of the slit-like crack under σ_{22}^0. The sum of the two surface dislocations defined on the upper and lower surfaces of the infinitisimally thin inclusion gives the factor of 2 in (37.21). The distribution of dislocations (37.21) has been suggested by Eshelby (1957); see also Bilby and Eshelby (1968).

In the above discussion we have derived the equivalent dislocation distribution from the equivalent inclusion for a given crack problem. It is also possible to find an equivalent inclusion from an equivalent dislocation distribution. For the Dugdale crack we have

$$2\alpha_{32} = \frac{2k_0}{\pi\mu}\log\frac{x_1(a_1^2-c^2)^{1/2}+c(a_1^2-x_1^2)^{1/2}}{x_1(a_1^2-c^2)^{1/2}-c(a_1^2-x_1^2)^{1/2}} \quad \text{for } c<|x_1|<a_1$$
$$= \frac{2k_0}{\pi\mu}\log\frac{x_1(a_1^2-c^2)^{1/2}+c(a_1^2-x_1^2)^{1/2}}{c(a_1^2-x_1^2)^{1/2}-x_1(a_1^2-c^2)^{1/2}} \quad \text{for } |x_1|<c \tag{37.22}$$

from (32.37) in the BCS model. The equivalent eigenstrain ϵ^*_{22} is obtained from (37.18). It is found that $a_2\epsilon^*_{22}$ is zero at $x_1 = \pm a_1$, infinite at $x_1 = \pm c$, and finite at $x_1 = 0$.

Impotent distribution of dislocations

Mura (1968) has pointed out that if

$$\beta^p_{ji} = -\beta^p_{ij}, \tag{37.23}$$

the displacement and stress fields caused by α_{hi} as defined by (37.11) or (37.15) become identically zero. Such a distribution of dislocations is called impotent. It is obvious from (3.23) and (3.25) that $\epsilon^*_{ij} = \epsilon^p_{ij} = (1/2)(\beta^p_{ij} + \beta^p_{ji}) = 0$ leads to $u_i = \sigma_{ij} = 0$.

As an example, consider the case

$$\beta^p_{21}(x_1) = -\beta^p_{12}(x_1). \tag{37.24}$$

Equation (37.11) leads to an impotent distribution of dislocations $\alpha_{31}(x_1) = -\beta^p_{21,1}(x_1)$. If β^p_{21} and β^p_{12} are uniform in $x_1 < 0$ and zero in $x_1 > 0$, (37.15) leads to $\alpha_{31} = \beta^p_{21}$ along the surface $x_1 = 0$ as shown in Fig. 37.3(d). Figure 37.3(a) deforms to Fig. 37.3(b) by β^p_{21}, while Fig. 37.3(a) deforms to Fig. 37.3(c) by β^p_{12}. The superimposition of the deformations due to β^p_{21} and β^p_{12} becomes Fig. 37.3(d). The elastic energy is minimum zero and the distribution of dislocations in Fig. 37.3(d) results in an impotent distribution of dislocations. The dislocation distribution shown in Fig. 37.3(b) is the polygonization.

Another example of impotent distribution of dislocations is shown in Fig. 37.4, where

$$\beta^p_{23}(x_1) = -\beta^p_{32},$$
$$\alpha_{33}(x_1) = -\beta^p_{23,1}(x_1), \tag{37.25}$$
$$\alpha_{22}(x_1) = \beta^p_{32,1}(x_1).$$

Planes perpendicular to the x_1-axis rotate clockwise by the dislocation array α_{22} while they rotate counterclockwise by α_{33}. The resulting

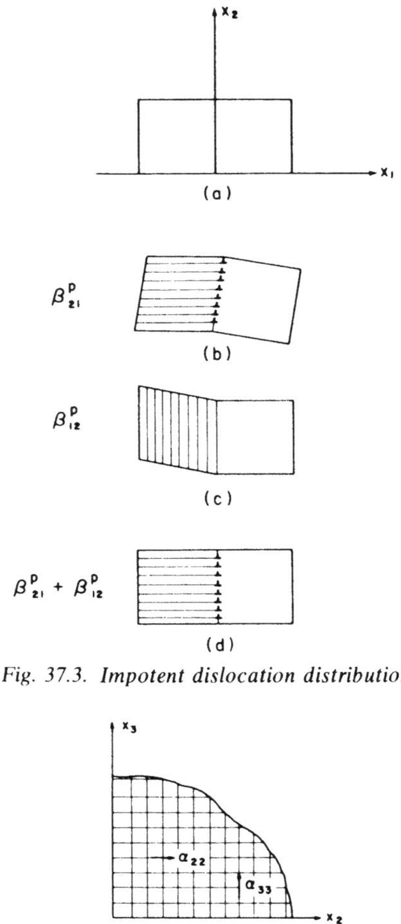

Fig. 37.3. Impotent dislocation distribution.

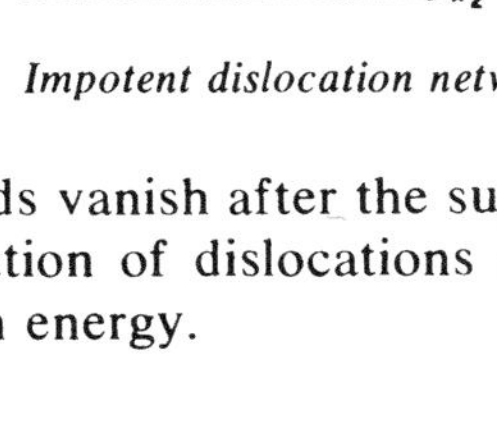

Fig. 37.4. Impotent dislocation network.

deformation and stress fields vanish after the superimposition of α_{22} and α_{33}. This impotent distribution of dislocations constructs a dislocation network of minimum strain energy.

38. Dislocation flux tensor

Using the analogy of heat conduction, we consider the growth rate of dislocations threading an arbitrary surface S; see Fig. 38.1. Let the dislocation density tensor change by an amount $\dot{\alpha}_{hi}$ per unit time. The

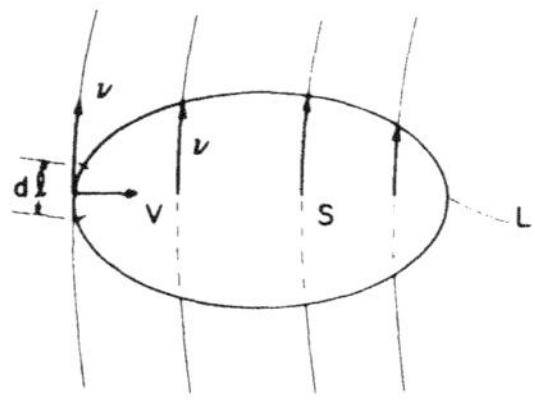

Fig. 38.1. Dislocations threading surface S increases by moving-in dislocations through L.

growth rate of the total Burgers vector of these dislocations is $\int_S \dot{\alpha}_{hi} n_h dS$, where $\mathbf{n}$ is the normal vector on surface S. The net transport of dislocations moving through the line element $d\mathbf{l}$ of the boundary S is $\boldsymbol{\nu} \cdot (d\mathbf{l} \times \mathbf{V})$ or $\epsilon_{lhj} V_l \nu_h dl_j$, where $\mathbf{V}$ is the dislocation velocity, and $\boldsymbol{\nu}$ the dislocation direction. The growth rate of the total Burgers vector on S is equal to the flux of dislocations moving through L, namely,

$$\int_S \dot{\alpha}_{hi} n_h dS = \int_L \epsilon_{lhj} V_{lhi} dl_j, \tag{38.1}$$

where

$$V_{lhi} = \sum V_l \nu_h b_i. \tag{38.2}$$

The Burgers vector for a single dislocation has been denoted by $\mathbf{b}$ in (38.2) and the summation applies to all dislocations. Mura (1963) has defined V_{lhi} as the dislocation flux (or velocity) tensor. Applying Stoke's theorem of integration to the right side of (38.1) it follows that

$$\int_S \dot{\alpha}_{hi} dS_h = \int_S \epsilon_{hlj} (\epsilon_{mnj} V_{mni})_{,l} dS_h. \tag{38.3}$$

Since this holds for any surface S, we have

$$\dot{\alpha}_{hi} = \epsilon_{hlj} \epsilon_{mnj} V_{mni,l}. \tag{38.4}$$

On the other hand, the motion of dislocations causes a plastic distortion rate. For example, $\dot{\beta}^p_{21}$ is caused by the $+x_1$-direction-motion of α_{31} and by the $-x_3$-direction-motion of α_{11}, namely

$$\dot{\beta}^p_{21} = V_{131} - V_{311}. \tag{38.5}$$

More generally,

$$\dot{\beta}^p_{ji} = -\epsilon_{jmn} V_{mni}. \tag{38.6}$$

When it is substituted into (38.4), we have

$$\dot{\alpha}_{hi} = -\epsilon_{hlj}\dot{\beta}^{P}_{ji,l}, \tag{38.7}$$

or

$$\alpha_{hi} = -\epsilon_{hlj}\beta^{P}_{ji,l} \tag{38.8}$$

by integrating with respect to time. Expression (38.8) is the same as (37.11).

We shall derive (38.6) by a different method, starting from a consideration of a single dislocation loop whose line element is moving with velocity **V**. The plastic distortion associated with this dislocation loop L can be expressed in the Fourier integral form,

$$\beta^{P}_{ji}(\mathbf{x}, t) = \iint_{-\infty}^{\infty} \bar{\beta}^{P}_{ji}(\boldsymbol{\xi}, \omega) \exp\{i(\boldsymbol{\xi} \cdot \mathbf{x} - \omega t)\} d\boldsymbol{\xi} d\omega, \tag{38.9}$$

where

$$\bar{\beta}^{P}_{ji}(\boldsymbol{\xi}, \omega) = (2\pi)^{-4} \iint_{-\infty}^{\infty} \beta^{P}_{ji}(\mathbf{x}, t) \exp\{-i(\boldsymbol{\xi} \cdot \mathbf{x} - \omega t)\} d\mathbf{x} dt. \tag{38.10}$$

Substituting (37.1) into (38.10), we have

$$\bar{\beta}^{P}_{ji}(\boldsymbol{\xi}, \omega) = -(2\pi)^{-4} b_i \int_{-\infty}^{\infty} dt \int_{S(t)} \exp\{-i(\boldsymbol{\xi} \cdot \mathbf{x} - \omega t)\} dS_j(\mathbf{x}). \tag{38.11}$$

When (38.11) is multiplied by $-i\omega$, we obtain

$$\begin{aligned} -i\omega\bar{\beta}^{P}_{ji}(\boldsymbol{\xi}, \omega) &= (2\pi)^{-4} b_i \int_{-\infty}^{\infty} dt \int_{S(t)} \frac{\partial}{\partial t} \exp\{-i(\boldsymbol{\xi} \cdot \mathbf{x} - \omega t)\} dS_j(\mathbf{x}) \\ &= (2\pi)^{-4} b_i \int_{-\infty}^{\infty} dt \left[\frac{\partial}{\partial t} \int_{S(t)} \exp\{-i(\boldsymbol{\xi} \cdot \mathbf{x} - \omega t)\} dS_j(\mathbf{x}) \right. \\ &\quad \left. - \int_{\partial S/\partial t} \exp\{-i(\boldsymbol{\xi} \cdot \mathbf{x} - \omega t)\} dS_j(\mathbf{x}) \right]. \end{aligned} \tag{38.12}$$

The first term in (38.12) vanishes by introducing proper initial conditions such as

$$S(t) = 0 \quad \text{at} \quad t = \pm\infty. \tag{38.13}$$

The increment of the slip surface S swept by $d\mathbf{l}$ is determined by the vector product **V** and $d\mathbf{l}$, that is,

$$\partial S_j / \partial t = \epsilon_{jnh} V_n dl_h. \tag{38.14}$$

Then, (38.12) can be written as

$$-i\omega\bar{\beta}^{P}_{ji}(\boldsymbol{\xi},\omega) = -(2\pi)^{-4}\epsilon_{jnh}\int_{-\infty}^{\infty} dt \int_{L} \exp\{-i(\boldsymbol{\xi}\cdot\mathbf{x}-\omega t)\} b_i V_n dl_h, \tag{38.15}$$

or, from (37.6), as

$$-i\omega\bar{\beta}^{P}_{ji}(\boldsymbol{\xi},\omega) = -(2\pi)^{-4}\epsilon_{jnh}\iint_{-\infty}^{\infty} \exp\{-i(\boldsymbol{\xi}\cdot\mathbf{x}-\omega t)\} V_n\alpha_{hi} d\mathbf{x}dt. \tag{38.16}$$

On the other hand, we have from (38.10),

$$-i\omega\bar{\beta}^{P}_{ji}(\boldsymbol{\xi},\omega) = (2\pi)^{-4}\iint_{-\infty}^{\infty} \dot{\beta}^{P}_{ji}(\mathbf{x},t)\exp\{-i(\boldsymbol{\xi}\cdot\mathbf{x}-\omega t)\} d\mathbf{x}dt. \tag{38.17}$$

Comparing (38.16) and (38.17), it holds that

$$\dot{\beta}^{P}_{ji} = -\epsilon_{jnh}V_n\alpha_{hi}, \tag{38.18}$$

or

$$\dot{\beta}^{P}_{ji} = -\epsilon_{jnh}V_{nhi}. \tag{38.19}$$

Although all these formulae have been obtained by considering the motion of a single dislocation loop, the result (39.19) also holds for continuous distributions of dislocations by the principle of superimposition.

The general theory of moving dislocations has been developed by Kosevich (1962–65), Mura (1963), Bross (1964, 1968), Zorski (1966), Amari (1968), Günther (1967, 1969), Kossecka (1969), Lardner (1969), Malén (1970), and Minagawa (1971). The earliest work on the corresponding field equations has been done by Höllander (1960), where Poisson's ratio has been assumed to be zero for mathematical simplicity.

**Line integral expression of displacement and plastic distortion fields*

The elastic distorsion and stress fields of a dislocation line L (or loop) are expressed by the line integrals along L in (7.9) and (7.15), while the displacement field is expressed by the surface integral on the slip surface S in (7.6). The displacement field can also be expressed by a line integral, if a creation history of the dislocation line is prescribed. First consider the displacement field (8.9) for a given $\beta^{P}_{ji}(\mathbf{x},t)$, where $\bar{\epsilon}^{*}_{mn}(\boldsymbol{\xi},\omega)$

is replaced by $\bar{\beta}^{p}_{nm}(\boldsymbol{\xi}, \omega)$,

$$u_i(\mathbf{x}, t) = -\iint_{-\infty}^{\infty} iC_{jlmn}\bar{\beta}^{p}_{nm}(\boldsymbol{\xi}, \omega)\xi_l N_{ij}(\boldsymbol{\xi}, \omega)D^{-1}(\boldsymbol{\xi}, \omega) \times \exp\{i(\boldsymbol{\xi} \cdot \mathbf{x} - \omega t)\}d\boldsymbol{\xi}d\omega, \tag{38.20}$$

where

$$\bar{\beta}^{p}_{nm}(\boldsymbol{\xi}, \omega) = (2\pi)^{-4}\iint_{-\infty}^{\infty} \beta^{p}_{nm}(\mathbf{x}, \mathbf{t}) \exp\{-i(\boldsymbol{\xi} \cdot \mathbf{x} - \omega t)\}d\mathbf{x}dt. \tag{38.21}$$

If $\dot{\beta}^{p}_{ji}$ is a plastic distortion rate associating with $V_n\alpha_{hi}$ through (38.18), then (38.20) can be written in terms of $V_n\alpha_{hi}$. From (38.17) and (38.18),

$$\bar{\beta}^{p}_{nm}(\boldsymbol{\xi}, \omega) = (2\pi)^{-4}(1/i\omega)\epsilon_{nkh}\iint_{-\infty}^{\infty} V_k(\mathbf{x}, t)\alpha_{hm}(\mathbf{x}, t) \times \exp\{-i(\boldsymbol{\xi} \cdot \mathbf{x} - \omega t)\}d\mathbf{x}dt. \tag{38.22}$$

When it is substituted into (38.20),

$$u_i(\mathbf{x}, t) = -(2\pi)^{-4}\epsilon_{nkh}\iiiint_{-\infty\,-\infty}^{\infty\,\infty} C_{jlmn}V_k(\mathbf{x}', t')\alpha_{hm}(\mathbf{x}', \mathbf{t}')\frac{\xi_l}{\omega}\frac{N_{ij}(\boldsymbol{\xi}, \omega)}{D(\boldsymbol{\xi}, \omega)} \times \exp\{i\boldsymbol{\xi} \cdot (\mathbf{x} - \mathbf{x}') - i\omega(t - t')\}d\boldsymbol{\xi}d\omega d\mathbf{x}'dt'. \tag{38.23}$$

When a discrete (single) dislocation with Burgers vector b_i is considered, expression (37.6) is used and the integral with respect to $\mathbf{x}'$ in (38.23) becomes a line integral along the dislocation line L. The time derivative of (38.23) is

$$\dot{u}_i(\mathbf{x}, t) = \epsilon_{nkh}\oint_L b_m dl_h(\mathbf{x}')\int_{-\infty}^{\infty} dt' C_{jlmn}V_k(\mathbf{x}', t')G_{ij,l}(\mathbf{x} - \mathbf{x}', t - t'), \tag{38.23.1}$$

where G_{ij} is the Green's function defined by (8.10).

When a uniform motion of the dislocation distribution with a constant velocity is considered, we can write

$$\alpha_{hi}(\mathbf{x}, t) = \alpha_{hi}(\mathbf{x} - \mathbf{v}t). \tag{38.24}$$

Putting

$$\mathbf{x}' - \mathbf{v}t' = \mathbf{y}, \quad d\mathbf{x}' = d\mathbf{y}, \tag{38.25}$$

the integration with respect to t' in (38.23) yields $2\pi\delta(\boldsymbol{\xi}\cdot\mathbf{V}-\omega)$. Then, (38.23) becomes

$$u_i(\mathbf{x}, t) = -(2\pi)^{-3}\epsilon_{nkh}\iint_{-\infty}^{\infty} C_{jlmn}V_k\alpha_{hm}(\mathbf{y})\frac{\xi_l}{\boldsymbol{\xi}\cdot\mathbf{V}}\frac{N_{ij}(\boldsymbol{\xi}, \boldsymbol{\xi}\cdot\mathbf{V})}{D(\boldsymbol{\xi}, \boldsymbol{\xi}\cdot\mathbf{V})}$$

$$\times\exp\{i\boldsymbol{\xi}\cdot(\mathbf{x}-\mathbf{y}-\mathbf{v}t)\}d\boldsymbol{\xi}d\mathbf{y}. \tag{38.26}$$

The above result can be directly applied to problems of the static dislocation distribution by a limiting process $\mathbf{V}\to 0$ and $V_k/V\to v_k/v$, where $\mathbf{v}$ is the direction of motion when the dislocation α_{hm} is introduced into the material. Then (38.26) becomes

$$u_i(\mathbf{x}) = -(2\pi)^{-3}\epsilon_{nkh}\iint_{-\infty}^{\infty} C_{jlmn}\alpha_{hm}(\mathbf{y})\frac{\xi_l v_k}{\boldsymbol{\xi}\cdot\mathbf{v}}\frac{N_{ij}(\boldsymbol{\xi})}{D(\boldsymbol{\xi})}\exp\{i\boldsymbol{\xi}\cdot(\mathbf{x}-\mathbf{y})\}d\boldsymbol{\xi}d\mathbf{y}. \tag{38.27}$$

It is interesting to note that the displacement field for a given static dislocation distribution $\alpha_{hm}(\mathbf{x})$ is not unique, depending on the choice of $\mathbf{v}$.

For a single dislocation line L, using (37.6) we have

$$u_i(\mathbf{x}) = -(2\pi)^{-3}\epsilon_{nkh}\int_{-\infty}^{\infty}\int_L C_{jlmn}b_m\frac{\xi_l v_k}{\boldsymbol{\xi}\cdot\mathbf{v}}\frac{N_{ij}(\boldsymbol{\xi})}{D(\boldsymbol{\xi})}\exp\{i\boldsymbol{\xi}\cdot(\mathbf{x}-\mathbf{y})\}d\boldsymbol{\xi}dl_h. \tag{38.28}$$

This expression has been obtained by Mura (1971).

The plastic distortion field, as well as the displacement field for a given static distribution of dislocation is not a state quantity. It is uniquely determined only when the history of creation of the dislocations is given. As with the displacement field, the history is assigned by a given uniform motion of dislocations in the $\mathbf{v}$ direction with an infinitely small velocity $\mathbf{V}$.

Substituting (38.22) into (38.9), we have

$$\beta^p_{ji}(\mathbf{x}, t) = (2\pi)^{-4}\epsilon_{jkh}\iiiint_{-\infty-\infty}^{\infty\ \infty}\frac{V_k(\mathbf{x}', t')}{i\omega}\alpha_{hi}(\mathbf{x}', t')\exp\{i\boldsymbol{\xi}\cdot(\mathbf{x}-\mathbf{x}')$$

$$-i\omega(t-t')\}d\boldsymbol{\xi}d\omega d\mathbf{x}'dt'. \tag{38.29}$$

For a uniformly moving dislocation distribution with a constant velocity

V, (38.29) becomes

$$\beta^{p}_{ji}(\mathbf{x},t) = (2\pi)^{-3}\epsilon_{jkh}\iint_{-\infty}^{\infty} \frac{V_k}{i\boldsymbol{\xi}\cdot\mathbf{V}}\,\alpha_{hi}(\mathbf{y})\exp\{i\boldsymbol{\xi}\cdot(\mathbf{x}-\mathbf{y}-\mathbf{V}t)\}d\boldsymbol{\xi}d\mathbf{y}. \tag{38.30}$$

For a static distribution of dislocations, after limiting process $\mathbf{V}\to 0$, $V_k/V \to v_k/v$, (38.30) becomes

$$\beta^{p}_{ji}(\mathbf{x}) = (2\pi)^{-3}\epsilon_{jkh}\iint_{\infty}^{\infty} \frac{v_k}{i\boldsymbol{\xi}\cdot\mathbf{v}}\,\alpha_{hi}(\mathbf{y})\exp\{i\boldsymbol{\xi}\cdot(\mathbf{x}-\mathbf{y})\}d\boldsymbol{\xi}d\mathbf{y}. \tag{38.31}$$

It is easy to show that two β^{p}_{ji} with different **v** yield the same dislocation density tensor when (38.31) is substituted into (38.8) and the divergency law

$$\alpha_{hi,h}(\mathbf{x}) = 0 \tag{38.32}$$

is taken into account. Equation (38.32) can be derived from (38.8). It implies that dislocation lines must close on themselves.

39. Energies and forces

In this section we consider the elastic strain energy of a static dislocation distribution, the interaction energy, and the forces acting on dislocations.

The elastic strain energy W^* caused by a plastic distortion β^{p}_{ji} has the same form as (13.3),

$$W^* = -(1/2)\int_{-\infty}^{\infty}\sigma_{ij}\beta^{p}_{ji}d\mathbf{x}, \tag{39.1}$$

where β^{p}_{ji} is associated with α_{hi} through (38.8).

For a dislocation loop L, (37.1) leads to

$$W^* = (1/2)\int_S \sigma_{ij}b_i n_j dS, \tag{39.2}$$

where S is the slip plane bounded by L. For example, the straight screw dislocation, shown in Fig. 4.1, has an elastic strain energy per unit length of dislocation line,

$$W^* = \frac{\mu b}{4\pi}\log[x]_0^{\infty} = \frac{\mu b}{4\pi}\log(R/r_0), \tag{39.3}$$

where R is the size of material and r_0 is the atomic distance of lattice (dislocation core radius).

If the straight dislocation is in an anisotropic medium, σ_{ij} in (39.2) is calculated from (36.16). Namely,

$$W^* = (1/2)C_{ijkl} \int_{r_0}^{R} r^{-1} dr \nu_h \sum_{klh} (\mathbf{m}, \mathbf{n}) b_i n_j$$

$$= (1/2)C_{ijkl} \nu_h \sum_{klh} (\mathbf{m}, \mathbf{n}) b_i n_j \log(R/r_0), \tag{39.4}$$

where S in (39.2) has been taken as the plane with a normal vector $\mathbf{n}$ as shown in Fig. 36.1.

The self-force on a dislocation loop is derived from the variation of W^* for an infinitesimal change of β^p_{ij} in (39.1). The variation becomes

$$\delta W^* = -\int_{-\infty}^{\infty} \sigma_{ij} \delta\epsilon^p_{ij} d\mathbf{x}. \tag{39.5}$$

The factor (1/2) disappears when a variation of (13.1) is taken and integration by parts is performed as in (13.2). The variation $\delta\epsilon^p_{ij}$ is caused by a vertical displacement $\delta\boldsymbol{\xi}$ of a line element dl of the dislocation loop L. It holds from (38.2) and (38.6) that

$$\delta\beta^p_{ji} = -\epsilon_{jmn} \delta\xi_m \alpha_{ni}. \tag{39.6}$$

Since $\delta\epsilon^p_{ij} = (1/2)(\delta\beta^p_{ji} + \delta\beta^p_{ij})$ and $\alpha_{ni} d\mathbf{x} = \nu_n b_i dl$, (39.5) becomes

$$\delta W^* = \int_L \epsilon_{jmn} \nu_n b_i \sigma_{ij} \delta\xi_m dl, \tag{39.7}$$

where $\boldsymbol{\nu}$ is the direction of the line element.

The Peach–Koehler (1950) force acting on the dislocation line element is defined by

$$f_m = -\epsilon_{mnj} \nu_n \sigma_{ij} b_i. \tag{39.8}$$

Then,

$$\delta W^* = -\int_L f_m \delta\xi_m dl. \tag{39.9}$$

The interaction energy between the dislocation L and a stress field σ^0_{ij} is defined in the same way as (13.13),

$$\Delta W = -\int_{-\infty}^{\infty} \sigma^0_{ij} \epsilon^p_{ij} d\mathbf{x}. \tag{39.10}$$

Its variation becomes

$$\delta(\Delta W) = -\int_{-\infty}^{\infty} \sigma_{ij}^{0}\delta\epsilon_{ij}^{p}d\mathbf{x}$$
$$= -\int_{L} f_{m}^{0}\delta\xi_{m}dl, \tag{39.11}$$

where f_m^0 has the same form as (38.9) with σ_{ij}^0. The irreversibility of plastic work done on dl by σ_{ij} and σ_{ij}^0 is

$$(f_m + f_m^0)V_m > 0, \tag{39.12}$$

where $\mathbf{V}$ is the velocity of dislocation line. The above expression can be extended to a continuous distribution of dislocations α_{ni}, moving with dislocation velocity tensor V_{mni}. The rate of plastic work per unit volume then becomes

$$-\epsilon_{mnj}(\sigma_{ij} + \sigma_{ij}^0)V_{mni} > 0. \tag{39.13}$$

The last result is important in investigating the mathematical theory of plasticity in the context of the dislocation theory, described in later sections.

Further investigations on the self-force f_m are presented by Asaro and Hirth (1973), Barnett (1976), and Gavazza and Barnett (1976), Minagawa (1970), Das et al. (1973), and Golebiewska-Lasota (1978), among others.

Dynamic consideration

Let us consider the following Lagrangian functional for a moving dislocation loop in an infinitely extended material:

$$\int_{t_0}^{t_1} \mathscr{L}dt = \int_{t_0}^{t_1} dt \int_{D} (\tfrac{1}{2}\rho\dot{u}_i\dot{u}_i - \tfrac{1}{2}\sigma_{ij}\beta_{ji})d\mathbf{x}, \tag{39.14}$$

where

$$\beta_{ji} = u_{i,j},$$
$$\sigma_{ij,j} = \rho\ddot{u}_i, \tag{39.15}$$
$$\sigma_{ij} = C_{ijkl}\beta_{lk}.$$

The above formulae are defined everywhere except on the slip surface S. The boundary of S (which is the dislocation line) is moving with velocity $\dot{\boldsymbol{\xi}}$. The variation of (39.14) becomes

$$\delta \int_{t_0}^{t_1} \mathscr{L} dt = \int_{t_0}^{t_1} dt \int_D (\rho \dot{u}_i \delta \dot{u}_i - \sigma_{ij} \delta \beta_{ji}) d\mathbf{x}$$

$$= \int_{t_0}^{t_1} dt \int_D \left[\rho \dot{u}_i \frac{\partial}{\partial t} (\delta u_i) - \sigma_{ij} \frac{\partial}{\partial x_j} (\delta u_i) \right] d\mathbf{x}. \tag{39.16}$$

Applying the Gauss theorem, we have

$$\int_D \sigma_{ij} \frac{\partial}{\partial x_j} (\delta u_i) d\mathbf{x} = \int_{S+\delta S} \sigma_{ij} n_j [\delta u_i] dS - \int_D \sigma_{ij,j} \delta u_i d\mathbf{x}$$

$$= \int_{\delta S} \sigma_{ij} n_j b_i \delta S - \int_D \sigma_{ij,j} \delta u_i d\mathbf{x}, \tag{39.17}$$

where δS is the variation of slip surface S, which is expressed by the vector product of a virtual displacement of the dislocation position $\delta\boldsymbol{\xi}$ and the dislocation line element $\boldsymbol{\nu} dl$,

$$n_j \delta S = \epsilon_{jlh} \delta \xi_l \nu_h dl. \tag{39.18}$$

In the above derivation, $[u_i]$ is the difference of u_i evaluated at the upper and lower surfaces of the slip plane: $[u_i] = b_i$ on S. After giving a vertical displacement $\delta \xi_l$ to the position of the dislocation, the variation $[\delta u_i] = \delta[u_i]$ is zero on S, and b_i on δS.

Furthermore, we have

$$\int_{t_0}^{t_1} \rho \dot{u}_i \frac{\partial}{\partial t} (\delta u_i) dt d\mathbf{x} = -\int_{\delta S} \rho \dot{u}_i b_i \dot{\xi}_j n_j \delta S - \int_{t_0}^{t_1} \rho \ddot{u}_i \delta u_i dt d\mathbf{x}. \tag{39.19}$$

Substituting (39.17), (39.19) into (39.16) we obtain

$$\delta \int_{t_0}^{t_1} \mathscr{L} dt = -\int_{t_0}^{t_1} dt \oint_L (\rho \dot{u}_i \dot{\xi}_j + \sigma_{ij}) b_i \epsilon_{jlh} \delta \xi_l \nu_h dl. \tag{39.20}$$

The term $\rho \dot{u}_i \dot{\xi}_j b_i \epsilon_{jlh} \nu_h$ is called the Lorentz force. Since $\dot{\xi}_j dt = \delta \xi_j$ and $\epsilon_{jlh} \delta \xi_j \delta \xi_l = 0$, (39.20) becomes

$$\delta \int_{t_0}^{t_1} \mathscr{L} dt = -\int_{t_0}^{t_1} dt \oint_L \sigma_{ij} b_i \epsilon_{jlh} \delta \xi_l \nu_h dl. \tag{39.21}$$

The Lorentz force does not contribute to the energy balance and the result is the same as the static case (39.7).

The elastic solution of a uniformly moving straight dislocation has been obtained by Frank (1949) and Eshelby (1949). Frank has shown that when a screw dislocation is moving with a constant velocity V, the total energy (the sum of the elastic strain energy and the kinetic energy) is given by $E_0/(1 - V^2/c^2)^{1/2}$, where E_0 is the elastic strain energy of the dislocation at rest, and c the shear wave velocity $(\mu/\rho)^{1/2}$. His calculation has been stimulated by the work done by Kontorova and Frenkel (1938) who studied a one-dimensional dislocation model and found a strikingly analogous form to that of a particle in special relativity. Frank's calculation was extended to an edge dislocation by Eshelby (1949). In this case no relativistic relation is found since the dilatational wave velocity is also involved.

The uniform motion of a straight dislocation is possible in the absence of an applied stress since $\sigma_{ij}b_i\epsilon_{jlh}\nu_h$ becomes zero at the dislocation position. It is not yet clear how the dislocation arrives at the constant velocity. Eshelby has extended his analysis to the Peierls dislocation (1940). No applied stress is necessary for uniform motion if the width of the dislocation is $2\bar{\zeta} = \{b/(1-\nu)\}D(V)/D(0)$ for an edge dislocation or $2\bar{\zeta} = b\beta$ for a screw dislocation, where $D(V) = -2\mu(2c^2/V^2)(\gamma - \alpha^4/\beta)$, $\gamma = (1 - V^2/a^2)^{1/2}$, $\beta = (1 - V^2/c^2)^{1/2}$, $\alpha = (1 - V^2/2c^2)^{1/2}$, $c = (\mu/\rho)^{1/2}$ and $a = \{(\lambda + 2\mu)/\rho\}^{1/2}$. It is interesting to note that the width of the edge dislocation vanishes when $D(V) = 0$, the solution of which is the Rayleigh wave velocity. The width of the screw dislocation vanishes at $V = c$. Eshelby's calculation has been extended by Leibfried and Dietze (1949) to a dislocation moving in the middle of a plate (direction of dislocation parallel to the plate surface). Weertman (1960, 1961, 1966) has completed Eshelby's analysis (1949) by calculating the elastic strain energy and the kinetic energy of the edge dislocation. He has found that the shear stress on the slip plane (except at the dislocation position) decreases with increasing dislocation velocity, vanishes at the Rayleigh velocity, and changes sign at velocities higher than the Rayleigh velocity. Because of this, edge dislocations of like sign attract each other (contrary to the usual situation) when the dislocations are traveling at velocities above the Rayleigh velocity. Such a critical velocity of a dislocation is called the threshold velocity. The threshold velocities have been calculated for various crystals by Teutonico (1961, 1962, 1963) and Weertman (1962). They started their stress analysis from the general solution obtained by Bullough and Bilby (1954) for a uniformly moving straight dislocation in an anisotropic media. If a dislocation of general shape is considered, the formula (38.23.1) is recommended.

The effect of free surfaces and the inhomogeneity of materials on dislocation motion has not been fully investigated. Weertman (1963) considered a dislocation moving on the interface between two isotropic media of different elastic constants and densities. The case where a dislocation is moving parallel to the interface has been solved by M.S. Lee (1972).

Although the energy of a moving dislocation becomes infinite as its velocity approaches the sound velocity, there are dislocation solutions of the elastic equations for greater dislocation velocities than c. This dislocation-like solution was first demonstrated by Eshelby (1956). He suggests its posible application to the propagation of diffusionless transformations by dislocations; however, no successful application has been reported. Callias and Markenscoff (1980) analyzed the stress field due to the general transient motion of a supersonic screw dislocation and in particular the wave-front behaviour.

The applied stress required for the supersonic dislocation motion becomes infinity. This difficulty is avoided by Weertman (1967, 1969) by considering a spread dislocation with a distribution function $B(x_1 - Vt)$. If B is a constant in the region $-A < (x_1 - Vt) < A$ and zero elsewhere, the equilibrium condition is satisfied for a uniformly applied stress. Weertman also has discussed atomic force law under which the equilibrium condition is satisfied without applied stress: see the review paper by Mura (1972). For a spread supersonic dislocation Clifton and Markenscoff (1981) estimated that the required stress to sustain the supersonic velocity would exceed the theoretical strength of the material.

Almost all calculations for moving dislocations have been limited to straight infinite dislocations. Very few calculations have been done for circular dislocation loops. Günther (1968) has obtained the elastic solution for a circular edge dislocation moving uniformly in the direction of the normal to the circular plane. In the region $V < c < a$ the elastic field shows double Lorentz contractions and in the region $c < a < V$ there arise two Mach cones radiated from the dislocation loop with angles $\pm\sin^{-1}(V/c)$ and $\pm\sin^{-1}(V/a)$ with respect to the direction of the motion. No disturbance is seen outside the domain bounded by the cones and the circular plane, where c and a are the shear and dilatational velocities.

Unlike the uniform motion of a straight dislocation with a subsonic velocity, an oscillating straight dislocation generally requires the application of an oscillating stress (forces vibration). Free vibrations, however, are possible under special circumstances. By using Eshelby's elastic solution (1949) for an oscillating screw dislocation and the equilibrium condition, Nabarro (1951) has obtained the amplitude and phase-difference for the position of the oscillating dislocation as functions of the amplitude and wave number of an incident sound wave. He also has calculated the scattering cross-section (the ratio of the rate of radiation to the incident energy flux). Same calculation is done for an edge dislocation by Kiusalaas and Mura (1964). On the other hand, Leibfried (1950) has pointed out that a dislocation moving through a flux of sound waves (which are a natural consequence of the thermal energy of the crystal) experiences a retarding force which is proportional to the dislocation velocity. The numerical value of his result, however, is criticized by Nabarro (1951), Lothe (1960), and Eshelby (1962), Eshelby has used a kink model for mathematical simplicity and found that a kink moving through an isotropic flux of elastic waves has a scattering cross-section proportional to the square of its width, and experiences a retarding force proportional to the product of its velocity and the energy density of the waves. Pegel (1966), Laub and Eshelby (1966), and Ninomiya and Ishioka (1967) have obtained dispersion curves for the free vibration of a straight dislocation. They found that free vibrations are possible for a certain range of frequencies and wavelengths, but that in general a suitable applied stress is required to maintain the vibration.

Eshelby (1953) has considered several cases of the accelerated motion of a screw dislocation. The dislocation is at rest for negative t, and thereafter moves 1) with a velocity $c^2t/(x_0^2+c^2t^2)^{1/2}$ and acceleration $c^2x_0^2/(x_0^2+c^2t^2)^{3/2}$, 2) with a constant acceleration, 3) with a constant velocity. He has calculated the uniform applied stresses required for these motions. He also has calculated the velocity of the dislocation when a constant stress is applied at $t=0$ and $t=t_1$ so that the total impulse is a constant value. Like Nabarro (1951), Eshelby also accounts for the Peierls stress law. According to the present author's opinion, no such attention is necessary as long as the equilibrium condition is being considered at $\bar{x}_1=0$, $x_2=b/2$. Some details of the elastic field caused by the uniform motion of a dislocation starting from rest are given by Ang and Williams (1959) and by Kiusalaas and Mura (1964). A solution for the general non-uniform motion has been obtained by Markenscoff (1980) for a screw dislocation and by Markenscoff and Clifton (1981) for an edge dislocation. The self-stress of a dislocation starting from rest and moving with a constant velocity has been obtained by Kiusalaas and Mura (1964) for a screw dislocation and Clifton and Markenscoff (1981) for an edge dislocation.

The interaction between a varying stress field and a moving screw dislocation has an electro-magnetic analogy as shown by Eshelby (1953) and Nabarro (1951). From this analogy, Nabarro predicts that a moving dislocation receives a Lorentz force (in addition to the Peach–Koehler force) similar to a moving-line density of charge. Since the Lorentz force has been predicted strictly from this analogy, its existence has been criticized by several authors. Although in some papers the Lorentz force is derived from a Lagrangian formalism, either the mathematics or the physics involved in these derivations is questionable.

The scattering of elastic waves by a dislocation is only one of the many energy dissipative mechanisms which have been suggested by a number of investigators. Eshelby (1949) has shown that the thermoelastic effect around a moving edge dislocation produces an irreversible heat flow causing energy dissipation. A more rigorous analysis of the thermoelastic dissipation due to an edge dislocation moving at an arbitrary speed has been performed by Weiner (1958). In order to produce a similar thermal effect for a moving screw dislocation, Mason (1960) has proposed a shear wave effect which would instantaneously raise the temperature of those phonons which have components along the compressed direction, while lowering the temperature of those which have components along the extended direction.

As an element of the dislocation line moves through the lattice, its potential energy varies in a roughly sinusoidal manner; Orowan (1940). Therefore, the traveling dislocation experiences an oscillatory force. The energy is dissipated through this oscillation by a mechanism similar to that mentioned in the last section. Hart (1955) has calculated the magnitude of this dissipation. Various other dissipative mechanisms have been studied by Lothe (1962), Koehler (1952) and Granato and Lücke (1956). Some atomistic aspects of thermally activated motions of dislocations have been investigated by Weiner (1969). Kröner (1955), Holländer (1960), Schaefer (1969), Günther (1967), Minagawa (1971), Golebiewska–Lasota (1979) among others discussed some analogies between the electromagnetic and moving dislocation fields. However, no practical applications have been found yet.

40. Plasticity

In this section the mathematical theory of plasticity is interpreted in terms of dislocation theory. The basic idea is that a stress field in continuum plasticity is the sum of the applied elastic stress (solution of a boundary value problem in elasticity) and the dislocation stress caused by dislocations created in the so-called plastic domains. The Mises yield criterion and the Prandtle–Reuss relation between stress and plastic strain rate are also derived from the theory of continuous distributions of dislocations.

Mathematical theory of plasticity

The fundamental equations of continuum plasticity are reviewed here before the dislocation theory is introduced. In the absence of body forces the equation of equilibrium is,

$$\sigma_{ij,j} = 0. \tag{40.1}$$

The traction boundary conditions are,

$$\sigma_{ij} n_j = F_i \text{ on } S, \tag{40.2}$$

where F_i is the applied force on surface S and n_i is the exterior unit normal on S. The strain ϵ_{ij} is the sum of elastic strain e_{ij} and the plastic strain ϵ^p_{ij},

$$\epsilon_{ij} = e_{ij} + \epsilon^p_{ij}. \tag{40.3}$$

The strain ϵ_{ij} is compatible and expressed by

$$\epsilon_{ij} = (1/2)(u_{i,j} + u_{j,i}), \tag{40.4}$$

where u_i is the displacement component. The material is in a plastic state when

$$(1/2)s_{ij}s_{ij} = k^2 = \bar{\sigma}^2/3, \tag{40.5}$$

where

$$s_{ij} = \sigma_{ij} - \delta_{ij}\sigma_{kk}/3 \tag{40.6}$$

is the reduced stress component (or deviatoric). k is the shearing yield stress for a perfectly elasto-plastic material and the shearing flow stress for a work-hardening material, while $\bar{\sigma}$ is the yield stress or flow stress in a simple tension test. $\bar{\sigma}$ (or k) is a constant for the perfectly elasto-plastic material (see Fig. 40.1a) while it is a function of plastic strain for the work-hardening material (see Fig. 40.1b). In this latter case we have

$$\bar{\sigma} = \bar{\sigma}\left(\int d\bar{\epsilon}^p\right), \tag{40.7}$$

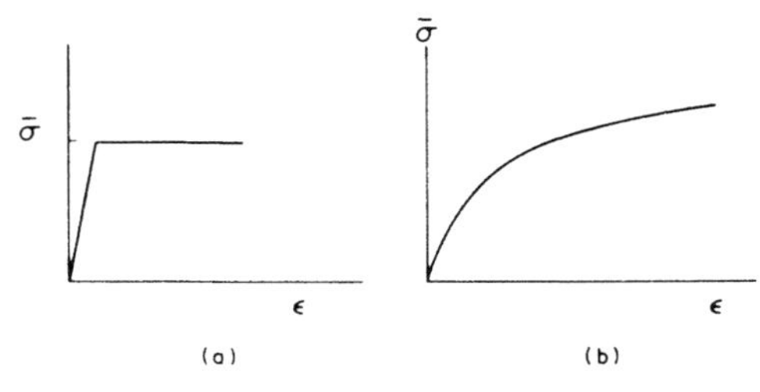

Fig. 40.1. Stress-strain curve in the simple tension test. (a) is perfectly elasto-plastic material. (b) is work-hardening material.

where

$$d\bar{\epsilon}^p = \{(2/3)d\epsilon_{ij}^p d\epsilon_{ij}^p\}^{1/2} \tag{40.8}$$

is the effective plastic strain increment. $\bar{\sigma}$ is also called the effective stress and, from (40.5),

$$\bar{\sigma} = \{(3/2)s_{ij}s_{ij}\}^{1/2}. \tag{40.9}$$

The condition (40.5) is called the Mises' yield criterion (1928). Finally, the constitutive equations are

$$\begin{aligned} de'_{ij} &= ds_{ij}/2\mu, \\ de_{kk} &= d\sigma_{kk}/3K, \\ d\epsilon_{ij}^p &= s_{ij}d\mu^*, \end{aligned} \tag{40.10}$$

where μ is the elastic shear, and K the bulk modulus. $d\mu^*$ is an unknown scaler quantity for the perfectly elasto-plastic material. For the work-hardening material, it is

$$d\mu^* = 3d\bar{\epsilon}^p/2\bar{\sigma} = 3(d\bar{\epsilon}^p/d\bar{\sigma})d\bar{\sigma}/2\bar{\sigma}. \tag{40.11}$$

The last result has been obtained from $d\epsilon_{ij}^p d\epsilon_{ij}^p = (d\mu^*)^2 s_{ij}s_{ij}$ and definitions (40.8) and (40.9). The slope $d\bar{\sigma}/d\bar{\epsilon}^p$ is obtained from the stress-strain curve in Fig. 40.1(b). The third equation in (40.10) has been proposed by Reuss (1930). It is a modification of the Lévy (1871) and Mises (1928) equation, $d\epsilon_{ij} = d\mu^* s_{ij}$. A special case of the Reuss equation has been obtained by Prandtle (1924) for the plane problems.

Dislocation theory

Using the dislocation flux tensor V_{nhi} in (38.19), we write the plastic strain rate as

$$\dot{\epsilon}_{ij}^p = (1/2)(\dot{\beta}_{ji}^p + \dot{\beta}_{ij}^p), \tag{40.12}$$

where

$$\dot{\beta}_{ji}^p = -\epsilon_{jnh}V_{nhi}. \tag{40.13}$$

The rate of plastic work per unit volume is

$$\delta W^p/dt = \sigma_{ij}\dot{\epsilon}_{ij}^p = -\epsilon_{lhj}\sigma_{ji}V_{lhi}. \tag{40.14}$$

The plastic deformation does not cause any volume change, that is,

$$\dot{\epsilon}_{ii}^p = -\epsilon_{lhi}V_{lhi} = 0. \tag{40.15}$$

since the plastic work is irreversible,

$$-\epsilon_{lhj}\sigma_{ji}V_{lhi} > 0. \tag{40.16}$$

Mura (1965) has shown that the two conditions (40.15) and (40.16) are satisfied when

$$V_{lhi} = d\mu^* g_{lhi} \tag{40.17}$$

with

$$g_{lhi} = -(1/2)\epsilon_{lhj}s_{ji}, \tag{40.18}$$

where $d\mu^*$ is a positive scaler quantity. g_{lhi} is the x_l-component of the force acting on α_{hi} dislocations. The magnitude of the force must be a constant k for the perfectly elasto-plastic materials. Equation (40.13) leads to

$$g_{lhi}g_{lhi} = (1/2)s_{ij}s_{ij} = k^2 \tag{40.19}$$

which is the Mises yield criterion (40.5)

When (40.12) and (40.13) are used with (40.17), we have

$$\dot{\epsilon}^p_{ij} = \dot{\beta}^p_{ij} = \dot{\beta}^p_{ji} = d\mu^* s_{ij}, \tag{40.20}$$

which corresponds to the Reuss equation in (40.10). $d\mu^*$ may be determined from the simple tension test for work-hardening materials as seen in the mathematical theory of plasticity. $d\mu^*$ is a history-depending quantity. Kröner (1963) has assumed it to be a function of dislocation loop density. The constitutive equations in plasticity have been further studied by Eisenberg (1970), Hahn and Jaunzemis (1973), Lambermont (1974), Smith (1970), Karihaloo (1975), Bodner (1978), Weng and Phillips (1978), and Werne and Kelly (1978), among others, from the viewpoint of continuous distributions of dislocations. Gilman (1966) has discussed the stress-strain law in terms of multiplication and motion of dislocations.

After a series of original works by Nye (1953), Bilby, Gardner and Smith (1958) and Kröner (1963), Mura (1967) has postulated that the stress field in continuum plasticity is the sum of the applied elastic stress and the dislocation stress. He also has shown that the dislocation density tensor α_{hi} can be expressed in terms of the stress components if the impotent dislocations are excluded, and the material is infinitely extended.

We start our discussion with (37.14), that is,

$$u_{i,j} = \beta_{ji} + \beta^p_{ji}. \tag{40.21}$$

When substituted into (37.11), we have

$$\alpha_{hi} = \epsilon_{hlj}\beta_{ji,l} = \epsilon_{hlj}e_{ji,l} + \epsilon_{hlj}\omega_{ji,l}, \tag{40.22}$$

where

$$\beta_{ji} = e_{ji} + \omega_{ji}. \tag{40.23}$$

The first term in (40.23) is the elastic strain and the second term is the rotational tensor ($\omega_{ji} = -\omega_{ij}$). The part of α_{hi} contributed from the second term in (40.22) is impotent. The displacement and stress fields due to $\epsilon_{hlj}\omega_{ji,l}$ can be considered to be equivalent to those caused by eigenstrain $\beta^*_{ji} = -\omega_{ji}$ [see (37.23)]. Therefore, the potent part of α_{hi} can be written as

$$\alpha_{hi} = \epsilon_{hlj}e_{ji,l}. \tag{40.24}$$

Since $e_{ji} = e_{ij} = C^{-1}_{ijmn}\sigma_{mn}$, (40.24) becomes

$$\alpha_{hi} = C^{-1}_{ijmn}\epsilon_{hlj}\sigma_{mn,l}. \tag{40.25}$$

The above result indicates that the potent dislocation distribution can be obtained if the corresponding stress distribution is known.

The basic idea of the dislocation theory in connection with the continuum theory of plasticity is that a stress field in an elasto-plastic medium is the sum of the (potent) dislocation stress and the elastic applied stress. Several examples are shown in the following subsections.

Plane strain problems

Consider an isotropic material subjected to an applied boundary force. The material is deformed to an elasto-plastic state of plane strain. In the plastic domains, the sum of the applied elastic stress and the dislocation stress satisfies the Mises yield criterion,

$$(\sigma_x - \sigma_y)^2/4 + \sigma_{xy}^2 = k^2. \tag{40.26}$$

When the material is perfectly elasto-plastic, the stress field satisfying the above condition may be written as (Hill 1960),

$$\begin{aligned} \sigma_x &= -p - k \sin 2\phi, \\ \sigma_y &= -p + k \sin 2\phi, \\ \sigma_{xy} &= k \cos 2\phi, \end{aligned} \tag{40.27}$$

where $p = -(\sigma_x + \sigma_y)/2$ and ϕ is the angle between the direction of the first shear line and the x-axis measured counter-clockwise. On the first and the second shear lines, the shear stress takes on the value of $\pm k$.

The stress is related to the elastic strain by Hooke's law in any domain (plastic or elastic),

$$\begin{aligned} e_x &= s_x/2\mu - p(1-2\nu)/E, \\ e_y &= s_y/2\mu - p(1-2\nu)/E, \\ e_{xy} &= s_{xy}/2\mu, \end{aligned} \tag{40.28}$$

where

$$\begin{aligned} s_x &= (\sigma_x - \sigma_y)/2, \\ s_y &= (\sigma_y - \sigma_x)/2, \\ s_{xy} &= \sigma_{xy}, \\ p &= -(\sigma_x + \sigma_y)/2. \end{aligned} \tag{40.29}$$

Equation (40.24) leads to

$$\begin{aligned} \alpha_{31} &= \partial e_{xy}/\partial x - \partial e_x/\partial y, \\ \alpha_{32} &= \partial e_y/\partial x - \partial e_{xy}/\partial y. \end{aligned} \tag{40.30}$$

The dislocation density is obtained from (40.30), (40.28), and (40.27). The components of α_{hi} in the directions of the shear lines 1 and 2 become

$$\begin{aligned} \alpha_1 &= \frac{k}{\mu}\frac{\partial \phi}{\partial s_2} + \frac{1-2\nu}{E}\frac{\partial p}{\partial s_2}, \\ \alpha_2 &= \frac{k}{\mu}\frac{\partial \phi}{\partial s_1} - \frac{1-2\nu}{E}\frac{\partial p}{\partial s_1}, \end{aligned} \tag{40.31}$$

where ds_1 and ds_2 are line elements of the first and second shear lines. The above result has been obtained from (40.25) by assuming the x-direction to be the first shear line and the y-direction to be the second shear line. The Burgers vector of α_1 has the direction of the first shear line and that of α_2 is in the second shear line.

The equations of equilibrium with (40.27) lead to the Hencky relations,

$$\begin{aligned} \frac{\partial p}{\partial s_1} + 2k\frac{\partial \phi}{\partial s_1} &= 0, \\ \frac{\partial p}{\partial s_2} - 2k\frac{\partial \phi}{\partial s_2} &= 0. \end{aligned} \tag{40.32}$$

Then, (40.31) becomes

$$\alpha_1 = \frac{2(2-\nu)}{E} k \frac{\partial \phi}{\partial s_2}$$
$$\alpha_2 = \frac{2(2-\nu)}{E} k \frac{\partial \phi}{\partial s_1}. \tag{40.33}$$

The above result has been derived by Mura (1967). A similar result has been obtained by Eisenberg (1970) and Weng and Phillips (1976).

Lines of discontinuity in the continuum theory of plasticity are defined as those lines across which the stress varies in a discontinuous manner. In the dislocation theory, the lines of discontinuity are the lines along which surface dislocations are distributed. Excluding the impotent dislocations, we write the surface dislocation density tensor as,

$$\alpha_{3i} = \epsilon_{3lk}[e_{ki}]n_l, \tag{40.34}$$

where $[e_{ki}] = e_{ki}$ (in domain 2) $- e_{ki}$ (in domain 1), and the normal n_l is directed toward domain 2 from domain 1. The above equation has been obtained from (37.15), (40.21), and (40.23). The traction force should be continuous on the line of discontinuity,

$$[\sigma_{ij}]n_j = 0. \tag{40.35}$$

Equations (40.34) and (40.35) lead to

$$\alpha_{31} = \frac{2-\nu}{E}[p]n_2,$$
$$\alpha_{32} = -\frac{2-\nu}{E}[p]n_1. \tag{40.36}$$

The resultant of α_{hi} along the line of discontinuity is

$$\alpha_s = \alpha_{31}\frac{dx_1}{ds} + \alpha_{32}\frac{dx_2}{ds} = -\frac{2-\nu}{E}[p], \tag{40.37}$$

where ds is the line element along the line of discontinuity, and $dx_1/ds = -n_2$, $dx_2/ds = n_1$. The Burgers vector of α_s is tangential to the line of discontinuity, (see Fig. 40.7).

Consider a half-space with a uniform pressure $P = (2+\pi)k$ along $|x| \le a$, $y = 0$. The first and second shear lines are shown by the solid and dashed lines in Fig. 40.2. The distribution of dislocations is easily found from (40.28). In the fan-shaped region in the right-hand side in Fig. 40.2,

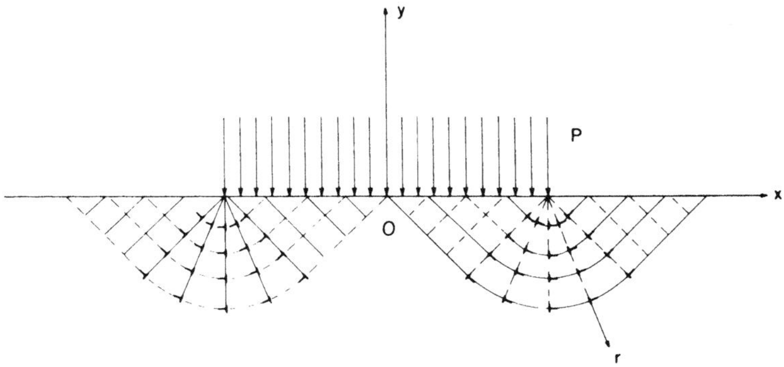

Fig. 40.2. Dislocation distributions in the fan-shaped plastic domains.

$\partial\phi/\partial s_1 = 1/r$, and

$$\alpha_1 = 0, \quad \alpha_2 = \frac{(2-\nu)}{E}\frac{2k}{r}. \tag{40.38}$$

Similarly,

$$\alpha_1 = \frac{(2-\nu)}{E}\frac{2k}{r}, \quad \alpha_2 = 0, \tag{40.39}$$

in the left-hand fan shaped region. In the above equations, r is the distance measured from the edge of the indenter.

In order to demonstrate that the stress field is the sum of the applied elastic stress and the dislocation stress, we consider the simpler example shown in Fig. 40.3. The half-infinite domain is loaded by a semi-infinite band of uniform pressure P on the surface. The applied elastic stress-components are

$$\begin{aligned} \sigma_r^A &= -\frac{P}{\pi}\left(\frac{\pi}{2} + \theta - \frac{1}{2}\sin 2\theta\right), \\ \sigma_\theta^A &= -\frac{P}{\pi}\left(\frac{\pi}{2} + \theta + \frac{1}{2}\sin 2\theta\right), \\ \sigma_{r\theta}^A &= \frac{P}{2\pi}(1 + \cos 2\theta), \end{aligned} \tag{40.40}$$

(see Timoshenko and Goodier 1951), where the polar coordinates (r, θ) are used. Let us assume that the plastic domain is $-\theta_0 \leq \theta \leq \theta_0$. Analogously with (40.39), the dislocations are distributed along the radii with

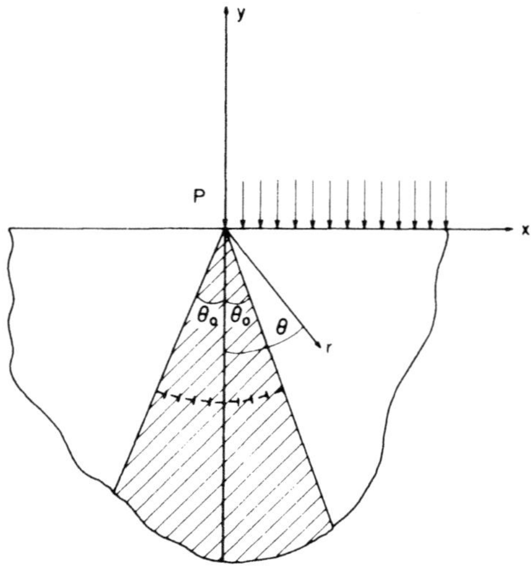

Fig. 40.3. A semi-infinite band of uniform pressure P on the surface of a half-space.

the density inversely proportional to r,

$$\alpha_{3r} = b/r. \tag{40.41}$$

The stress field of the distributed dislocations can be obtained by integrating over the plastic domain $|\theta| \leq \theta_0$ the solution for a single dislocation (Dundurs and Mura (1964)) multiplied by the density (40.39). The result is

$$\begin{aligned} \sigma_r^D &= \frac{\mu b}{2\pi}(2f + f''), \\ \sigma_\theta^D &= \frac{\mu b}{2\pi} 2f, \\ \sigma_{r\theta}^D &= -\frac{\mu b}{2\pi} f', \end{aligned} \tag{40.42}$$

where

$$f(\theta) = (2\theta_0 + \sin 2\theta_0)(\sin 2\theta + 2\theta) - \pi \sin 2\theta_0 \cos 2\theta + 2\pi\theta_0 \qquad \text{for } -\pi/2 \leq \theta \leq -\theta_0,$$

$$f(\theta) = (2\theta_0 + \sin 2\theta_0)(\sin 2\theta + 2\theta) + \pi \cos 2\theta_0 \sin 2\theta - 2\pi\theta \qquad \text{for } -\theta_0 \leq \theta \leq \theta_0,$$

$$f(\theta) = (2\theta_0 + \sin 2\theta_0)(\sin 2\theta + 2\theta) + \pi \sin 2\theta_0 \cos 2\theta - 2\pi\theta_0 \quad \text{for } \theta_0 \le \theta \le \pi/2, \tag{40.43}$$

and $f' = df/d\theta$.

The total stress field is

$$\sigma_r = \sigma_r^A + \sigma_r^D, \quad \sigma_\theta = \sigma_\theta^A + \sigma_\theta^D, \quad \sigma_{r\theta} = \sigma_{r\theta}^A + \sigma_{r\theta}^D. \tag{40.44}$$

The Mises yield criterion becomes

$$(1/2)(\sigma_r - \sigma_\theta)^2 + \sigma_{r\theta}^2 \le k^2. \tag{40.45}$$

The equality should hold in $-\theta_0 \le \theta \le \theta_0$, and the inequality in the rest of the body. The condition (40.45) is satisfied if (Mura 1968)

$$b = k/\mu(1 + \cos 2\theta_0), \quad P = 2k(2\theta_0 + \sin 2\theta_0 + \pi \cos 2\theta_0)/(1 + \cos 2\theta_0). \tag{40.46}$$

The above relations determine the plastic domain and the necessary distribution of dislocations for a given P. The relation between p and θ_0 and the stress field (40.44) agree with the classical solution obtained by Sokolovsky (1950).

Vilmann and Mura (1979) have considered a two-dimensional distribution of dislocations as shown in Fig. 40.4, and have modified the Griffith fracture theory for application to ductile materials. From slip line theory, the stress field in the fan-shaped plastic domain is

$$\sigma_{11} = k\{(1 + 3\pi/2 + 2\theta) - \sin 2\theta\}, \quad \sigma_{22} = k\{(1 + 3\pi/2 + 2\theta) + \sin 2\theta\}, \quad \sigma_{12} = k \cos 2\theta. \tag{40.47}$$

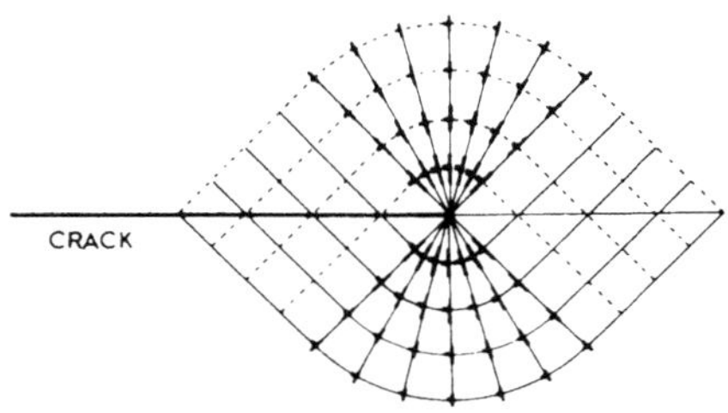

Fig. 40.4. Dislocation distributions at the crack tip.

From (40.24) and (40.47) the dislocation density in the radial direction is

$$\alpha_r = \frac{1-\nu}{\mu}\frac{2k}{r}, \tag{40.48}$$

where r is the distance from the vertex. The Burgers vector of α_r is the radial direction. The application of this result to fatigue has been considered by Mura and Vilmann (1981).

When the applied load is reduced to zero, an additional dislocation distribution, Fig. 40.5(b), is superimposed on the old distribution, Fig. 40.5(a). When the load is applied again, the resulting dislocation distribution is the sum of those shown in Figs. (a), (b), and (c). The sum of the dislocation distributions (b) and (c) is a dipole distribution which can be observed in fatigue tests. The second unloading adds the distribution (b) to the last one and the subsequent loading adds the distribution (c) to

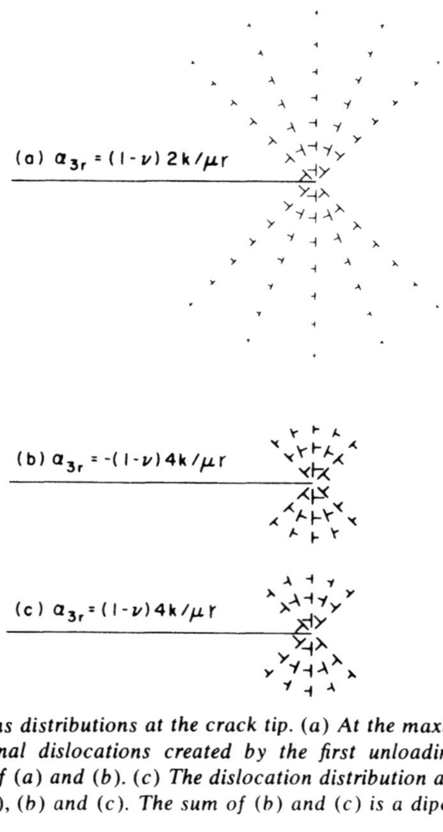

Fig. 40.5. Dislocations distributions at the crack tip. (a) At the maximum stress in the first loading. (b) Additional dislocations created by the first unloading. The resulting distribution is the sum of (a) and (b). (c) The dislocation distribution at the second loading is the sum of (a), (b) and (c). The sum of (b) and (c) is a dipole distribution.

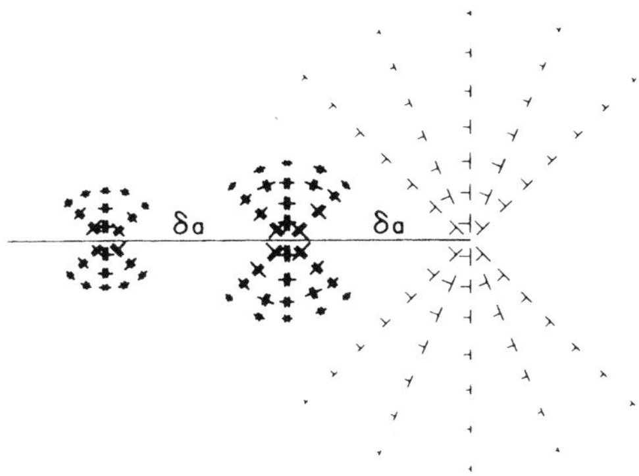

Fig. 40.6. Dislocation distributions of a propagating fatigue crack.

the preceding one and so on. Thus, the strength of the dipoles increases by the repeating loading. After some number of cyclic loadings, the crack tip advances by δa as shown in Fig. 40.6. The domain of the dipole distribution increases with the growth of the crack.

Figure 40.7 shows surface dislocations along the lines of discontinuity when the wedge becomes fully plastic under pressure P. As seen in Prager and Hodge (1951), we have

$$[p] = -2k \sin \gamma \tag{40.49}$$

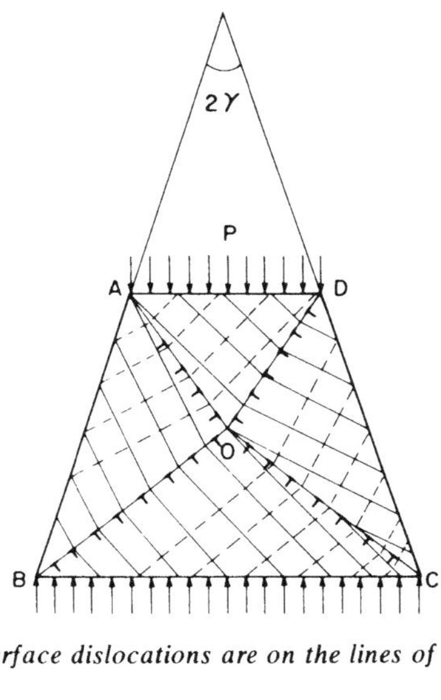

Fig. 40.7. Surface dislocations are on the lines of discontinuity.

along AOC and DOB, where 2γ is the wedge angle. From (40.37) the density of the surface dislocations is obtained as

$$\alpha_s = \frac{2-\nu}{E} 2k \sin\gamma. \tag{40.50}$$

These distributions of dislocations shown in the above examples have been obtained from the statically admissible stress fields which are essentially independent of the displacement and strain fields (kinematically admissible fields). However, the present results of dislocation distributions give possible patterns of plastic deformation. Knowing the source of the dislocations (crack tips, for instance), one can evaluate possible plastic strains. These strains are the product of the dislocation density and the distance from the source.

Beams and cylinders

Nye (1953) has introduced the concept of a dislocation density and has expressed the curvature of a beam caused by dislocations in terms of the dislocation density tensor. He has considered the case where the dislocation stress disappears. Read (1957), Bilby, Gardner and Smith (1958) have dealt with the case where the stress does not vanish. They have obtained a relation among the stress gradient, curvature, and the dislocation density tensor. In this book, a somewhat different approach is used as illustrated by the following examples.

Consider an infinitely long beam in the x_1-direction with a height of $2h$. Under pure bending, all quantities depend only on the x_2-variable, and only the normal stress and strain in the x_1-direction are considered. From (40.13)

$$\dot{\beta}^p_{11} = -V_{231}; \tag{40.51}$$

this shows that all dislocations are of the edge type with Burgers vector in the x_1-direction and that their motion takes place in the x_2-direction, i.e., the plastic domain expands in the x_2-direction. Although this motion of dislocations is of a clime type, it can be regarded as the sum of two slips of two dislocations whose Burgers vectors are $\pm 45°$ with respect to the x_1-direction. The sum of the two Burgers vectors is equal to the Burgers vector in the x_1-direction.

When the edge dislocations with density b_1 are distributed uniformly in a strip domain $c < x_2 < h$ in the infinitely extended material, the dislocation stress is, from (4.13) and (4.14),

$$\sigma_{11} = \frac{-\mu b_1}{2\pi(1-\nu)} \int_{-\infty}^{\infty} dx_1' \int_c^h \frac{\bar{y}(3\bar{x}^2 + \bar{y}^2)}{(\bar{x}^2 + \bar{y}^2)^2} dx_2'$$

$$= -\frac{\mu b_1}{1-\nu}(h-c) \quad \text{for } x_2 \geq h,$$

$$= -\frac{\mu b_1}{1-\nu}(2x_2 - h - c) \quad \text{for } h \geq x_2 \geq c,$$

$$= \frac{\mu b_1}{1-\nu}(h-c) \qquad \text{for } c \geq x_2, \tag{40.52}$$

where $\bar{x} = x_1 - x_1'$, and $\bar{y} = x_2 - x_2'$. The other stress components disappear. If the dislocations are distributed in the two strips as shown in Fig. 40.8(a), the stress field becomes

$$\sigma_{11} = \frac{-2\mu b_1}{1-\nu}(x_2 - c) \quad \text{for } h \geq x_2 \geq c,$$

$$\sigma_{11} = 0 \qquad \text{for } c \geq |x_2|, \tag{40.53}$$

$$\sigma_{11} = \frac{-2\mu b_1}{1-\nu}(x_2 + c) \quad \text{for } -c \geq x_2 \geq -h.$$

Although (40.53) has been constructed from the solution for the infinite body, it can be used as the solution of the beam with the plastic domains $h \geq |x_2| \geq c$ if the linear stress

$$\sigma_{11} = M_0 x_2 / I \tag{40.54}$$

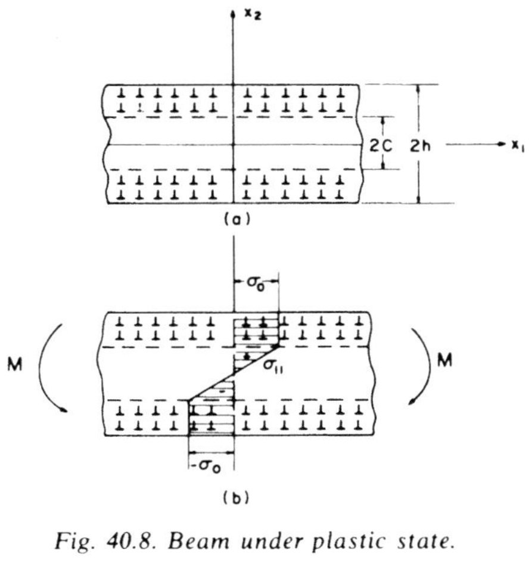

Fig. 40.8. Beam under plastic state.

is added to (40.53), where

$$M_0 = 2\mu b_1(h-c)^2(2h+c)/3(1-\nu),\qquad I = 2h^3/3. \tag{40.55}$$

The stress (40.54) is determined such that the sum of (40.53) and (40.54) yields zero moment. This must be true since the beam is free from external tractions.

When the beam is perfectly elasto-plastic with the yield stress σ_0 and is bent by moment M (applied stress Mx_2/I), the total stress must be $\pm\sigma_0$ in the plastic domains, that is,

$$\frac{-2\mu b_1}{1-\nu}(x_2 \mp c) + \frac{M_0 x_2}{I} + \frac{M x_2}{I} = \pm\sigma_0. \tag{40.56}$$

The above conditions are satisfied when

$$M_0/I + M/I = 2\mu b_1/(1-\nu),\qquad 2\mu b_1 c/(1-\nu) = \sigma_0,\quad b_1 = \alpha_{31} = (1-\nu^2)\sigma_0/cE. \tag{40.57}$$

Then, the total stress becomes the well-known distribution as shown in Fig. 40.8(b), and the moment necessary for this plastic state is

$$M = \sigma_0(h^2 - c^2)/3 \tag{40.58}$$

which agrees with the classical solution.

The dislocation distribution shown in this example can also be derived from (40.25). From the stress distribution in Fig. 40.8(b), α_{31} becomes zero in the plastic domains, and equals $-\sigma_0/cE$ in the elastic domain. If the uniform distribution σ_0/cE is superimposed everywhere, the resulting distribution becomes identical to the one shown in Fig. 40.8(b). The uniform distribution does not cause any stress field. E is Young's modulus. The factor $(1-\nu^2)$ in the α_{31}-expression in (40.57) arises from the deviation of the beam theory from the plane strain state.

Similar calculations can be done for the torsion of a solid circular cylinder (Mura 1970). Assuming that all dislocations are of the same sign and parallel to the x_3-axis (the axis of the cylinder), (40.13) leads to

$$\dot\beta^p_{13} = -V_{233},\qquad \dot\beta^p_{23} = V_{133}, \tag{40.59}$$

or, equivalently, to

$$\dot\beta^p_{\theta 3} = V_{r33}. \tag{40.60}$$

This shows that the plastic deformation is caused by the radial motion of α_{33}. In Fig. 40.9 screw dislocations $\alpha_{33} = b_3$ are distributed uniformly in the plastic domain. The dislocation stress is obtained from (4.8) Integrating for this dislocation distribution we have

$$\begin{aligned} \sigma_{\theta z} &= -\mu b_3(r^2 - c^2)/2r \quad \text{for } c \le r \le a, \\ &= 0 \qquad\qquad\qquad\quad \text{for } r \le c. \end{aligned} \tag{40.61}$$

If the cylinder is free from external tractions, the following stress must be added to (40.61) so that the total twisting moment disappears:

$$\sigma_{\theta z} = M_0 r/J, \tag{40.62}$$

where

$$\begin{aligned} M_0 &= \pi\mu b_3(a^2 - c^2)^2/4, \\ J &= \pi a^4/2. \end{aligned} \tag{40.63}$$

When the cylinder is a perfectly elasto-plastic material with the yield shear stress k twisted by moment M, the total stress must be equal to k in $c \le r \le a$, that is,

$$-\mu b_3(r - c) + M_0 r/J + Mr/J = k \quad c \le r \le a, \tag{40.64}$$

where (40.61) has been approximated by $-\mu b_3(r - c)$ by assuming $c \approx a \approx r$. The above condition is satisfied when

$$\begin{aligned} M_0/J + M/J &= \mu b_3, \\ \mu b_3 c &= k. \end{aligned} \tag{40.65}$$

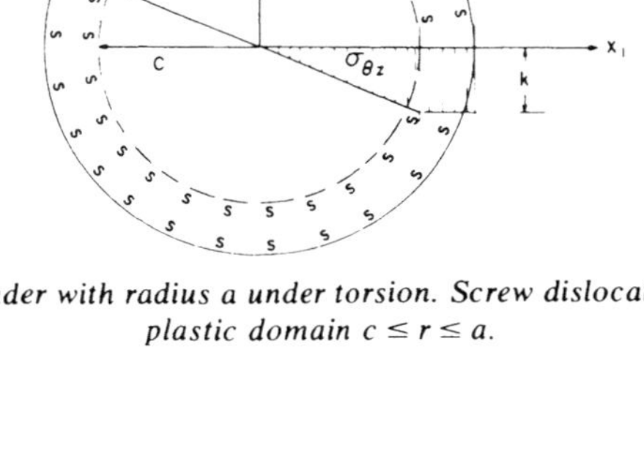

Fig. 40.9. A solid cylinder with radius a under torsion. Screw dislocations are distributed in plastic domain $c \le r \le a$.

The second equation in (40.65) determines the dislocation density b_3, and the first equation yields M necessary for having plastic domain $c \le r \le a$. The total stress distribution under M is shown in Fig. 40.9. According to the elementary strength of materials, the torsion of a circular cylinder is described by

$$\beta^{p}_{3\theta} = \theta r \tag{40.66}$$

or

$$\begin{aligned} \beta^{p}_{32} &= -\theta x_2, \\ \beta^{p}_{32} &= \theta x_1. \end{aligned} \tag{40.66$'$}$$

Equation (37.11) leads to the dislocation distribution

$$\alpha_{11} = \alpha_{22} = \theta. \tag{40.67}$$

It can be proved that the dislocation distribution (40.67) is equivalent to $\alpha_{33} = -2\theta$ if the impotent distributions of dislocations are superimposed on (40.67). The impotent distributions are

$$\begin{aligned} \beta^{p}_{31} &= \theta x_2, \\ \beta^{p}_{13} &= -\theta x_2, \end{aligned} \tag{40.68}$$

and

$$\begin{aligned} \beta^{p}_{32} &= -\theta x_1, \\ \beta^{p}_{23} &= \theta x_1. \end{aligned} \tag{40.69}$$

Then, (40.68) gives

$$\begin{aligned} \alpha_{11} &= -\theta, \\ \alpha_{33} &= -\theta, \end{aligned} \tag{40.70}$$

and (40.69) leads to

$$\begin{aligned} \alpha_{22} &= -\theta, \\ \alpha_{33} &= -\theta. \end{aligned} \tag{40.71}$$

Therefore, the sum of (40.67), (40.70), and (40.71) yields

$$\alpha_{33} = -2\theta \tag{40.72}$$

which is the dislocation distribution assumed after (40.60).

Shioya and Shioiri (1976) have explained the non-uniform yield-zone pattern in twisted mild steel bars with circular cross-sections in terms of continuous distributions of dislocations. Flinn (1970), and Jimma and

Kawai (1979) have investigated the deformation mechanism in fabrication processes by continuous distributions of dislocations. Jimma and Masuda (1977) have studied longitudinal and torsional waves in a linearly work-hardening material by using moving dislocations. Smith (1958) has proposed a model of a planar network of moving dislocations at shock fronts in explosive metals. Ohnami et al. (1972) and Shiozawa and Ohnami (1974) have studied flow stresses of metals and X-ray diffraction by means of the geometrical property of continuous distributions of dislocations. An electron microscopic study of dislocation walls has been done by Yamamoto et al. (1977). Owen (1971) has assumed the stress of a dislocation dipole to be the work-hardening in the material and has derived the corresponding stress-strain relations of composite materials.

41. Dislocation model for fatigue crack initiation

The initiation of fatigue cracks is one of the most important stages in the fatigue fracture process of metals. A large number of metallographic observations has been carried out to elucidate the micromechanisms responsible for crack initiation. The state of the art is found in the recent excellent review articles by Grosskrentz (1971), and Laird and Duquette (1972). The site of crack initiation varies depending on the microstructure of the material and the applied stresses. Among the possible sites of crack initiation, the persistent slip band is a preferential one for pure, single-phased metals and some poly-phased metals under a low strain cycling. The cyclic strain is concentrated along the slip band, accompanied by extrusion or intrusion. Laird (1976), Mughrabi (1980) among others have investigated the nature of microscopic mechanisms of persistent slip bands.

Most of the models proposed to account for the formation of extrusions and intrusions, are based upon the Mott assumption (1958) that dislocations move along different paths under forward and reverse loadings. These models, however, cannot explain the most important experimental fact in fatigue: the 'ratcheting' of plastic deformation by a cyclic loading. Lin and Ito (1969) have proposed a model for ratcheting in which a speciment is subjected to a special initial stress state. Zarka et al. (1980) also have proposed a phenomenological model for the ratcheting. Recently, Tanaka and Mura (1981) have proposed a more reasonable model for ratcheting by considering two adjacent layers of dislocation pile-ups. Each layer has a different sign. Their model can also yield the Coffin–Manson type law for the crack initiation and the

Petch type equation for the grain size dependency of fatigue strength. This section will introduce the theory of Tanaka and Mura.

Let us consider an applied shear stress pattern as shown in Fig. 41.1. Figures 41.2 and 41.3 show models of dislocation pile-ups. Dislocations are piled up on layer I by the loading (forward) force and on layer II by the unloading (reverse) force. The signs of the dislocations are opposite in the two layers. The dislocations in Fig. 41.2 are created at the surface of the specimen and moved to the interior of a grain. The dislocations in Fig. 41.3 are created inside a grain and moved to the grain boundary.

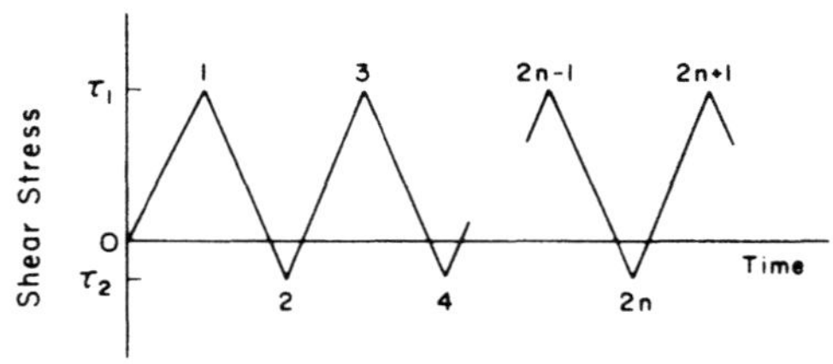

Fig. 41.1. Applied shear stress pattern.

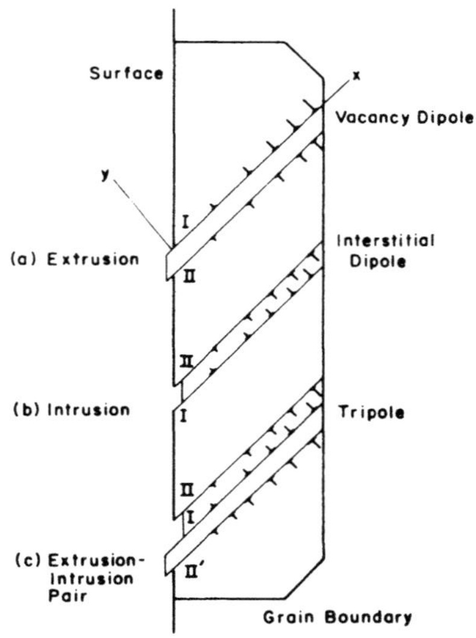

Fig. 41.2. Dislocation distribution in a most favorably oriented grain and the formation of extrusion and intrusion.

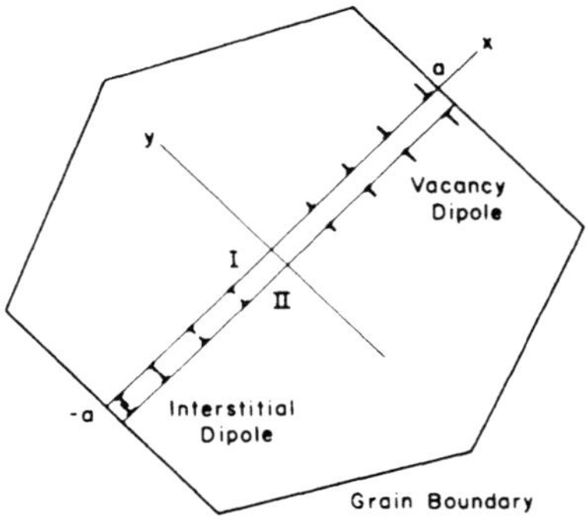

Fig. 41.3. Dislocation distribution in a most favorably oriented grain.

We calculate the plastic strain, number of dislocations, the associated energy, and the stress field caused by the dislocation motion. Under the first loading τ_1 greater than the frictional stress k, the dislocation distribution with density $D_1(x)$ is produced on layer I as shown in Fig. 41.3. (The following analysis is developed for the dislocation pile-up in Fig. 41.3, but the result also applies to the case shown by Fig. 41.2 when the plane $x = 0$ is assumed to be a free surface.) By assuming k to be constant, we express the equilibrium condition of dislocations as

$$\tau_1^D + \tau_1 - k = 0, \tag{41.1}$$

where τ_1^D is the dislocation stress (the back stress) given by

$$\tau_1^D = A \int_{-a}^{a} \frac{D_1(\xi)d\xi}{x - \xi}. \tag{41.2}$$

The domain of the dislocation distribution is $-a < x < a$, and

$$A = \mu b/2\pi(1 - \nu), \tag{41.3}$$

where b is the Burgers vector, μ the shear modulus, and ν Poisson's ratio. The solution of (41.1) is found by the method in Appendix 4. When the solution is assumed to be unbounded at $x = \pm a$, it becomes

$$D_1(x) = (\tau_1 - k)x/\pi A(a^2 - x^2)^{1/2}. \tag{41.4}$$

The total number of dislocations existing between $x = 0$ and a is

$$N_1 = \int_0^a D_1(x)dx = (\tau_1 - k)a/\pi A. \tag{41.5}$$

The average plastic strain caused by these dislocations is

$$\gamma_1 = \int_{-a}^{a} bD_1(x)xdx = (\tau_1 - k)ba^2/2A. \tag{41.6}$$

The multiplication of the number of pile-up layers in a unit area by the γ_1 value yields the macroscopic plastic strain. For simplicity, γ_1 is called the plastic strain.

The stored energy of dislocations is

$$U_1 = -\frac{1}{2}\int_{-a}^{a} \tau_1^D bD_1(x)xdx = \gamma_1(\tau_1 - k)/2. \tag{41.7}$$

When the applied stress is reversed from τ_1 to τ_2, the dislocations with opposite signs are piled up on layer II as shown in Figs. 41.2 and 41.3. We assume that the dislocations in layer I do not move by the unloading. This assumption of the irreversibility of dislocation motion will be discussed later. By denoting the density of dislocations on layer II by $D_2(x)$ and the back stress due to $D_2(x)$ by τ_2^D, we express the equilibrium condition in layer II as

$$\tau_2^D + \tau_1^D + \tau_2 + k = 0. \tag{41.8}$$

The distance between layers I and II is assumed to be very small compared with the pile-up length. Then, τ_1^D on layer II can be regarded as the same as on layer I. Equations (41.1) and (41.8) yield

$$\tau_2^D - (\Delta\tau - 2k) = 0, \tag{41.9}$$

where $\Delta\tau = \tau_1 - \tau_2$. Only when $\Delta\tau$ is larger than $2k$ can the dislocations on layer II start to move from $x = 0$ and pile up at $x = \pm a$. The dislocation density $D_2(x)$, the total number of dislocations between $x = 0$ and $x = a$ on layer II, and the plastic strain increment γ_2 are obtained as

$$\begin{aligned} D_2(x) &= -(\Delta\tau - 2k)x/\pi A(a^2 - x^2)^{1/2}, \\ N_2 &= -(\Delta\tau - 2k)a/\pi A, \\ \gamma_2 &= -(\Delta\tau - 2k)ba^2/2A. \end{aligned} \tag{41.10}$$

The stored energy of dislocations $D_2(x)$ becomes

$$U_2 = -\gamma_2(\Delta\tau - 2k)/2. \tag{41.11}$$

The pile-up of opposite dislocations on layer II causes a positive back stress on layer I. This back stress enhances the pile-up of dislocations on layer I during the next loading. Similarly the dislocations on layer I enhance the pile-up of dislocations on layer II during the subsequent unloading. This is the ratcheting mechanism proposed by Tanaka and Mura (1981). After repeated loading and unloading, the extrusion or intrusion is created on the surface of the specimen as shown in Fig. 41.2.

The increment of dislocation $D_k(x)$, the dislocation number N_k, the plastic strain increment γ_k, the back stress increment τ_k^D, and the stored energy U_k at the k-th step of the loading and unloading processes are obtained in a similar manner. They are

$$\begin{aligned} &D_k(x) = (-1)^{k+1}\Delta D(x),\ N_k = (-1)^{k+1}\Delta N,\ \gamma_k = (-1)^{k+1}\Delta\gamma, \\ &\quad \tau_k^D = (-1)^{k+1}(2k - \Delta\tau),\ U_k = \Delta U, \end{aligned} \tag{41.12}$$

where

$$\begin{aligned} \Delta\tau &= \tau_1 - \tau_2, \\ \Delta D(x) &= (\Delta\tau - 2k)x/\pi A(a^2 - x^2)^{1/2}, \\ \Delta N &= (\Delta\tau - 2k)a/\pi A, \\ \Delta\gamma &= (\Delta\tau - 2k)ba^2/2A, \\ \Delta U &= \Delta\gamma(\Delta\tau - 2k)/2. \end{aligned} \tag{41.13}$$

The index k takes $2n$ at the minimum stress after n cycles and $2n + 1$ at the maximum stress after n cycles. The total values of these quantities obtained in (41.12) or (41.13) are

$$\begin{aligned} D_{\text{I}}(x) &= \sum_{n=0}^{n} D_{2n+1}(x) = D_1(x) + n\Delta D(x), \\ N_{\text{I}} &= \sum_{n=0}^{n} N_{2n+1} = N_1 + n\Delta N, \\ U_{\text{I}} &= \sum_{n=0}^{n} U_{2n+1} = U_1 + n\Delta U, \end{aligned} \tag{41.14}$$

for the dislocation pile-up on layer I, and

$$\begin{aligned} D_{\text{II}}(x) &= \sum_{n=1}^{n} D_{2n}(x) = -n\Delta D(x), \\ N_{\text{II}} &= \sum_{n=1}^{n} N_{2n} = -n\Delta N, \\ U_{\text{II}} &= \sum_{n=1}^{n} U_{2n} = n\Delta U, \end{aligned} \tag{41.15}$$

for the dislocation pile-up on layer II. $D_I(x)$ is written as

$$D_I(x) = T_I x/\pi A(a^2 - x^2)^{1/2}, \tag{41.16}$$

where

$$T_I = \tau_I - k + n(\Delta\tau - 2k). \tag{41.17}$$

The stress field due to the dislocation distribution $D_I(x)$ is the same as that for the shear crack under shear stress T_I (see Bilby and Eshelby, 1968). The stress intensity factor at the pile-up tip is

$$K_I = T_I(\pi a)^{1/2} = \{\tau_I - k + n(\Delta\tau - 2k)\}(\pi a)^{1/2}. \tag{41.18}$$

Similarly, the stress intensity factor at the tip of layer II becomes

$$K_{II} = T_{II}(\pi a)^{1/2} = -n(\Delta\tau - 2k)(\pi a)^{1/2}. \tag{41.19}$$

A tensile stress, built up at the tip between the two layers I and II, could become large enough to create the nucleus of a crack. Since the densities of pile-up dislocations on the two layers are about the same for long-life fatigue, except for the sign, the tensile stress σ_{xx} at $x = a$ and at half the distance between the two layers becomes

$$\sigma_{xx} = \sigma_{xx}^{DI} + \sigma_{xx}^{DII} = 3\sqrt{a}\,n(\Delta\tau - 2k)/\sqrt{2h}, \tag{41.20}$$

where h is the distance between the two layers. Embryonic cracks can be formed from this tensile stress as shown in Figs. 41.4 and 41.5. Formation of the intrusion causes the stress concentration and the intrusion can also be a crack embryo. The depth of the intrusion equals

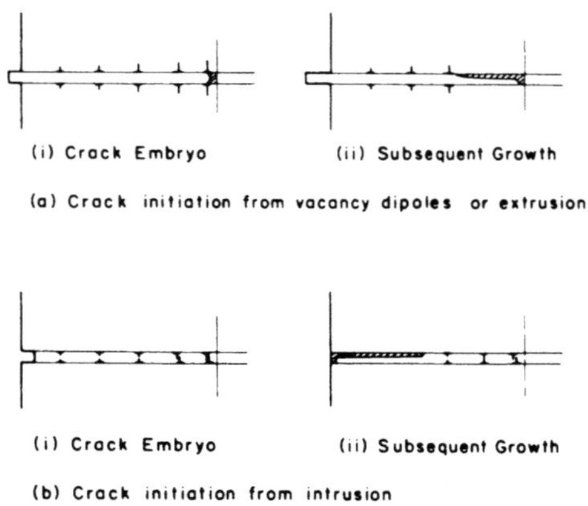

Fig. 41.4. Two types of crack initiation.

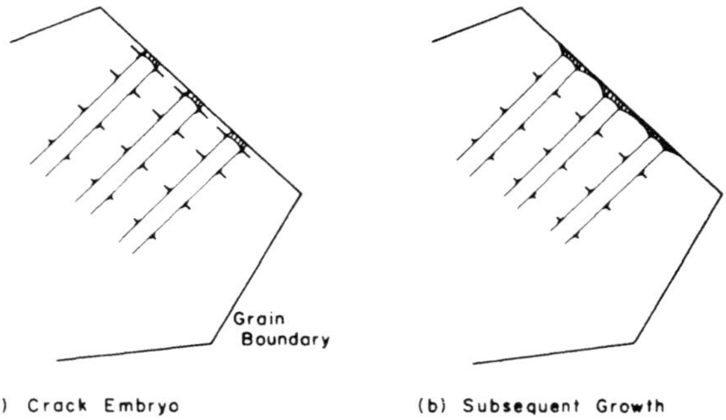

Fig. 41.5. Grain boundary crack initiated by stacked pileups of vacancy dipoles.

the total number of accumulated dislocations which is $N_{\mathrm{I}} = N_{\mathrm{II}} = n\Delta N$, multiplied by the Burgers vector.

The condition of the growth of the crack embryo will be treated from the viewpoint of energy balance. We assume that when the stored strain energy due to dislocations accumulated after n cycles becomes equal to the surface energy, the layers of dislocation dipoles are transformed into a free surface. The life of the crack initiation n_c is now defined as the number of stress cycles when the following energy condition is satisfied:

$$U = U_{\mathrm{I}} + U_{\mathrm{II}} = 2n_c\Delta U = 4aW_s, \tag{41.21}$$

where W_s is the specific fracture energy for a unit area. The right-hand side of (41.21) is $2aW_s$ for the pile-up shown in Fig. 41.2. Equation (41.21) leads to

$$\begin{aligned} n_c &= 4aW_s/(\Delta\tau - 2k)\Delta\gamma \\ &= 8AW_s/ba(\Delta\tau - 2k)^2 \\ &= 2bW_sa^3/A(\Delta\gamma)^2. \end{aligned} \tag{41.22}$$

The last result,

$$n_c(\Delta\gamma)^2 = 2bW_sa^3/A, \tag{41.23}$$

is similar to the Coffin (1954)–Manson (1954) law. They have found the life law for the complete fracture of smooth specimens under high strain cycling. Later, a similar law was confirmed by Lukás et al. (1974) to be valid for low-strain and long-life fatigue. The life up to the initiation of the crack on a grain size order is about fifty percent of the total life of fatigue (see Taira et al. 1979). Therefore, the Coffin–Manson relation seems to be valid for the initiation of cracks in long-life fatigue. The strain energy of dislocations is accumulated in the same amount in each loading and unloading except the first loading. The amount of stored energy does not correspond to the total area of the hysteresis loop, but only to the shaded area shown in Fig. 41.6. The energy corresponding to the remaining area of the loop is the work dissipated against the friction stress. Martin (1961) derived the Coffin–Manson relation by regarding the segment of the plastic work associated with workhardening as an accumulating damage.

Tomkins (1968) also derived the Coffin–Manson law from a different view-point. He assumed that the crack growth per cycle is equal to the product of plastic strain and plastic zone size. Integrating the growth rate equation from an initial crack length to a final length, he obtained the Coffin–Manson law.

The second equation in (41.22) can be written as

$$\Delta\tau = 2k + (8AW_s/bn_c)^{1/2}a^{-1/2} \tag{41.24}$$

which is of the Petch-type equation for constant initial life. The Petch-type relation has been reported for the long-life fatigue of several metals (see Taira et al. 1979, Thompson and Backofen 1971, Ebner and Backofen 1959).

The most striking advantage of the Tanaka and Mura model based on the irreversibility assumption for dislocation motion is that the progress of the ratcheting of a plastic deformation in the slip band can be calculated in each loading and unloading. Two neighboring layers, where the motion of dislocations with opposite signs takes place, can be regarded as the zone of strain localization akin to the persistent slip bands found in low strain fatigue. Several experiments have been

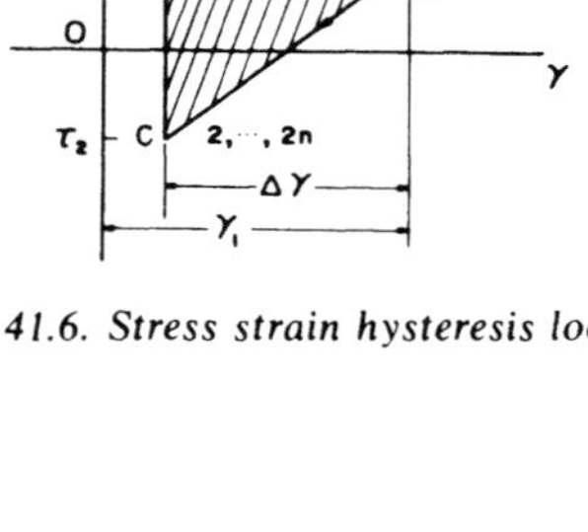

Fig. 41.6. Stress strain hysteresis loop.

reported on the slip movement in such bands and most of them indicate a certain irreversibility of plastic deformation. Keith and Gilman (1959) have observed in their etch pit study that dislocation movements were irreversible under cyclic stressing except for a small motion of dislocation loops in lithium fluoride crystals. More macroscopically, Charsley and Desvaux (1969), and Desvaux and Charsley (1969) have found that a partial reversal of tensile slip steps occurred during compression, but no total reversal was seen in single crystals with wavy and planar slip modes.

7

Material properties and related topics

In this chapter, we demonstrate how inclusion problems are applied to estimate macroscopic plastic properties of aggregates, i.e. particle bearing materials and polycrystalline materials. In these materials, internal stress usually develops as a result of plastic deformation which is inherently heterogeneous due to the heterogeneity of the constituent medium. These internal stresses can be calculated in a straightforward manner by using the methods of Chapters 2 and 3. If one finds a way to connect these internal stresses with the applied stress needed to produce the plastic deformation, one can eventually predict the macroscopic plastic properties.

42. Macroscopic average

We begin with the somewhat general properties of the internal stresses of inclusions, not explicitly mentioned in the preceding chapters.

Average of internal stresses

$$\int_D \sigma_{ij} dD = \int_D \sigma_{ik}\delta_{jk} dD \tag{42.1}$$

is rewritten as

$$\int_D \sigma_{ij} dD = \int_D \sigma_{ik} x_{j,k} dD, \tag{42.2}$$

since $x_{j,k} = \delta_{jk}$. Integrating by parts and noting $\sigma_{ik,k} = 0$ in D and $\sigma_{jk}n_k = 0$ on its boundary $|D|$, one can see that the above equation is always equal to zero, that is,

$$\int_D \sigma_{ij} dD = \int_{|D|} \sigma_{ik} x_j n_k dS - \int_D \sigma_{ik,k} x_j dD = 0. \tag{42.3}$$

Note that this is valid without specifying the elastic constants or the origin of the internal stress.

Macroscopic strains

Suppose that the eigenstrain ϵ^*_{ij} is given to an aggregate Ω of inclusions, producing an internal stress field σ_{ij} and strain field ϵ_{ij}. Let us calculate the average of ϵ_{ij} over the whole body, $\langle\epsilon_{ij}\rangle_D$, defined by

$$\langle\epsilon_{ij}\rangle_D = (1/V)\int_D \epsilon_{ij} dD = (1/V)\int_D (e_{ij} + \epsilon^*_{ij}) dD, \tag{42.4}$$

where V is the volume of the body and e_{ij} the elastic strain. If the elastic constants are uniform, (42.4) is rewritten as

$$\langle\epsilon_{ij}\rangle_D = (1/V) C^{-1}_{ijkl} \int_D \sigma_{kl} dD + (1/V)\int_D \epsilon^*_{ij} dD, \tag{42.5}$$

which becomes

$$\langle\epsilon_{ij}\rangle_D = (1/V)\int_D \epsilon^*_{ij} dD \tag{42.6}$$

from (42.3). Since ϵ^*_{ij} is defined only in Ω, the integral domain in (42.6) can be changed to Ω. By defining the volume fraction f of the inclusions as Ω/V, (42.6) is written

$$\langle\epsilon_{ij}\rangle_D = f\langle\epsilon^*_{ij}\rangle_\Omega = f\epsilon^*_{ij}; \tag{42.7}$$

the last result applies when ϵ^*_{ij} is uniform in Ω. Expression (42.6) can be used to describe macroscopic strains induced by dislocation motions.

When the inclusions have elastic moduli different from those of the matrix, the inhomogeneous inclusions are simulated by the equivalent inclusions defined in Section 22. That is, when the eigenstrain of the inclusion is ϵ^p_{ij}, we have

$$\sigma_{ij} = C_{ijkl}(\epsilon_{kl} - \epsilon^{**}_{kl}) = C^*_{ijkl}(\epsilon_{kl} - \epsilon^p_{kl}) \qquad \text{in } \Omega, \tag{42.8}$$

where

$$\epsilon^{**}_{kl} = \epsilon^p_{kl} + \epsilon^*_{kl}, \tag{42.9}$$

$$\epsilon_{kl} = S_{klmn}\epsilon^{**}_{mn}, \tag{42.10}$$

and C^*_{ijkl} is the elastic moduli of the inhomogeneous inclusions. Then (42.6) and (42.7) are written, respectively, as

$$\langle \epsilon_{ij} \rangle_D = (1/V) \int_D (C^{-1}_{ijkl} \sigma_{kl} + \epsilon^{**}_{ij}) dD = (1/V) \int_\Omega \epsilon^{**}_{ij} dD \tag{42.11}$$

and

$$\langle \epsilon_{ij} \rangle_D = f \epsilon^{**}_{ij}. \tag{42.12}$$

Hence, the macroscopic strain is directly related to the eigenstrain of the equivalent inclusion.

In the above procedure, it is assumed that the stress due to a particular inclusion is not disturbed by other inclusions. This approximation is permitted when $f \ll 1$. However, it is a poor approximation when f is large.

Tanaka-Mori's theorem

For simplicity, let us consider a homogeneous body containing an inclusion Ω not necessarily ellipsoidal (see Fig. 42.1). At this stage of discussion, the shape of the inclusion is not specified. Give to the inclusion a uniform eigenstrain ϵ^*_{ij}. Let the internal stress due to ϵ^*_{ij} be σ^∞_{ij}, implying that σ^∞_{ij} is calculated by assuming the body to be infinitely extended. σ^∞_{ij} outside Ω is safely given by

$$\sigma^\infty_{ij}(\mathbf{x}) = -C_{ijkl} \int_\Omega C_{pqmn} \epsilon^*_{mn} G_{kp,ql}(\mathbf{x} - \mathbf{x}') d\mathbf{x}' \tag{42.13}$$

from (3.25), where $\epsilon^*_{mn} = 0$ for points exterior to Ω. Let us integrate σ^∞_{ij} over the domain $V_2 - V_1$ as shown in Fig. 42.1. V_2 and V_1 are ellipsoidal and $V_2 \supset V_1 \supset \Omega$.

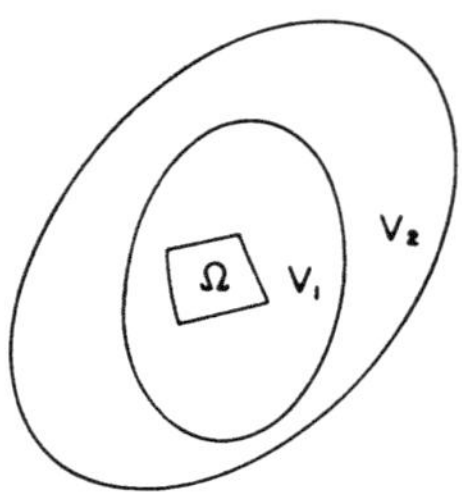

Fig. 42.1. Inclusion Ω with an arbitrary shape in an infinite domain.

$$\int_{V_2-V_1} \sigma_{ij}^{\infty} d\mathbf{x} = -C_{ijkl} \int_{V_2} d\mathbf{x} \int_{\Omega} C_{pqmn} \epsilon_{mn}^* G_{kp,ql}(\mathbf{x}-\mathbf{x}') d\mathbf{x}'$$
$$+ C_{ijkl} \int_{V_1} d\mathbf{x} \int_{\Omega} C_{pqmn} \epsilon_{mn}^* G_{kp,ql}(\mathbf{x}-\mathbf{x}') d\mathbf{x}'. \qquad (42.14)$$

When we change the order of integration in (42.14),

$$\int_{V_2-V_1} \sigma_{ij}^{\infty} d\mathbf{x} = -\int_{\Omega} d\mathbf{x}' \left\{ C_{ijkl} \int_{V_2} C_{pqmn} \epsilon_{mn}^* G_{kp,ql}(\mathbf{x}-\mathbf{x}') d\mathbf{x} \right\}$$
$$+ \int_{\Omega} d\mathbf{x}' \left\{ C_{ijkl} \int_{V_1} C_{pqmn} \epsilon_{mn}^* G_{kp,ql}(\mathbf{x}-\mathbf{x}') d\mathbf{x} \right\}. \qquad (42.15)$$

Since

$$-\int_{V_2} C_{pqmn} \epsilon_{mn}^* G_{kp,ql}(\mathbf{x}-\mathbf{x}') d\mathbf{x}$$

gives the total distortion at $\mathbf{x}'$ when the uniform eigenstrain is defined in V_2 and we have assumed V_2 to be an ellipsoid, this distortion is independent of $\mathbf{x}'$ and is determined by the shape of V_2, as long as $\mathbf{x}' \in V_2$. This condition is satisfied since $\mathbf{x}' \in \Omega \subset V_2$. A similar remark applies to the second term in (42.15). Thus, (42.15) is rewritten as

$$\int_{V_2-V_1} \sigma_{ij}^{\infty} d\mathbf{x} = \Omega C_{ijkl} [S_{klmn}(V_2) \epsilon_{mn}^* - S_{klmn}(V_1) \epsilon_{mn}^*], \qquad (42.16)$$

[see (11.1) and (11.15)], where Ω is the volume of the inclusion, and $S_{klmn}(V_2)$ and $S_{klmn}(V_1)$ are calculated by (17.14) as

$$S_{klmn} = (1/8\pi) C_{pqmn} \int_{S^2} \bar{\xi}_q \{ \bar{\xi}_l N_{kp}(\bar{\boldsymbol{\xi}}) + \bar{\xi}_k N_{lp}(\bar{\boldsymbol{\xi}}) \} D^{-1}(\bar{\boldsymbol{\xi}}) dS(\bar{\zeta}). \qquad (42.17)$$

Equation (42.16) leads to interesting conclusions: The volume integral of σ_{ij}^{∞} over $V_2 - V_1$ is proportional to Ω; it is independent of the absolute position and size of V_2 and V_1 as long as $V_2 \supset V_1$; it depends only on the shape of V_2 and V_1; and it vanishes when V_2 and V_1 are similar in shape and have the same orientation, that is, when $a_1/b_1 = a_2/b_2 = a_3/b_3$, and

$$\begin{aligned} &V_2\colon x_1^2/a_1^2 + x_2^2/a_2^2 + x_3^2/a_3^2 \le 1, \\ &V_1\colon (x_1 - x_1^0)^2/b_1^2 + (x_2 - x_2^0)^2/b_2^2 + (x_3 - x_3^0)^2/b_3^2 \le 1. \end{aligned} \qquad (42.18)$$

The last conclusion (42.16) is valid even when the eigenstrain in Ω is non-uniform and the elastic moduli of Ω differ from those of the matrix.

When Ω is an ellipsoidal inclusion we can have the limiting situation, $V_1 \rightarrow \Omega$. Then

$$\int_{V_2-\Omega} \sigma_{ij}^{\infty} d\mathbf{x} = \Omega C_{ijkl}[S_{klmn}(V_2)\epsilon_{mn}^{*} - S_{klmn}(\Omega)\epsilon_{mn}^{*}], \quad V_2 \supset \Omega. \tag{42.19}$$

If V_2 and Ω are similar in shape, (42.19) vanishes,

$$\int_{V_2-\Omega} \sigma_{ij}^{\infty} d\mathbf{x} = 0, \quad V_2 \supset \Omega. \tag{42.20}$$

Similarly, for the corresponding strain,

$$\int_{V_2-\Omega} \epsilon_{ij}^{\infty} d\mathbf{x} = 0, \quad V_2 \supset \Omega. \tag{42.20.1}$$

The above properties of the integration in the domain around an ellipsoidal inclusion have been obtained by Tanaka and Mori (1972). Brown and Stobbs (1971) have noted (42.20) by using explicit form of σ_{ij}^{∞} due to a spherical inclusion in an isotropic medium. It should be noted that (42.20) is valid for an inhomogeneous inclusion with arbitrary and non-uniform eigenstrains: (42.15) becomes zero even if $\epsilon_{mn}^{*} = \epsilon_{mn}^{*}(\mathbf{x}')$ when V_2 and V_1 have the same ellipsoidal shape and orientation (see Fig. 42.2).

Image stress

Let us consider a finite body D containing an ellipsoidal inclusion Ω with eigenstrain ϵ_{ij}^{*}. In order to utilize the property (42.20), we assume that D is similar to Ω in terms of its orientation and shape. This assumption does not restrict the problem when Ω is very small compared with D.

Let us find the internal stress σ_{ij} due to Ω. In the first approximation, σ_{ij} is taken as σ_{ij}^{∞} defined previously. The integration of σ_{ij}^{∞} over the whole body can be written as

$$\int_{D} \sigma_{ij}^{\infty} dD = \int_{\Omega} \sigma_{ij}^{\infty} dD + \int_{D-\Omega} \sigma_{ij}^{\infty} dD. \tag{42.21}$$

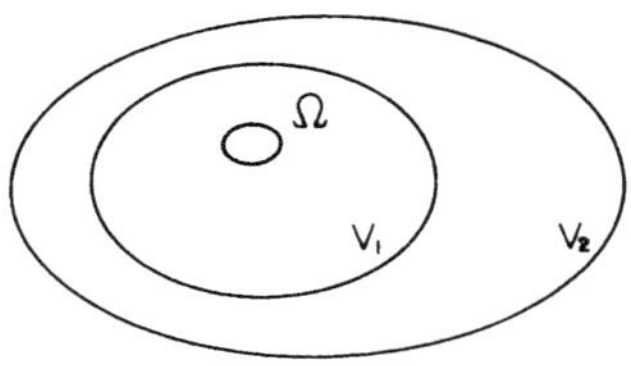

Fig. 42.2. Domains Ω, V_1 and V_2 have the same shape and orientation.

Because of (42.20), the second term vanishes; thus

$$\int_D \sigma_{ij}^{\infty} dD = \int_\Omega \sigma_{ij}^{\infty} dD. \tag{42.22}$$

Equation (42.22) shows that σ_{ij}^{∞} does not possess the property (42.3) of internal stresses in a finite body and is based on the fact that σ_{ij}^{∞} does not satisfy the traction free condition $\sigma_{ij}^{\infty} n_j = 0$ on $|D|$. Thus, one must add the correction term σ_{ij}^{I} to σ_{ij}^{∞} to obtain the exact solution for the internal stress σ_{ij},

$$\sigma_{ij} = \sigma_{ij}^{\infty} + \sigma_{ij}^{I}, \tag{42.23}$$

where $\sigma_{ij,j}^{I} = 0$ in D and $\sigma_{ij}^{I} n_j + \sigma_{ij}^{\infty} n_j = 0$ on $|D|$. σ_{ij}^{I} is called the image stress. The determination of this image stress is generally a difficult problem. However, one can estimate its average value from (42.3), (42.23), and (42.22). This average is

$$\langle \sigma_{ij}^{I} \rangle_D = (1/V) \int_D \sigma_{ij}^{I} dD = -(1/V) \int_D \sigma_{ij}^{\infty} dD = -(1/V) \int_\Omega \sigma_{ij}^{\infty} dD, \tag{42.24}$$

where V is the volume of D.

When the eigenstrain ϵ_{ij}^{*} is uniform in Ω, (42.24) becomes

$$\langle \sigma_{ij}^{I} \rangle_D = -(\Omega/V) \sigma_{ij}^{\infty}(\Omega) = -(\Omega/V) C_{ijkl} \{ S_{klmn}(\Omega) \epsilon_{mn}^{*} - \epsilon_{kl}^{*} \}. \tag{42.25}$$

$\langle \sigma_{ij}^{I} \rangle_D$ is small compared to the magnitude of σ_{ij}^{∞}. Nevertheless, when σ_{ij}^{I} and σ_{ij}^{∞} are integrated over the whole body, they attain values of the same order.

We can obtain a result similar to (42.25) for strain. Let

$$\epsilon_{ij} = \epsilon_{ij}^{\infty} + \epsilon_{ij}^{I}, \tag{42.26}$$

where ϵ_{ij}^{I} is the image strain due to the presence of the free surface. Equation (42.20.1) leads to

$$\langle \epsilon_{ij}^{\infty} \rangle_D = (1/V) \int_D \epsilon_{ij}^{\infty} dD = (1/V) \int_\Omega \epsilon_{ij}^{\infty} dD = (\Omega/V) S_{ijkl}(\Omega) \epsilon_{kl}^{*}, \tag{42.27}$$

while from (42.6) we have

$$\langle \epsilon_{ij} \rangle_D = \langle \epsilon_{ij}^{\infty} \rangle_D + \langle \epsilon_{ij}^{I} \rangle_D = (\Omega/V) \epsilon_{ij}^{*}. \tag{42.28}$$

Therefore, combining these results, we have

$$\langle \epsilon_{ij}^{I} \rangle_D = -(\Omega/V) \{ S_{ijkl}(\Omega) \epsilon_{kl}^{*} - \epsilon_{ij}^{*} \} \tag{42.29}$$

which can be obtained directly from (42.25) by inspection.

The importance of the image stress and strain has been discussed by

Eshelby (1954, 1956) in connection with macroscopic deformations and average changes in the lattice constants.

Random distribution of inclusions—Mori and Tanaka's theory

Seldom is only a single precipitate formed in a material. For example, when the volume fraction of precipitates is 0.001 and their size is 100 Å, some 10^{15} precipitates are present in a unit volume. Even if the stress field due to an individual precipitate were known, including the image stress term, it would take many years to add up the stresses due to all these precipitates. Moreover, the positions of individual precipitates are generally not known.

However, the average stress in the matrix and in the precipitates (inclusions) as well as the elastic energy are easily calculated when the precipitates are randomly distributed. Sometimes all these quantities are needed to discuss the macroscopic properties.

For simplicity, homogeneous ellipsoidal inclusions, with uniform eigenstrain ϵ^*_{ij}, are treated first. Let us formally define the average stresses in the matrix as $\langle\sigma_{ij}\rangle_M$ and those in the inclusions as $\langle\sigma_{ij}\rangle_I$. Equation (42.3) can then be expressed as

$$f\langle\sigma_{ij}\rangle_I + (1-f)\langle\sigma_{ij}\rangle_M = 0, \tag{42.30}$$

where f is the volume fraction of the inclusions. Now let us introduce a new single inclusion. The number of inclusions is so large that this addition does not affect f. Since random distribution of the inclusions is assumed, we will insert the new inclusion randomly in the matrix. Then, within the inclusion (or within any inclusion) the stress becomes

$$\sigma_{ij} = \sigma^\infty_{ij} + \langle\sigma_{ij}\rangle_M, \tag{42.31}$$

where σ^∞_{ij} is the stress calculated for a single inclusion present in an infinite medium:

$$\sigma^\infty_{ij} = C_{ijkl}(S_{klmn}\epsilon^*_{mn} - \epsilon^*_{kl}), \tag{42.32}$$

and $\langle\sigma_{ij}\rangle_M$ is the stress already present in the matrix. Since the new inclusion can be inserted at any place in the matrix, the average of σ_{ij} must be equal to $\langle\sigma_{ij}\rangle_I$. The average of (42.31) over all the positions available for inclusions is

$$\langle\sigma_{ij}\rangle_I = \sigma^\infty_{ij} + \langle\sigma_{ij}\rangle_M. \tag{42.33}$$

Substituting (42.33) into (42.30), the following equations are obtained:

$$\langle\sigma_{ij}\rangle_M = -f\sigma_{ij}^{\infty}, \tag{42.34}$$

$$\langle\sigma_{ij}\rangle_I = \sigma_{ij}^{\infty} - f\sigma_{ij}^{\infty}. \tag{42.35}$$

It should be noted that $-f\sigma_{ij}^{\infty}$ in the above equations is the average of the sum of σ_{ij}^{∞} and σ_{ij}^{I} caused by all the inclusions.

The elastic strain energy per unit volume of the body is written as

$$W^* = -(1/2)\epsilon_{ij}^* \int_{\Omega} \sigma_{ij} dD/V = -(1/2)\epsilon_{ij}^*\langle\sigma_{ij}\rangle_I f$$

$$= -(1/2)f(1-f)\sigma_{ij}^{\infty}\epsilon_{ij}^* \tag{42.36}$$

from (42.35) and (25.2) or (13.3). It is extremely useful that the average stress and the elastic energy can be written in such simple forms when the inclusions are randomly distributed. This fact has been found by Mori and Tanaka (1973). Brown (1973) has considered the same problem and has obtained a slightly different result.

A parallel discussion can be given in the situation when the eigenstrain is not uniform in an inclusion. In such a case, (42.33) can be written as

$$\langle\sigma_{ij}\rangle_I = \langle\sigma_{ij}^{\infty}\rangle_I + \langle\sigma_{ij}\rangle_M \tag{42.37}$$

with

$$\langle\sigma_{ij}^{\infty}\rangle_I = (1/\Omega)\int_{\Omega} \sigma_{ij}^{\infty} dD, \tag{42.37.1}$$

where σ_{ij}^{∞} is the stress for a single inclusion present in an infinite medium and Ω is the volume of the inclusion. Then, (42.30) and (42.37) lead to

$$\langle\sigma_{ij}\rangle_M = -f\langle\sigma_{ij}^{\infty}\rangle_I, \tag{42.38}$$

$$\langle\sigma_{ij}\rangle_I = (1-f)\langle\sigma_{ij}^{\infty}\rangle_I. \tag{42.38.1}$$

From (42.31) the stress inside a representative inclusion can be written as

$$\sigma_{ij} = \sigma_{ij}^{\infty} - f\langle\sigma_{ij}^{\infty}\rangle_I. \tag{42.39}$$

When the inclusions are inhomogeneous with eigenstrains ϵ_{ij}^p, the treatment may be slightly modified. Let us assume, for simplicity, that ϵ_{ij}^p is uniform in the inhomogeneous inclusions. The equation (42.30) is still valid. The average stress in the matrix can be written as

$$\langle\sigma_{ij}\rangle_M = C_{ijkl}e_{kl}^0, \tag{42.40}$$

where e^0_{kl} is to be determined. Let us introduce a new single inhomogeneous inclusion into the matrix, as before. Then, the stress in this inclusion can be determined by the equivalent inclusion method (22.13),

$$\sigma_{ij} = C_{ijkl}(e^0_{kl} + S_{klmn}\epsilon^{**}_{mn} - \epsilon^{**}_{kl}) = C^*_{ijkl}(e^0_{kl} + S_{klmn}\epsilon^{**}_{mn} - \epsilon^p_{kl}), \tag{42.41}$$

where C^*_{ijkl} are the elastic moduli of the inclusions. The stresses σ_{ij} in (42.41) can be assumed to be $\langle\sigma_{ij}\rangle_I$. Then, (42.30) leads to

$$fC_{ijkl}(e^0_{kl} + S_{klmn}\epsilon^{**}_{mn} - \epsilon^{**}_{kl}) + (1-f)C_{ijkl}e^0_{kl} = 0 \tag{42.42}$$

or

$$fC_{ijkl}(S_{klmn}\epsilon^{**}_{mn} - \epsilon^{**}_{kl}) + C_{ijkl}e^0_{kl} = 0. \tag{42.43}$$

Equations (42.41) and (42.43) are necessary and sufficient to determine 12 unknowns e^0_{ij} and ϵ^{**}_{ij}.

If $f \ll 1$, the following approximation is admissible. Setting $\epsilon^{**}_{ij} = \epsilon^*_{ij} + \epsilon^{*\prime}_{ij}$,

$$(\sigma^\infty_{ij} \equiv) C_{ijkl}(S_{klmn}\epsilon^{*\prime}_{mn} - \epsilon^{*\prime}_{kl}) = C^*_{ijkl}(S_{klmn}\epsilon^{*\prime}_{mn} - \epsilon^p_{kl}) \tag{42.44}$$

and

$$(\sigma'_{ij} \equiv) C_{ijkl}(e^0_{kl} + S_{klmn}\epsilon^*_{mn} - \epsilon^*_{kl}) = C^*_{ijkl}(e^0_{kl} + S_{klmn}\epsilon^*_{mn}), \tag{42.45}$$

where $\epsilon^{*\prime}_{ij}$ and ϵ^*_{ij} are the respective equivalent eigenstrains for ϵ^p_{ij} and e^0_{ij}.

It is easy to solve (42.44) for $\epsilon^{*\prime}_{ij}$ under given ϵ^p_{ij}. We assume that $\epsilon^*_{ij} \ll \epsilon^{*\prime}_{ij}$. Then, (42.43) leads to

$$\langle\sigma_{ij}\rangle_M = C_{ijkl}e^0_{kl} \approx -f\sigma^\infty_{ij}. \tag{42.46}$$

Substituting e^0_{ij} obtained from (42.46) into (42.45), we can derive approximate values of ϵ^*_{ij}. Then, (42.41) can be written as

$$\langle\sigma_{ij}\rangle_I = \sigma^\infty_{ij} + \sigma'_{ij}. \tag{42.47}$$

It is apparent that when $f \ll 1$, (42.38) is also a good approximation for the situation where ϵ^p_{ij} is not uniform in an inclusion. The discussion of this problem is similar to that developed above.

43. Work-hardening of dispersion hardened alloys

Metallurgists invented a dispersion strengthened alloy whose matrix is a plastically soft and ductile material and throughout which particles of plastically strong second phases are dispersed. Sometimes these particles are called nonshearable, implying that the particles cannot be cut by the crystal dislocations (glide dislocations). These particles thus hinder the dislocation motion and thereby raise the yield stress. Macroscopic yielding in a dispersion strengthened alloy is explained by the Orowan mechanism (see Ashby 1966); dislocations bowing out between particles. The critical stress for this bowing out can be calculated by using the results (36.15) or (36.25). For recent developments of this problem, the reader is referred to Bacon et al. (1973) and Scattergood and Bacon (1975). When dislocations bow out, dislocation loops, called Orowan loops, are left around the particles. The internal stresses from these Orowan loops affect the motion of the dislocations. Macroscopically, this effect is manifested by work-hardening. One approach used to estimate work-hardening is the balancing of forces acting on the glide dislocations bowing out between two neighboring particles, where one force comes from the applied stress, one from the glide dislocation itself, and one from the internal stresses caused by the Orowan loops. An approximate method using this assumption has been employed by (Hart 1972). However, this type of approach depends critically on the model adopted, the position of the loops around a particle, as well as the number of the particles considered. A particular loop may hinder or assist the motion of a dislocation, depending on the relative position of the loop and the dislocation. It is, therefore, desirable to take another approach, with which the estimation of work-hardening can be made imdependently of the choice of a detailed model. The following sections are devoted to this purpose and are mainly applications of the analysis of the elastic fields produced by an ellipsoidal inclusions.

Work-hardening in simple shear

Let us consider the situation where the matrix deforms plastically but the inclusions do not. For simplicity, the shape of the inclusions is assumed to be spherical and the plastic deformation in the matrix is assumed to be uniform. As an example, the plastic deformation is assumed to occur in simple shear; the slip plane is thus perpendicular to the x_3 direction, while the slip direction is parallel to the x_1 direction. Let the plastic strain in the matrix be $\epsilon_{31}^{p} = \gamma_p/2$. ($\gamma_p$ is the glide strain in the slip system.) Superimpose $\epsilon_{31}^{p} = -\gamma_p/2$ uniformly throughout the specimen. Since the uniform plastic strain over the entire body does not produce internal stress, the internal stress induced by the prescribed plastic deformation in the matrix is calculated by considering the situation where

$$\begin{aligned} \epsilon^{p}_{31} &= 0, && \text{in the matrix,} \\ \epsilon^{p}_{31} &= -\gamma_p/2, && \text{in the inclusions.} \end{aligned} \tag{43.1}$$

From (22.13a) and (22.19), the stress in an inclusion is given by

$$\sigma^{B}_{31} = \mu\mu^*(1-2S)\gamma_p/g \tag{43.2}$$

when the inclusion is isolated in an infinite matrix. Here μ and μ^* are the shear moduli of the matrix and the inclusion, respectively,

$$g = 2(\mu^* - \mu)S + \mu, \tag{43.3}$$

and

$$S = S_{3131} = (4-5\nu)/\{15(1-\nu)\}, \tag{43.4}$$

where ν is the Poisson ratio of the matrix. From (43.1), (43.2) and (25.3), the elastic strain energy due to an inclusion is calculated as

$$W^* = (1/2)V_0\mu\mu^*(1-2S)\gamma_p^2/g, \tag{43.5}$$

where V_0 is the volume of the inclusion. When many inclusions coexist with a volume fraction f, the elastic strain energy per unit volume can be approximated by

$$\bar{W}^* = (1/2)f\mu\mu^*(1-2S)\gamma_p^2/g, \tag{43.6}$$

provided f is much smaller than 1. The otherwise uniform applied stress, $\sigma^0_{31} = \sigma^0$, is disturbed by the inhomogeneity in the elastic constant of the inclusions. The applied stress in the inclusion is calculated as

$$\sigma^{A}_{31} = \sigma^0\{1 + (\mu^* - \mu)(1-2S)/g\} \tag{43.7}$$

from (22.7). Thus, the change in the external potential per unit volume $\bar{V}^*$ is given by (25.40),

$$\bar{V}^* = -\sigma^0\gamma_p + f\sigma^0\{1 + (\mu^* - \mu)(1-2S)/g\}\gamma_p. \tag{43.8}$$

The Gibbs free energy (25.41) per unit volume of the specimen, $\Delta\bar{W}$, is the sum of $\bar{W}^*$ and $\bar{V}^*$,

$$\Delta\bar{W} = (1/2)f\mu\mu^*(1-2S)\gamma_p^2/g - \sigma^0\gamma_p + f\sigma^0\{1 + (\mu^* - \mu)(1-2S)/g\}\gamma_p. \tag{43.9}$$

The specimen is in stable equilibrium when $\partial\Delta\bar{W}/\partial\gamma_p = 0$ and $\partial^2\Delta\bar{W}/\partial\gamma_p^2 > 0$. That is, the stable state is achieved by

$$\sigma^0\{(1-f) - f(\mu^* - \mu)(1-2S)/g\} = f\mu\mu^*(1-2S)\gamma_p/g. \tag{43.10}$$

Since $f \ll 1$ is assumed, (43.10) can be written as

$$\sigma^0 = f\mu\mu^*(1-2S)\gamma_p/g, \tag{43.11}$$

which represents the equation relating the applied stress and the plastic strain (the constitutive equation) as is schematically shown by the dotted line in Fig. 43.1.

Since a finite stress is needed to start the plastic deformation, one has to raise the curve to the solid line,

$$\sigma^0 = \sigma_y + f\mu\mu^*(1-2S)\gamma_p/g. \tag{43.12}$$

The physical reason for this is as follows: When the increase in γ_p is $\delta\gamma_p$, the increase in $\bar{W}^*$ is $f\mu\mu^*(1-2S)\gamma_p\delta\gamma_p/M$, and the work done by the applied stress (negative of the external potential increase, δU) is $\sigma^0\delta\gamma_p$. At the same time, the increase in γ_p in the matrix accompanies the energy dissipation $\delta D^* \approx \sigma_y\delta\gamma_p$. The energy conservation law, $\delta U = \delta\bar{W}^* + \delta D^*$, yields the solid line in Fig. 43.2, which indicates linear work-hardening after yielding. Yielding by the Orowan mechanism is typical of energy dissipation. Thus, σ_y can be identified as the critical stress for the operation of the Orowan mechanism. The above method of treating the plastic deformation of dispersion-strengthened alloys appeared in the paper by Tanaka and Mori (1970). Experimentally, linear work-hardening is usually observed in a dispersion-hardened alloy when the strain is small and the deformation occurs at a low temperature, where the diffusion effect can be ignored. An example (Tokushige, 1976) is given in Fig. 43.2, in which the solid line is the experimentally observed stress-strain curve of a Cu–SiO_2 single crystal and the dotted line is calculated from (43.11). The SiO_2 particle is known to have a

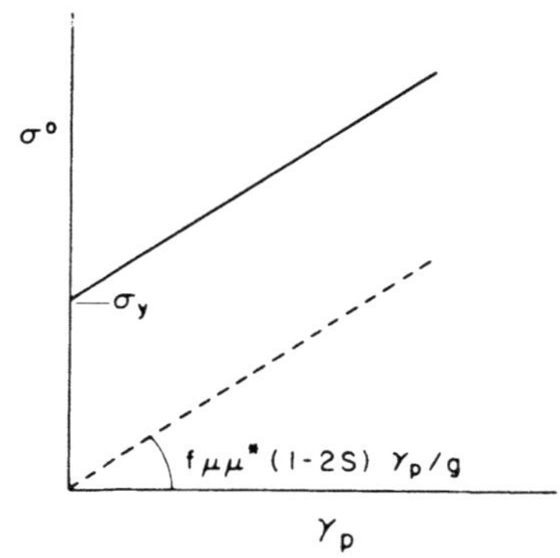

Fig. 43.1. Theoretical lines of linear work hardening due to particles with volume fraction f.

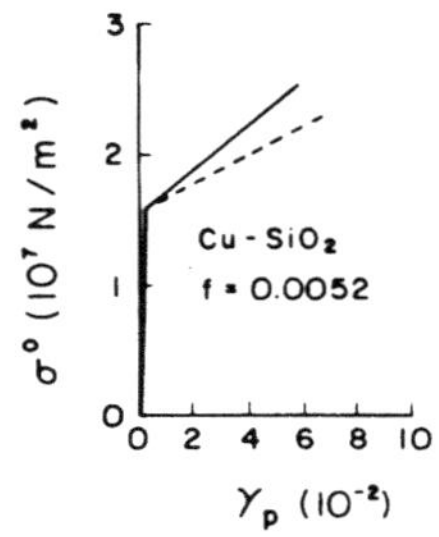

Fig. 43.2. The solid line is the experimentally observed stress-strain curve and the dotted line is calculated from (43.11) (after Tokushige 1976).

spherical shape in this alloy. Thus, $S = (4 - 5\nu)/\{15(1 - \nu)\}$. Here $f = 0.0052$, $\mu = 4.61 \times 10^{10}\,\mathrm{N/m^2}$, $\mu^* = 3.13 \times 10^{10}\,\mathrm{N/m^2}$ and $\nu = 0.33$. The calculation accounts for 65% of the observed hardening.

It is easy to extend the above calculation to other shapes of inclusions as well as other modes of plastic deformation such as uniaxial deformation as shown already by (25.58) (Tanaka and Mori, 1970). The elastic anisotropy of the matrix and the inclusions can also be taken into account (Lin et al., 1973). The anisotropy in plastic deformation induced by the asymmetrical arrangement of inclusions in a composite material has been discussed with an approach similar to the above method (Tanaka et al., 1973).

Brown and Stobbs (1971), and Brown and Clarke (1975) have taken a different approach to derive the hardening rate, a method essentially identical to (43.11), although their paper contains some misleading interpretations. These have been corrected later by Mori and Tanaka (1973). From (42.46) and (43.2), the average internal stress in the matrix is given as

$$\langle\sigma_{31}\rangle_M = -f\mu\mu^*(1 - 2S)\gamma_p/g, \tag{43.13}$$

after the matrix has been plastically deformed by γ_p. Thus, to continue the plastic deformation, one has to raise the applied stress by

$$\Delta\sigma_{31}^0 = -\langle\sigma_{31}\rangle_M \tag{43.14}$$

to overcome the internal stress. Note that (43.14) is identical to (43.11). Both the stress fields due to the inclusions in the infinite matrix and the image stresses due to the free surface effect contribute to $\langle\sigma_{31}\rangle_M$. $-\langle\sigma_{ij}\rangle_M$ is sometimes called the back stress. It is clear that $\langle\sigma_{ij}\rangle_M$ hinders the progress of plastic deformation, and aids the plastic flow in the reverse direction. Of course, one can see this from the preceding energy consideration. This property should be manifested in the Bauschinger effect. Conversely, the existence and magnitude of the back stress can be examined from the measurement of the Bauschinger effect. Some studies have been done along this line (Atkinson et al., 1974, Gould et al., 1974, and Mori and Narita, 1975). The magnitude of the back stress, $-\langle\sigma_{31}\rangle_M$, determined from the difference of the flow curves between forward and reverse strainings, is well accounted for by (43.13) when strain is small.

Let us examine (43.10) and (43.11) further. These equations account for the interaction among inclusions and for the free surface effect. The procedure mentioned in the preceding section, (42.40) through (42.47), can be used to improve these approximations. Then, the average internal stress in an inclusion is evaluated as

$$\langle\sigma_{31}\rangle_I = \mu\mu^*(1-2S)\gamma_p\{(1-f)-f(\mu^*-\mu)(1-2S)/g\}g, \tag{43.15}$$

and $\bar{W}^*$ now is

$$\bar{W}^* = (1/2)f\mu\mu^*(1-2S)\gamma_p^2\{(1-f)-f(\mu^*-\mu)(1-2S)/g\}/g. \tag{43.16}$$

Writing $\Delta\bar{W}$ with (43.8) and (43.16), we obtain (43.11) instead of (43.10), as an equation satisfying the equilibrium condition.

Dislocations around an inclusion

Here we discuss the dislocation distribution around an inclusion when $\epsilon_{31}^p = \gamma_p/2$ is uniform in the matrix and $\epsilon_{ij}^p = 0$ in the inclusion. For simplicity, assume

$$\begin{cases} \beta_{31}^p = \gamma_p & \text{in the matrix,} \\ \beta_{ji}^p = 0 & \text{in the inclusions.} \end{cases} \tag{43.17}$$

Since there is a jump in β_{ji}^p at the matrix-inclusion interface, the surface dislocations α_{hi} defined by (37.16) exist at the interface. Let $\mathbf{n}$ be the unit normal vector at the interface directed toward the matrix. Then,

$$\begin{aligned} \alpha_{11} &= -\gamma_p n_2, \\ \alpha_{21} &= \gamma_p n_1. \end{aligned} \tag{43.18}$$

All other components are zero. This means that the infinitesimal glide loops, which have the Burgers vectors parallel to the loop planes, are present at the interface (see Fig. 37.2 and Fig. 43.3(a)). The total Burgers vector of the surface dislocations crossing the unit length of the line element perpendicular to $\boldsymbol{\nu}$ and lying on the interface, becomes

$$B_1 = \alpha_{11}\nu_1 + \alpha_{21}\nu_2 = \gamma_p(1-n_3^2)^{1/2}, \tag{43.19}$$

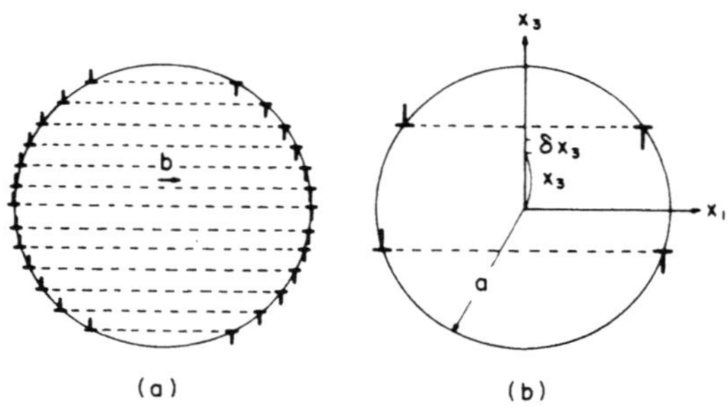

Fig. 43.3. Surface dislocations around a particle.

since

$$\begin{aligned}\nu_1 &= -n_2/(n_1^2+n_2^2)^{1/2},\\ \nu_2 &= n_1/(n_1^2+n_2^2)^{1/2},\\ \nu_3 &= 0, \quad n_1^2+n_2^2+n_3^2=1.\end{aligned} \tag{43.20}$$

This result (43.19) implies that the density of the dotted lines in Fig. 43.3(a) is a constant equal to γ_p along the x_3-axis.

With the dislocation distribution defined previously, another physical explanation can be given for the work-hardening derived in this section. Physically, the plastic deformation is caused by the motion of the crystal dislocations with a finite Burgers vector. Thus, the situation described by (43.17) does not generally occur literally, say, at the atomic dimension. All that (43.17) implies in the case of crystal plasticity is that slip occurs randomly, and macroscopically uniformly. Here, "macroscopic" may be defined in such a manner that uniformity is discussed in comparison with the inclusion dimension. For example, (43.19) shows that the number n of the (crystal) Orowan loops surrounding a particle is, on the average, given by

$$n = 2a\gamma_p/b, \tag{43.21}$$

and that these loops are randomly distributed around the particle. Here, b is the magnitude of the crystal dislocations Burgers vector responsible for slip, and a is the radius of the spherical inclusion. This situation is schematically shown in Fig. 43.3(b). When the inclusions are randomly distributed, it is apparent that the position of the Orowan loops around a particular inclusion is also random. Let Ndx_3 be the number of the Orowan loops per unit volume, situated between x_3 and x_3+dx_3 with respect to the center of an inclusion (see Fig. 43.3). N is called the

Orowan loop density. Then, uniformity of the plastic deformation in the matrix means that N is constant related to γ_p by

$$\int_{-a}^{a} N dx_3 = nN_p \tag{43.22}$$

or

$$\gamma_p = Nb/N_p, \tag{43.23}$$

where N_p is the number of inclusions per unit volume.

Now let us examine the stress field due to these Orowan loops. Consider the average of the internal stress in the matrix. Since there are many loops, this average remains almost unchanged when a loop is transferred from one particle to another. We repeat this until some particles are uniformly covered by the loops (other particles may not have any loop), then, divide these loops into the loops with an infinitesimal Burgers vector and redistribute these infinitesimal loops to all the particles uniformly. It is clear that, in the final stage, the infinitesimal loops are distributed uniformly as the surface dislocations described in (43.18) or (43.21), and that the average internal stress in the matrix is the same as that in the original distribution of the Orowan loops with a finite Burgers vector. In short, the work-hardening discussed in this section is that brought about by the average internal stress due to the Orowan loops. For the discrete dislocation approach, it is difficult to conduct a quantitative calculation. Such difficulty can be avoided by using the stress field of an inclusion, as has been suggested by Mori and Tokushige (1977).

When the situation arises that the Orowan loop distribution density N is not constant but is a function of x_3, the average internal stress in the matrix may still be calculated by the above procedure. The average internal stress in the matrix can be reproduced by giving the eigenstrain

$$\epsilon_{31}^{p} = -N(x_3)b/2N_p \tag{43.24}$$

to every particle. Then, the solutions of Sections 19 and 20 can be used together with (42.37.1). Such a situation will be discussed in the next section.

Uniformity of plastic deformation

One of the reasons that the straightforward evaluation of work-hardening in dispersion strengthened materials is possible is the assumption of uniform plastic deformation throughout the matrix. In the context of the approach taken in the present section, this assumption is justified, as shown below.

For simplicity, let us assume $f \ll 1$ and that the plastic strain in the matrix ϵ_{31}^{p} varies only along the x_3-direction, that is,

$$\epsilon_{31}^{p} = \gamma_p(x_3)/2. \tag{43.25}$$

Neglecting the interaction among the inclusions and the disturbance of the applied stress $\sigma_{31}^0 = \sigma^0$ due to the inhomogeneity of the inclusions, one can write the Gibbs free energy per unit volume of the specimen as

$$\bar{W} = \frac{1}{2}\int_{D_3} fA\mu\{\gamma_p(x_3)\}^2 dx_3 - \int_{D_3} \sigma^0\gamma_p(x_3)dx_3 \tag{43.26}$$

with

$$A = \mu^*(1-2S)/g. \tag{43.27}$$

Here, energy dissipation is not considered and D_3 indicates the integral domain with respect to x_3. The first term in (43.26) represents the elastic strain energy (see (43.6)). The second term arises from the external potential. It is seen that (43.26) takes a minimum when

$$fA\mu\gamma_p(x_3) = \sigma^0. \tag{43.28}$$

That is, the stable state is achieved when the specimen is uniformly covered by the plastic domain in which the matrix deforms plastically with constant γ_p determined by σ^0 in (43.28). As mentioned before, uniform slip cannot occur in a literal sense, since crystalline materials deform plastically by movement of crystal dislocations. Therefore, equation (43.28) should be interpreted in such a manner that the slip lines, each of which is produced by a single dislocation movement, cover the specimen uniformly, i.e., the slip line spacing becomes as small as possible. With this conclusion and the consideration of the dislocation configuration around the inclusion discussed previously, the procedure to evaluate the work hardening is justified. The above procedure and conclusions are identical to those given by Mori and Mura (1976).

44. Diffusional relaxation of internal and external stresses

Let us consider the elastic deformation of an infinite body due to σ_{31}^0. When the body contains an inhomogeneity, the stress field is disturbed. When the inhomogeneous particle is harder than the matrix, the interface boundary of the matrix (dotted line) and the particle (solid line) deform as shown in Fig. 44.1, if the two domains deform freely without

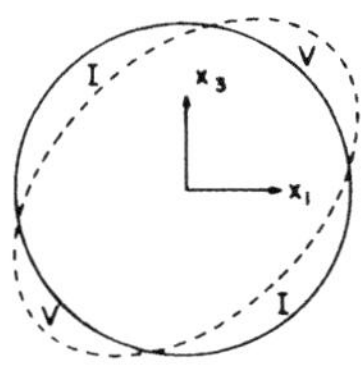

Fig. 44.1. When an inhomogeneous particle is harder than a matrix, the interface boundary of the matrix (dotted line) and the particle (solid line) deform as shown here by σ^0_{31} if no constraint exists from the surroundings.

any constraint from the surroundings. The internal stress (stress disturbance) can be interpreted as the eigenstress caused by the eigenstrain corresponding to the misfit (the difference between the dotted domain and solid domain in Fig. 44.1). If interfacial diffusion takes place on the particle surface, mass must be transported from region I to region V. The misfit would therefore disappear, and the internal stress and elastic energy would decrease; this is diffusional relaxation. Similarly, diffusional relaxation takes place when the matrix is subjected to plastic deformation but the inclusion is not.

In this section, some experimental evidence of diffusional relaxation will be mentioned. Next, the relaxation rate will be discussed and the creep deformation will be considered as a balance between the accumulation and the relaxation of the internal stress. The diffusional relaxation of the external stress will be also discussed.

Relaxation of the internal stress in a plastically deformed dispersion strengthened alloy

Suppose that plastic deformation $\epsilon^p_{31} = \gamma_p/2$ occurs only in the matrix. The internal stress in this state is caused by the eigenstrain given in (43.1), which can be understood from Fig. 44.1. If a complete relaxation occurs by a diffusional process which transports mass from region I to region V in Fig. 44.1, then the eigenstrain in the inclusion would be reduced by

$$\Delta\epsilon^p_{31} = \gamma_p/2 \tag{44.1}$$

and the resulting eigenstrain in the inclusion would become zero. In that case, work-hardening is lost by the amount given by (43.11); the flow stress decreases correspondingly. Figure 44.2 shows the flow stress decrease of Cu–SiO_2 crystals ($f = 0.0052$) work-hardened at 77°K for 20 min. annealing at 473°K (Mori and Tokushige 1977). The dotted line is

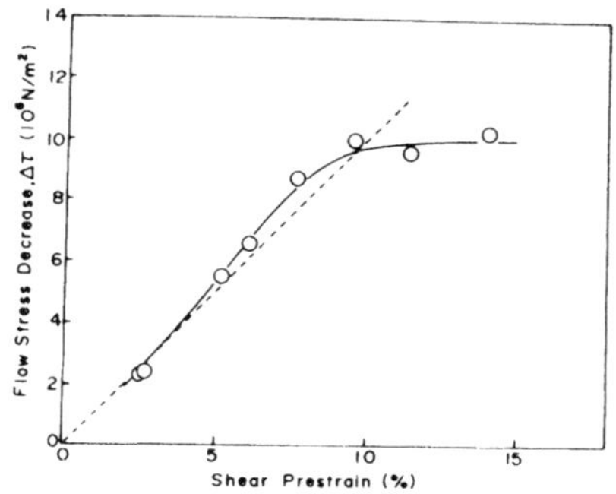

Fig. 44.2. The flow stress decreases by diffusion after prestraining (after Mori and Tokushige 1977).

drawn according to (43.11). Except for the high strain region, the agreement between the prediction and the observation is good. Metallurgically, the deviation at the high strain region indicates that some relaxation processes which do not need diffusion have occurred.

The diffusional relaxation mentioned in this section can be further checked. The additional eigenstrain, (44.1), should be observed by the macroscopic deformation. By using (42.8), (42.12) and (22.19), the macroscopic strain increment $\langle\epsilon_{31}\rangle_D$ due to $\Delta\epsilon^p_{31}$ in the inclusions can be calculated as

$$\langle\epsilon_{31}\rangle_D = f\epsilon^{**}_{31} \tag{44.2}$$

with

$$\epsilon^{**}_{31} = \mu^*\Delta\epsilon^p_{31}/g, \tag{44.3}$$

where g is defined in (43.3). From (43.11) and (44.1) through (44.3), the loss of the work hardening, $\Delta\sigma^0$, is related to the macroscopic strain observed in relaxation annealing. It becomes

$$\Delta\sigma^0 = 2\mu(1-2S)\langle\epsilon_{31}\rangle_D. \tag{44.4}$$

This has been tested in Cu–SiO_2 crystals as shown in Fig. 44.3 (Mori and Narita 1975). The specimens were prestrained and annealed. The change in flow stress, $\Delta\sigma^0$, and the change in specimen length (corresponding to $\langle\epsilon_{31}\rangle_D$) due to annealing were measured. We assume that the annealing causes $\Delta\epsilon^p_{31}$ in (44.3). The theoretical values from (44.4) and the corresponding measured values are plotted by the filled circles and the triangles, respectively. These are in good agreement. The open circles represent the change in the back stress, $-\langle\sigma_{31}\rangle_M$, due to annealing. These

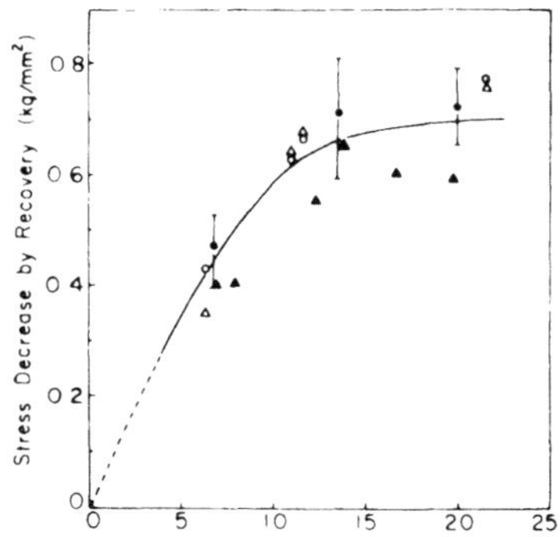

Fig. 44.3. The decrease in the back stress due to the recovery against the prestrain (after Mori and Narita 1975). The axis of abscissa is shear prestrain (0/0).

changes were determined by Bauschinger tests before and after annealing. $-\langle\sigma_{31}\rangle_M$ is calculated as half of the difference of the flow stresses between forward and reverse straining. The changes in $-\langle\sigma_{31}\rangle_M$ also compare well with $\Delta\sigma^0$. The saturation tendency of $\Delta\sigma^0$ with prestraining is consistent with Fig. 44.2. Since the specimens were considered to have annealed fully, the change in back stress by annealing is equal to the back stress present before annealing. The saturation of the back stresses at high prestrains in Fig. 44.3 agrees with an observation by Atkinson et al. (1974).

Diffusional relaxation process, climb rate of an Orowan loop

There are three types of diffusion: volume, interfacial and pipe diffusion. The last one is caused by mass transport along the core of a dislocation, where diffusion is fast with a low activation energy because of the severely disturbed atomic arrangement. In this section we employ the pipe diffusion as a cause of climb of the Orowan loops. The interfacial diffusion will be examined in connection with stress relaxation in later sections.

The climb process of the Orowan loop is explained by Fig. 44.4. If atoms flow from the left-hand side of an Orowan loop to the right-hand side through its own pipe, the Orowan loop will climb up. Let us calculate the number of atoms, δN, diffusing when an Orowan loop with the Burgers vector $\mathbf{b}$ climbs from x_3 to $x_3 + \delta x_3$. The particles are assumed to be spherical with radius a. Since the thickness of an extra plane defined by the edge component of a dislocation is expressed by the magnitude of $\boldsymbol{\nu} \times \mathbf{b} = (b \sin\theta)$ (see Fig. 44.5), where $\boldsymbol{\nu}$ is the unit vector

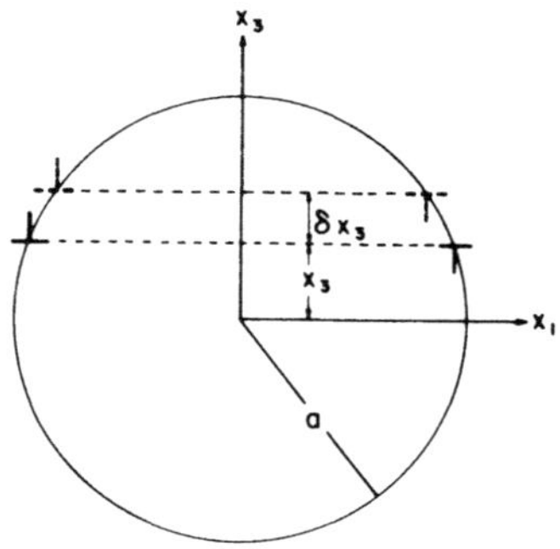

Fig. 44.4. The climb process of the Orowan loop.

along the dislocation, δN is calculated as

$$\delta N = (1/\Omega)\delta x_3 \int_0^{\pi} b \sin\theta (a^2 - x_3^2)^{1/2} d\theta = 2b(a^2 - x_3^2)^{1/2}\delta x_3/\Omega, \qquad (44.5)$$

where Ω is the atomic volume. The elastic energy of the Orowan loop W^* is given by

$$W^* = \alpha\mu b^2 2\pi (a^2 - x_3^2)^{1/2} \qquad (44.6)$$

(see, for example, Barnett 1976). Here, α is a constant of the order of unity and μ is the shear modulus. Thus, the change in the elastic energy δW^* is

$$\delta W^* = -2\pi\alpha\mu b^2 x_3 \delta x_3/(a^2 - x_3^2)^{1/2} \qquad (44.7)$$

when the loop climbs up by δx_3. The diffusion distance of the atoms, l, is approximated to be a constant given by

$$l = \pi(a^2 - x_3^2)^{1/2} \qquad (44.8)$$

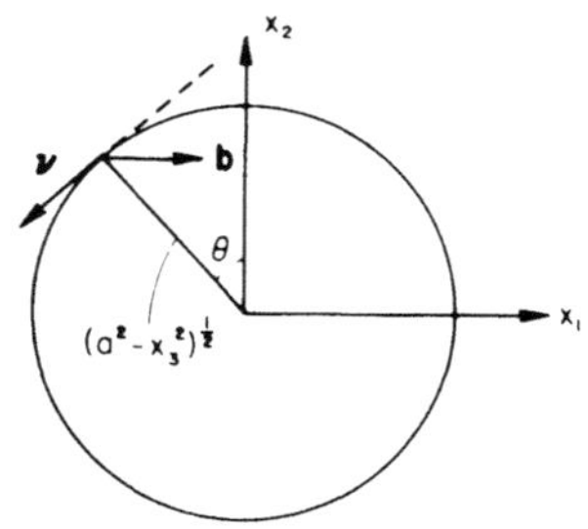

Fig. 44.5. Explanation for equation (44.5).

which is the average over the possible travelling distance. One can define the force F acting on an atom as $-(\delta W^*/l\delta N)$. Thus,

$$F = \alpha\mu b\Omega x_3(a^2 - x_3^2)^{-3/2}. \tag{44.9}$$

The Einstein relation gives the drift velocity v of an atom as

$$v = \alpha\mu b\Omega x_3(a^2 - x_3^2)^{-3/2}D_p/kT, \tag{44.10}$$

where D_p is the pipe diffusion constant along the dislocation, k is the Boltzmann constant, and T is the temperature. Let us define by h^2 the cross section of the dislocation pipe. During the time interval δt, atoms $\delta N = 2h^2 v\delta t/\Omega$ flow and consequently the Orowan loop climbs by δx_3 related to δN by (44.5). Thus, the climb rate, $V = dx_3/dt$, of the Orowan loops is given by

$$V = \frac{dx_3}{dt} = Ax_3(a^2 - x_3^2)^{-2} \tag{44.11}$$

with

$$A = \alpha\mu\Omega h^2 D_p/kT. \tag{44.12}$$

The above treatment of the climb rate of an Orowan loop is developed by Mori and Tokushige (1977). The treatment can be easily extended to the case when the inclusions are ellipsoidal (see Okabe and Mori 1979). Hazzledine and Hirsch (1974) have taken a different approach to analyze the climb process of the Orowan loops.

Suppose that the matrix deforms with a strain rate of $\dot{\epsilon}_{31}^p = \dot{\gamma}_p/2$. If slip occurs uniformly, from (43.23) the Orowan loop density distribution N increases by $\dot{\gamma}_p N_p/b$ per unit time, N_p being the number of inclusions per unit volume. At the same time, the loop density decreases with a rate of $-\partial(NV)/\partial x_3$. Thus, in total, one has a differential equation,

$$\frac{\partial N}{\partial t} = \dot{\gamma}_p N_p/b - \frac{\partial(NV)}{\partial x_3}. \tag{44.13}$$

When $\dot{\gamma}_p = 0$,

$$\frac{\partial N}{\partial t} = -\frac{\partial(NV)}{\partial x_3} \tag{44.14}$$

corresponds to annealing. It can be shown that the solution of (44.14) with the initial condition of $N(x_3, 0) = N_0$ is

$$N(x_3, t) = N_0 \frac{z(a^2 - x_3^2)^2}{x_3(a^2 - z^2)^2}, \tag{44.15}$$

where z is the root of

$$At = a^4 \log(x_3/z) - (x_3^2 - z^2)(4a^2 - x_3^2 - z^2)/4. \tag{44.16}$$

Mori and Mura (1978) have solved (44.16) numerically to obtain $N(x_3, t)$ which may be approximated by $N = N_0(m_0 + m_2x_3^2/a^2)$ if proper constants m_0 and m_2 are chosen. N_0 is related to the prestrain Υ_p which has induced work-hardening.

From (43.24), one can express the eigenstrain ϵ_{31}^p needed to calculate the back stress $-\langle\sigma_{31}\rangle_M$ in a polynomial form. Thus, by using the result of Section 19 and (22.13), one can calculate σ_{31}^∞, by which $-\langle\sigma_{31}\rangle_M$ is obtained from (42.38) by assuming $f \ll 1$. Mori and Mura (1978) have numerically calculated the back stress decrease of a work hardened Cu-SiO_2 alloy and compared it to the experimentally observed softening on annealing. Although the calculation reasonably accounts for the observation, it predicts the temperature range at which significant softening would occur to be narrower than the observed range. This has been corrected by Mori and Osawa (1979) who have taken the distribution of the particle size in an actual alloy into consideration. It is to be noted that the climb rate of the loop and thus the loop distribution brought about by the relaxation are sensitive to the particle size.

Recovery creep of a dispersion strengthened alloy

If the Orowan loop density distribution is in balance between loss due to climb, and gain due to the accumulation by plastic deformation, one would expect stationary creep. The Orowan loop density distribution corresponding to a stationary creep rate $\dot{\gamma}_p$ is obtained from

$$\frac{\partial(NV)}{\partial x_3} = \dot{\gamma}_p N_p/b, \tag{44.17}$$

since $\partial N/\partial t = 0$ in (44.13). Since V is given by (44.11), it is clear that the following equation satisfies (44.17):

$$N = \dot{\gamma}_p(N_p/b)(a^2 - x_3^2)^2/A \tag{44.18}$$

with A given by (44.12). From (43.24) and (44.18), the eigenstrain given to a particle for calculation of $\langle\sigma_{ij}\rangle_M$ is written as

$$\epsilon_{31}^p = -\dot{\gamma}_p(a^2 - x_3^2)^2/2A \tag{44.19}$$

which is in the form of (19.1). Thus, using the results of Section 19 (Asaro and Barnett 1975, Mura and Kinoshita 1978) and the equivalency

condition developed in Section 22, one finds the stress in the inclusion as

$$\sigma_{31}^{\infty} = \dot{\gamma}_p a^4 (B_0 + B_2^{pq} x_p x_q / a^2 + B_4^{pqrs} x_p x_q x_r x_s / a^4)/(2A), \tag{44.20}$$

when the inclusion is in an infinite matrix. The coefficients B_0, B_2^{pq} and B_4^{pqrs} are all computable and expressed in terms of the elastic constants of the matrix and the inclusion. Thus, by (42.38) and (42.37.1) one obtains the average of the internal stress in the matrix as

$$\langle \sigma_{31} \rangle_M = -\mu f \dot{\gamma}_p a^4 (A_0 - 2A_2 + A_4)/(2A), \tag{44.21}$$

where A_0, A_2 and A_4 are given by Okabe et al. (1980). Therefore, the applied stress $\sigma_{31}^0 = \sigma^0$ needed to maintain stationary creep deformation is given by

$$\sigma^0 = \sigma^D + \mu f \dot{\gamma}_p a^4 (A_0 - 2A_2 + A_4)/(2A), \tag{44.22}$$

where σ^D is the flow stress of the matrix (in absence of any particle).

The experimental observations relevant to (44.22), such as the existence of the stationary creep and the linear dependence of $\dot{\gamma}_p$ on σ^0 have recently been made (Mori and Osawa 1979, Okabe and Mori 1979, and Okabe et al. 1980). Figure 44.6 shows the observed stationary creep

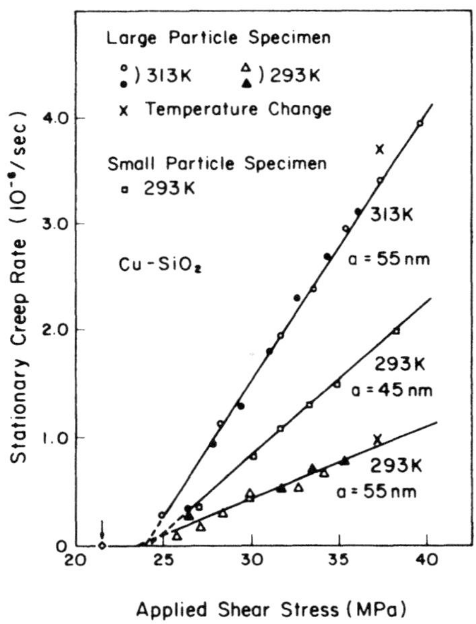

Fig. 44.6. Theoretical lines and experimental values of the stationary creep with parameters of particle sizes and testing temperatures.

rate plotted against the applied stress in a Cu–SiO_2 alloy (single crystals, $f = 0.005$ [Okabe et al. 1980]). In a certain range of σ^0, $\dot{\gamma}_p$ is linearly related to σ^0, in accordance with (44.22). One can also examine the particle size dependence of creep rates. The ratio of the slope of the creep rate for the specimen with $a = 45$ nm to that for the specimen with $a = 55$ nm is obtained as 2.17, which is in excellent agreement with $(55)^4/(45)^4 = 2.23$, as required by (44.22). Finally, let us compare the calculation with the absolute magnitude of the slope in Fig. 44.6. By conducting numerical calculations, one obtains

$$\sigma^0 = \sigma^D + 0.5838\mu f \dot{\gamma}_p a^4/(2A). \tag{44.23}$$

Here, μ is the shear modulus of the matrix (Cu, 4.61×10^{10} N/m^2). In the calculation, Poisson's ratio of Cu has been taken to be 0.33, and the shear modulus and Poisson's ratio of SiO_2, 3.13×10^{10} N/m^2 and 0.18, respectively. By assuming $\alpha = 1/2$, $\Omega = 11.8 \times 10^{-30}$ m^2, $h^2 = b^2$, $b = 2.56 \times 10^{-10}$ m, $D_p = b^2 \times 10^{12} \exp(-Q/kT)/s$, and $Q = 0.56$ ev, one obtains, from (44.23),

$$\dot{\gamma}_p/(\sigma^0 - \sigma^D) = 1.09 \times 10^{-7}/Mp_a s, \tag{44.24}$$

for $a = 55$ nm at $T = 293$°K. The activation energy Q has been determined from Fig. 44.6. The experimental value of the slope corresponding to this situation is about $0.66 \times 10^{-7}/Mp_a s$. This argument is satisfactory. It has been similarly found that (44.22) accounts for the observations of stationary creep in Al–Si crystals (Mori and Osawa 1979). Okabe and Mori (1979) have extended the above analysis to the situation of rod shaped inclusions in conjunction with the experiment on a Cu-mullite alloy. Again, in this case a satisfactory agreement is found between the calculation and the observation.

Interfacial diffusional relaxation

Relaxation of the internal stress induced by plastic deformation in a material with plastically strong inclusions has been discussed on the basis of the diffusion of atoms through the cores of the dislocations at the matrix-inclusion interface. This is an adequate picture, since the existence of these dislocations at the interface is responsible for the internal stress. However, if the stress is brought about by some other reason, it is better to seek another path of diffusion such as interfacial diffusion. Interfacial diffusion especially applies in the situation where external stress is present in a specimen which has a so-called incoherent inclusion. The origin of the stress is independent of the situation at the

interface and the atomic arrangement at the matrix-inclusion interface is in disorder, so the diffusion along the interface is presumably fast. Even if an interface is described by an array of dislocations, such as a tilt or twist grain boundary, and atoms are considered to diffuse through pipes of these dislocations, the diffusional process is adequately treated as interfacial diffusion. In the following, we will discuss the relaxation of an external stress due to diffusion at the matrix-inclusion interface.

For simplicity, a cylindrical inclusion of radius a lying parallel to the x_2 axis is considered. The matrix and the inclusion are elastically isotropic with the respective shear moduli of μ and μ^*. The Poisson ratio of the matrix is ν. Let the applied stress far from the inclusion be $\sigma_{31}^0 = \sigma^0$. Let atomic diffusion take place along the matrix-inclusion interface. After diffusion, the resulting stress field can be calculated by considering a uniform eigenstrain of $\Delta\epsilon_{31}^p$ in the inclusion. $\Delta\epsilon_{31}^p$ is a misfit strain caused by either an inclusion geometry change or a matrix geometry change due to the interfacial diffusion. These changes are illustrated by Figs. 44.7(a) and 44.7(b), both of which give the same misfit $\Delta\epsilon_{31}^p$. The dotted line in Fig. 44.7(a) is the shape of the inclusion in the stress free state, while the solid line in the figure is the void in the matrix after the inclusion is removed. This is the situation when the diffusing species are the material constituting the inclusion. Figure 44.7(b) applies to the situation when the diffusing species are the matrix material.

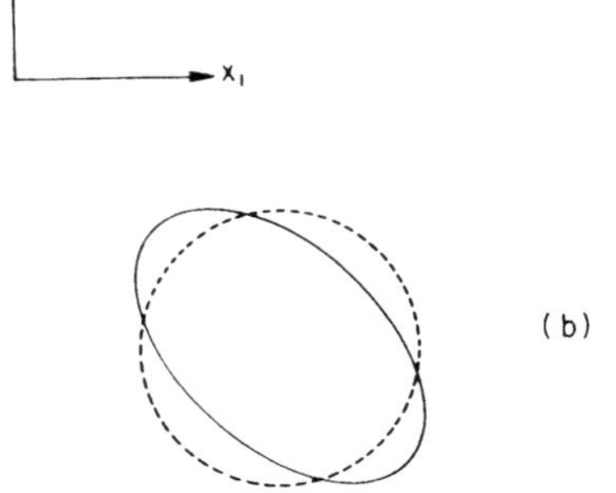

Fig. 44.7. Change of the misfit due to the interfacial diffusion.

The eigenstress $\Delta\sigma_{ij}$ caused by $\Delta\epsilon_{ij}^{p}$ in a cylinder is calculated from (22.11) as

$$\Delta\sigma_{31} = -2\mu\mu^*(1-2S)\Delta\epsilon_{31}^{p}/g, \tag{44.25}$$

where

$$S = S_{3131} = (3-4\nu)/\{8(1-\nu)\} \tag{44.26}$$

and

$$g = 2(\mu^* - \mu)S + \mu. \tag{44.27}$$

The increase in the elastic strain energy, ΔW^*, is given by (25.31) ($\sigma_{ij}^{B} = \Delta\sigma_{31}$):

$$\Delta W^* = -\Delta\sigma_{ij}\Delta\epsilon_{ij}^{p}V_0/2 = 2\mu\mu^*(1-2S)(\Delta\epsilon_{31}^{p})^2V_0/g, \tag{44.28}$$

where V_0 is the volume of the inclusion per unit length, πa^2. The applied stress in the inclusion σ_{ij}^{A} is calculated as

$$\sigma_{31}^{A} = \sigma^0\mu^*/g. \tag{44.29}$$

Thus, the change in the external potential due to $\Delta\epsilon_{31}^{p}$ is obtained from (25.32) as

$$\Delta V = -\sigma_{ij}^{A}\Delta\epsilon_{ij}^{p}V_0 = -2\sigma^0\mu^*\Delta\epsilon_{31}^{p}V_0/g. \tag{44.30}$$

The total energy change or the Gibbs free energy change ΔW due to $\Delta\epsilon_{31}^{p}$ is the sum of (44.28) and (44.30),

$$\Delta W = 2\mu\mu^*(1-2S)(\Delta\epsilon_{31}^{p})^2V_0/g - 2\sigma^0\mu^*\Delta\epsilon_{31}^{p}V_0/g. \tag{44.31}$$

Thus, the misfit strain $\overline{\Delta\epsilon_{31}^{p}}$ which minimizes the Gibbs free energy is found to be

$$\overline{\Delta\epsilon_{31}^{p}} = \sigma^0/\{2\mu(1-2S)\}. \tag{44.32}$$

In this situation, the internal stress $\overline{\Delta\sigma_{31}}$ is given by

$$\overline{\Delta\sigma_{31}} = -\sigma^0\mu^*/g \tag{44.33}$$

from (44.25). Thus, the total stress in the inclusion vanishes,

$$\sigma_{31}^{A} + \overline{\Delta\sigma_{31}} = 0 \tag{44.34}$$

from (44.29) and (44.33). It should be noted that the equilibrium condition, (44.34), is a special case of more general conditions. Let the applied stress in an inclusion be σ_{ij}^{A} and the internal stress be $\Delta\sigma_{ij}$ which is brought about by the eigenstrain $\Delta\epsilon_{ij}^{p}$ due to the interfacial diffusion.

Then, the Gibbs free energy is, generally,

$$\Delta W = -\Delta\sigma_{ij}\Delta\epsilon_{ij}^{p} V_0/2 - \sigma_{ij}^{A}\Delta\epsilon_{ij}^{p} V_0 \tag{44.35}$$

which is minimized when

$$-(\Delta\sigma_{ij} + \sigma_{ij}^{A})\delta(\Delta\epsilon_{ij}^{p}) = 0, \tag{44.36}$$

since $\Delta\sigma_{ij}$ is linearly related to $\Delta\epsilon_{ij}^{p}$. In diffusion, mass is conserved. It can be expressed by no-dilation,

$$\Delta\epsilon_{ii}^{p} = 0. \tag{44.37}$$

From (44.36) and (44.37), it is found that at equilibrium the total stress in the inclusion becomes hydrostatic,

$$\Delta\sigma_{ij} + \sigma_{ij}^{A} = \delta_{ij}\sigma \tag{44.38}$$

which includes (44.34) as a special case. It is further noted that the above discussion and conclusion are identical to those developed for a misfitting incoherent particle in Section 26.

One can discuss the kinetics of the above diffusional relaxation in a simple manner. From Fig. 44.1, the number of atoms which have diffused to accomplish the formation of misfit $\Delta\epsilon_{31}^{p}$ is calculated as

$$N = 2V_0\Delta\epsilon_{31}^{p}/\pi\Omega \tag{44.39}$$

per unit length of the cylindrical inclusion. Here, Ω is the atomic volume. N is the volume of region I divided by Ω, where the major and minor axes of the ellipse in Fig. 44.1 are $a(1+\Delta\epsilon_{31}^{p})$ and $a(1-\Delta\epsilon_{31}^{p})$, respectively. When the relaxation is incomplete and the misfit strain changes by $\delta(\Delta\epsilon_{31}^{p})$, the change in the Gibbs free energy is given by

$$\delta(\Delta W) = -(\overline{\Delta\epsilon_{31}^{p}} - \Delta\epsilon_{31}^{p})4\mu^{*}\mu(1-2S)V_0\delta(\Delta\epsilon_{31}^{p})/g \tag{44.40}$$

from (44.31) and (44.32). The number of the transported atoms for this process is given from (44.39) as

$$\delta N = 2V_0\delta(\Delta\epsilon_{31}^{p})/\pi\Omega. \tag{44.41}$$

Since the diffusion distance l is $\sim\pi a/2$, the force F acting on an atom is calculated as

$$F = -\delta(\Delta W)/l\delta N = (\overline{\Delta\epsilon_{31}^{p}} - \Delta\epsilon_{31}^{p})4\mu^{*}\mu(1-2S)\Omega/ag \tag{44.42}$$

from (44.40) and (44.41). The Einstein relation gives the drift velocity v of an atom as

$$v = FD_s/kT, \tag{44.43}$$

where D_s is the interfacial diffusion constant, k the Boltzmann constant and T the temperature. Let the thickness of the interfacial diffusional layer be h. Then, the number of the atoms which diffuse during the time interval δt is given by

$$\delta N = 4hv\delta t/\Omega, \tag{44.44}$$

since the cross section of the diffusion path is $4h$ per unit length of the cylinder. From (44.41) through (44.44), one obtains the rate of the misfit strain as

$$d(\Delta\epsilon^{\rho}_{31})/dt = (\overline{\Delta\epsilon^{\rho}_{31}} - \Delta\epsilon^{\rho}_{31})/\tau, \tag{44.45}$$

where the relaxation time τ is given by

$$\tau = \frac{kT}{hD_s}\frac{agV_0}{8\pi\mu^*\mu(1-2S)\Omega} \tag{44.46}$$

or by

$$\tau = \left\{\frac{1}{2\mu^*} + \frac{(3-4\nu)}{2\mu}\right\}\frac{a^3kT}{4hD_s\Omega} \tag{44.47}$$

from (44.26) and (44.27), and $V_0 = \pi a^2$. Since ϵ^{ρ}_{31} is related to the eigenstrain of the equivalent inclusion, ϵ^*_{31}, in a manner similar to (44.3), and since the macroscopic deformation is expressed by (44.2) when the volume fraction of the inclusion f is small, one expects the specimen to respond to the applied stress by

$$d\gamma/dt = (\bar{\gamma} - \gamma)/\tau \tag{44.48}$$

with τ given by (44.46) or (44.47). Here γ is the macroscopic shear strain caused by interfacial diffusion, and

$$\begin{aligned} \gamma &= f\mu^*\Delta\epsilon^{\rho}_{31}/g, \\ \bar{\gamma} &= f\mu^*\overline{\Delta\epsilon^{\rho}_{31}}/g = f\mu^*\sigma^0/\{2\mu(1-2S)g\}, \end{aligned} \tag{44.49}$$

from (44.32). It is clear that, due to diffusional relaxation, the specimen behaves as a standard linear solid, (44.48) and (44.49), the relaxation strength of which is by definition (Zener 1948) given by

$$\Delta\mu = f\mu^*/\{(1-2S)g\}. \tag{44.50}$$

The preceding treatment of relaxation from interfacial diffusion followed Mori, Okabe and Mura (1979) who also have discussed spherical inclusions. Koeller and Raj (1978) have considered relaxation in a situation exactly the same as above. However, because they take a different approach, the result is not exactly the same, although the conclusions are similar in nature.

Okabe et al. (1981) have measured internal friction of Al–S; alloys to examine equation (44.47). The relaxation time has a third power dependence of the particle size of Si. The internal friction peak occurs when the frequency ω_p times τ is unity. This relation is plotted in Fig. 44.8, where $D_s = D_0 \exp(-Q/kT_p)$. The slope of the lines in the figure agrees with (44.47).

McClintock (1968) has suggested that cavities nucleate on interfaces by tearing the inclusion away from the ductile matrix or by the cracking of non-deformable inclusions. The incompatibility (misfit) between the inclusion and the matrix is expressed by the distribution of dislocations (43.18). These dislocations are dissipated by punching from the inclusion to initiate the interfacial cavities or cracks. This idea was first proposed by Ashby (1966) and further developed by Argon et al. (1975), Argon (1976), Raj and Ashby (1971, 1972, 1975), Raj (1975, 1976) and Chuang et al. (1979). It is interesting to note that condition (44.34) is equivalent to the condition of cavity formation.

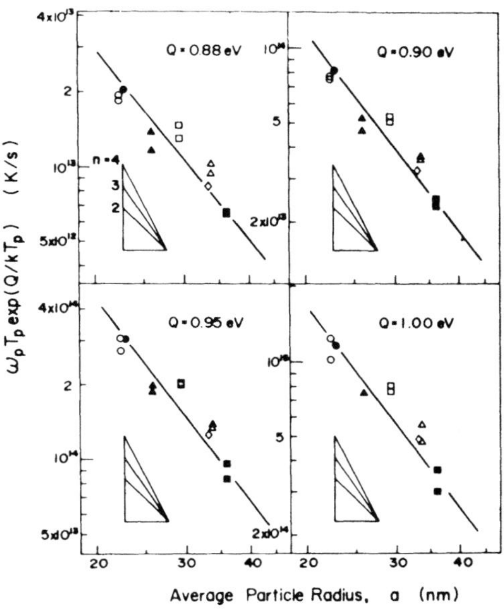

Fig. 44.8. Particle size dependency of peak temperature T_p.

45. Average elastic moduli of composite materials

The evaluation of average elastic moduli (the shear modulus and bulk modulus) of composite materials or polycrystals is one of the classical problems in micromechanics. The pioneer works on this subject have been done by Voigt (1889) and Reuss (1929). The Voigt approximation gives upper bounds and the Reuss approximation gives lower bounds of the average elastic moduli, as proved by Boas and Schmid (1934), Bishop and Hill (1951), and Hill (1952) (see also Bruggeman 1934, Huber and Schmid 1934, and Boas 1935).

The Voight approximation

Let us consider a composite material which consists of inhomogeneities with various shear moduli $\mu_1, \mu_2, \ldots, \mu_n$ and a matrix with the shear modulus μ_0 (see Fig. 45.1). The average (or effective) shear modulus $\bar{\mu}$ of the composite material will be investigated here. The volume fractions of the inhomogeneities are denoted by $c_1, c_2, \ldots, c_n$ and that of the remainder (matrix) by c_0. The shapes of the inhomogeneities are arbitrary.

Let us assume that the average elastic strain of the composite material is ϵ_{12}^0 when a shear stress is applied. Voigt (1889) approximated the average stress of the composite material as

$$\bar{\sigma} = \sum_{r=0}^{n} 2c_r\mu_r\epsilon_{12}^0 \tag{45.1}$$

which assumes that all the composite elements are subjected to the same

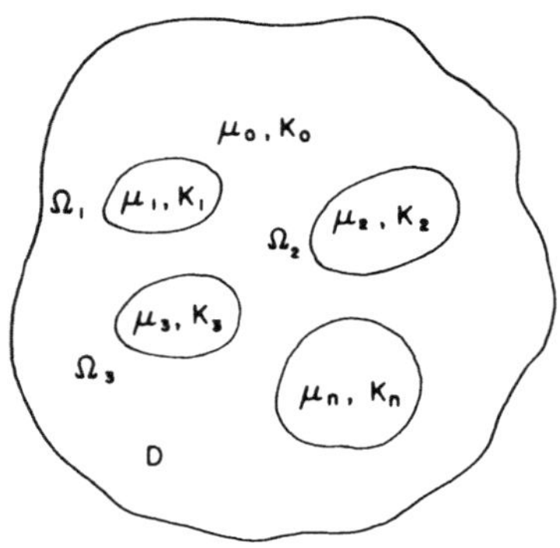

Fig. 45.1. Composite material containing inhomogeneities $\Omega_1, \Omega_2 \ldots$ μ: shear modulus K: bulk modulus.

uniform strain ϵ^0_{12}. The definition of the average shear modulus is

$$\bar{\mu} = \bar{\sigma}/\bar{\gamma}, \tag{45.2}$$

where $\bar{\gamma} = 2\epsilon^0_{12}$. (45.3)

Then, the Voigt average shear modulus is

$$\bar{\mu} = \sum_{r=0}^{n} c_r \mu_r. \tag{45.4}$$

The Voigt approximation can also be applied to anisotropic inhomogeneities. Since $\sigma_{12} = 2C_{1212}\epsilon^0_{12}$ and $\bar{\sigma} = \int_D \sigma_{12} dD$, where D is the unit volume, the average of C_{1212} becomes

$$\bar{C}_{1212} = \sum_{r=0}^{n} c_r C^{(r)}_{1212}, \tag{45.5}$$

where $C^{(r)}_{1212}$ is the elastic modulus in the r-th inhomogeneity with reference to the same coordinate system in which ϵ^0_{12} is measured.

The average shear modulus $\bar{\mu}$ of a polycrystal can be evaluated from the anisotropic elastic moduli C_{ijkl} of constituent single crystals. The single crystals are assumed to be randomly oriented with respect to the x, y, z axes shown in Fig. 45.2. The average shear modulus is isotropic and becomes, from (45.4),

$$\bar{\mu} = (1/8\pi^2) \int_0^{2\pi} d\beta \int_0^{\pi} \sin\theta d\theta \int_0^{2\pi} C_{1212} d\phi, \tag{45.6}$$

where C_{1212} is referred to the x, y, z axes and (β, θ, ϕ) are the Euler angles. A random orientation of a single crystal is represented by a point

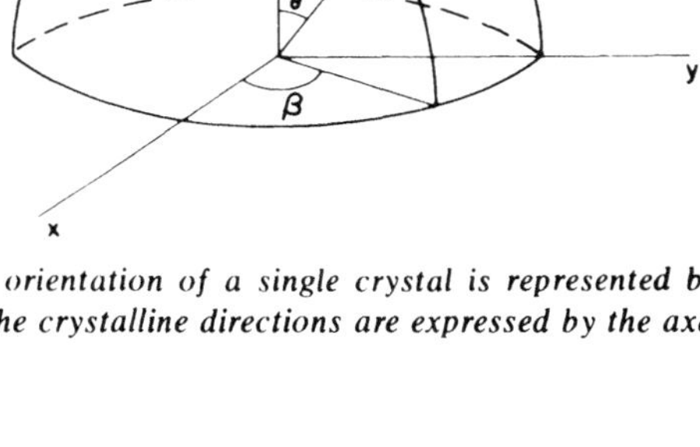

Fig. 45.2. A random orientation of a single crystal is represented by a point of the unit sphere. The crystalline directions are expressed by the axes 1, 2, 3.

on the unit sphere shown in Fig. 45.2. The orthogonal crystallographic directions of the crystal are expressed by the 1, 2 and 3 directions in the figure. The location of the point on the unit sphere indicates the 3-direction, and angle ϕ from the median indicates the 1-direction, where the plane containing the 1- and 2-directions is tangential to the unit sphere. The surface element of the unit sphere is $\sin\theta d\beta d\theta$ and the total area is 4π. Angle ϕ varies from 0 to 2π. Therefore, $(1/8\pi^2)\sin\theta d\beta d\theta d\phi$ in (45.6) corresponds to c_r in (45.5). The elastic modulus C_{1212} in the x, y, z coordinate system and C_{IJKL} in the 1, 2, 3 coordinate system are related by

$$C_{1212} = a_{1I}a_{2J}a_{1K}a_{2L}C_{IJKL}, \tag{45.7}$$

where

$$a_{IJ} = \begin{pmatrix} \cos\theta\cos\beta\cos\phi - \sin\beta\sin\phi, & -\cos\theta\cos\beta\sin\phi - \sin\beta\cos\phi, & \sin\theta\cos\beta \\ \cos\theta\sin\beta\cos\phi + \cos\beta\sin\phi, & -\cos\theta\sin\beta\sin\phi + \cos\beta\cos\phi, & \sin\theta\sin\beta \\ -\sin\theta\cos\phi, & \sin\theta\sin\phi, & \cos\theta \end{pmatrix} \tag{45.8}$$

The average bulk modulus $\bar{K}$ can be obtained in a similar way. The bulk moduli of the inhomogeneities are denoted by $K_1, K_2, \ldots, K_n$ and that of the matrix by K_0. The Voigt approximation is obtained by assuming that all the elements of the composite material have the same dilatation ϵ^0_{ii} as the average. Then, the average hydrostatic stress $\bar{\sigma}$ is evaluated by

$$\bar{\sigma} = \sum_{r=0}^{n} c_r K_r \bar{\theta}, \tag{45.9}$$

where

$$\bar{\theta} = \epsilon^0_{ii}. \tag{45.10}$$

By definition, the average bulk modulus is

$$\bar{K} = \bar{\sigma}/\bar{\theta}. \tag{45.11}$$

Therefore, we have

$$\bar{K} = \sum_{r=0}^{n} c_r K_r. \tag{45.12}$$

The average bulk modulus of a polycrystal is evaluated in the same way. Since $\sigma_{ii} = C_{iijj}\bar{\theta}/3$, and $\bar{\sigma} = (1/3)\int_D \sigma_{ii} dD$, we have

$$\bar{\sigma} = (1/9)\int_D C_{iijj}\bar{\theta}dD, \tag{45.13}$$

where D is the unit volume. The quantity C_{iijj} is an invariant and is independent of β, θ, ϕ in (45.6). Therefore, (45.13) is simply $\bar{\sigma} = (1/9)C_{iijj}\bar{\theta}$, which leads to

$$\bar{K} = (1/9)C_{iijj}. \tag{45.14}$$

The Reuss approximation

Reuss (1929) has assumed that all the elements of the composite material (Fig. 45.1) are subjected to a uniform stress equal to the average stress. According to Reuss' assumption, the average shear strain for a given shear stress $\bar{\sigma} = \sigma_{12}^0$ is expressed by

$$\bar{\gamma} = \sum_{r=0}^{n} c_r\bar{\sigma}/\mu_r. \tag{45.15}$$

By definition (45.2), the average shear modulus is obtained as

$$\bar{\mu} = \left(\sum_{r=0}^{n} c_r/\mu_r\right)^{-1}. \tag{45.16}$$

Similarly, the average bulk modulus becomes

$$\bar{K} = \left(\sum_{r=0}^{n} c_r/K_r\right)^{-1}. \tag{45.17}$$

For a polycrystal, we have

$$\bar{\mu} = \left[(1/2\pi^2)\int_0^{2\pi} d\beta \int_0^{\pi} \sin\theta d\theta \int_0^{2\pi} s_{1212} d\phi\right]^{-1}, \tag{45.18}$$

where s_{ijkl} is the elastic compliance and $\gamma = 2\epsilon_{12} = 4s_{1212}\sigma_{12}^0$. The average bulk modulus obtained by the Reuss method is identical to the Voigt result for the following reason: since $\theta = \epsilon_{ii} = s_{iijj}\bar{\sigma}$ for the hydrostatic stress $\bar{\sigma}$ and s_{iijj} is an invariant, it follows that $\bar{\theta} = s_{iijj}\bar{\sigma}$ and $\bar{K} = 1/s_{iijj} = C_{iijj}/9$.

Table 45.1 (Hill 1952) contains numerical results of $\bar{\mu}$ and $\bar{K}$ for various polycrystal metals. The Voigt and Reuss approximations for $\bar{\mu}$ are denoted by G_V and G_R, respectively. The Voigt and Reuss approximations for $\bar{K}$ are identical and are denoted by K_V or K_R. The average values of Young's modulus and Poisson's ratio are evaluated from μ and K by using the well known relations (A2.2). C_{ij} in the table are the elastic constants of the single crystals. Table 45.2 shows the experimental observation. According to Hill $(1/2)(G_R + G_V)$ or $(G_R G_V)^{1/2}$ are good approximations since

$$G_R < G < G_V, \tag{45.19}$$

where the true value of $\bar{\mu}$ is denoted by G.

Table 45.1 (after Hill 1952)

	C_{11}	C_{12}	C_{44}	G_R	G_V	$K_R = K_V$	E_R	E_V	ν_R	ν_V
Al	1.08	0.622	0.284	0.26	0.26	0.78	0.71	0.71	0.349	0.348
Cu	1.70	1.23	0.75	0.40	0.54	1.39	1.09	1.44	0.369	0.328
Au	1.86	1.57	0.42	0.24	0.31	1.67	0.69	0.87	0.431	0.413
α-Fe	2.37	1.41	1.16	0.74	0.89	1.73	1.93	2.29	0.313	0.280

Unit = 10^{12} dyn/cm^2.

Table 45.2 (after Hill 1952)

	G	K	E	ν
Al	0.265	0.74	0.71	0.34
Cu	0.436	1.33	1.18	0.35
Au	0.278	1.66	0.79	0.42
α-Fe	0.808	1.59	2.07	0.285

Unit = 10^{12} dyn/cm^2.

Hill's theory

Hill (1952) has shown that the Voigt approximation and the Reuss approximation are the upper and lower bounds of the true average elastic moduli.

Let the average stress and strain in an aggregate of single crystals under an applied force X_i be denoted by $\bar{\sigma}_{ij}$ and $\bar{\epsilon}_{ij}$. $\bar{\sigma}_{ij}$ and $\bar{\epsilon}_{ij}$ are uniform quantities which can be calculated from the boundary values. Namely, $\bar{\sigma}_{ij}$ is estimated from

$$\bar{\sigma}_{ij} n_j = X_i \quad \text{on } S, \tag{45.19.1}$$

where n_j is the unit exterior normal of the material surface S. $\bar{\epsilon}_{ij}$ is obtained from the equation

$$\begin{aligned} \bar{\epsilon}_{ij} &= (1/2)(\bar{u}_{i,j} + \bar{u}_{j,i}) \quad \text{in } D, \\ \bar{u}_i &= u_i^0 \quad \text{on } S, \end{aligned} \tag{45.19.2}$$

where $\bar{u}_i$ is the average displacement which is a linear function of the space coordinates.

Denoting the average elastic moduli by $\bar{C}_{ijkl}$, we can write

$$\bar{\sigma}_{ij} = \bar{C}_{ijkl}\bar{\epsilon}_{kl} \tag{45.19.3}$$

or

$$\bar{\epsilon}_{ij} = \bar{s}_{ijkl}\bar{\sigma}_{kl}, \tag{45.19.4}$$

where $\bar{s}_{ijkl}$ is the average elastic compliance tensor [see (A2.18)′]. The actual stress, strain, and displacement are denoted by σ_{ij}, ϵ_{ij} and u_i, respectively. They must satisfy the same boundary conditions as in (45.19.1) and (45.19.2),

$$\left.\begin{aligned} \sigma_{ij}n_j &= X_i \\ u_i &= u_i^0 \end{aligned}\right\} \text{ on } S. \tag{45.19.5}$$

Since $\sigma_{ij,j} = 0$, $\bar{\sigma}_{ij,j} = 0$, we have

$$\begin{aligned} \int_D \sigma_{ij}\epsilon_{ij}dD &= \int_D \sigma_{ij}u_{i,j}dD = \int_S X_i u_i^0 dS, \\ \int_D \bar{\sigma}_{ij}\bar{\epsilon}_{ij}dD &= \int_D \bar{\sigma}_{ij}\bar{u}_{i,j}dD = \int_S X_i u_i^0 dS. \end{aligned} \tag{45.19.6}$$

Therefore, it holds that

$$\bar{\sigma}_{ij}\bar{\epsilon}_{ij} = (1/V)\int_D \sigma_{ij}\epsilon_{ij}dD, \tag{45.19.7}$$

where V is the volume of D. New stress σ_{ij}^0 and strain ϵ_{ij}^0 are defined by the following: σ_{ij}^0 is the stress that would exist in a crystal with a local orientation and having the strain $\bar{\epsilon}_{ij}$. ϵ_{ij}^0 is the strain that would be produced in such a crystal by stress $\bar{\sigma}_{ij}$. Namely,

$$\begin{aligned} \sigma_{ij}^0 &= C_{ijkl}\bar{\epsilon}_{kl}, \\ \epsilon_{ij}^0 &= s_{ijkl}\bar{\sigma}_{kl}, \end{aligned} \tag{45.19.8}$$

where C_{ijkl} and s_{ijkl} represent the elastic modulus and the compliance of the single crystal. We have

$$\sigma_{ij} = C_{ijkl}\epsilon_{kl}, \quad \epsilon_{ij} = s_{ijkl}\sigma_{kl}. \tag{45.19.9}$$

From (45.19.8) and (45.19.9), $\sigma_{ij}^0\epsilon_{ij} = \sigma_{ij}\bar{\epsilon}_{ij}$ and $\sigma_{ij}\epsilon_{ij}^0 = \bar{\sigma}_{ij}\epsilon_{ij}$, and therefore,

$$\begin{aligned} \sigma_{ij}\epsilon_{ij} + (\sigma_{ij} - \sigma_{ij}^0)(\epsilon_{ij} - \bar{\epsilon}_{ij}) &= \sigma_{ij}^0\bar{\epsilon}_{ij} + 2(\epsilon_{ij} - \bar{\epsilon}_{ij})\sigma_{ij}, \\ \sigma_{ij}\epsilon_{ij} + (\sigma_{ij} - \bar{\sigma}_{ij})(\epsilon_{ij} - \epsilon_{ij}^0) &= \bar{\sigma}_{ij}\epsilon_{ij}^0 + 2(\sigma_{ij} - \bar{\sigma}_{ij})\epsilon_{ij}. \end{aligned} \tag{45.19.10}$$

The second terms in the left-hand sides in (45.19.10) are positive since they can be written as $C_{ijkl}(\epsilon_{kl}-\bar{\epsilon}_{kl})(\epsilon_{ij}-\bar{\epsilon}_{ij})$ and $s_{ijkl}(\sigma_{kl}-\bar{\sigma}_{kl})(\sigma_{ij}-\bar{\sigma}_{ij})$, respectively. It is also show that

$$\begin{aligned}\int_D \sigma_{ij}(\epsilon_{ij}-\bar{\epsilon}_{ij})dD &= \int_S X_i(u_i^0-u_i^0)dS = 0,\\ \int_D (\sigma_{ij}-\bar{\sigma}_{ij})\epsilon_{ij}dD &= \int_S (X_i-X_i)u_i^0 dS = 0,\end{aligned} \tag{45.19.11}$$

by integrating by parts. Then, (45.19.10) leads to

$$\begin{aligned}V\bar{\sigma}_{ij}\bar{\epsilon}_{ij} &\le \bar{\epsilon}_{ij}\int_D \sigma_{ij}^0 dD,\\ V\bar{\sigma}_{ij}\bar{\epsilon}_{ij} &\le \bar{\sigma}_{ij}\int_D \epsilon_{ij}^0 dD,\end{aligned} \tag{45.19.12}$$

where (45.19.7) has been used for the derivation. Thus, we have, finally,

$$\begin{aligned}\bar{C}_{ijkl} &\le (1/V)\int_D C_{ijkl}dD\\ \bar{s}_{ijkl} &\le (1/V)\int_D s_{ijkl}dD.\end{aligned} \tag{45.19.13}$$

This result indicates that the Voigt approximations give upper bounds and the Reuss approximations yield lower bounds, since $4\bar{s}_{1212}=1/\bar{C}_{1212}$, and $\bar{s}_{iijj}=9/\bar{C}_{iijj}$.

Eshelby's method

Let us consider again the composite material shown in Fig. 45.1. We evaluate the average shear strain $\bar{\gamma}$ when an average shear stress is given as $\sigma_{12}^0=\bar{\sigma}$. When the relation between $\bar{\gamma}$ and $\bar{\sigma}$ is obtained, the average shear modulus is determined by (45.2).

The stress in the composite material varies from place to place. Here it is denoted by $\sigma_{12}^0+\sigma_{12}$, where σ_{12} is the disturbance due to inhomogeneities. Eshelby's equivalent inclusion method in Section 22 is employed. The composite material is simulated by a homogeneous material having the same elastic moduli as those of the matrix and containing inclusions Ω_r with eigenstrain ϵ_{12}^*, where $r=1,2,\ldots,n$. Ω_r is the r-th inclusion with the same volume and location as the r-th inhomogeneity. The equivalency equation (22.5′) is

$$\begin{aligned}\sigma_{12}^0+\sigma_{12} &= 2\mu_0(\epsilon_{12}^0+\epsilon_{12}-\epsilon_{12}^*)\\ &= 2\mu_r(\epsilon_{12}^0+\epsilon_{12}),\end{aligned} \quad \text{in } \Omega_r \tag{45.20}$$

where

$$\epsilon_{12} = 2S_{1212}\epsilon^*_{12}, \quad \sigma^0_{12} = 2\mu_0\epsilon^0_{12}, \tag{45.21}$$

and S_{ijkl} is Eshelby's tensor; $S_{1212} = (4 - 5\nu_0)/15(1 - \nu_0)$ for spherical inclusions. The equivalent eigenstrain ϵ^*_{12} in Ω_r is obtained from (45.20) as

$$\epsilon^*_{12} = (\mu_0 - \mu_r)\epsilon^0_{12}/g_r, \tag{45.22}$$

where

$$g_r = \mu_0 + 2S_{1212}(\mu_r - \mu_0). \tag{45.23}$$

The following quantities in Ω_r are easily calculated by substituting (45.22) into (45.21) and (45.20):

$$\begin{aligned}
\epsilon_{12} &= 2S_{1212}(\mu_0 - \mu_r)\epsilon^0_{12}/g_r, \\
\sigma_{12} &= 2(2S_{1212} - 1)(\mu_0 - \mu_r)\mu_0\epsilon^0_{12}/g_r, \\
\epsilon^0_{12} + \epsilon_{12} &= \mu_0\epsilon^0_{12}/g_r, \\
\sigma^0_{12} + \sigma_{12} &= 2\mu_r\mu_0\epsilon^0_{12}/g_r.
\end{aligned} \tag{45.24}$$

According to Tanaka-Mori's theorem (42.20), we have

$$\int_{D-\Omega_r} \sigma_{12}dD = 0 \tag{45.25}$$

and, therefore,

$$\int_D \sigma_{12}dD = \sum_{r=1}^{n} c_r(2S_{1212} - 1)(\mu_0 - \mu_r)\sigma^0_{12}/g_r. \tag{45.26}$$

This does not, however, satisfy the condition for stress disturbance

$$\int_D \sigma_{12}dD = 0. \tag{45.27}$$

Therefore, we add the uniform stress

$$-\sum_{r=1}^{n} c_r(2S_{1212} - 1)(\mu_0 - \mu_r)\sigma^0_{12}/g_r \tag{45.28}$$

to σ_{12} in (45.24). Consequently, the uniform strain

$$-\sum_{r=1}^{n} c_r(2S_{1212} - 1)(\mu_0 - \mu_r)\sigma^0_{12}/2\mu_0 g_r \tag{45.29}$$

is added to $\epsilon_{12}^0 + \epsilon_{12}$ in (45.24). The average strain is

$$\bar{\gamma} = 2 \int_D (\epsilon_{12}^0 + \epsilon_{12}) dD = \sum_{r=1}^{n} c_r \bar{\sigma}/g_r - \sum_{r=1}^{n} c_r (2S_{1212} - 1)(\mu_0 - \mu_r)\bar{\sigma}/\mu_0 g_r + c_0\gamma_0, \tag{45.30}$$

where γ_0 is the average strain in the matrix, and

$$\gamma_0 = \bar{\sigma}/\mu_0. \tag{45.31}$$

By using definition (45.2) and

$$c_0 + \sum_{r=1}^{n} c_r = 1, \tag{45.32}$$

we have

$$\bar{\mu} = \mu_0 \Big/ \left[1 + \sum_{r=1}^{n} c_r(\mu_0 - \mu_r)/\{\mu_0 + 2S_{1212}(\mu_r - \mu_0)\} \right]. \tag{45.33}$$

Similarly, the average bulk modulus is obtained as

$$\bar{K} = K_0 \Big/ \left[1 + \sum_{r=1}^{n} c_r(K_0 - K_r)/\{K_0 + (1/3)S_{iijj}(K_r - K_0)\} \right]. \tag{45.34}$$

The Reuss approximation is a special case, $2S_{1212} = 1$. The Voight approximation can also be obtained by $2S_{1212} = 0$ as well as by expanding the denominator, and assuming that the second term is much less than 1.

The results (45.33) and (45.34) can be obtained from the energy consideration of Section 25. The elastic strain energy of the composite material is written as (25.11). On the other hand, we can write it as

$$W^* = (1/2)\bar{\sigma}^2/\bar{\mu}. \tag{45.34.1}$$

Substituting (45.22) and (45.34.1) into (25.11), we have

$$(1/2)\bar{\sigma}^2/\bar{\mu} = (1/2)\bar{\sigma}^2/\mu_0 + (1/2)\bar{\sigma}^2 \sum_{r=1}^{n} c_r(\mu_0 - \mu_r)/\mu_0 g_r. \tag{45.34.2}$$

Then, it leads to (45.33).

Self-consistent method

When the equivalent inclusion method is applied to obtain the stress disturbance of inhomogeneities, we assume the existence of a matrix which has the shear and bulk moduli μ_0 and K_0. This assumption, however, does not hold in some cases. Polycrystals, for instance, have no matrix, although they are aggregates of particles with different elastic moduli. In composite materials as shown in Fig. 45.1, the meaning of the matrix becomes rather vague when the volume fraction of inhomogeneities is increased, particularly when $c_0 \to 0$. Besides this, the interaction among inhomogeneities becomes more prominent as the number of particles increases. Kröner (1958) and Budiansky and Wu (1962) have proposed the self-consistent method in order to avoid the difficulties mentioned above.

Let us consider a composite material as shown in Fig. 45.1. The average strain and stress under an applied shear stress field are denoted by ϵ_{12}^0 and σ_{12}^0, respectively. Hooke's law is

$$\sigma_{12}^0 = 2\bar{\mu}\epsilon_{12}^0, \tag{45.35}$$

where $\bar{\mu}$ is the average shear modulus. The stress disturbance due to the inhomogeneities is denoted by σ_{12}. The self-consistent method assumes that the composite material can be simulated by a homogeneous material which has a shear modulus $\bar{\mu}$, Poisson's ratio $\bar{\nu}$, and an eigenstrain ϵ_{12}^* in Ω_r $(r = 1, 2, \ldots, n)$. The equivalency equation (22.5′) becomes

$$\begin{aligned} \sigma_{12}^0 + \sigma_{12} &= 2\bar{\mu}(\epsilon_{12}^0 + \epsilon_{12} - \epsilon_{12}^*) \\ &= 2\mu_r(\epsilon_{12}^0 + \epsilon_{12}), \end{aligned} \quad \text{in } \Omega_r \tag{45.36}$$

where

$$\epsilon_{12} = 2S_{1212}\epsilon_{12}^*. \tag{45.37}$$

The Poisson's ratio appearing in S_{1212} is the average Poisson's ratio $\bar{\nu}$. ϵ_{12}^* satisfying (45.36) is obtained as

$$\epsilon_{12}^* = (\bar{\mu} - \mu_r)\epsilon_{12}^0/\bar{g}_r, \tag{45.38}$$

where

$$\bar{g}_r = \bar{\mu} + 2S_{1212}(\mu_r - \bar{\mu}). \tag{45.39}$$

The strain and stress in Ω_r are obtained as

$$\begin{aligned} \epsilon_{12}^0 + \epsilon_{12} &= \bar{\mu}\epsilon_{12}^0/\bar{g}_r, \\ \sigma_{12}^0 + \sigma_{12} &= 2\mu_r\bar{\mu}\epsilon_{12}^0/\bar{g}_r. \end{aligned} \tag{45.40}$$

The average values are

$$\begin{aligned} \bar{\gamma} &= 2\int_D (\epsilon_{12}^0 + \epsilon_{12})dD = \sum_{r=1}^{n} c_r\bar{\mu}\bar{\gamma}/\bar{g}_r + c_0\gamma_0, \\ \bar{\sigma} &= \int_D (\sigma_{12}^0 + \sigma_{12})dD = \sum_{r=1}^{n} c_r\mu_r\bar{\mu}\bar{\gamma}/\bar{g}_r + c_0\sigma_0, \end{aligned} \tag{45.41}$$

where σ_0 and γ_0 are the average shear stress and strain in the original matrix with shear modulus μ_0:

$$\sigma_0 = \mu_0 \gamma_0. \tag{45.42}$$

Eliminating σ_0 from (45.41) and (45.42), we have

$$\bar{\sigma} = \mu_0 \bar{\gamma} + \sum_{r=1}^{n} c_r(\mu_r - \mu_0)\bar{\mu}\bar{\gamma}/\bar{g}_r, \tag{45.43}$$

and, therefore,

$$\bar{\mu} = \mu_0 + \sum_{r=1}^{n} c_r(\mu_r - \mu_0)\bar{\mu}/\{\bar{\mu} + 2S_{1212}(\mu_r - \bar{\mu})\}. \tag{45.44}$$

Similarly, the average bulk modulus is obtained as

$$\bar{K} = K_0 + \sum_{r=1}^{n} c_r(K_r - K_0)\bar{K}/\{\bar{K} + (1/3)S_{iijj}(K_r - \bar{K})\}. \tag{45.45}$$

In the case of spherical inclusions, we have

$$\begin{aligned} S_{1212} &= (4 - 5\bar{\nu})/15(1 - \bar{\nu}) \\ S_{iijj} &= (1 + \bar{\nu})/(1 - \bar{\nu}). \end{aligned} \tag{45.46}$$

The relations in (A2.2) give

$$\bar{\nu} = (3\bar{K} - 2\bar{\mu})/2(\bar{\mu} + 3\bar{K}). \tag{45.47}$$

Equations (45.44) and (45.45) with (45.46) and (45.47) are simultaneous equations to determine $\bar{\mu}$ and $\bar{K}$ for given c_r, μ_r, K_r $(r = 0, 1, \ldots, n)$. These equations are valid even if $c_0 = 0$. In this case, μ_0 and K_0 are taken as zero in (45.44) and (45.45). Equations in (45.41) for $c_0 = 0$ lead to

$$1 = \sum_{r=1}^{n} c_r \mu_r/\bar{g}_r, \quad 1 = \sum_{r=1}^{n} c_r \bar{\mu}/\bar{g}_r. \tag{45.48}$$

The above relations were pointed out by Budiansky (1965).

T. T. Wu (1966) has solved simultaneously (45.44) and (45.45) and has investigated the shape effect of inclusions when Eshelby's tensor S_{ijkl} for spheroids is used. He concludes that the disk-shaped inclusions give by far the most significant increase in Young's modulus. Walpole (1967) has extended this theory to include the anisotropic case and has proposed a convergent series of successive approximations. Kneer (1965) and Morris (1970) have calculated elastic moduli of polycrystals where the orientation distribution of single crystals is expanded in a series of generalised spherical harmonics. Hershey's paper (1954) may be one of the oldest ones, after Voigt and Reuss, dealing with the elasticity of a polycrystalline in terms of the elasticity of the individual grains.

But the equivalent inclusion method, including the self-consistent method, is not accurate enough when one wants to know, for instance, the effect of the distribution pattern (morphology) of unidirectional fibers in a reinforced composite. In order to calculate the interaction among composite elements, boundary value problems of elasticity must first be solved. Exact analyses along this line have been attempted by Adams and Doner (1967), C. H. Chen (1970), Chen and Cheng (1967), and Sendeckyj (1970, 1971), among others.

Upper and lower bounds

Hashin and Shtrikman (1962, 1963) have proposed a variational principle for finding upper and lower bounds of the average elastic moduli of a composite material. Their method gives generally better bounds than the Voigt and Reuss bounds. Their method has been generalized and modified by Hill (1963), Walpole (1966, 1969, 1970), Willis (1977), Willis and Acton (1976), Kröner (1977), and Laws and McLaughlin (1979), among others.

Let us consider an inhomogeneous system with various elastic moduli C_{ijkl}. The overall (average) elastic moduli are denoted by $\bar{C}_{ijkl}$. The strain and stress state under an applied load is simulated by the equivalent inclusion method. The inhomogeneous system is simulated by a homogeneous system with eigenstrain ϵ^*_{ij} and elastic moduli C^0_{ijkl}. The average strain is denoted by ϵ^0_{ij}. The equivalency equation (22.5′) becomes

$$C^0_{ijkl}(\epsilon^0_{kl} + \epsilon_{kl} - \epsilon^*_{kl}) = C_{ijkl}(\epsilon^0_{kl} + \epsilon_{kl}), \qquad (45.49)$$

where ϵ_{kl} is the strain disturbance due to the inhomogeneities, expressed by (3.24) in terms of $C^0_{klmn}\epsilon^*_{mn}$. Let us define a polarization stress $\boldsymbol{\sigma}^*$ by

$$\sigma^*_{ij} \equiv C^0_{ijkl}\epsilon^*_{kl}. \qquad (45.50)$$

Expression (3.24) can be written symbolically as

$$\boldsymbol{\epsilon} = \boldsymbol{\Gamma}\boldsymbol{\sigma}^*. \qquad (45.51)$$

Then, (45.49) is written as

$$\boldsymbol{\sigma}^* + (\mathbf{C} - \mathbf{C}^0)(\boldsymbol{\epsilon}^0 + \boldsymbol{\Gamma}\boldsymbol{\sigma}^*) = 0 \qquad (45.52)$$

or

$$(\mathbf{C} - \mathbf{C}^0)^{-1}\boldsymbol{\sigma}^* + (\boldsymbol{\epsilon}^0 + \boldsymbol{\Gamma}\boldsymbol{\sigma}^*) = 0. \qquad (45.52')$$

By the use of definition

$$(f, g) = (1/V) \int_D f(\mathbf{x})g(\mathbf{x})d\mathbf{x}, \tag{45.53}$$

Hashin and Shrikman (1962) have proposed the variational principle

$$\delta[(1/2)(\boldsymbol{\sigma}^*, (\mathbf{C}-\mathbf{C}^0)^{-1}\boldsymbol{\sigma}^*) + (1/2)(\boldsymbol{\sigma}^*, \boldsymbol{\Gamma}\boldsymbol{\sigma}^*) + (\boldsymbol{\sigma}^*, \boldsymbol{\epsilon}^0)] = 0, \tag{45.54}$$

where the comparison function is $\boldsymbol{\sigma}^*$. The stationary equation for the variation becomes (45.52′) and its stationary value is

$$(1/2)(\boldsymbol{\sigma}^*, \boldsymbol{\epsilon}^0) = (1/2) \int_D C^0_{ijkl}\epsilon^*_{kl}\epsilon^0_{ij}dD = (1/2) \int_D \sigma^0_{kl}\epsilon^*_{kl}dD, \tag{45.55}$$

where σ^0_{ij} is defined by

$$\sigma^0_{ij} = C^0_{ijkl}\epsilon^0_{kl}. \tag{45.56}$$

The stationary value of the functional can be written, from (25.11), as

$$W^* - W^0, \tag{45.57}$$

where W^* is the elastic strain energy and W^0 is $(1/2)\int_D \sigma^0_{ij}\epsilon^0_{ij}dD$.

It can also be shown that the stationary value is a minimum when $\mathbf{C}-\mathbf{C}^0$ is positive definite, and a maximum when $\mathbf{C}-\mathbf{C}^0$ is negative. That is,

$$(1/2)(\boldsymbol{\epsilon}^0, \bar{\mathbf{C}}\boldsymbol{\epsilon}^0) - (1/2)(\boldsymbol{\epsilon}^0, \mathbf{C}^0\boldsymbol{\epsilon}^0) \le (1/2)(\boldsymbol{\sigma}^*, (\mathbf{C}-\mathbf{C}^0)^{-1}\boldsymbol{\sigma}^*) + (1/2)(\boldsymbol{\sigma}^*, \boldsymbol{\Gamma}\boldsymbol{\sigma}^*) + (\boldsymbol{\sigma}^*, \boldsymbol{\epsilon}^0) \tag{45.58}$$

for a positive definite $\mathbf{C}-\mathbf{C}^0$, and

$$(1/2)(\boldsymbol{\epsilon}^0, \bar{\mathbf{C}}\boldsymbol{\epsilon}^0) - (1/2)(\boldsymbol{\epsilon}^0, \mathbf{C}^0\boldsymbol{\epsilon}^0) \ge (1/2)(\boldsymbol{\sigma}^*, (\mathbf{C}-\mathbf{C}^0)^{-1}\boldsymbol{\sigma}^*) + (1/2)(\boldsymbol{\sigma}^*, \boldsymbol{\Gamma}\boldsymbol{\sigma}^*) + (\boldsymbol{\sigma}^*, \boldsymbol{\epsilon}^0) \tag{45.59}$$

for a negative definite $\mathbf{C}-\mathbf{C}^0$. When $\boldsymbol{\sigma}^*$ is approximated as a linear functional of $\boldsymbol{\epsilon}^0$ in the variational method (45.54), the inequalities (45.58) and (45.59) can be used for estimating the lower and upper bounds of $\bar{\mathbf{C}}$ by choosing two sets of $\mathbf{C}^0$ so that $\mathbf{C}-\mathbf{C}^0$ becomes positive or negative definite, respectively. The minimum or maximum of the functional in (45.54) depends on the operator $(\mathbf{C}-\mathbf{C}^0)^{-1}+\boldsymbol{\Gamma}$. Willis (1977) has shown that the operator is positive definite so long as $\mathbf{C}-\mathbf{C}^0$ is positive definite at each point of D, and is negative definite so long as $\mathbf{C}-\mathbf{C}^0$ is negative definite at each point of D, as follows. Suppose that $\mathbf{C}-\mathbf{C}^0$ is positive definite. Then,

$$(\mathbf{x}, (\mathbf{C}-\mathbf{C}^0)\mathbf{x}) > 0 \tag{45.60}$$

for any non-zero $\mathbf{x}$. Let $(\mathbf{C}-\mathbf{C}^0)\mathbf{x}=\mathbf{y}$. Then, $\mathbf{x}=(\mathbf{C}-\mathbf{C}^0)^{-1}\mathbf{y}$; therefore,

$$(\mathbf{x},(\mathbf{C}-\mathbf{C}^0)\mathbf{x})=(\mathbf{x},\mathbf{y})=(\mathbf{y},\mathbf{x})=(\mathbf{y},(\mathbf{C}-\mathbf{C}^0)^{-1}\mathbf{y})>0. \tag{45.61}$$

This means that $(\mathbf{C}-\mathbf{C}^0)^{-1}$ is positive definite if $\mathbf{C}-\mathbf{C}^0$ is positive definite. We know that for any $\boldsymbol{\epsilon}^*$ on $\boldsymbol{\sigma}^*$, equation

$$\boldsymbol{\sigma}=\mathbf{C}^0(\boldsymbol{\epsilon}-\boldsymbol{\epsilon}^*)=\mathbf{C}^0\boldsymbol{\epsilon}-\boldsymbol{\sigma}^* \tag{45.62}$$

satisfies the condition $\sigma_{ij,j}=0$ and $(\boldsymbol{\sigma},\boldsymbol{\epsilon})=0$ as long as $\boldsymbol{\epsilon}$ is obtained from (45.51). We have, therefore,

$$(\boldsymbol{\sigma}^*,\boldsymbol{\Gamma}\boldsymbol{\sigma}^*)=(\boldsymbol{\sigma}^*,\boldsymbol{\epsilon})=((\mathbf{C}^0\boldsymbol{\epsilon}-\boldsymbol{\sigma}),\boldsymbol{\epsilon})=(\mathbf{C}^0\boldsymbol{\epsilon},\boldsymbol{\epsilon})>0. \tag{45.63}$$

This means, then, that $\boldsymbol{\Gamma}$ is positive definite.

If $\mathbf{C}-\mathbf{C}^0$ is negative definite, we may consider the following approach. From (45.50) and (45.62),

$$\boldsymbol{\epsilon}^*=(\mathbf{C}^0)^{-1}\boldsymbol{\sigma}^*,\qquad \boldsymbol{\epsilon}=(\mathbf{C}^0)^{-1}\boldsymbol{\sigma}+\boldsymbol{\epsilon}^*. \tag{45.64}$$

Hill (1963) calls $\boldsymbol{\epsilon}^*$ the strain polarization. It now follows that

$$\begin{aligned}(\boldsymbol{\sigma}^*,\boldsymbol{\Gamma}\boldsymbol{\sigma}^*)=(\mathbf{C}^0\boldsymbol{\epsilon}^*,\boldsymbol{\epsilon})&=(\mathbf{C}^0\boldsymbol{\epsilon}^*,((\mathbf{C}^0)^{-1}\boldsymbol{\sigma}+\boldsymbol{\epsilon}^*))\\&=(\boldsymbol{\epsilon}^*,\mathbf{C}^0\boldsymbol{\epsilon}^*)+(\mathbf{C}^0\boldsymbol{\epsilon}^*,(\mathbf{C}^0)^{-1}\boldsymbol{\sigma})\\&=(\boldsymbol{\epsilon}^*,\mathbf{C}^0\boldsymbol{\epsilon}^*)+((\mathbf{C}^0\boldsymbol{\epsilon}-\boldsymbol{\sigma}),(\mathbf{C}^0)^{-1}\boldsymbol{\sigma})\\&=(\boldsymbol{\epsilon}^*,\mathbf{C}^0\boldsymbol{\epsilon}^*)-(\boldsymbol{\sigma},(\mathbf{C}^0)^{-1}\boldsymbol{\sigma}),\end{aligned} \tag{45.65}$$

since $(\boldsymbol{\epsilon},\boldsymbol{\sigma})=0$. Consider the following identities:

$$\begin{aligned}\mathbf{C}^0(\mathbf{C}^0)^{-1}&=\mathbf{C}\mathbf{C}^{-1},\\ \mathbf{C}^0\mathbf{C}^{-1}&=\mathbf{C}^0\mathbf{C}^{-1}.\end{aligned} \tag{45.66}$$

The difference of the above two identities leads to

$$(\mathbf{C}-\mathbf{C}^0)^{-1}\mathbf{C}^0=(\mathbf{C}^0)^{-1}((\mathbf{C}^0)^{-1}-\mathbf{C}^{-1})^{-1}-\mathbf{I}, \tag{45.67}$$

where $\mathbf{I}$ is the identity. Thus,

$$(\boldsymbol{\sigma}^*,(\mathbf{C}-\mathbf{C}^0)^{-1}\mathbf{C}^0\boldsymbol{\epsilon}^*)=(\boldsymbol{\sigma}^*,(\mathbf{C}^0)^{-1}((\mathbf{C}^0)^{-1}-\mathbf{C}^{-1})^{-1}\boldsymbol{\epsilon}^*)-(\boldsymbol{\sigma}^*,\boldsymbol{\epsilon}^*), \tag{45.68}$$

and, therefore,

$$(\boldsymbol{\sigma}^*,(\mathbf{C}-\mathbf{C}^0)^{-1}\boldsymbol{\sigma}^*)=(\boldsymbol{\epsilon}^*,((\mathbf{C}^0)^{-1}-\mathbf{C}^{-1})^{-1}\boldsymbol{\epsilon}^*)-(\boldsymbol{\epsilon}^*,\mathbf{C}^0\boldsymbol{\epsilon}^*). \tag{45.69}$$

The sum of (45.65) and (45.69) provides

$$\begin{aligned}(\boldsymbol{\sigma}^*,(\mathbf{C}-\mathbf{C}^0)^{-1}\boldsymbol{\sigma}^*)+(\boldsymbol{\sigma}^*,\boldsymbol{\Gamma}\boldsymbol{\sigma}^*)=&-(\boldsymbol{\epsilon}^*,(\mathbf{C}^{-1}-(\mathbf{C}^0)^{-1})^{-1}\boldsymbol{\epsilon}^*)\\&-(\boldsymbol{\sigma},(\mathbf{C}^0)^{-1}\boldsymbol{\sigma}).\end{aligned} \tag{45.70}$$

If $(\mathbf{C}-\mathbf{C}^0)$ is negative definite, then $(\mathbf{C}^{-1}-(\mathbf{C}^0)^{-1})^{-1}$ is positive definite. Therefore, the above equation (45.70) proves that the functional of $\boldsymbol{\sigma}^*$ in (45.59) is negative definite.

Some numerical examples are shown in Figs. 45.3 and 45.4 for a composite material containing spherical inhomogeneities. The ratios between the elastic isotropic moduli of the spheres and the matrix are taken as $\mu_1/\mu_0 = 100$, $K_1/K_0 = 100$. Poisson's ratio is taken as $\nu_1 = \nu_0 = 0.3$. The average shear modulus and bulk modulus are shown as functions of the volume fraction c_1 of spheres. For comparison, the Voigt bounds (45.4) and (45.12), the Reuss bounds (45.16) and (45.17), Eshelby's approximations (45.33) and (45.34), and the self-consistent approximations (45.44) and (45.45) are also shown in the figures.

Other related works

Many papers have been published for predicting elastic properties of fibrous composites. The review papers by Hashin (1964) and by Chamis and Sendeckyj (1968), for instance, provide comprehensive lists of the

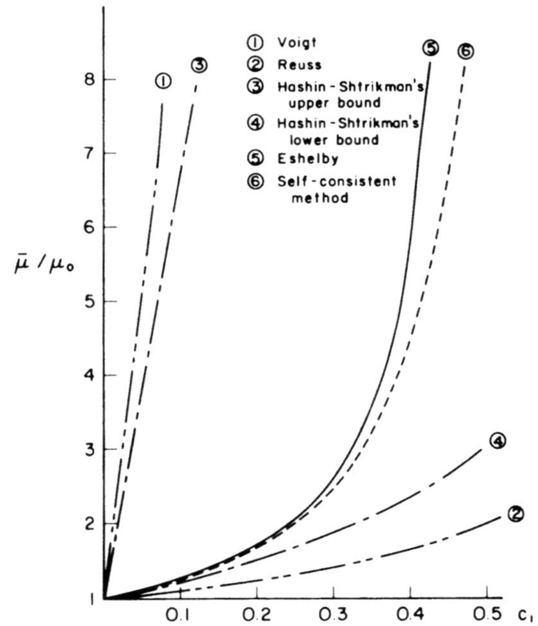

Fig. 45.3. The effective shear modulus versus the volume fraction of spherical inhomogeneities, $\mu_1/\mu_0 = 100$.

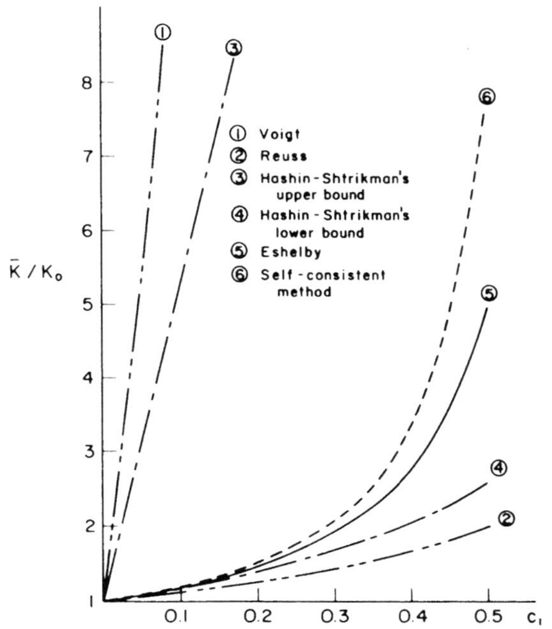

Fig. 45.4. The effective bulk modulus versus the volume fraction of spherical inhomogeneities, $K_1/K_0 = 100$.

literature. Various theories have been proposed which can be classified, according to Chamis and Sendeckyj, as netting analysis, mechanics of materials, self-consistent model, variational, exact, statistical, discrete element, semi-empirical models, and theories accounting for micro-structure.

For many steady-state and transient dynamic problems of structural mechanics, it is helpful and sometimes required to know the dispersion characteristics of free harmonic waves, i.e. the dependence of the phase velocity as the wavelength. For waves propagating in the direction of the fibers, a fiber-reinforced composite is essentially a system of wave guides. Achenback and Herrmann (1968) have derived the dispersion curves for transverse time-harmonic waves. Further dispersion curves are obtained by means of a smoothing operation (Sun, Achenbach and Herrmann 1968), and by variational methods (Kohn, Krumhansl and Lee 1972, Wheeler and Mura 1973, Nemat-Nasser and Minagawa 1975, and Nemat-Nasser and Horgan 1980).

The self-consistent imbedding approximation has been applied to rocks and porous media by many authors: Walsh (1968, 1969), Korringa

(1973), Anderson et al. (1974), O'Connelll and Budiansky (1974), Budiansky and O'Connell (1974), Edil et al. (1975), and Korringa and Thompson (1977).

Another important field of micromechanics is the mechanics of granular materials (soil, powder, rockfill). The works of Drucker and Prager (1952), Shield (1953, 1954), Mindlin (1954), and Mandl and Luque (1970), are phenomenological continuum theories. The microstructural continuum theories for granular materials have been developed by Mandel (1947), Spencer (1964), Horne (1965), Mogami (1965), Oda (1972), and Mehrabadi and Cowin (1978), Mróz (1980), among others (see Cowin and Satake 1978). An interesting application of the Cosserat continuum (1909) to granular materials has been given by Oshima (1953), Mindlin (1964), Eringen (1966), and Satake (1978). For finite plastic deformations of porous metals and geotechnical materials such as cohesionsless or cohesive soils, Nemat-Nasser and Shokooh (1980) have developed a theory which accounts for plastic volumetric changes, pressure sensitivity, microscopic frictional effects, and a nonassociative flow rule.

46. Plastic behavior of polycrystalline metals and composites

Taylor (1938) has evaluated the flow stress of a polycrystal in tension from the critical shear stress required to cause slip in a single crystal. He regards the polycrystal as an aggregate of randomly oriented FCC crystals that are assumed to be rigid-plastic. Hershey (1954) and T. H. Lin (1957) have considered the influence of elastic strains on the Taylor theory. They have assumed that the total strain is the same in each grain (single crystal) and outlined a procedure which would predict slipping in different slip systems within a grain with an increase in the prescribed total strain. Later, Kröner (1961), Budiansky and Wu (1962), Lin (1964), Hill (1965), and Hutchinson (1970), among others, have abandoned the assumption of a homogenous strain throughout the crystal and used instead the self-consistent method. These and related topics are reviewed in this chapter.

Taylor's analysis

The first realistic model for calculating the polycrystal uniaxial stress-strain relation from that of the single crystal was proposed by Taylor (1938). An aggregate of randomly oriented crystals under tension is assumed to be rigid-plastic. The associated average stress is denoted by $\bar{\sigma}_{ij}$ and the average plastic strain by $\bar{\epsilon}^p_{ij}$.

Suppose that grains of the polycrystal are FCC single crystals which have various orientations. Consider an arbitrary single grain. The relative orientations of the four slip planes of the FCC crystal are those found in the faces of a regular tetrahedron, and the three possible slip directions in each plane coincide with the edges of the triangular faces. The orientation of the tetrahedron *abcd*, relative to the cubic crystal axes x, y, z is shown in Fig. 46.1. Relative to an arbitrary set of Cartesian axes 1, 2, 3, the nth slip system ($n = 1, 2, \ldots, 12$) may be specified by the unit vector $m_i^{(n)}$ normal to the slip plane, and the unit vector $n_i^{(n)}$ in the slip direction.

Consider a uniform stress state σ_{ij} in a crystal of a given orientation. The resolved shear stress on the nth slip system is

$$\begin{aligned} \tau^{(n)} &= \sigma_{ij} m_j^{(n)} n_i^{(n)} \\ &= \sigma_{ij} \alpha_{ij}^{(n)}, \end{aligned} \tag{46.1}$$

and hence its rate is

$$\dot{\tau}^{(n)} = \dot{\sigma}_{ij} \alpha_{ij}^{(n)}, \tag{46.1'}$$

where

$$\alpha_{ij}^{(n)} = \tfrac{1}{2}(m_i^{(n)} n_j^{(n)} + m_j^{(n)} n_i^{(n)}). \tag{46.2}$$

The plastic strain in the grain due to the slips that have occurred along all the slip system is

$$\epsilon^p_{ij} = \sum_{n=1} \gamma^{(n)} \alpha_{ij}^{(n)}, \tag{46.3}$$

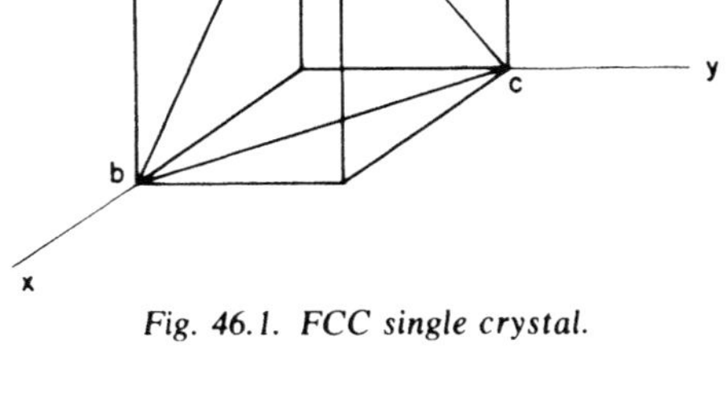

Fig. 46.1. FCC single crystal.

and its time derivative

$$\dot{\epsilon}^p_{ij} = \sum_{n=1} \dot{\gamma}^{(n)} \alpha^{(n)}_{ij}, \tag{46.4}$$

where $\gamma^{(n)}$ is the amount of plastic shear strain associated with $\tau^{(n)}$, the summation being taken over all active slip systems.

Taylor assumes that the plastic strain and its time rate are homogeneous and independent of the orientations of grains; in other words,

$$\epsilon^p_{ij} = \bar{\epsilon}^p_{ij}, \quad \dot{\epsilon}^p_{ij} = \dot{\bar{\epsilon}}^p_{ij}. \tag{46.5}$$

The plastic work done on all the crystals is

$$\bar{\sigma}_{ij}\dot{\bar{\epsilon}}^p_{ij} = \int_D \sigma_{ij}\dot{\epsilon}^p_{ij} dD = \int_D \sum_{n=1} \tau^{(n)}\dot{\gamma}^{(n)} dD, \tag{46.6}$$

where D is taken as a unit volume. Taylor further assumes that the strain hardening law chosen to represent the single crystal behavior, is

$$\tau_k = \tau^{(1)} = \tau^{(2)} = \cdots = F\left(\sum_n |\gamma^{(n)}|\right); \tag{46.7}$$

that is, the flow stress on each system in a grain is equal and is a function of $\Sigma_n |\gamma^{(n)}|$. Taylor's analysis follows the following steps. For a given $\dot{\bar{\epsilon}}^p_{ij}$, $\dot{\gamma}^{(n)}$ are calculated from

$$\dot{\bar{\epsilon}}^p_{ij} = \sum_n \dot{\gamma}^{(n)} \alpha^{(n)}_{ij} \tag{46.8}$$

and

$$\sum_n |\dot{\gamma}^{(n)}| = \text{minimum}, \tag{46.9}$$

where $\alpha^{(n)}_{ij}$ varies from grain to grain. The orientations of the grains are equally distributed in a stereographic projection. After $\dot{\gamma}^{(n)}$ are found, they are substituted into (46.7) to evaluate $\tau^{(n)}$. The average stress $\bar{\sigma}_{ij}$ is calculated from (46.6) by using $\dot{\bar{\epsilon}}^p_{ij}$, $\dot{\gamma}^{(n)}$ and $\tau^{(n)}$. The integral on the right-hand side in (46.6) is replaced by the integral shown in (45.6), since the orientations of the grains are equally distributed on the unit sphere associated with the Euler angles. A numerical example calculated by Taylor is shown in Fig. 46.2, where the load-extension curve ($P \sim \epsilon$) for a polycrystal aluminum is calculated from the shear stress–strain ($S \sim s$) curve of a single crystal aluminum. Points indicated by $\odot$ are data for the single crystal, $\times$ and $+$ show the experimental values of the polycrystal, and points $\bigcirc$ are the calculated values.

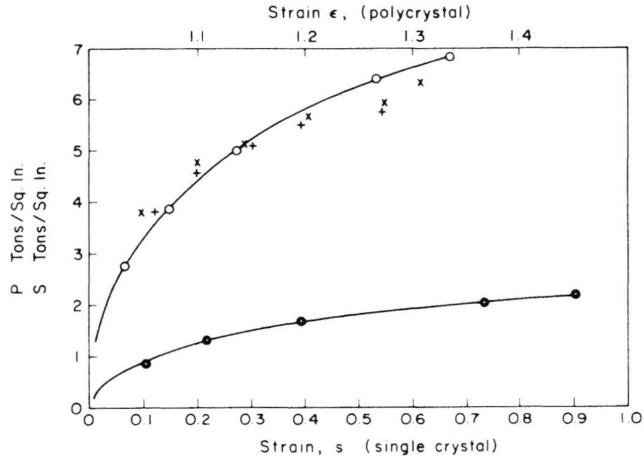

Fig. 46.2. Theoretical prediction of polycrystal stress-strain curve from a single crystal stress-strain curve (after G.I. Taylor 1938).

The minimum principle (46.9) has been proved by Bishop and Hill (1951) in the following manner. Let $d\epsilon_{ij}$ be a prescribed increment of strain in a single crystal and let σ_{ij} be a stress which will produce this strain by activating a set of shears $d\gamma^{(n)}$ with critical shear stress τ_k. Let $d\gamma^{*(n)}$ be any set of shears which are geometrically equivalent to the prescribed strain, but which are not necessarily operable by any stress satisfying the yield conditions; that is,

$$d\epsilon_{ij} = \sum_n d\gamma^{(n)}\alpha_{ij}^{(n)} = \sum_n d\gamma^{*(n)}\alpha_{ij}^{(n)} \tag{46.10}$$

and

$$\begin{aligned} &\sum_n \tau^{(n)} d\gamma^{(n)} = \tau_k \sum_n |d\gamma^{(n)}|, \\ &\sum_n \tau^{(n)} d\gamma^{*(n)} \leq \tau_k \sum_n |d\gamma^{*(n)}|. \end{aligned} \tag{46.11}$$

Multiplying (46.10) by σ_{ij}, we have, from (46.1),

$$\sum_n \tau^{(n)} d\gamma^{(n)} = \sum_n \tau^{(n)} d\gamma^{*(n)}. \tag{46.12}$$

Equations (46.11) and (46.12) lead to

$$\sum_n |d\gamma^{(n)}| < \sum_n |d\gamma^{*(n)}| \tag{46.13}$$

which is equivalent to the statement (46.9).

Taylor's hardening law (46.7) has been further studied and modified by Hutchinson (1964), Hill and Rice (1972), Havner and Shalaby (1977, 1978), Shalaby and Havner (1978), and Havner et al. (1979).

Self-consistent method

Taylor's analysis has been modified by using the idea of the self-consistent method explained in Section 45. The stress σ_{ij} and strain ϵ_{ij} in a grain are different from the average stress $\bar{\sigma}_{ij}$ and the average strain $\bar{\epsilon}_{ij}$. The difference of the plastic strain ϵ^p_{ij} and its average value $\bar{\epsilon}^p_{ij}$ is considered the cause of the non-uniformity. The misfit $\epsilon^p_{ij} - \bar{\epsilon}^p_{ij}$ is considered as the eigenstrain used for the spherical inclusion problem. The strain caused by the misfit can be written as $S_{ijkl}(\epsilon^p_{kl} - \bar{\epsilon}^p_{kl})$, where S_{ijkl} is the Eshelby tensor. If the material has no grains and is homogeneous everywhere, the strain is $\epsilon_{ij} = C^{-1}_{ijkl}\bar{\sigma}_{kl} + \bar{\epsilon}^p_{ij}$. However, when the material has the misfit, we have

$$\epsilon_{ij} = C^{-1}_{ijkl}\bar{\sigma}_{kl} + \bar{\epsilon}^p_{ij} + S_{ijkl}(\epsilon^p_{kl} - \bar{\epsilon}^p_{kl}). \tag{46.14}$$

The first term on the right-hand side of (46.14) is the average elastic strain. The last term is composed of the elastic and plastic strain disturbances, the sum of which is the total strain disturbance due to the misfit between ϵ^p_{ij} and $\bar{\epsilon}^p_{ij}$. On the other hand, it must hold that

$$\epsilon_{ij} = C^{-1}_{ijkl}\sigma_{kl} + \epsilon^p_{ij}, \tag{46.15}$$

where σ_{ij} is the stress in a grain. The overall average of σ_{ij} is $\bar{\sigma}_{ij}$. The overall average of ϵ_{ij} is $\bar{\epsilon}_{ij}$, and that of ϵ^p_{ij} is $\bar{\epsilon}^p_{ij}$. Multiplying the elastic moduli to the equation resulting from (46.14) = (46.15), we have

$$\sigma_{ij} = \bar{\sigma}_{ij} + C_{ijkl}\{S_{klmn}(\epsilon^p_{mn} - \bar{\epsilon}^p_{mn}) - (\epsilon^p_{kl} - \bar{\epsilon}^p_{kl})\} \tag{46.16}$$

or its time derivative

$$\dot{\sigma}_{ij} = \dot{\bar{\sigma}}_{ij} + C_{ijkl}\{S_{klmn}(\dot{\epsilon}^p_{mn} - \dot{\bar{\epsilon}}^p_{mn}) - (\dot{\epsilon}^p_{kl} - \dot{\bar{\epsilon}}^p_{kl})\}. \tag{46.16'}$$

The twelve slip systems are considered in a FCC single crystal of a given orientation. The resolved shear stress on the nth slip system can be expressed by (46.1). The plastic strain in the grain can also be given by (46.3).

$\tau^{(n)}$ and $\gamma^{(n)}$ are related by the plastic shear stress–shear strain relation of the single crystal,

$$\tau = \tau(\gamma) \quad \text{or } \dot{\tau} = \tau'\dot{\gamma}. \tag{46.17}$$

$\tau^{(n)}$ is the value of τ, and $\gamma^{(n)}$ is the value of γ for the nth slip system.

Equations (46.16′), (46.1′), (46.3), (46.17), and the following concept of average are fundamental ingredients in finding $\bar{\sigma}_{ij}$ as a function of $\bar{\epsilon}^p_{ij}$. The average of ϵ^p_{ij} for all possible orientations of single crystals (grains) is $\bar{\epsilon}^p_{ij}$. Namely,

$$\bar{\epsilon}^p_{ij} = (\epsilon^p_{ij})_{\text{average}}. \tag{46.18}$$

$\bar{\sigma}_{ij}$ starts from zero. $\sigma_{ij} = \bar{\sigma}_{ij}$ until $\tau^{(n)}$ (resolved shear stress) in some grains reaches the elastic limit τ_y. If $\bar{\sigma}_{ij}$ is increased by $\dot{\bar{\sigma}}_{ij}$, then ϵ^p_{ij} is increased by $\dot{\epsilon}^p_{ij}$ from zero value. In order to find $\dot{\epsilon}^p_{ij}$ we must solve (46.16′) by assuming $\dot{\bar{\epsilon}}^p_{mn} = 0$ (initial value). Since $\dot{\sigma}_{ij}$ is also a function of $\dot{\epsilon}^p_{ij}$, it is convenient to rewrite (46.16′) in another form. Multiplying (46.16′) by $\alpha^{(n)}_{ij}$ (since we know which grains and which slip systems are plastically active at this moment, we know $\alpha^{(n)}_{ij}$) we have, from (46.1′),

$$\dot{\tau}^{(n)} = \dot{\bar{\sigma}}_{ij}\alpha^{(n)}_{ij} + \alpha^{(n)}_{ij}C_{ijkl}\{S_{klmn}(\dot{\epsilon}^p_{mn} - \dot{\bar{\epsilon}}^p_{mn}) - (\dot{\epsilon}^p_{kl} - \dot{\bar{\epsilon}}^p_{kl})\}. \tag{46.19}$$

Furthermore, from (46.4) and (46.17) we have

$$\tau'\dot{\gamma}^{(n)} = \dot{\bar{\sigma}}_{ij}\alpha^{(n)}_{ij} + \alpha^{(n)}_{ij}C_{ijkl}\left\{S_{klmn}\left(\sum_{n=1}\dot{\gamma}^{(n)}\alpha^{(n)}_{mn} - \dot{\bar{\epsilon}}^p_{mn}\right) - \left(\sum_{n=1}\dot{\gamma}^{(n)}\alpha^{(n)}_{kl} - \dot{\bar{\epsilon}}^p_{kl}\right)\right\}, \tag{46.20}$$

where $n = 1, 2, \ldots, 12$ when all slip systems are plastically active. Some of the n slip systems are not active in the initial stage after the elastic limit. The unknown quantities are $\dot{\gamma}^{(n)}$ in (46.20) ($\dot{\bar{\epsilon}}^p_{ij} = 0$). The solutions of $\dot{\gamma}^{(n)}$ are substituted into (46.4) to obtain $\dot{\epsilon}^p_{ij}$. $\dot{\bar{\epsilon}}^p_{ij}$ is not zero for the calculation of the second stage when $\bar{\sigma}_{ij}$ is further increased. $\dot{\bar{\epsilon}}^p_{ij}$ is calculated from $\dot{\epsilon}^p_{ij}$ by

$$\dot{\bar{\epsilon}}^p_{ij} = (1/8\pi^2)\int_0^{2\pi} d\beta \int_0^{\pi} \sin\theta d\theta \int_0^{2\pi} \dot{\epsilon}^p_{ij} d\phi, \tag{46.21}$$

where the Euler angles (β, θ, ϕ) (Fig. 45.2) relate the orientation of stress axes (1, 2, 3) to that of the crystal axes (x, y, z). The new increment $\dot{\epsilon}^p_{ij}$ provides an increment of $\dot{\gamma}^{(n)}$ in (46.20) by giving new $\dot{\bar{\sigma}}_{ij}$ and $\dot{\bar{\epsilon}}^p_{ij}$, where $\dot{\bar{\sigma}}_{ij}$ is arbitrarily chosen and $\dot{\bar{\epsilon}}^p_{ij}$ is calculated by (46.21) in the preceding calculation. The same step-by-step calculation is continued. $\bar{\epsilon}_{ij}$ is easily calculated by

$$\bar{\epsilon}_{ij} = C^{-1}_{ijkl}\bar{\sigma}_{kl} + \bar{\epsilon}^p_{ij}. \tag{46.22}$$

A more general plastic shear stress-strain relation is

$$\dot{\tau}^{(n)} = \sum_{m=1} h^{(nm)}\dot{\gamma}^{(m)} \tag{46.23}$$

which may be used for the left-hand side of (46.20). In this calculation, we must keep in mind the loading and unloading conditions as well as the condition that $\dot{\tau}^{(n)}$ and $\dot{\gamma}^{(n)}$ must have the same sign.

The method of solving (46.20) is essentially the one proposed by Kröner (1961), Budiansky and T. T. Wu (1962), T. H. Lin (1964) and Hutchinson (1970). The method proposed by Hill (1965) is slightly different. Equation (46.15) is written as

$$\sigma_{ij} = C_{ijkl}(\epsilon_{kl} - \epsilon^{p}_{kl})$$

or (46.24)

$$\dot{\sigma}_{ij} = C_{ijkl}(\dot{\epsilon}_{kl} - \dot{\epsilon}^{p}_{kl}).$$

When (46.24) is multiplied by $\alpha^{(n)}_{ij}$, we have

$$\dot{\tau}^{(n)} = \alpha^{(n)}_{ij} C_{ijkl}(\dot{\epsilon}_{kl} - \dot{\epsilon}^{p}_{kl}). \tag{46.25}$$

Substituting (46.23) and (46.4) into (46.25), we obtain

$$\sum_{m=1} h^{(nm)}\dot{\gamma}^{(m)} + \alpha^{(n)}_{ij} C_{ijkl} \sum_{m=1} \dot{\gamma}^{(m)}\alpha^{(m)}_{kl} = \alpha^{(n)}_{ij} C_{ijkl}\dot{\epsilon}_{kl}. \tag{46.26}$$

By solving (46.26) to express $\dot{\gamma}^{(n)}$ in terms of $\dot{\epsilon}_{kl}$, we have

$$\dot{\gamma}^{(n)} = f^{(n)}_{kl}\dot{\epsilon}_{kl}. \tag{46.27}$$

Then, (46.24) becomes

$$\dot{\sigma}_{ij} = C_{ijkl}\left(\dot{\epsilon}_{kl} - \sum_{n=1} f^{(n)}_{pq}\alpha^{(n)}_{kl}\dot{\epsilon}_{pq}\right). \tag{46.28}$$

If (46.28) is substituted into (46.16′), we have

$$C_{ijkl}\left(\dot{\epsilon}_{kl} - \sum_{n=1} f^{(n)}_{pq}\alpha^{(n)}_{kl}\dot{\epsilon}_{pq}\right) = \dot{\bar{\sigma}}_{ij} + C_{ijkl}\{S_{klmn}(\dot{\epsilon}^{p}_{mn} - \dot{\bar{\epsilon}}^{p}_{mn}) - (\dot{\epsilon}^{p}_{kl} - \dot{\bar{\epsilon}}^{p}_{kl})\}. \tag{46.29}$$

From (46.29) we can calculate $\dot{\epsilon}^{p}_{ij}$ in terms of given values of $\dot{\bar{\sigma}}_{ij}$ and the initial value of $\dot{\bar{\epsilon}}^{p}_{ij}$ in the preceding calculation. The new $\dot{\bar{\epsilon}}^{p}_{ij}$ is calculated from $\dot{\epsilon}^{p}_{ij}$ by (46.21).

The early work by Lin (1957) used (46.20) without the second term, in which he assumed $\dot{\epsilon}^{p}_{ij} = \dot{\bar{\epsilon}}^{p}_{ij}$. Figure 46.3 shows the numerical example obtained by Hutchinson (1970), showing the comparison of various theoretical tensile stress-plastic strain curves for FCC polycrystals composed of isotropic and non-hardening single crystal grains. The Lin model refers to Lin's early work (1957). The K. B. W. model stands for the works by Kröner, Budiansky and Wu. These curves are obtained by assuming that the single crystals have no work-hardening. In this case

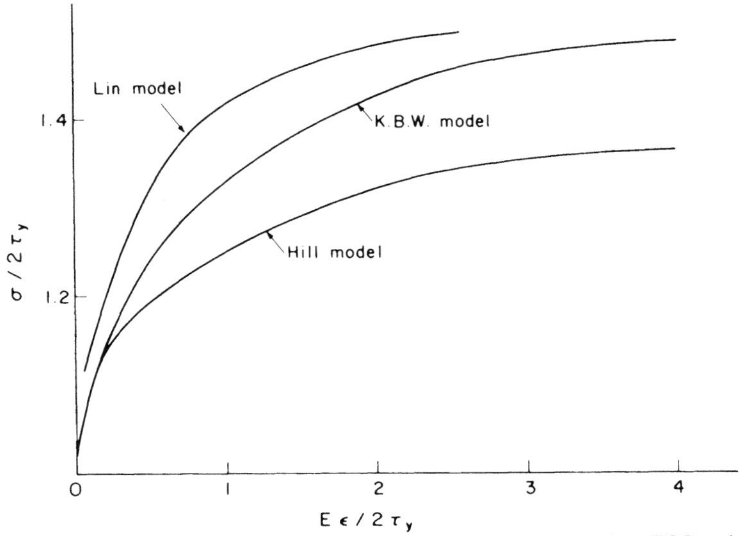

Fig. 46.3. Various theoretical predictions of stress-strain curves of a FCC polycrystal composed of isotropic and non-hardening single crystal grains (after Hutchinson 1970).

the left-hand side in (46.20) is zero after yielding. The yield shear stress is denoted by τ_y.

Equation (46.20) can also be used for determining the yield surface of a polycrystal. Lin and Ito (1966) and Lin (1971) have assumed that a polycrystal consists of identical basic cubic blocks of 64 cube-shaped crystals having different orientations. The calculation is similar to the cubic inclusion problem. Further interesting studies on the yield surface have been done by Mandel (1966), Bui (1970), Zarka (1972, 1973), and Weng and Phillips (1977), among others.

For the first original work to determine the yield function from multi-slip of an isotropic aggregate of FCC crystals the reader is referred to the two papers by Bishop and Hill (1951). Recently, Bassani (1977) has examined the yield function of FCC (and BCC) metals whose texture gives rise to transversely isotropic plastic properties.

Chen and Argon (1979) employed the self-consistent theory to consider the effect of grain boundary sliding in creeping polycrystals and composites where the sliding boundaries are treated as disk-shaped viscous inhomogeneities and the matrix (grain interior) follows a power law in creep. Raj and Ashby (1971) and Raj (1975) considered a similar problem and investigated the nucleation of intergranular cracks during steady state and cyclic creep. Their analysis is based upon stress concentrations at inclusions which are produced by grain boundary sliding.

Ghahremani (1980) has investigated the effect of grain boundary sliding on anelasticity of polycrystalline materials. Recently, Weng (1981) has employed the self-consistent method for predicting the creep and recovery strain of an aluminum alloy.

Embedded weakened zone

If a material contains a weakened zone, Ω, the zone can be simulated by an ellipsoidal inhomogeneous inclusion. For simplicity it is assumed that the principal axes of the ellipsoid are parallel to the coordinate axes. The two phases Ω and the matrix have different shear moduli μ^* and μ, respectively, and different resistances for plastic deformation. It is assumed that an applied stress σ^0_{12} causes uniform plastic strain ϵ^α_{12} in Ω, and uniform plastic strain ϵ^β_{12} in the matrix. The stress disturbance σ_{ij} can be obtained from (22.19) where ϵ^p_{ij} is replaced by $(\epsilon^\alpha_{12} - \epsilon^\beta_{12})$. For a single weakened zone, the stress in Ω is obtained as

$$\sigma_{12} = -(1 - 2S_{1212})\frac{2\mu^*(\epsilon^\alpha_{12} - \epsilon^\beta_{12}) + (1 - \mu^*/\mu)\sigma^0_{12}}{1 + 2(\mu^*/\mu - 1)S_{1212}}. \tag{46.30}$$

The strain disturbance ϵ_{ij} in Ω is

$$\epsilon_{12} = 2S_{1212}\frac{(\mu^*/\mu)(\epsilon^\alpha_{12} - \epsilon^\beta_{12}) + (1 - \mu^*/\mu)\sigma^0_{12}/2\mu}{1 + 2(\mu^*/\mu - 1)S_{1212}}. \tag{46.31}$$

The total stress and strain in Ω are

$$\begin{aligned}\sigma^t_{12} &= \sigma^0_{12} + \sigma_{12},\\ \epsilon^t_{12} &= \epsilon^0_{12} + \epsilon_{12},\end{aligned} \tag{46.32}$$

where

$$\sigma^0_{12} = 2\mu\epsilon^0_{12}. \tag{46.33}$$

Then, we have

$$\sigma^t_{12} = -2\mu(1/2S_{1212} - 1)\epsilon^t_{12} + \sigma^0_{12}/2S_{1212}. \tag{46.34}$$

Eshelby's tensor S_{1212} becomes 7/30 for a spherical Ω and 15/30 for a penny-shaped Ω. The above equation has been obtained by Rudnicki (1977). σ^t_{12} versus ϵ^t_{12} is shown in Fig. 46.4 by the straight line ACDB for a given σ^0_{12}. If the constitutive equation of the weakened zone is expressed by curve OCP, point C is uniquely determined for a given σ^0_{12} as the

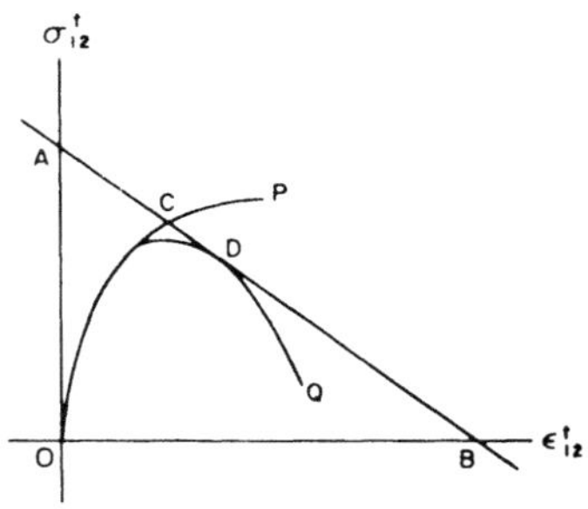

Fig. 46.4. Stress-strain state inside an inclusion is represented by an intersection of AB and OP (or OQ).

intersection of the straight line and the curve. On the other hand, if the constitutive equation is expressed by curve ODQ, for instance, the plastic flow inside Ω becomes unstable, perhaps cracks will be initiated to accommodate deformation. Rudnicki's theory is used by Rice and Rudnicki (1979) in a study of earthquake premonitory processes. It is argued that the fluid coupling effects serve to stabilize the weakened rock against rapid fracture, and give rise instead to a precursory period of accelerating, but initially quasi-static straining which ultimately leads to dynamic instability.

47. Viscoelasticity of composite materials

Eshelby (1957) has pointed out that the elastic theory of inclusions can be extended to viscoelastic materials. The viscoelastic creep of composite materials, for instance, can be evaluated by the viscoelastic theory of inclusions.

Homogeneous inclusions

Consider an infinitely extended viscoelastic material, and let an ellipsoidal sub-domain Ω of the material be subjected to an eigenstrain $\epsilon^*_{ij}(\mathbf{x}, t)$. The inclusion Ω and the matrix $D - \Omega$ are assumed to have the same viscoelastic properties. The stress and strain are denoted by σ_{ij} and ϵ_{ij}, respectively. Equations (2.1), (2.2) and (2.10) still hold,

$$\epsilon_{ij} = e_{ij} + \epsilon^*_{ij}, \tag{47.1}$$

where $\epsilon^*_{ij} = 0$ in $D - \Omega$. Instead of Hooke's law, we now consider

$$\sigma_{ij}(\mathbf{x}, t) = L_{ijkl}(t) e_{kl}(\mathbf{x}, 0) + \int_0^\infty L_{ijkl}(t - \tau) \frac{de_{kl}(\mathbf{x}, \tau)}{d\tau}\, d\tau \tag{47.2}$$

or

$$e_{ij}(\mathbf{x}, t) = M_{ijkl}(t)\sigma_{kl}(\mathbf{x}, 0) + \int_0^\infty M_{ijkl}(t-\tau)\frac{d\sigma_{kl}(\mathbf{x}, \tau)}{d\tau}\,d\tau \tag{47.3}$$

assuming that $e_{ij}(\mathbf{x}, t)$, $\sigma_{ij}(\mathbf{x}, t)$, $L_{ijkl}(t)$ and $M_{ijkl}(t)$ are all zero for $t<0$. L_{ijkl} and M_{ijkl} are called the tensorial relaxation and creep functions, respectively, and have the same properties of symmetry with respect to the indices as the elastic moduli and compliances. Let the Laplace transform (A3.21) of a function $f(t)$ be indicated by a bar,

$$\bar{f}(s) = \int_0^\infty f(t)\exp(-st)dt, \quad s>0. \tag{47.4}$$

The Laplace transform of (47.1) and (47.2) are obtained as

$$\bar{\epsilon}_{ij} = \bar{e}_{ij} + \bar{\epsilon}^*_{ij}, \tag{47.5}$$

and

$$\bar{\sigma}_{ij} = s\bar{L}_{ijkl}\bar{e}_{kl}, \tag{47.6}$$

where

$$\int_0^\infty M(t-\tau)\exp(-st)dt = \bar{M}\exp(-s\tau)$$

and

$$\int_0^\infty \frac{de}{d\tau}\exp(-s\tau)d\tau = -e(0) + s\bar{e}$$

are used. Equations (2.10) for equilibrium are transformed to

$$\bar{\sigma}_{ij,j} = 0 \tag{47.7}$$

and equations (2.2) become

$$\bar{\epsilon}_{ij} = (1/2)(\bar{u}_{i,j} + \bar{u}_{j,i}). \tag{47.8}$$

The system of equations (47.5) ~ (47.8) becomes completely equivalent to the elastic case, where the elastic moduli C_{ijkl} correspond to $s\bar{L}_{ijkl}$.

For isotropic materials, we have

$$L_{ijkl}(t) = \lambda(t)\delta_{ij}\delta_{kl} + \mu(t)(\delta_{ik}\delta_{jl} + \delta_{il}\delta_{jk}) \tag{47.9}$$

and

$$\bar{L}_{ijkl} = \bar{\lambda}\delta_{ij}\delta_{kl} + \bar{\mu}(\delta_{ik}\delta_{jl} + \delta_{il}\delta_{jk}) \tag{47.9'}$$

(see Fung 1965). λ and μ in (A2.1) correspond to $s\bar{\lambda}$ and $s\bar{\mu}$ for the viscoelasticity. When an ellipsoidal inclusion has a uniform eigenstrain ϵ_{ij}^*, the Laplace transform of the stress in Ω is obtained from (11.20), by changing μ to $s\bar{\mu}$ and ν to $\bar{\nu} = (1/2)\bar{\lambda}/(\bar{\lambda} + \bar{\mu})$,

$$\begin{aligned}\bar{\sigma}_{11}/2s\bar{\mu} = &\left[\frac{a_1^2}{8\pi(1-\bar{\nu})}\left\{\frac{1-\bar{\nu}}{1-2\bar{\nu}}3I_{11} + \frac{\bar{\nu}}{1-2\bar{\nu}}(I_{21}+I_{31})\right\}\right.\\ &\left.+\frac{1-2\bar{\nu}}{8\pi(1-\bar{\nu})}\left\{\frac{1-\bar{\nu}}{1-2\bar{\nu}}I_1 - \frac{\bar{\nu}}{1-2\bar{\nu}}(I_2+I_3)\right\} - \frac{1-\bar{\nu}}{1-2\bar{\nu}}\right]\overline{\epsilon_{11}^*}\\ &+\left[\frac{a_2^2}{8\pi(1-\bar{\nu})}\left\{\frac{1-\bar{\nu}}{1-2\bar{\nu}}I_{12} + \frac{\bar{\nu}}{1-2\bar{\nu}}(3I_{22}+I_{32})\right\}\right.\\ &\left.-\frac{1-2\bar{\nu}}{8\pi(1-\bar{\nu})}\left\{\frac{1-\bar{\nu}}{1-2\bar{\nu}}I_1 - \frac{\bar{\nu}}{1-2\bar{\nu}}(I_2-I_3)\right\} - \frac{\bar{\nu}}{1-2\bar{\nu}}\right]\overline{\epsilon_{22}^*}\\ &+\left[\frac{a_3^2}{8\pi(1-\bar{\nu})}\left\{\frac{1-\bar{\nu}}{1-2\bar{\nu}}I_{13} + \frac{\bar{\nu}}{1-2\bar{\nu}}(3I_{33}+I_{23})\right\}\right.\\ &\left.-\frac{1-2\bar{\nu}}{8\pi(1-\bar{\nu})}\left\{\frac{1-\bar{\nu}}{1-2\bar{\nu}}I_1 - \frac{\bar{\nu}}{1-2\bar{\nu}}(I_3-I_2)\right\} - \frac{\bar{\nu}}{1-2\bar{\nu}}\right]\overline{\epsilon_{33}^*},\\ \bar{\sigma}_{12}/2s\bar{\mu} = &\left\{\frac{a_1^2+a_2^2}{8\pi(1-\bar{\nu})}I_{12} + \frac{1-2\bar{\nu}}{8\pi(1-\bar{\nu})}(I_1+I_2) - 1\right\}\overline{\epsilon_{12}^*},\end{aligned} \tag{47.10}$$

and other components are obtained by the cyclic permutation of (1, 2, 3). The inverse transform of $\bar{\sigma}_{ij}$ is, from (A3.22),

$$\sigma_{ij}(\mathbf{x}, t) = (1/2\pi i)\int_{\gamma-i\infty}^{\gamma+i\infty} \bar{\sigma}_{ij}(\mathbf{x}, s)\exp(st)ds. \tag{47.11}$$

The above result can be easily extended to more general eigenstrains and anisotropic materials. The Laplace transform of the displacement is obtained from (6.2) as

$$\bar{u}_i = -i(2\pi)^{-3}\int_{-\infty}^{\infty}\int_{\Omega} s\bar{L}_{jlmn}\overline{\epsilon_{mn}^*}\xi_l(\bar{N}_{ij}/\bar{D})\exp\{i\boldsymbol{\xi}\cdot(\mathbf{x}-\mathbf{x}')\}d\boldsymbol{\xi}d\mathbf{x}', \tag{47.12}$$

where $\bar{N}_{ij}$ and $\bar{D}$ are the cofactors and the determinant of the matrix having the ij element $s\bar{L}_{imjn}\xi_m\xi_n$. The above equation has been used by Kuo and Mura (1973) for solving a circular twist disclination in viscoelastic materials.

Inhomogeneous inclusions

The equivalent inclusion method in Section 22 can be used for the viscoelastic creep of composite materials. The constitutive equation of the matrix is assumed to be (47.2). On the other hand, the inhomogeneities behave as

$$\sigma_{ij}^{0}(t) + \sigma_{ij}(\mathbf{x}, t) = L_{ijkl}^{*}(t)\{\epsilon_{kl}^{0}(0) + \epsilon_{kl}(\mathbf{x}, 0)\} + \int_{0}^{\infty} L_{ijkl}^{*}(t-\tau)\frac{d}{d\tau}\{\epsilon_{kl}^{0}(\tau) + \epsilon_{kl}(\mathbf{x}, \tau)\}d\tau, \tag{47.13}$$

where

$$\sigma_{ij}^{0}(t) = L_{ijkl}(t)\epsilon_{kl}^{0}(0) + \int_{0}^{\infty} L_{ijkl}(t-\tau)\frac{d}{d\tau}\epsilon_{kl}^{0}(\tau)d\tau \tag{47.14}$$

is the applied stress at infinity, and $\epsilon_{ij}^{0}(t)$ is the corresponding strain. σ_{ij} in (47.2) and (47.13) is the stress disturbance due to the inhomogeneities. The equivalent inclusion method is to simulate σ_{ij} by the stress caused by homogeneous inclusions with proper ϵ_{ij}^{*}. The equivalency is obtained by the equation (47.13) = (47.14) + (47.2). The Laplace transform of this equation is

$$s\overline{L_{ijkl}^{*}}(\bar{\epsilon}_{kl}^{0} + \bar{\epsilon}_{kl}) = s\bar{L}_{ijkl}(\bar{\epsilon}_{kl}^{0} + \bar{\epsilon}_{kl} - \overline{\epsilon_{kl}^{*}}) \tag{47.15}$$

which can be obtained directly from (22.5). $\bar{\epsilon}_{kl}$ or $\bar{\sigma}_{ij}$ is a linear function of $\overline{\epsilon_{ij}^{*}}$. Equation (47.15) determines $\overline{\epsilon_{ij}^{*}}$ necessary for the equivalent inclusion. The stress state can be completely determined after ϵ_{ij}^{*} for a given $\sigma_{ij}^{0}(t)$ is evaluated.

Laws and McLaughlin (1978) have estimated the viscoelastic creep compliances of composite materials and have given numerical solutions for two common composite materials, namely an isotropic dispersion of spheres, and an uni-directional fibre reinforced material. Viscous liquids can be treated in the same manner as in viscoelastic materials if velocity is used in place of displacement. Bilby, Eshelby and Kundu (1975), and Howard and Brierley (1976) have found numerically the change of a viscous inhomogeneity inside a viscous liquid of different viscosity. Hashin (1965) has obtained the macroscopic viscoelastic heterogeneous media in terms of effective relaxation moduli and creep compliances. Recently, Budiansky and Hutchinson (1980) have determined the growth-rate and shape for an isolated void growing in a self-similar way in an infinite block of incompressible power-law, viscous material (see also Budiansky, Hutchinson and Lutsky 1980). Taya and Seidel (1981) have obtained the void growth rate when a viscous material contains many voids. They consider the interaction among the voids by using Mori-Tanaka's method.

Waves in an infinite medium

Consider an infinite medium which consists of an isotropic matrix and ellipsoidal inhomogeneities Ω with volume fraction c. The matrix and the inhomogeneities are assumed to be different standard linear solids. Setting $D \equiv \partial/\partial t$, the stress–strain relations are written as

$$\left.\begin{aligned} P_1(D)'\sigma_{ij} &= Q_1(D)'e_{ij} \\ P_2(D)\sigma_{kk} &= Q_2(D)e_{kk} \end{aligned}\right\} \text{ in the matrix} \tag{47.16}$$

and

$$\left.\begin{aligned} P_1^*(D)'\sigma_{ij} &= Q_1^*(D)'e_{ij} \\ P_2^*(D)\sigma_{kk} &= Q_2^*(D)e_{kk} \end{aligned}\right\} \text{ in } \Omega, \tag{47.17}$$

where $'\sigma_{ij}$ and $'e_{ij}$ are the reduced stress and strain defined by (A2.6), and $P(D)$ and $Q(D)$ are polynomials of D.

Consider an applied strain $\epsilon_{ij}^0 = \bar{\epsilon}_{ij}^0 \exp(i\omega t)$. The strain and stress disturbances due to the inhomogeneities are also assumed to have the form $\epsilon_{ij} = \bar{\epsilon}_{ij} \exp(i\omega t)$ and $\sigma_{ij} = \bar{\sigma}_{ij} \exp(i\omega t)$. If the inhomogeneities are simulated by inclusions with eigenstrain $\epsilon_{ij}^* = \bar{\epsilon}_{ij}^* \exp(i\omega t)$, the equivalency equations become

$$\begin{aligned} &\{Q_1^*(i\omega)/P_1^*(i\omega)\}('\bar{\epsilon}_{ij}^0 + '\bar{\epsilon}_{ij}) = \{Q_1(i\omega)/P_1(i\omega)\}('\bar{\epsilon}_{ij}^0 + '\bar{\epsilon}_{ij} - '\bar{\epsilon}_{ij}^*), \\ &\{Q_2^*(i\omega)/P_2^*(i\omega)\}(\bar{\epsilon}_{kk}^0 + \bar{\epsilon}_{kk}) = \{Q_2(i\omega)/P_2(i\omega)\}(\bar{\epsilon}_{kk}^0 + \bar{\epsilon}_{kk} - \bar{\epsilon}_{kk}^*). \end{aligned} \tag{47.18}$$

The above equations determine $\bar{\epsilon}_{ij}^*$ for a given set of $\bar{\epsilon}_{ij}^0$ since $\bar{\epsilon}_{ij}$ is a linear function of $\bar{\epsilon}_{ij}^*$ as shown by (11.15), that is,

$$\bar{\epsilon}_{ij} = S_{ijkl}\bar{\epsilon}_{kl}^*. \tag{47.19}$$

Comparing with (22.17), the following correspondence holds:

$$\begin{aligned} 2\mu^* &= Q_1^*(i\omega)/P_1^*(i\omega), \quad & K^* &= Q_2^*(i\omega)/P_2^*(i\omega), \\ 2\mu &= Q_1(i\omega)/P_1(i\omega), & K &= Q_2(i\omega)/P_2(i\omega). \end{aligned} \tag{47.20}$$

Eshelby's tensor S_{ijkl} in (47.19) contains Poisson's ratio of the matrix, which is expressed as

$$\nu = (3K - 2\mu)/2(3K + \mu) \tag{47.21}$$

from (A2.2). K and μ are, of course, functions of $P(i\omega)$ and $Q(i\omega)$ through (47.20).

The average (or effective) shear modulus $\bar{\mu}$ and bulk modulus $\bar{K}$ are determined from (45.33) and (45.34), where $\mu_0 = \mu$, $K_0 = K$, $\mu_r = \mu^*$, $K_r = K^*$ and $c_r = c$. In the above discussion we have assumed that the state of deformation is quasi-static. When a dynamic state of the wave propagation with angular frequency ω is considered, we assume that the material will respond by the complex moduli $\bar{\mu}$ and $\bar{K}$ obtained in the above discussion. Then, the equations of motion become

$$\bar{\mu}\bar{u}_{i,jj} + (\bar{\mu}/3 + \bar{K})\bar{u}_{j,ji} + \rho\omega^2\bar{u}_i = 0 \tag{47.22}$$

or

$$(\bar{\lambda} + 2\bar{\mu})\bar{u}_{j,ji} - \bar{\mu}\epsilon_{ijk}\epsilon_{kmn}\bar{u}_{n,mj} + \rho\omega^2\bar{u}_i = 0, \tag{47.22'}$$

where ρ is the average density of the material and $\bar{\lambda} + 2\bar{\mu} = \bar{K} + 4\bar{\mu}/3$. The solutions for (47.22) are the real parts of

$$\bar{u}_i = An_i \exp(i\omega[t \pm \{\rho/(\bar{K} + 4\bar{\mu}/3)\}^{1/2} n_k x_k]) \tag{47.23}$$

and

$$\bar{u}_i = C_i \exp(i\omega[t \pm (\rho/\bar{\mu})^{1/2} n_k x_k]), \tag{47.24}$$

where A, C_i, n_i are arbitrary constants and $C_j n_j = 0$. Equation (47.23) represents a plane dilatational wave and equation (47.24) represents a plane shear wave. The real parts of the factors in front of $n_k x_k$ are the inverse of wave velocities, v_D, v_S; whereas the imaginary parts are attenuation factors α_D, α_S. We have

$$v_D = [Re\{\rho/(\bar{K} + 4\bar{\mu}/3)\}^{1/2}]^{-1}, \quad \alpha_D = -\omega Im\{\rho/(\bar{K} + 4\bar{\mu}/3)\}^{1/2}, \tag{47.25}$$

for dilational waves, and

$$v_S = [Re(\rho/\bar{\mu})^{1/2}]^{-1}, \quad \alpha_S = -\omega Im(\rho/\bar{\mu})^{1/2}, \tag{47.26}$$

for shear waves.

Walsh (1968, 1969) has calculated the wave velocities and attenuations in a partially melted rock, where the melted zone is simulated by a penny-shaped inclusion of liquid. The seismic velocities of cracked rocks also are calculated by Anderson et al. (1974), O'Connell and Budiansky (1974), and Budiansky and O'Connell (1976), among others.

48. Elastic wave scattering

The subject of multiple scattering of waves is of interest in many fields of engineering and science. In acoustics, it has important practical applications in studies of the distribution of flaws in solids, fiber reinforced composites, porous media, rocks, earth, underwater signal transmissions, etc.

The basic scattering problem is shown in Fig. 48.1. An incident

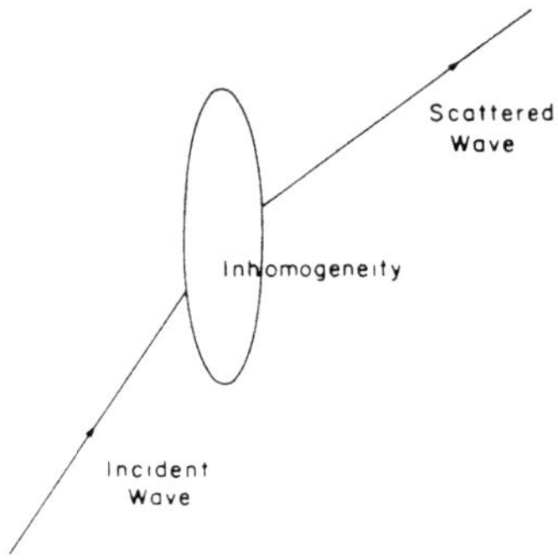

Fig. 48.1. Scattering of elastic wave due to an inhomogeneity.

plane wave with angular frequency ω and wave vector $\mathbf{k}$ is expressed by

$$\begin{aligned} u_i^0(\mathbf{x}, t) &= u_i^0(\mathbf{x}) \exp(-i\omega t), \\ u_i^0(\mathbf{x}) &= u_i^A \exp(i\mathbf{k} \cdot \mathbf{x}), \end{aligned} \tag{48.1}$$

the scattered wave by

$$u_i^s(\mathbf{x}, t) = u_i^s(\mathbf{x}) \exp(-i\omega t) \tag{48.2}$$

and the total wave amplitude by

$$u_i(\mathbf{x}) = u_i^0(\mathbf{x}) + u_i^s(\mathbf{x}). \tag{48.3}$$

The equations of motion are

$$C_{ijkl} u_{k,lj}(\mathbf{x}) + \rho\omega^2 u_i(\mathbf{x}) = 0 \tag{48.4}$$

in the matrix and

$$C^*_{ijkl} u_{k,lj}(\mathbf{x}) + \rho^*\omega^2 u_i(\mathbf{x}) = 0 \tag{48.5}$$

in the inhomogeneity Ω, where the elastic moduli and the density in Ω are distinguished by an asterisk, *. Equation (48.5) can be written as

$$C_{ijkl} u_{k,lj}(\mathbf{x}) + \rho\omega^2 u_i(\mathbf{x}) + X_i(\mathbf{x}) = 0, \tag{48.6}$$

where

$$X_i(\mathbf{x}) = \Delta C_{ijkl} u_{k,lj}(\mathbf{x}) + \Delta\rho\omega^2 u_i(\mathbf{x}) \tag{48.7}$$

and

$$\begin{aligned} C^*_{ijkl} &= C_{ijkl} + \Delta C_{ijkl}, \\ \rho^* &= \rho + \Delta\rho. \end{aligned} \tag{48.8}$$

The quantity $X_i(\mathbf{x})$ is equivalent to a body force distributed in the domain Ω. The solution of a unit body force is known as the steady-state elastic wave Green's function (9.37). Then, for the distributed body force X_i we have

$$u_i(\mathbf{x}) = u_i^0(\mathbf{x}) + \int_\Omega \{\Delta C_{mjkl} u_{k,lj}(\mathbf{x}') + \Delta\rho\omega^2 u_m(\mathbf{x}')\} g_{im}(\mathbf{x}-\mathbf{x}')d\mathbf{x}' \tag{48.9}$$

or

$$u_i(\mathbf{x}) = u_i^0(\mathbf{x}) + \Delta\rho\omega^2 \int_\Omega u_m(\mathbf{x}') g_{im}(\mathbf{x}-\mathbf{x}')d\mathbf{x}' + \Delta C_{mjkl} \int_\Omega u_{k,l}(\mathbf{x}') g_{im,j}(\mathbf{x}-\mathbf{x}')d\mathbf{x}' \tag{48.10}$$

after integration by parts.

For an isotropic material, $g_{im}(\mathbf{x}-\mathbf{x}')$ is known from (9.40). The far-field scattered amplitudes can be calculated by using the following asymptotic equalities:

$$\begin{aligned} &1/|\mathbf{x}-\mathbf{x}'| \approx 1/x, \\ &|\mathbf{x}-\mathbf{x}'| \approx x - (\mathbf{x}'\cdot\bar{\mathbf{x}}), \\ &\frac{\partial^2}{\partial x_i'\partial x_j'} \frac{\exp(i\alpha|\mathbf{x}-\mathbf{x}'|)}{|\mathbf{x}-\mathbf{x}'|} \approx -\frac{\alpha^2\bar{x}_i\bar{x}_j}{x}\exp(i\alpha x - i\alpha\mathbf{x}'\cdot\bar{\mathbf{x}}), \end{aligned} \tag{48.11}$$

where $\bar{\mathbf{x}} = \mathbf{x}/x$.

From (9.40), we have

$$\begin{aligned} g_{ij}(\mathbf{x}-\mathbf{x}') &\approx \frac{1}{4\pi\rho\omega^2 x}[\beta^2(\delta_{ij}-\bar{x}_i\bar{x}_j)\exp(i\beta x - i\beta\mathbf{x}'\cdot\bar{\mathbf{x}}) \\ &\quad + \alpha^2\bar{x}_i\bar{x}_j\exp(i\alpha x - i\alpha\mathbf{x}'\cdot\bar{\mathbf{x}})], \\ g_{ij,k}(\mathbf{x}-\mathbf{x}') &\approx -\frac{\partial}{\partial x_k'} g_{ij}(\mathbf{x}-\mathbf{x}') \\ &= \frac{i\bar{x}_k}{4\pi\rho\omega^2 x}[\beta^3(\delta_{ij}-\bar{x}_i\bar{x}_j)\exp(i\beta x - i\beta\mathbf{x}'\cdot\bar{\mathbf{x}}) \\ &\quad + \alpha^3\bar{x}_i\bar{x}_j\exp(i\alpha x - i\alpha\mathbf{x}'\cdot\bar{\mathbf{x}})]. \end{aligned} \tag{48.12}$$

When the expressions in (48.12) are substituted into (48.10), we have

$$u_i^s(\mathbf{x}) \approx (\delta_{im} - \bar{x}_i\bar{x}_m) f_m(\beta) x^{-1}\exp(i\beta x) + \bar{x}_i\bar{x}_m f_m(\alpha) x^{-1}\exp(i\alpha x), \tag{48.13}$$

where

$$f_m(\alpha) = \frac{\alpha^2}{4\pi\rho\omega^2}\left[\Delta\rho\omega^2\int_\Omega u_m(\mathbf{x}')\exp(-i\alpha\mathbf{x}'\cdot\bar{\mathbf{x}})d\mathbf{x}' + i\alpha\bar{x}_j\Delta C_{mjkl}\int_\Omega u_{k,l}(\mathbf{x}')\exp(-i\alpha\mathbf{x}'\cdot\bar{\mathbf{x}})d\mathbf{x}'\right] \tag{48.14}$$

which is called the f-vector. Since $u_m(\mathbf{x}')$ and $u_{k,l}(\mathbf{x}')$ in the integrands of the integrals in (48.14) are unknown, an iterative procedure is employed. The first approximation is obtained by substituting $u^0_m(\mathbf{x}')$ and $u^0_{k,l}(\mathbf{x}')$ for these unknown functions. This first approximation is called the first Born (or Rayleigh-Gaus) approximation (see Gubernatis et al., 1977).

The Born approximation becomes progressively worse as changes in the material properties become large, and is limited to $k|\mathbf{x}-\mathbf{x}'|<1$ (long wavelength). This has been concluded by Gubernatis et al. (1977) by comparing the approximation with the exact solution for a spherical inhomogeneity obtained by Ying and Truell (1956), and Einspruch et al. (1960). More convenient perturbation methods, however, have been developed by Datta (1977), Gubernatis (1979), Mal and Knopoff (1967), and Gubernatis and Domany (1979), among others.

Another approximation method is based upon the eigenfunction expansions. According to Morse and Feshbach (1958), and Pao and Mow (1971), the dynamic displacement amplitude $u_i(\mathbf{x})$ for an isotropic material can be expressed as

$$u_i(\mathbf{x}) = \Phi_{,i} - i\epsilon_{ijk}x_j\psi_{,k} - i(x_i\chi_{,j} - x_j\chi_{,i})_{,j}. \tag{48.15}$$

When it is substituted into the wave equation (47.22′) (u_i for $\bar{u}_i$), the wave equation is satisfied if

$$\begin{aligned} &\Phi_{,ii} + \alpha^2\Phi = 0,\\ &\psi_{,ii} + \beta^2\psi = 0,\\ &\chi_{,ii} + \beta^2\chi = 0, \end{aligned} \tag{48.16}$$

are satisfied.

For the spherical coordinates the eigenfunctions of the scalar Helmholtz equations (48.16) are products of spherical Bessel functions and spherical harmonics $Y_{lm}(\theta, \phi)$. For non-spherical defects, the expansion method has been used in several different ways to obtain approximate solutions as seen in the papers by Waterman (1969, 1971), Varatharajulu and Pao (1976), Waterman (1976), and Visscher (1978).

Recently Gubernatis, Domany, and Krumhansl (1980) have published an excellent review paper on the elastic wave scattering theory with applications to non-destructive evaluation. For randomly distributed inhomogeneities we refer to the papers by Bose and Mal (1973), McCoy (1973), Sobezyk (1976), Datta (1977), and Varadan and Varadan (1979), among others.

49. Interaction between dislocations and inclusions

The term inclusion is interpreted here in a broader sense to include inhomogeneity and crack. Dispersed second-phase particles and non-metallic inclusions exert a large influence upon the mechanical properties of materials. In addition to increasing the yield strength by raising the stress necessary to move dislocations through the matrix, the presence of dispersed phases and inclusions affects the fracture behavior of these materials by providing sites for crack formation via particle or particle-interface cracking at the tip of blocked slip bands.

Inclusions and dislocations

Consider an infinite elastic medium containing a circular cylindrical inclusion of radius a (Fig. 49.1). The shear modulus and Poisson's ratio of the matrix are denoted by μ_1 and ν_1 and the corresponding constants of the inclusion by μ_2 and ν_2. The stress field of a dislocation located at $(\xi, 0)$ is found by Dundurs (1967) for a screw dislocation and by Dundurs and Mura (1964) and Dundurs and Sendeckyj (1965) for an edge dislocation. The evaluation of the Peach-Koeler force requires the stress component σ_{yz} for the screw dislocation and σ_{yx} for the edge dislocation.

For the screw dislocation, the stress at $(x, 0)$ is

$$\sigma_{yz} = \frac{\mu_1 b}{2\pi}\left\{\frac{1}{x-\xi} + \frac{\mu_2 - \mu_1}{\mu_2 + \mu_1}\left(\frac{1}{x - a^2/\xi} - \frac{1}{x}\right)\right\}, \tag{49.1}$$

where $x > a$.

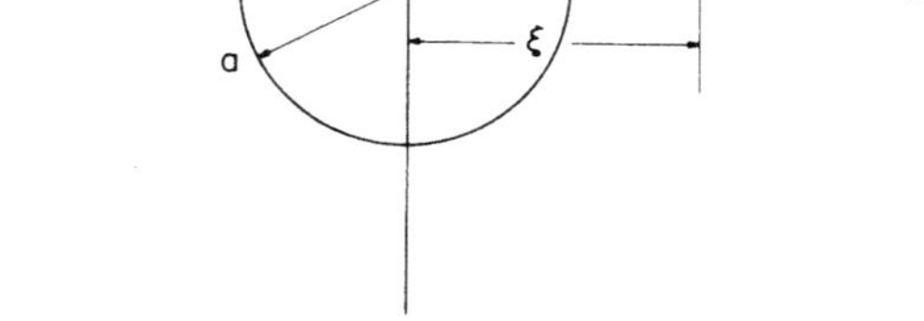

Fig. 49.1. A dislocation near a cylindrical inhomogeneity.

For the edge dislocation, the stress at $(x, 0)$ becomes, for $x > a$,

$$\sigma_{yx} = \frac{\mu_1 b}{2\pi(1-\nu_1)}\left[\frac{1}{x-\xi} - \frac{A+B}{2(x-a^2/\xi)} - A\,\frac{(\xi/a)^2-1}{(\xi/a)^3}\right.$$
$$\times\left\{1 - \frac{(\xi/a)^2-1}{\xi/a}\,\frac{a}{x-a^2/\xi}\right\}$$
$$\left.\times\frac{a}{(x-a^2/\xi)^2} + \frac{A+B}{2x} + (B-A)\,\frac{a}{2(\xi/a)x^2} - A\,\frac{a^2}{x^3}\right], \tag{49.2}$$

where b is the Burgers vector, and

$$\begin{aligned} A &= (1-\mu_2/\mu_1)/(1+\mu_2\kappa_1/\mu_1),\\ B &= (\kappa_2-\mu_2\kappa_1/\mu_1)/(\kappa_2+\mu_2/\mu_1),\\ \kappa_1 &= 3-4\nu_1, \quad \kappa_2 = 3-4\nu_2. \end{aligned} \tag{49.3}$$

More complicated cases have been considered by several authors. When a circular inclusion has a slipping interface, the elastic interaction between an edge dislocation and the inclusion is investigated by Dundurs and Gangadharan (1969). The elastic interaction between a circular inclusion in a semi-infinite body and a screw dislocation is given by Tamate (1973). Sendeckyj (1970) has given the elastic field of a screw dislocation near an arbitrary number of circular inclusions. Tamate (1968) has investigated the behavior of a screw dislocation near a partially bonded bimetallic interface. The case for an edge dislocation is considered by Tamate and Kurihara (1970). A dislocation near or on a surface layer is treated by Weeks et al. (1968), Lee and Dundurs (1973), Masumura and Glicksman (1977) for the edge case. Gavazza and Barnett (1974), and Willis et al. (1972) have found the general solution for the interaction between a spherical inclusion and a screw dislocation. If the inclusion and a dislocation are far apart, the long-range interaction approximation is possible as shown by Barnett (1971), Comninou and Dundurs (1972), Yoo and Ohr (1972), Lin and Mura (1973), Yoo (1974), and Heinisch and Sines (1976). The interaction between a crack and a dislocation has been investigated by Tamate and Sekine (1972), and Rice and Thomson (1974).

If the inclusion is a half-space, we take $\xi/a \to 1$, $\{(\xi/a)^2 - 1\}a \to 2\xi$, $a - a^2/\xi \to \xi$ and $x - a^2/\xi \to x + \xi$ in (49.1) and (49.2) (see Fig. 49.2). Then, (49.1) and (49.2) become

$$\sigma_{yz} = \frac{\mu_1 b}{2\pi}\left(\frac{1}{x-\xi} + \frac{\mu_2-\mu_1}{\mu_2+\mu_1}\,\frac{1}{x+\xi}\right) \tag{49.4}$$

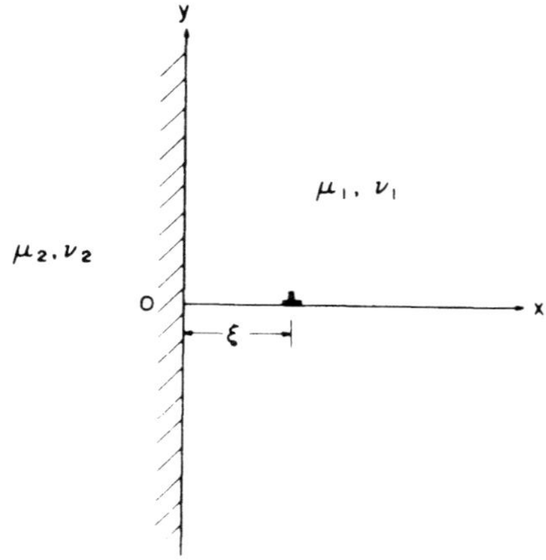

Fig. 49.2. A dislocation near a two-phase material.

for the screw dislocation, and

$$\sigma_{yx} = \frac{\mu_1 b}{2\pi(1-\nu_1)}\left\{\frac{1}{x-\xi} - \frac{A+B}{2(x+\xi)} - 2A\xi\frac{(x-\xi)}{(x+\xi)^3}\right\} \tag{49.5}$$

for the edge dislocation.

The above solutions agree with those obtained by Head (1953). The Peach-Koehler force can be obtained by $\sigma_{zy}b$ or $\sigma_{yx}b$ at $x = \xi$, where the term $1/(x-\xi)$ is omitted for the calculation.

The problem of a linear array of length L containing n dislocations piled up along the x-axis against the circular inclusion under the application of an applied stress σ^0_{zy} (for screw) or σ^0_{yx} (for edge) may be formulated as follows. For static equilibrium the force acting on any one dislocation in the pile-up due to all other pile-up dislocations must balance the force due to the applied stress. This requires that $\sigma_{zy} + \sigma^0_{zy} = 0$ or $\sigma_{yx} + \sigma^0_{yx} = 0$,

$$\frac{\mu_1 b}{2\pi}\left\{\sum_{\substack{j=0\\ j\neq i}}^{n-1}\frac{1}{x_i - x_j} + \frac{\mu_2-\mu_1}{\mu_2+\mu_1}\sum_{j=0}^{n-1}\left(\frac{1}{x_i - a^2/x_j} - \frac{1}{x_i}\right)\right\} + \sigma^0_{zy} = 0,$$

$$\text{for } i = 1, 2, \ldots, n-1. \tag{49.6}$$

for the screw dislocation array. A similar equation holds for the edge dislocation array.

When n is large, the discrete dislocations in the array can be replaced by a continuous distribution of dislocations with density $f(\xi)$.

Equation (49.6) becomes

$$\frac{\mu_1 b}{2\pi}\int_a^{a+L}\left\{\frac{1}{x-\xi}+\frac{\mu_2-\mu_1}{\mu_2+\mu_1}\left(\frac{1}{x-a^2/\xi}-\frac{1}{x}\right)\right\}f(\xi)d\xi+\sigma^0_{zy}=0$$

$$\text{for } a<x\leqq L. \quad (49.7)$$

Solving (49.6) yields values of $x_1, x_2, \ldots, x_{n-1}$ and n for a given constant σ^0_{zy}. Equation (49.7) determines $f(x)$ and L for a given σ^0_{zy}.

Solutions of these equations are obtained by many researchers (see a review paper by Chou and Li 1969). Eshelby, Frank and Nabarro (1951) have solved the discrete dislocations pile-up problem in the homogeneous medium ($\mu_2=\mu_1$, $\nu_1=\nu_2$),

$$D\sum_{\substack{j=0\\ j\neq i}}^{n-1}\frac{1}{x_i-x_j}-\sigma=0 \quad \text{for } i=1,2,\ldots,n-1, \tag{49.8}$$

where

$$D=\mu_1 b/2\pi, \qquad \sigma=-\sigma^0_{yz}, \tag{49.9}$$

for the screw dislocation case, and

$$D=\mu_1 b/2\pi(1-\nu_1), \qquad \sigma=-\sigma^0_{yx}, \tag{49.10}$$

for the edge dislocation case. Equilibrium equation (49.8) holds for the $(n-1)$ locations. The leading dislocation at x_0 is a locked dislocation. Without losing generality, we can assume that $x_0=0$ and an obstacle is locking the leading dislocation. Eshelby et al. obtained $x_1, x_2, \ldots, x_{n-1}$ as the roots of the generalized Laguerre polynomial $L^{(1)}_{n-1}(x)$ by taking $D/2\sigma$ as a unit of length. The stress σ_{tip} acting on the locked dislocation is obtained as $(n-1)\sigma$, from the virtual work argument of Cottrell (1949). When the leading dislocation is displaced by δx_0, the other $(n-1)$ dislocations are also displaced by δx_0 without losing the equilibrium since only the relative location of the dislocations is essential for the equilibrium. The work done by the applied stress σ on the $(n-1)$ dislocations during the virtual displacement δx_0 is $\sigma b(n-1)\delta x_0$ which must be equal to the work done by the locking force $\sigma_{\text{tip}}\, b$ on the leading dislocation. Namely, $\sigma b(n-1)\delta x_0=\sigma_{\text{tip}} b\delta x_0$, which leads to $\sigma_{\text{tip}}=\sigma(n-1)$.

Stroh (1954, 1955) and Y. T. Chou (1967) have calculated details of the stress field and energy associated with the Eshelby, Frank and

Nabarro problem. A modification of the Eshelby, Frank and Nabarro problem is made by Head and Thomson (1962) by considering a variation of the applied stress and by Y. T. Chou (1967) by taking a variation of the strength of the locked dislocation. J. C. M. Li (1969) has investigated the case of extended dislocations. The discrete dislocations in the Eshelby, Frank and Nabarro problem can also be replaced by a continuous distribution of dislocations as done by Leibfried (1951). He has solved the integral equation

$$D \int_0^L \frac{f(\xi)}{x-\xi}\, d\xi - \sigma = 0 \quad \text{for } 0 < x \leqq L, \tag{49.11}$$

and has obtained

$$f(x) = (\sigma/\pi D)\{(L-x)/x\}^{1/2}. \tag{49.12}$$

The total number of dislocations is found to be

$$n = \int_0^L f(x)dx = \sigma L/2D, \tag{49.13}$$

in agreement with the discrete dislocation case.

The pile-up problem in a two-phase medium (two half spaces) requires solving an integral equation of the type (see Fig. 49.3)

$$D \int_0^L \left\{\frac{1}{x-\xi} - \frac{\alpha}{x+\xi} - \frac{\beta\xi(x-\xi)}{(x+\xi)^3}\right\} f(\xi)d\xi - \sigma = 0 \quad \text{for } 0 < x \leqq L \tag{49.14}$$

or

$$D\left\{\sum_{\substack{j=0\\ j\neq i}}^{n-1} \frac{1}{x_i - x_j} - \sum_{j=0}^{n-1} \frac{\alpha}{x_i + x_j} - \sum_{j=0}^{n-1} \frac{\beta x_j(x_i - x_j)}{(x_i + x_j)^3}\right\} - \sigma = 0 \quad \text{for } i = 1, 2, \ldots, n-1, \tag{49.15}$$

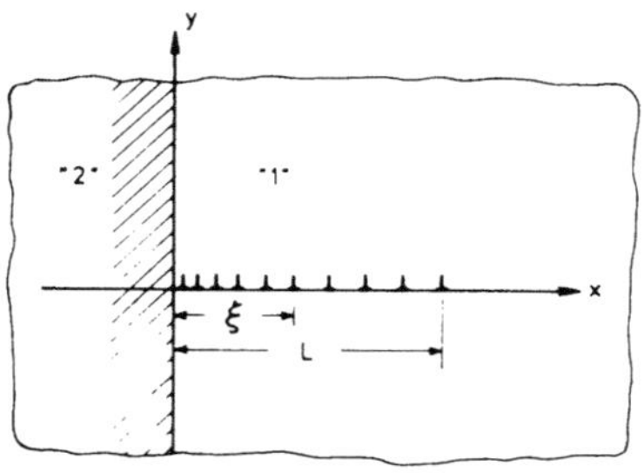

Fig. 49.3. Pile-up dislocations at the interface.

where definitions (49.9) and (49.10) are used, and

$$\begin{aligned} \alpha &= (\mu_1 - \mu_2)/(\mu_1 + \mu_2), \quad & \beta &= 0 \quad & &\text{(for screw)}, \\ \alpha &= (1/2)(A + B), & \beta &= 2A & &\text{(for edge)}. \end{aligned} \tag{49.16}$$

The solution of (49.14) is found by Y. T. Chou (1965), Barnett and Tetelman (1966), Barnett (1967), and Smith (1967) for the screw dislocation case, and by Kuang and Mura (1968) for the edge dislocation case. The distribution density of the screw dislocations is

$$f(x) = \frac{2\sigma}{b\mu_1 \sin(\pi s/2)} \sinh\{s \cosh^{-1}(L/x)\}, \quad 0 \leqq x \leqq L, \tag{49.17}$$

where

$$s = (2/\pi) \sin^{-1}\{\mu_1/(\mu_1 + \mu_2)\}^{1/2}. \tag{49.18}$$

The number of screw dislocations in the pile-up is then found to be

$$n = (\sigma L s/D)/\sin(s\pi). \tag{49.19}$$

The stress at the tip of the pile-up becomes

$$\sigma_{yz} = \frac{\sigma \cos(s\pi/2)}{\sin^2(s\pi)} \left\{ \left(\frac{2L}{r}\right)^s - 2\cos(s\pi/2) \right\}, \tag{49.20}$$

where r is the distance from the tip in the x-direction. The order of stress singularity at $r = 0$ depends on the shear moduli of the two phases. The value of s is 1/2 for the homogeneous ($\mu_1 = \mu_2$) case.

The corresponding quantities for the case of edge dislocation pile-up are too complicated to record here. Kuang and Mura (1968) showed that these quantities are depending on the roots s of the equation

$$\cos(\pi s) + \alpha - \beta(s-1)^2 = 0. \tag{49.21}$$

Numerical examples of the edge dislocation distributions are shown in Fig. 49.4 where $\Gamma = \mu_2/\mu_1$. These distributions are very close to the screw dislocation distributions given by (49.17).

The nature of the stress singularity at the tip of the edge dislocation pile-up can be investigated by the following alternative method. Equation (49.14) is replaced by

$$D \int_0^\infty \left\{ \frac{1}{x-\xi} - \frac{\alpha}{x+\xi} - \frac{\beta\xi(x-\xi)}{(x+\xi)^3} \right\} f(\xi)\,d\xi - \sigma + H(x) = 0 \qquad \text{for } 0 < (x/L) < \epsilon, \tag{49.22}$$

where $\epsilon \ll 1$ and $H(x)$ is nonsingular as $x \to 0^+$.

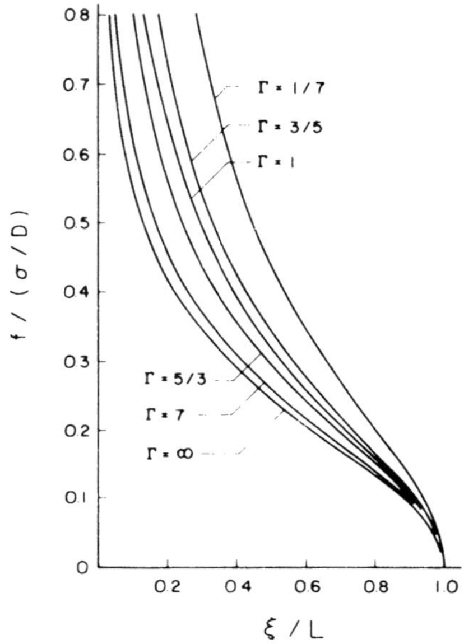

Fig. 49.4. Distributions of pile-up edge dislocations.

Assuming that $f(\xi)$ is of the form

$$f(\xi) = f_0/\xi^{1-k} \quad (0 < k \leq 1), \tag{49.23}$$

we get

$$f_0 \pi D x^{k-1}\{\cos(\pi k) - \alpha + \beta k^2\}/\sin(\pi k) - \sigma + \bar{H}(x) = 0, \tag{49.24}$$

where $\bar{H}(x)$ is not singular as $x \to 0^+$. Since only the first term in (49.24) is singular as $x \to 0^+$, for identity to hold, the following relation must be satisfied:

$$-\cos(\pi k) + \alpha - \beta k^2 = 0. \tag{49.25}$$

The above equation is equivalent to (49.21) when $s - 1 = k$. The stress at the tip of the pile-up can be evaluated from the same integral in (49.22) except $x < 0$. The integral gives

$$\sigma_{12} = -\frac{(\mu_2)^2}{\mu_1}\frac{f_0}{\sin(\pi k)}\left\{\frac{2k+1}{1+(3-4\nu_1)\mu_2/\mu_1} - \frac{2k-1}{(3-4\nu_2)+\mu_2/\mu_1}\right\}\Big/ r^{1-k}, \tag{49.26}$$

where $r = |x|$. The order of stress singularity at the tip is the same as that of dislocation distribution (49.25). Equations (49.22) ~ (49.24) are presented by Keer (1968) in a private communication.

Barnett (1969) has shown that

$$
\begin{aligned}
&0 < (1-k) \leq 1/2 && \text{if } \mu_2/\mu_1 \geq (3-4\nu_2)/(3-4\nu_1), \\
&(1/2) \leq (1-k) < 1 && \text{if } \mu_2/\mu_1 \leq (3-4\nu_2)/(3-4\nu_1).
\end{aligned}
\tag{49.27}
$$

The relation between k and Dundurs constants defined by (6.15) is discussed by Dundurs and Lee (1970).

The problem of screw dislocation pile-up against a circular cylinder requires solving (49.7). The exact solution is given by Barnett and Tetelman (1966, 1967). Toya (1976) has investigated the interfacial debonding caused by a screw dislocation pile-up at a circular cylindrical rigid inclusion.

Cracks in two-phase materials

The equilibrium conditions for pile-up dislocations can also be interpreted as the conditions for a fracture when the conditions for the free (crack) surface are identical to the equilibrium conditions for the dislocation array under an applied stress. Tucker (1973) has used the method of Kuang and Mura (1968, 1969) (the Mellin transform and the Wiener-Hopf technique) and analyzed the behavior of Griffith, Zener-Stroh and Bullough-Gilman cracks in bi-metals. The effect of a circular inclusion on the stresses around a slit-like crack is investigated by Tamate (1968).

When a crack is located along the interface of a bi-metal, the well-known elastic solutions (e.g. Williams 1959, Erdogan 1963, England 1965, Rice and Sih 1965, Clements 1971) characteristically involve oscillatory singularities. Comninou (1977) has reconsidered the problem using a continuous distribution of crack dislocations and found a non-oscillatory solution. She assumed that the crack is not completely open and that its faces are in frictionless contact near the crack tips (see also Comninou 1978, Dundurs and Comninou 1979, Comninou and Schmueser 1979). It has been shown by Achenbach et al. (1979) that the introduction of a small cohesive zone of a certain characteristic in the vicinity of the crack tip also serves to remove the corresponding oscillatory singularity. Recently, Hayashi and Nemat-Nasser (1981) have considered a branched crack at the interface of two dissimilar materials.

The analysis of Comninou is as follows. Consider a crack of length $2L$ lying in the interface of two elastic solids with shear moduli μ_1 and μ_2 and Poisson's ratio ν_1, ν_2 as shown in Fig. 49.5. Under the action of

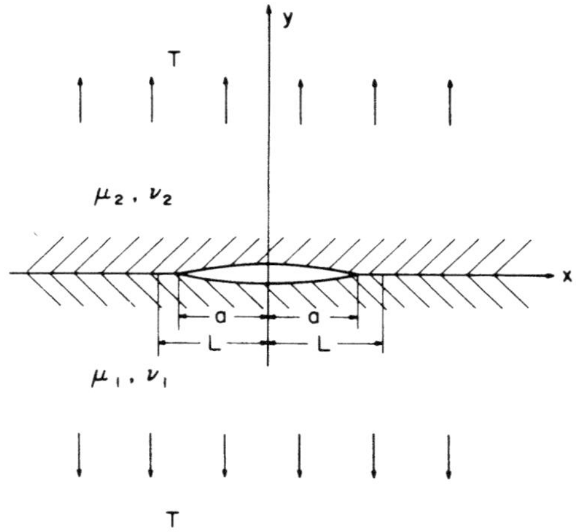

Fig. 49.5. A crack at an interface of a two-phase material.

uniform tension T applied at infinity, in the direction normal to the interface, the crack opens in the interval $(-a, a)$, where a is an unknown to be determined in the course of solution. We assume that a portion of the crack is closed and its two sides are in frictionless contact in the intervals $(-L, -a)$ and (a, L). The boundary conditions are that the shear tractiom must vanish in $(-L, L)$ and that the normal traction must vanish in $(-a, a)$. To enforce these conditions we take advantage of the known stress fields for edge dislocations and consider a distribution $B_x(x)$ of glide dislocations in the interval $(-L, L)$, and a distribution $B_y(x)$ of climb dislocations in the interval $(-a, a)$.

When a discrete edge dislocation is located at point ξ on the interface between two solids, the induced interface tractions are (see Comninou 1977)

$$\begin{aligned} \sigma_{xy}(x, 0) &= Cb_x/\pi(x-\xi), \\ \sigma_{yy}(x, 0) &= -\beta Cb_x\delta(x-\xi), \end{aligned} \tag{49.28}$$

for a glide dislocations, and

$$\begin{aligned} \sigma_{xy}(x, 0) &= \beta Cb_y\delta(x-\xi), \\ \sigma_{yy}(x, 0) &= Cb_y/\pi(x-\xi), \end{aligned} \tag{49.29}$$

for a climb dislocation, where $\delta(x-\xi)$ is the Dirac delta function, and b_x

and b_y are the components of the Burgers vector. Moreover,

$$C = \frac{2\mu_2(1+\alpha)}{(\kappa_1+1)(1-\beta^2)} = \frac{2\mu_2(1-\alpha)}{(\kappa_2+1)(1-\beta^2)}, \tag{49.30}$$

where α and β are the Dundurs constants (6.15).

The boundary conditions on the traction mentioned before lead to the integral equations which determine B_x and B_y. They are

$$C\beta B_y(x)[H(x+a)-H(x-a)] + \frac{C}{\pi}\int_{-L}^{L} \frac{B_x(\xi)}{x-\xi}\,d\xi = 0, \quad -L < x < L, \tag{49.31}$$

$$T - C\left[\beta B_x(x) - \frac{1}{\pi}\int_{-a}^{a} \frac{B_y(\xi)}{x-\xi}\,d\xi\right] = 0, \quad -a < x < a, \tag{49.32}$$

where $H(x)$ is the Heaviside unit step function. The condition for single-valued displacements for $|x| > L$ requires that

$$\int_{-L}^{L} B_x(\xi)d\xi = 0, \tag{49.33}$$

$$\int_{-a}^{a} B_y(\xi)d\xi = 0. \tag{49.34}$$

Denoting by $g(x)$ the gap between the solids and by $h(x)$ their relative slip, we have

$$\begin{aligned} g(x) &= u_y^{(2)}(x,0) - u_y^{(1)}(x,0), \\ h(x) &= u_x^{(2)}(x,0) - u_x^{(1)}(x,0). \end{aligned} \tag{49.35}$$

The definition of continuous distribution of dislocations (38.8) gives

$$B_x(x) = -dh/dx, \quad B_y(x) = -dg/dx. \tag{49.36}$$

Thus, we can assume that $B_y(x)$ is continuous in $(-a, a)$ and bounded at the ends $-a, a$. Then, (49.31) can be viewed as the Hilbert integral equation (A4.1) for the unknown function $B_x(x)$ and solved formally by treating $B_y(x)$ as known. Assuming $B_x(x)$ is unbounded at $-L$ and L, as the asymptotic analysis (Comninou 1977) indicates, $B_x(x)$ is obtained as (A.4.3), where $a_1 = -L$, $b_1 = L$, $c_{1+q} = -L$, $C_{2+q} = L$, $p = 1$, $q = 0$. The solution, using our condition (49.33), becomes

$$B_x(\bar{x}) = -\frac{(\beta/\pi)}{(1-\bar{x}^2)^{1/2}}\int_{-\bar{a}}^{\bar{a}} \frac{(1-t^2)^{1/2}}{(t-\bar{x})} B_y(t)dt, \quad -1 < \bar{x} < 1, \tag{49.37}$$

where

$$\bar{x} = x/L, \quad \xi/L = t, \quad a/L = \bar{a}. \tag{49.38}$$

Condition (49.34) is satisfied by symmetry. Since B_y is bounded, $B_x(x)$ is square root singular at L and $-L$.

Substituting $B_x(\bar{x})$ from (49.37) into (49.32), we have

$$-\frac{\beta^2}{(1-\bar{x}^2)^{1/2}}\int_{-\bar{a}}^{\bar{a}}\frac{(1-t^2)^{1/2}}{t-\bar{x}}B_y(t)dt+\int_{-\bar{a}}^{\bar{a}}\frac{1}{t-\bar{x}}B_y(t)dt=\pi T/C,$$
$$-\bar{a}<\bar{x}<\bar{a} \qquad (49.39)$$

which is a generalized Cauchy singular equation to determine $B_y(t)$. A numerical solution can be obtained by the method developed by Erdogan and Gupta (1972).

The singular behavior of the interface tractions can easily be established. The tractions at the interface are

$$\begin{aligned}\sigma_{xy}&=\frac{C}{\pi}\int_{-L}^{L}\frac{B_x(\xi)}{x-\xi}d\xi, \qquad |x|>L\\ &=-\frac{\beta C}{\pi}\frac{\operatorname{sgn}\bar{x}}{(\bar{x}^2-1)^{1/2}}\int_{-\bar{a}}^{\bar{a}}\frac{B_y(t)(1-t^2)^{1/2}}{t-\bar{x}}dt, \quad \bar{x}>1,\end{aligned} \qquad (49.40)$$

and

$$\begin{aligned}\sigma_{yy}(x,0)&=T+\frac{C}{\pi}\int_{-a}^{a}\frac{B_y(\xi)}{x-\xi}d\xi, \qquad |x|>L\\ &=T-C\beta B_x(x)+\frac{C}{\pi}\int_{-a}^{a}\frac{B_y(\xi)}{x-\xi}d\xi, \quad L>|x|>a.\end{aligned} \qquad (49.41)$$

It can be seen that $\sigma_{xy}(x,0)$ has square-root singularities at L and $-L$ and $\sigma_{yy}(x,0)$ is bounded at $L^+, -L^+$ but has square-root singularities at $L^-, -L^-$ because of the term $-C\beta B_x(x)$. Recently, Mak et al. (1980) have shown that both the normal and the shear stresses are singular at the crack tip when the interface adhesive zones have no-slip.

Finally, the solution of (49.39) must satisfy the following conditions: for no interpenetration,

$$g(x)>0, \quad |x|<a, \qquad (49.42)$$

and for contact,

$$\sigma_{yy}(x,0)\le 0, \quad a<|x|<L. \qquad (49.42')$$

Comninou showed a numerical example for the case $a/L=1-10^{-4}$ and $\beta=0.4854$. Her result shows that $\sigma_{xy}(x,0)(x^2/L^2-1)^{1/2}/T=1.050$, $\sigma_{yy}(x,0)/T=23.36$ at $x/L=1.0000$. These values cannot be obtained from the classical solution (e.g. England 1965) due to its oscillatory singularities. However, her solution is very close to England's solution when $x/L > 1.0002$. England's solution is

$$\sigma_{xy}(x, 0) = \frac{TL^{1/2}}{[2(x-L)]^{1/2}}(2\epsilon \cos \epsilon s - \sin \epsilon s), \quad x > L, \tag{49.43}$$

where

$$\epsilon = \frac{1}{2\pi} \log \frac{1-\beta}{1+\beta}, \quad s = \epsilon \log \frac{2L}{x-L}. \tag{49.44}$$

The crack extension force $\mathcal{G}_1$ defined by (34.6) becomes almost identical in both of the solutions and it can be approximated by

$$\mathcal{G}_1 \approx \pi K_2^2/4C, \tag{49.45}$$

where

$$K_2 = TL^{1/2}(1+4\epsilon^2)^{1/2}, \quad \epsilon \neq 0. \tag{49.46}$$

Comninou's consideration has been extended to a penny-shaped crack at the interface of two bonded dissimilar half-spaces by Keer et al. (1978). By considering contact zones, these solutions have no oscillatory singularities. The solutions obtained by Mossakovskii and Rykba (1964), Keer (1967), Willis (1972), and Lowengrub and Sneddon (1974) have the oscillatory singularities since their penny-shaped cracks are of Griffith type. However, for cracks in tension fields the oscillatory zone is very small compared to the radius of the crack and these solutions give good approximations outside the zone. The oscillatory zone is no longer small for a crack under shear fields. The solution of Lowengrub and Sneddon (1974) for a crack under internal pressure p gives

$$\sigma_{zz}(\rho, 0^+) = (K/\pi)\{S_1(\rho) + (\gamma/2)S_2(\rho)\} \tag{49.47}$$

at the interface $z = 0$, where ρ is the dimensionless radius,

$$\begin{aligned} K &= p/(\kappa_2 + \Gamma)^{1/2}(1 + \kappa_1\Gamma)^{1/2}, \\ \Gamma &= \mu_2/\mu_1, \quad \kappa = 3 - 4\nu, \\ \gamma &= \frac{1}{2\pi} \log \frac{\kappa_1\Gamma + 1}{\Gamma + \kappa_2}, \end{aligned} \tag{49.48}$$

and

$$\begin{aligned} S_1(\rho) &= -\frac{1}{\rho}\frac{\partial}{\partial\rho}\int_0^1 \frac{x^2\cos(\gamma\theta)}{(\rho^2 - x^2)^{1/2}}\,dx, \\ S_2(\rho) &= -\frac{1}{\rho}\frac{\partial}{\partial\rho}\int_0^1 \frac{x\sin(\gamma\theta)}{(\rho^2 - x^2)^{1/2}}\,dx, \\ \theta &= \frac{1}{2}\log\frac{1+x}{1-x}. \end{aligned} \tag{49.49}$$

50. Eigenstrains in lattice theory

In lattice theory (e.g. Born and Huang 1954, Maradudin 1958) a change of potential energy caused by a deformation is expressed by a function of atomic displacements measured from a perfect lattice state. For harmonic approximation the potential energy is expressed by a quadratic form of the atomic displacement components. The coefficients of the polynomial are chosen such that a uniform displacement (rigid motion) does not create any potential energy of deformation. The potential energy, however, does not vanish when a part of the material is displaced plastically and reformed into a perfect crystal. When a perfect crystal is deformed into another perfect crystal accompanying only a plastic shape change, no change of potential energy is expected. Therefore the classical lattice theory should be modified so that the potential energy does vanish for such a plastic deformation. A similar question about the incompleteness of the classical theory has been raised by M. Kuriyama (1967) from a different viewpoint.

It is shown that such a difficulty in the classical lattice theory simply can be eliminated when the concept of eigenstrains in continuum mechanics is introduced into the lattice theory. In continuum mechanics the eigenstrain has played an important role in analysis of elastic fields caused by line and point imperfections, inclusions, inhomogeneities, etc. The introduction of the concept of eigenstrains into lattice theory provides similar advantages and eliminates the above-mentioned difficulty in classical lattice theory.

The equation of motion of atom l in the classical lattice theory with the harmonic approximation is

$$M\ddot{u}_i(l) = -\sum_{l'} \phi_{ij}(ll')u_j(l'), \tag{50.1}$$

where the summation with respect to l' is referred to all atoms, M is the mass of an atom, $u_i(l)$ the displacement component from the perfect lattice state of atom l, indicating the position, and $\phi_{ij}(ll')$ are the force constants between l and l' atoms. The double indices appearing in the equations follow the summation convention from 1 to 3. The force constants have the following properties, since the right-hand side in equation (50.1) is zero for a constant displacement,

$$\sum_{l'} \phi_{ij}(ll') = \sum_{l} \phi_{ij}(ll') = 0. \tag{50.2}$$

Furthermore, the symmetry condition leads to

$$\phi_{ij}(ll') = \phi_{ji}(ll') = \phi_{ij}(l'l). \tag{50.3}$$

The summations with respect to l' in (50.1) and (50.2) are taken for all atoms including l. It is obvious from (50.2) that $\Sigma_{l'}\, \phi_{ij}(ll')u_j(l) = 0$. Then (50.1) can be written as

$$M\ddot{u}_i(l) = -\sum_{l'} \phi_{ij}(ll')[u_j(l') - u_j(l)]. \tag{50.4}$$

If we put

$$u_j(l') - u_j(l) = \beta_{kj}(ll')x_k(l'l), \tag{50.5}$$

$$x_k(l'l) = x_k(l') - x_k(l), \tag{50.6}$$

where $x_k(l)$ is a coordinate component of l atom in the perfect lattice state, we write (50.4)

$$M\ddot{u}_i(l) = -\sum_{l'} \phi_{ij}(ll')\beta_{kj}(ll')x_k(l'l). \tag{50.7}$$

It is important to keep in mind that (50.1) and (50.7) are equivalent only when $\beta_{kj}(ll')$ is compatible, namely, only when (50.5) holds. However, (50.5) does not always hold. If the material is involved in any inelastic phenomena, for instance, in plastic deformation, (50.5) does not hold and instead

$$u_j(l') - u_j(l) = [\beta_{kj}(ll') + \beta^*_{kj}(ll')]x_k(l'l). \tag{50.8}$$

Now we can call β_{kj} and β^*_{kj} the elastic distortion and eigendistortion, respectively, and their symmetric parts are called the elastic strain and eigenstrain, respectively. These definitions are strictly analogous to those of continuum mechanics, and the total strain is the sum of the elastic strain and eigenstrain. We assert that the fundamental equation of motion is (50.7) rather than (50.1), and condition (50.8) is subjected to equation (50.7). Equation (50.7) becomes (50.1) only when $\beta^*_{kj} = 0$. The plastic deformation which transforms a perfect crystal into another perfect crystal, accompanying a permanent shape change, can be described by zero elastic distortion (or strain), that is, $\beta_{kj} = 0$. In most inelastic problems u_i and β_{kj} are unknown quantities to be determined, while the components of β^*_{kj} are a priori given by the nature of the problems. For plastic deformations, $\beta_{kj}(ll')x_k(l'l)$ and $\beta^*_{kj}(ll')x_k(l'l)$ are the elastic and plastic displacements, respectively. The last displacement corresponds to the magnitude of glide which is a multiple of the lattice spacing. Analytically, however, it is expressed by a continuous function of coordinates of a generalized function. $\beta_{kj}(ll')$ and $\beta^*_{kj}(ll')$ are defined between atoms l and l' and are equal to $\beta_{kj}(l'l)$ and $\beta^*_{kj}(l'l)$, respectively.

When $\beta_{kj}(ll')$ is eliminated from (50.7) and (50.8), we have

$$M\ddot{u}_i(l) = -\sum_{l'} \phi_{ij}(ll')u_j(l') + \sum_{l'} \phi_{ij}(ll')\beta^*_{kj}(ll')x_k(l'l). \tag{50.9}$$

Comparing (50.9) and (50.1) it can be seen that an extra term appears as a fictitious body force acting on atom l.

The potential energy in the classical lattice theory is within the harmonic approximation

$$\Phi = \frac{1}{2}\sum_{l,l'} \phi_{ij}(ll')u_i(l)u_j(l') \tag{50.10}$$

or

$$\Phi = -\frac{1}{2}\sum_{l'-l} \phi_{ij}(ll')[u_i(l') - u_i(l)][u_j(l') - u_j(l)]. \tag{50.10.1}$$

The last expression (50.10.1) has been obtained by the use of (50.2) and the equality

$$\sum_{l'-l} \phi_{ij}(ll')u_i(l')u_j(l) + \sum_{l'-l} \phi_{ij}(ll')u_i(l)u_j(l') = \sum_{l,l'} \phi_{ij}(ll')u_i(l)u_j(l'), \tag{50.10.2}$$

where the summations $\Sigma_{l'-l}$ and $\Sigma_{l',l}$ are combinational and permutational, respectively.

Our modified theory asserts that the potential energy is not associated with the total displacement u_i but the elastic displacement defined by $\beta_{ki}(ll')x_k(l'l)$. Then, the new expression of the potential energy becomes, instead of (50.10),

$$\Phi = -\frac{1}{2}\sum_{l'-l} \phi_{ij}(ll')\beta_{ki}(ll')x_k(l'l)\beta_{mj}(ll')x_m(l'l). \tag{50.11}$$

If a component of stress tensor is defined as

$$\sigma_{ik}(ll') = -b^{-3}\phi_{ij}(ll')\beta_{mj}(ll')x_m(l'l)x_k(l'l) \tag{50.12}$$

(50.11) can be written as

$$\Phi = \frac{b^3}{2}\sum_{l'-l} \sigma_{ik}(ll')\beta_{ki}(ll'). \tag{50.13}$$

The definition of stress (50.12) has been chosen such that it agrees with one in the continuum mechanics for limiting case $b \to 0$. Similarly to continuum mechanics, the potential energy (the elastic strain energy caused by β^*_{ki}) can be written as

$$\Phi = -\frac{b^3}{2} \sum_{l'-l} \sigma_{ik}(ll')\beta^*_{ki}(ll') \tag{50.14}$$

for the static case. It is proved as follows. Equation (50.11) is written from (50.8) as

$$\Phi = -\frac{1}{2} \sum_{l'-l} \phi_{ij}(ll')\beta_{mj}(ll')x_m(l'l)[u_i(l') - u_i(l) - \beta^*_{ki}(ll')x_k(l'l)]. \tag{50.15}$$

For the static case (50.7) and (50.3) lead to

$$\sum_{l} \phi_{ij}(ll')\beta_{mj}(ll')x_m(l'l) = 0, \quad \sum_{l'} \phi_{ij}(ll')\beta_{mj}(ll')x_m(l'l) = 0 \tag{50.16}$$

and therefore (50.15) can be simplified to

$$\Phi = \frac{1}{2} \sum_{l'-l} \phi_{ij}(ll')\beta_{mj}(ll')x_m(l'l)x_k(l'l)\beta^*_{ki}(ll') \tag{50.17}$$

which is equivalent to (50.14) due to (50.12). Equation (50.14) has the same form as (13.3).

A uniformly moving screw dislocation

A straight screw dislocation along the x_3 axis is uniformly moving to the x_1 direction with a constant velocity v. At time $t = 0$, $l_{1/2,1/2}$, $l_{1/2,-1/2}$ atoms (see Figure 50.1) are in a quarter way jumping the next potential well. For mathematical simplicity, only the nearest neighborhood interaction is considered for a cubic crystal. Now we assume that

$$\begin{aligned} \phi_{33}(l_{m,n}l_{m+1,n}) &= -A, \\ \phi_{33}(l_{m,n}l_{m,n+1}) &= -B, \\ \phi_{33}(l_{m,n}l_{m,n}) &= 2(A+B). \end{aligned} \tag{50.18}$$

Since the plastic displacement $\beta^*_{23}(ll')x_2(l'l)$ must be b, we take $\beta^*_{23}(ll') = 1$ between $l = l_{k+1/2,1/2}$ and $l' = l_{k+1/2,-1/2}$ atoms (k = integer between $-\infty$ and vt/b). Then, our fundamental equation (50.9) becomes

$$\begin{aligned} M\ddot{u}_3(l_{m,n}) = {} & A[u_3(l_{m+1,n}) + u_3(l_{m-1,n})] + B[u_3(l_{m,n+1}) + u_3(l_{m,n-1})] \\ & - 2(A+B)u_3(l_{m,n}) + Bb \sum_{k=-\infty}^{vt/b} (\delta_{m,k+1/2}\delta_{n,1/2} - \delta_{m,k+1/2}\delta_{n,-1/2}), \end{aligned} \tag{50.19}$$

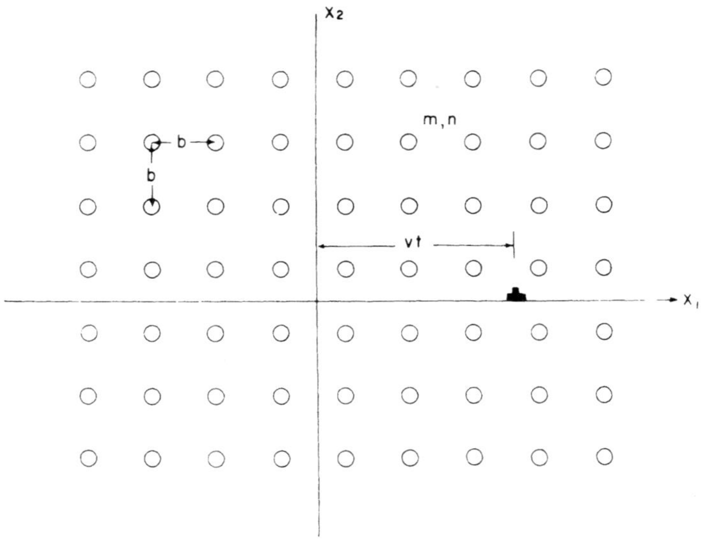

Fig. 50.1. A uniformly moving screw dislocation with velocity v. The atom positions are denoted by integers l, m and n.

where $\delta_{m,n}$ is the Kronecker delta. Although vt/b has been assumed as an integer, the analytic nature of our final solution can extend the solution to any value of vt/b.

Since

$$\delta_{m,k\pm1/2} = \frac{1}{2\pi}\int_{-\pi}^{\pi} \exp\{i(m-k\mp\tfrac{1}{2})\xi_1\}d\xi_1, \tag{50.20}$$

the solution of (50.19) is assumed as

$$u_3(l_{m,n}) = \frac{1}{(4\pi)^2}\int\int_{-\pi}^{\pi} \bar{u}_3(\xi_1,\xi_2)\exp\left\{i\left(m-\frac{vt}{b}\right)\xi_1 + in\xi_2\right\}d\xi_1 d\xi_2. \tag{50.21}$$

$\bar{u}_3(\xi_1, \xi_2)$ is easily obtained as a solution of an algebraic equation when (50.20) and (50.21) are substituted into (50.19). Then, we have

$$u_3(l_{m,n}) = \pm\frac{b}{4} - \frac{b}{\pi^2}\int\int_{0}^{\pi/2} \frac{\sin\xi_2}{\sin\xi_1}\frac{\sin(X\xi_1)\sin(2n\xi_2)}{(A/B)\sin^2\xi_1 + \sin^2\xi_2 - V^2\xi_1^2}d\xi_1 d\xi_2, \tag{50.22}$$

where

$$X = 2(m - vt/b), \quad V^2 = Mv^2/Bb^2 = v^2/c^2, \tag{50.23}$$

and c is the sound velocity of shear waves. The plus and minus signs of the first term in the right-hand side in (50.22) are taken for $n > 0$ and $n < 0$, respectively. When $V = 0$, (50.22) agrees with the result obtained by Maradudin (1958). When $A = B$ and $\xi_1/b = \bar{\xi}_1$, $\xi_2/b = \bar{\xi}_2$, $mb = x_1$, $nb = x_2$, $b \to 0$, (50.22) leads to the classical solution of continuum theory,

$$u_3(x_1, x_2) = (b/2\pi)\tan^{-1}(x_2/x_1'), \quad x_1' = (x_1 - vt)/(1 - V^2)^{1/2}. \tag{50.24}$$

The displacement due to applied stress $\sigma_{23} = \sigma$ is obtained as

$$u_3(l_{m,n}) = nb^2\sigma/B. \tag{50.25}$$

The total displacement w is

$$w = (50.22) + (50.25). \tag{50.26}$$

The displacement field is anti-symmetric about $x_2 = 0$ plane. The displacement of atoms just above the slip plane is obtained by taking $n = 1/2$. The numerical values are shown in Figure 50.2 for various values of V in case $A/B = 1$ and $B = \mu b$ (μ = shear modulus), where σ/μ is chosen such that

$$w/(\tfrac{1}{2}b) = 0.25 \quad \text{at } X = 1 \tag{50.27}$$

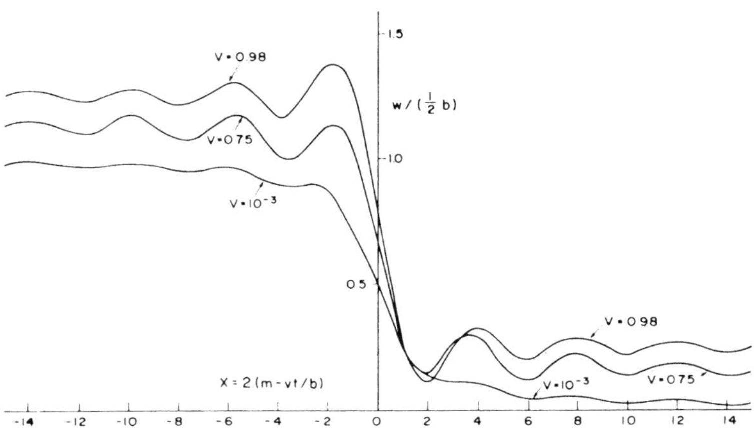

Fig. 50.2. The displacement of atoms just above the slip plane for various velocities of the dislocation, where $V = v/c$. The dislocation is in the position $X = 1$.

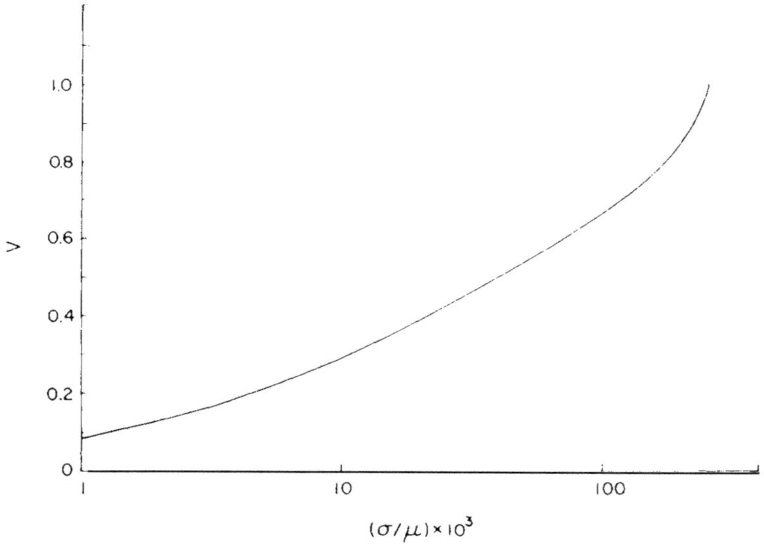

Fig. 50.3. Applied stress necessary to keep the dislocation in motion with velocity V.

is satisfied. The last condition means that the atoms at the dislocation position are in a quarter way jumping to the next potential well. The applied stress σ satisfying condition (50.27) becomes a function of V as shown in Figure 50.3.

In continuum theory no applied stress is necessary to maintain the dislocation in the uniform motion for $V < 1$ and the displacement profile of the upper slip plane is $w/(\frac{1}{2}b) = 1$ for $X < 0$ and $w = 0$ for $0 < X$.

Comparing the result shown in Figure 50.2 with the result of one-dimensional chain model obtained by Earmme and Weiner (1974), the present theory predicts the damping of oscillation of atoms in the trail of the dislocation and also predicts an oscillation of atoms ahead of the dislocation.

The theory in this section can be applied equally to vacancies and moving cracks as shown by Mura (1978).

Appendix 1

Einstein summation convention

If an index occurs twice in any one term, summation is taken from 1 to 3. For instance,

$$\begin{aligned}C_{ijkl}e_{kl} &= C_{ij1l}e_{1l} + C_{ij2l}e_{2l} + C_{ij3l}e_{3l}\\ &= (C_{ij11}e_{11} + C_{ij12}e_{12} + C_{ij13}e_{13})\\ &\quad + (C_{ij21}e_{21} + C_{ij22}e_{22} + C_{ij23}e_{23})\\ &\quad + (C_{ij31}e_{31} + C_{ij32}e_{32} + C_{ij33}e_{33}),\end{aligned} \tag{A1.1}$$

$$\sigma_{ij,j} = \sigma_{i1,1} + \sigma_{i2,2} + \sigma_{i3,3}. \tag{A1.2}$$

Kronecker delta

$$\delta_{ij} = \begin{cases}1 & \text{for } i = j\\ 0 & \text{for } i \neq j\end{cases}. \tag{A1.3}$$

Namely, $\delta_{11} = \delta_{22} = \delta_{33} = 1$, $\delta_{12} = \delta_{23} = \delta_{31} = \delta_{21} = \delta_{32} = \delta_{13} = 0$. We have

$$\delta_{ii} = 3. \tag{A1.4}$$

Permutation tensor

$$\epsilon_{ijk} = \begin{cases}1 & \text{for the even permutation of 1, 2, 3}\\ -1 & \text{for the odd permutation of 1, 2, 3.}\\ 0 & \text{for other cases.}\end{cases} \tag{A1.5}$$

Namely, $\epsilon_{123} = \epsilon_{231} = \epsilon_{312} = 1$, $\epsilon_{132} = \epsilon_{213} = \epsilon_{321} = -1$,

$$\epsilon_{112} = \epsilon_{223} = \epsilon_{333} = \cdots = 0.$$

We have

$$\epsilon_{ijk}\epsilon_{imn} = \delta_{jm}\delta_{kn} - \delta_{jn}\delta_{km}. \tag{A1.6}$$

Appendix 2

The elastic moduli

The elastic moduli for isotropic materials are

$$C_{ijkl} = \lambda\delta_{ij}\delta_{kl} + \mu\delta_{ik}\delta_{jl} + \mu\delta_{il}\delta_{jk}, \tag{A2.1}$$

where λ and μ are Lamé's constants. The following relations hold among Young's modulus E, Poisson's ratio ν, the bulk modulus K, and the shear modulus μ:

$$\begin{aligned} &\lambda = 2\mu\nu/(1-2\nu),\ E = 2(1+\nu)\mu,\ \nu = \lambda/2(\lambda+\mu), \\ &K = (2\mu+3\lambda)/3 = E/3(1-2\nu),\ \nu = (3K-2\mu)/2(3K+\mu). \end{aligned} \tag{A2.2}$$

Hooke's law becomes

$$\begin{aligned} \sigma_{11} &= 2\mu e_{11} + \lambda e_{kk}, \\ \sigma_{22} &= 2\mu e_{22} + \lambda e_{kk}, \\ \sigma_{33} &= 2\mu e_{33} + \lambda e_{kk}, \\ \sigma_{12} &= 2\mu e_{12}, \\ \sigma_{23} &= 2\mu e_{23}, \\ \sigma_{31} &= 2\mu e_{31}, \end{aligned} \tag{A2.3}$$

or

$$\sigma_{ij} = 2\mu e_{ij} + \delta_{ij}\lambda e_{kk}. \tag{A2.4}$$

The inverse of (A2.3) are,

$$\begin{aligned} e_{11} &= \{\sigma_{11} - \nu(\sigma_{22}+\sigma_{33})\}/E, \\ e_{22} &= \{\sigma_{22} - \nu(\sigma_{33}+\sigma_{11})\}/E, \\ e_{33} &= \{\sigma_{33} - \nu(\sigma_{11}+\sigma_{22})\}/E, \end{aligned}$$

$$e_{12} = \sigma_{12}/2\mu, \tag{A2.5}$$
$$e_{23} = \sigma_{23}/2\mu,$$
$$e_{31} = \sigma_{31}/2\mu.$$

When the reduced strain and stress are defined as

$$'e_{ij} = e_{ij} - \delta_{ij}e_{kk}/3,$$
$$'\sigma_{ij} = \sigma_{ij} - \delta_{ij}\sigma_{kk}/3. \tag{A2.6}$$

Hooke's law can be written as

$$'\sigma_{ij} = 2\mu' e_{ij},$$
$$\sigma_{kk} = 3Ke_{kk}. \tag{A2.7}$$

The elastic moduli for cubic crystals are

$$C_{ijkl} = \lambda\delta_{ij}\delta_{kl} + \mu\delta_{ik}\delta_{jl} + \mu\delta_{il}\delta_{jk} + \mu'\delta_{ijkl}, \tag{A2.8}$$

where all $\delta_{ijkl} = 0$, except $\delta_{1111} = \delta_{2222} = \delta_{3333} = 1$, and

$$\lambda = C_{12},\ \mu = C_{44},\ \mu' = C_{11} - C_{12} - 2C_{44}. \tag{A2.9}$$

These C_{ij} are the Voigt constants and are related to C_{ijkl} as shown in Table A2.1 for general cases. For instance, $C_{1122} = C_{12}$, $C_{1123} = C_{14}$. Since $C_{ij} = C_{ji}$, the number of independent constants is generally 21.

Table A2.1. Relation between Constant C_{ijkl} and the Voigt Elastic Constants C_{ij}

ij \ kl	11	22	33	23	31	12
11	C_{11}	C_{12}	C_{13}	C_{14}	C_{15}	C_{16}
22	C_{21}	C_{22}	C_{23}	C_{24}	C_{25}	C_{26}
33	C_{31}	C_{32}	C_{33}	C_{34}	C_{35}	C_{36}
23	C_{41}	C_{42}	C_{43}	C_{44}	C_{45}	C_{46}
31	C_{51}	C_{52}	C_{53}	C_{54}	C_{55}	C_{56}
12	C_{61}	C_{62}	C_{63}	C_{64}	C_{65}	C_{66}

Let us make the notational change,

$$e_{11} = e_1,\ e_{22} = e_2,\ e_{33} = e_3,\ e_{23} = e_4/2,\ e_{31} = e_5/2,\ e_{12} = e_6/2. \tag{A2.10}$$

and

$$\sigma_{11} = \sigma_1,\ \sigma_{22} = \sigma_2,\ \sigma_{33} = \sigma_3,\ \sigma_{23} = \sigma_4,\ \sigma_{31} = \sigma_5,\ \sigma_{12} = \sigma_6. \tag{A2.11}$$

Hooke's law is written as

$$\begin{aligned}
\sigma_1 &= C_{11}e_1 + C_{12}e_2 + C_{13}e_3 + C_{14}e_4 + C_{15}e_5 + C_{16}e_6,\\
\sigma_2 &= C_{21}e_1 + C_{22}e_2 + C_{23}e_3 + C_{24}e_4 + C_{25}e_5 + C_{26}e_6,\\
\sigma_3 &= C_{31}e_1 + C_{32}e_2 + C_{33}e_3 + C_{34}e_4 + C_{35}e_5 + C_{36}e_6,\\
\sigma_4 &= C_{41}e_1 + C_{42}e_2 + C_{43}e_3 + C_{44}e_4 + C_{45}e_5 + C_{46}e_6,\\
\sigma_5 &= C_{51}e_1 + C_{52}e_2 + C_{53}e_3 + C_{54}e_4 + C_{55}e_5 + C_{56}e_6,\\
\sigma_6 &= C_{61}e_1 + C_{62}e_2 + C_{63}e_3 + C_{64}e_4 + C_{65}e_5 + C_{66}e_6.
\end{aligned} \tag{A2.12}$$

For isotropic materials all $C_{ij} = 0$, except for

$$\begin{aligned}
C_{11} &= C_{22} = C_{33},\\
C_{12} &= C_{13} = C_{23},\\
C_{44} &= C_{55} = C_{66} = (C_{11} - C_{12})/2.
\end{aligned} \tag{A2.13}$$

For cubic crystals (Al, Cu, Au, Fe, Pb, Ni, Ag, W, UO_2, etc.) all $C_{ij} = 0$, except for

$$\begin{aligned}
C_{11} &= C_{22} = C_{33},\\
C_{12} &= C_{13} = C_{23},\\
C_{44} &= C_{55} = C_{66},
\end{aligned} \tag{A2.14}$$

when the coordinate system is chosen along the crystalline directions. For hexagonal crystals (Mg, Zn, Co, Cd, Al_2O_3, etc.) all $C_{ij} = 0$, except for

$$\begin{aligned}
&C_{11} = C_{22},\ C_{33},\\
&C_{12},\ C_{13} = C_{23},\\
&C_{44} = C_{55},\ C_{66} = (C_{11} - C_{12})/2,
\end{aligned} \tag{A2.15}$$

when the x_3 direction is chosen along the hexad axis. This case is transversely isotropic.

For tetragonal crystals (Sn, Zr, MgCl, etc.) all $C_{ij} = 0$, except for

$$\begin{aligned}
&C_{11} = C_{22},\ C_{33},\\
&C_{12},\ C_{13} = C_{23},\\
&C_{44} = C_{55},\ C_{66},
\end{aligned} \tag{A2.16}$$

when the x_3 direction is chosen along the hexad axis. For orthorhombic crystals (S, U, etc.), all $C_{ij} = 0$, except for

$$C_{11},\ C_{22},\ C_{33},$$
$$C_{12},\ C_{13},\ C_{23}, \tag{A2.17}$$
$$C_{44},\ C_{55},\ C_{66},$$

when the coordinate system is chosen along the crystalline directions.

The numerical values of C_{ij} for various materials are found in Simmons and Wang (1971).

The elastic strain can be expressed in the stress by

$$e_i = s_{ij}\sigma_j, \tag{A2.18}$$

or

$$e_{ij} = s_{ijkl}\sigma_{kl}, \tag{A2.18$'$}$$

where s_{ij} is called the elastic compliance. C_{ij} is sometimes called the elastic stiffness or the elastic constants.

There is a relation

$$s_{66} = 2(s_{11} - s_{12}) \tag{A2.19}$$

when

$$C_{66} = (C_{11} - C_{12})/2. \tag{A2.20}$$

Appendix 3

Fourier series and integrals

Every function which is piecewise smooth in the interval $-L \leq x \leq L$ and periodic with the period $2L$ may be expanded in a Fourier series; that is,

$$f(x) = \frac{1}{2} a_0 + \sum_{n=1}^{\infty} \left(a_n \cos \frac{\pi n x}{L} + b_n \sin \frac{\pi n x}{L} \right), \tag{A3.1}$$

where

$$\begin{aligned} a_0 &= \frac{1}{L} \int_{-L}^{L} f(x) dx, \\ a_n &= \frac{1}{L} \int_{-L}^{L} f(x) \cos \frac{\pi n x}{L} \, dx, \\ b_n &= \frac{1}{L} \int_{-L}^{L} f(x) \sin \frac{\pi n x}{L} \, dx. \end{aligned} \tag{A3.2}$$

The convergence of the Fourier series is uniform in every closed interval in which the function is continuous.

Equations (A3.1) and (A3.2) may also be written in the form

$$f(x) = \sum_{n=-\infty}^{\infty} \alpha_n \exp(i n \pi x / L), \tag{A3.3}$$

$$\alpha_n = \frac{1}{2L} \int_{-L}^{L} f(x) \exp(-i n \pi x / L) dx. \tag{A3.4}$$

To prove this, we derive (A3.1) from (A3.3). Substituting (A3.4) into

(A3.3), we have

$$f(x) = \sum_{n=-\infty}^{\infty} \frac{1}{2L} \int_{-L}^{L} f(t)\exp(-in\pi t/L)dt \exp(in\pi x/L)$$

$$= \sum_{n=-\infty}^{\infty} \left[\frac{1}{2L}\int_{-L}^{L} f(t)\cos(n\pi t/L)dt \cos(n\pi x/L) + \frac{1}{2L}\int_{-L}^{L} f(t)\sin(n\pi t/L)dt \sin(n\pi x/L)\right] \tag{A3.5}$$

since the terms $\cos(n\pi t/L)\sin(n\pi x/L)$ cancel the terms $\cos(-n\pi t/L)$ $\sin(-n\pi x/L)$. Furthermore, the identities $\cos(n\pi t/L)\cos(n\pi x/L) = \cos(-n\pi t/L)\cos(-n\pi x/L)$, etc. give the equivalency between (A3.5) and (A3.1).

It seems desirable to let L go to ∞, since then it is no longer necessary to require that f be continued periodically. We assume that $f(x)$ is piecewise smooth in every finite interval and that at the discontinuities, the value of the function is the arithmetic mean of the right-hand and left-hand limits. The further assumption that the integral $\int_{-\infty}^{\infty} |f(x)|dx$ exists is added. Setting $\pi/L = \delta$, (A3.3) and (A3.4) become

$$f(x) = \sum_{n=-\infty}^{\infty} \alpha_n \exp(in\delta x), \tag{A3.6}$$

$$\alpha_n = \frac{\delta}{2\pi}\int_{-L}^{L} f(x)\exp(-in\delta x)dx. \tag{A3.7}$$

Furthermore, we set $n\delta = \xi$, $\delta = d\xi$ and let $L \to \infty$ and $\delta \to 0$. Then, (A3.6) can be written as

$$f(x) = \int_{-\infty}^{\infty} \bar{f}(\xi)\exp(i\xi x)d\xi, \tag{A3.8}$$

where

$$\bar{f}(\xi) = \frac{1}{2\pi}\int_{-\infty}^{\infty} f(x)\exp(-i\xi x)dx. \tag{A3.9}$$

The expression (A3.8) with (A3.9) is called the Fourier integral form of $f(x)$. $\bar{f}(\xi)$ is called the Fourier transform of $f(x)$.

Dirac's delta function and Heaviside's unit function

Dirac's delta function $\delta(x)$ has the property

$$\int_{-\infty}^{\infty} \delta(x)F(x)dx = F(0) \tag{A3.10}$$

for any suitably continuous function $F(x)$. $\delta(x)$ may be defined by

$$\lim_{n\to\infty}\frac{\sin nx}{\pi x}=\delta(x). \tag{A3.11}$$

The Fourier transform of $\delta(x)$ is obtained from (A3.9) as

$$\bar{\delta}(\xi)=(1/2\pi). \tag{A3.12}$$

The Fourier integral form of $\delta(x)$ is, therefore,

$$\delta(x)=(1/2\pi)\int_{-\infty}^{\infty}\exp(i\xi x)d\xi. \tag{A3.13}$$

Although the integral in (A3.13) does not converge, the formality (A3.13) holds. $\delta(x)$ is the generalized function (see Lighthill 1964).

Heaviside's unit (or step) function $H(x)$ is defined by

$$H(x)=\begin{cases}1 & \text{for } x>0\\ 0 & \text{for } x<0\end{cases}. \tag{A3.14}$$

The Fourier transform of $H(x)$ becomes, from (A3.9),

$$\bar{H}(\xi)=1/2\pi i\xi \tag{A3.15}$$

if one can justify to throw out the term $\exp(-i\xi x)$ at $\xi=\infty$. According to Lighthill (p. 33, 1964), its correct expression is

$$\bar{H}(\xi)=\exp(-\tfrac{1}{2}\pi i\ \mathrm{sgn}\ \xi)/2\pi|\xi|, \tag{A3.16}$$

where sgn ξ is defined as

$$\mathrm{sgn}\ \xi=\begin{cases}1 & \text{for } \xi>0\\ -1 & \text{for } \xi<0\end{cases}. \tag{A3.17}$$

Laplace transformation

Combining (A3.8) and (A3.9), we have

$$f(t)=\frac{1}{2\pi}\int_{-\infty}^{\infty}e^{i\xi t}d\xi\int_{-\infty}^{\infty}f(\tau)e^{-i\xi\tau}d\tau, \tag{A3.18}$$

where t is used for x.

If

$$\begin{aligned} f(t)&=0 && \text{for } t<0,\\ &=e^{-\gamma t}F(t) && \text{for } t>0, \end{aligned} \tag{A3.19}$$

where $\gamma > 0$, (A3.18) becomes

$$F(t) = \frac{e^{\gamma t}}{2\pi} \int_{-\infty}^{\infty} e^{i\xi t} d\xi \int_{0}^{\infty} F(\tau) e^{-(\gamma + i\xi)\tau} d\tau$$

$$= \frac{1}{2\pi i} \int_{\gamma - i\infty}^{\gamma + i\infty} e^{zt} dz \int_{0}^{\infty} F(\tau) e^{-z\tau} d\tau, \tag{A3.20}$$

where $z = \gamma + i\xi$ is used.

The Laplace transform of a function $F(t)$ is defined as

$$\bar{F}(s) = \int_{0}^{\infty} F(t) e^{-st} dt \quad s > 0. \tag{A3.21}$$

When s is formally extended to a complex variable z, (A3.20) can be written as

$$F(t) = \frac{1}{2\pi i} \int_{\gamma - i\infty}^{\gamma + i\infty} e^{zt} \bar{F}(z) dz \tag{A3.22}$$

which is called the inversion integral.

If $\bar{F}(s)$ is obtained as an analytic function in the semi-infinite plane $R(z) < \gamma$ except at poles $z_n (n = 1, 2, \ldots)$ and

$$\begin{aligned} &|\bar{F}(x \pm i\beta_N)| < \delta_N \\ &|\bar{F}(-\beta_N + iy)| < M \end{aligned} \qquad (-\beta_N \leq x \leq \gamma, |y| \leq \beta_N) \tag{A3.23}$$

where $\delta_N \to 0$, $\beta_N \to \infty$ for $N \to \infty$, (A3.22) becomes

$$F(t) = \sum_{n=1} \rho_n(t), \tag{A3.24}$$

where ρ_n is the residue of $e^{zt}\bar{F}(z)$ at z_n. All poles of $e^{zt}\bar{F}(z)$ are located in the plane $R(z) < \gamma$.

Appendix 4

Dislocations pile-up

The Hilbert integral equation for an unknown $f(x)$,

$$\int_D \frac{f(t)}{x-t}\,dt + \sigma(x) = 0, \tag{A4.1}$$

has been investigated by Muskhelishvili (1953) and illustrated for the dislocation problems by Head and Louat (1955). In the above equation $\sigma(x)$ is a given piecewise function. It is understood that the Cauchy principal value of the integral (see Whittaker and Watson 1962, p. 75) is to be taken to avoid divergence at $t = x$.

Let us assume that D consists of a finite number of intervals of the x axis $(a_1, b_1), (a_2, b_2), \ldots, (a_p, b_p)$. Suppose that at q of the end-points of the segments, denoted by $c_1, c_2, \ldots, c_q$, $f(x)$ is to remain bounded, and that at the remaining $2p$-q end-points, denoted by $c_{q+1}, c_{q+2}, \ldots, c_{2p}$, $f(x)$ may be unbounded. Let

$$\begin{aligned} R_1(x) &= \prod_{k=1}^{q} (x - c_k), \\ R_2(x) &= \prod_{k=q+1}^{2p} (x - c_k), \end{aligned} \tag{A4.2}$$

where $R_1 = 1$ when $q = 0$, and $R_2 = 1$ when $2p < q + 1$. Then, if $p - q \geq 0$, solutions of (A4.1), bounded at $c_1, \ldots, c_q$, always exist and are given by

$$f(x) = \frac{1}{\pi^2}\left\{\frac{R_1(x)}{R_2(x)}\right\}^{1/2} \int_D \left\{\frac{R_2(t)}{R_1(t)}\right\}^{1/2} \frac{\sigma(t)}{x-t}\,dt + \left\{\frac{R_1(x)}{R_2(x)}\right\}^{1/2} Q_{p-q-1}(x), \tag{A4.3}$$

where $Q_{p-q-1}(x)$ is an arbitrary polynomial of degree not greater than $p - q - 1$. It is identically zero for $p = q$ and $p = q + 1$.

If $p-q<0$, a unique solution, bounded at $c_1, \ldots, c_q$ exists if and only if $\sigma(x)$ satisfies the conditions

$$\int_D \left\{\frac{R_2(t)}{R_1(t)}\right\}^{1/2} t^n \sigma(t)dt = 0, \quad n = 0, 1, \ldots, (q-p-1), \tag{A4.4}$$

and if this is so the solution is given by (A4.3) with $Q_{p-q-1}=0$. Moreover at a bounded end-point, $f(x)$ vanishes. The Cauchy principal values of the integrals are taken in all these formulae.

Let us consider an example. $\sigma=\sigma^0=\text{constant}$, $D=(-c, c)$, and $f(x)$ is unbounded at $-c$ and c. Then, $a_1=-c$, $b_1=c$, $p=1$, $q=0$, $c_{q+1}=-c$, and $c_{q+2}=c$. We have $R_1=1$, and $R_2(x)=(x+c)(x-c)$. Since $p-q>0$, (A4.3) becomes

$$\begin{aligned} f(x) &= \frac{\sigma^0}{\pi^2}\frac{1}{(c^2-x^2)^{1/2}}\int_{-c}^{c}\frac{(c^2-t^2)^{1/2}}{x-t}dt + \frac{Q_0}{(x^2-c^2)^{1/2}} \\ &= \frac{\sigma^0}{\pi}\frac{x}{(c^2-x^2)^{1/2}} + \frac{Q_0}{(x^2-c^2)^{1/2}}. \end{aligned} \tag{A4.5}$$

Q_0 is an arbitrary constant.

Another example will be considered: $D=(-a, a)$, $\sigma=\sigma^0$ in $|x|<c$, $\sigma=\sigma^0-k$ in $c<|x|<a$, and $f(x)$ is bounded at $-a$ and a, where σ^0 and k are given constants. Then, $a_1=-a$, $b_1=a$, $p=1$, $q=2$, $c_1=-a$, and $c_2=a$, therefore $R_1(x)=(x+a)(x-a)$, $R_2=1$. Since $p-q<0$,

$$\begin{aligned} f(x) &= (1/\pi^2)(a^2-x^2)^{1/2}\int_{-a}^{a}\frac{\sigma dt}{(a^2-t^2)^{1/2}(x-t)} \\ &= (1/\pi^2)(a^2-x^2)^{1/2}\left\{\int_{-a}^{-c}\frac{(\sigma^0-k)dt}{(a^2-t^2)^{1/2}(x-t)} + \int_{-c}^{c}\frac{\sigma^0 dt}{(a^2-t^2)^{1/2}(x-t)}\right. \\ &\quad \left. + \int_{c}^{a}\frac{(\sigma^0-k)dt}{(a^2-t^2)^{1/2}(x-t)}\right\}. \end{aligned} \tag{A4.6}$$

The condition (A4.4) becomes

$$\int_{-a}^{a}\frac{\sigma}{(a^2-t^2)^{1/2}}dt = 0. \tag{A4.7}$$

Further calculations for these integrals lead to

$$\begin{aligned} f(x) &= \frac{k}{\pi^2}\log\frac{x(a^2-c^2)^{1/2}+c(a^2-x^2)^{1/2}}{x(a^2-c^2)^{1/2}-c(a^2-x^2)^{1/2}} \quad \text{for } c<|x|<a \\ &= \frac{k}{\pi^2}\log\frac{x(a^2-c^2)^{1/2}+c(a^2-x^2)^{1/2}}{c(a^2-x^2)^{1/2}-x(a^2-c^2)^{1/2}} \quad \text{for } |x|<c \end{aligned} \tag{A4.8}$$

and

$$c/a = \cos(\pi\sigma^0/2k). \tag{A4.9}$$

The last result (A4.9) has been obtained from (A4.7).

References

H.I. Aaronson, C. Laird and K. R. Kinsman, Mechanisms of diffusional growth of precipitate crystals, in *Phase Transformations*, ed. H.I. Aaronson, American Society for Metals, (1970) 313–395.

J.D. Achenbach and G. Herrmann, Dispersion of free harmonic waves in fiber reinforced composites, *AIAA J.*, 6 (1968) 1832–1836.

J.D. Achenbach, Dynamic effects in brittle fracture, *Mechanics Today*, ed. S. Nemat–Nasser 1 (1974) 1–57.

J.D. Achenbach and M.R. Kanninen, Crack tip plasticity in dynamic fracture mechanics, in *Fracture Mechanics*, eds. N. Perrone, H. Liebowitz, D. Mulville and W. Pilkey, The Univ. Press of Virginia, (1978) 649–670.

J.D. Achenbach, L.M. Keer, R.P. Khetan, and S.H. Chen, Loss of adhesion at the tip of an interface crack, *J. Elasticity*, 9 (1979) 397–424.

D.F. Adams and D.R. Doner, Longitudinal shear loading of a unidirectional composite, *J. Compos. Mater.*, 1 (1967) 4–17.

K. Aderogba, Some results for two-phase elastic planes, *Proc. Roy. Soc. Edinb.*, A71 (1972) 249–262.

K. Aderogba, On eigenstresses in dissimilar media, Phil. *Mag.*, 35 (1977) 281–292.

S. Amari, A geometrical theory of moving dislocations and an elasticity, in *Unifying Study of Basic Problems in Engineering and Physical Sciences by Means of Geometry* 4, ed. K. Kondo, Gakujutsu Bunken Fukyukai, Tokyo, (1968) 284–294.

Amelinckx, *The Direct Observation of Dislocations*, Academic Press, (1964). *American Institute of Physics Handbook*, 2nd edition, McGraw-Hill, (1972).

D.L. Anderson, B. Minster and D. Cole, The effect of oriented cracks on seismic velocities, *J. Geophys. Res.*, 79 (1974) 4011–4015.

A.E. Andreykiv and M.M. Stadnik, Propagation of a plane crack with a piece-wise smooth contour, *Sov. Appl. Mech.*, 10 (1974) 1078–1083.
D.D. Ang and M.L. Williams, The dynamic stress field due to an extensional dislocation, in *4th Midwestern Conference on Solid Mechanics*, The Univ. of Texas, (1959).
K. Anthony, U. Essmann, A. Seeger and H. Träuble, Dislocations and the Cosserat-continuum with incompatible rotations, in *Mechanics of Generalized Continua*, ed. E. Kröner, Springer, (1968) 355–358.
K. Anthony, Die Theorie der disklinationen, *Arch. Ration. Mech. Anal.*, 39 (1970) 43–88.
A.J. Ardell, R.B. Nicholson and J.D. Eshelby, On the modulated structures of aged Ni-Al alloys, *Act. Metall.*, 14 (1966) 1295–1309.
A.S. Argon and J. Im, Separation of second phase particles in spheroidized 1045 steel, Cu-0.6 pct. Cr alloy, and maraging steel in plastic straining, *Metall. Trans.*, A6 (1975) 839–851.
A.S. Argon, J. Im and R. Safoglu, Cavity formation from inclusions in ductile fracture, *Metall. Trans.*, A6 (1975) 825–837.
A.S. Argon, Formation of cavities from nondeformable second-phase particles in low temperature ductile fracture, *J. Engng. Mater. Tech.*, 98 (1976) 60–68.
A.S. Argon, Stresses in and around slender elastic rods and platelets of different modulus in an infinite elastic medium under uniform strain at infinity, *Fibre Sci. and Tech.*, 9 (1976) 265–275.
R.J. Asaro and J.P. Hirth, Planar dislocation interactions in anisotropic media with applications to nodes, *J. Phys. F: metal physics*, 3 (1973) 1659–1671.
R.J. Asaro, J.P. Hirth, D.M. Barnett and J. Lothe, A further synthesis of sextic and integral theories for dislocations and line forces in anisotropic media, *Phys. stat. sol.* (b), 60 (1973) 261–271.
R.J. Asaro, Somigliana dislocations and internal stresses; with application to second phase hardening, *Int. J. Engng. Sci.*, 13 (1975) 271–286.
R.J. Asaro and D.M. Barnett, The non-uniform transformation strain problem for an anisotropic ellipsoidal inclusion, *J. Mech. Phys. Solids*, 23 (1975) 77–83.
R.J. Asaro and D.M. Barnett, Application of the geometrical theorems for dislocations in anisotropic elastic media, in *Computer Simulation for Materials Applications*, eds. R.J. Arsenault, J.R. Beeler, Jr. and J.A. Simmons, AIME, Nuclear Metallurgy Series No. 20 (Part 2) (1976) 313–324.

M.F. Ashby, The theory of the critical shear stress and work hardening of dispersion hardened crystals, in *Oxide Dispersion Hardening*, Gordon and Breach, (1966) 143–205.

M.F. Ashby, Work hardening of dispersion-hardened crystals, *Phil. Mag.*, 14 (1966) 1157–1178.

C. Atkinson, A simple model of a relaxed expanding crack, *Arkiv Fysik.*, 35 (1967) 469–476.

C. Atkinson and J.D. Eshelby, The flow of energy into the tip of a moving crack, *Int. J. Fracture Mech.*, 4 (1968) 3–8.

C. Atkinson and T.R. Kay, A simple model of relaxation at a crack tip, *Act. Metall.*, 19 (1971) 679–683.

C. Atkinson and M.F. Kanninen, A simple representation of crack tip plasticity: The inclined strip yield model, *Int. J. Fracture*, 13 (1977) 151–163.

J.D. Atkinson, L.M. Brown and W.M. Stobbs, The work hardening of copper-silica IV. The Bauschinger effect and plastic relaxation, *Phil. Mag.*, 30 (1974) 1247–1280.

D.J. Bacon, U.F. Kochs and R.O. Scattergood, The effect of dislocation self-interaction on the Orowan stress, *Phil. Mag.*, 28 (1973) 1241–1263.

D.J. Bacon, D.M. Barnett and R.O. Scattergood, Anisotropic Continuum Theory of Lattice Defects, *Progress in Materials Science Series*, 23 (1978) 51–262.

B.R. Baker, Dynamic stress created by a moving crack, *J. Appl. Mech.*, 29 (1962) 449–458.

G.I. Barenblatt, On finiteness conditions in the mechanics of continuous media. Static problems of the theory of elasticity, *J. Appl. Math. Mech.*, 24 (1960) 453–461.

G.I. Barenblatt, The mathematical theory of equilibrium cracks in brittle fracture, *Adv. in Appl. Mech.*, 7 (1962) 55–129.

D.M. Barnett and A.S. Tetelman, The stress distribution produced by screw dislocation pile-up at rigid circular cylindrical inclusions, *J. Mech. Phys. Solids*, 14 (1966) 329–348.

D.M. Barnett, The effect of shear modulus on the stress distribution produced by a planar array of screw dislocations near a bi-metallic interface, *Act. Metall.*, 15 (1967) 589–594.

D.M. Barnett and A.S. Tetelman, The stresses produced by a screw dislocation pile-up at a circular inclusion of finite rigidity, *Can. J. Phys.*, 45 (1967) 841–863.

D.M. Barnett, Edge dislocation stress concentrations in two-phase media, *Scrip. Metall.*, 3 (1969) 543–546.

D.M. Barnett, On nucleation of coherent precipitates near edge dislocations, *Scrip. Metall.*, 5 (1971) 261–266.

D.M. Barnett and L.A. Swanger, The elastic energy of a straight dislocation in an infinite anisotropic elastic medium, *Phys. stat. sol.* (b), 48 (1971) 419–428.

D.M. Barnett, The precise evaluation of derivatives of the anisotropic elastic Green's functions, *Phys. stat. sol.* (b), 49 (1972) 741–748.

D.M. Barnett and R.J. Asaro, The fracture mechanics of slit-like cracks in anisotropic elastic media, *J. Mech. Phys. Solids*, 20 (1972) 353–366.

D.M. Barnett, R.J. Asaro, S.D. Gavazza, D.J. Bacon and R.O. Scattergood, The effects of elastic anisotropy on dislocation line tension, *J. Phys. F: metal physics*, 2 (1972) 854–864.

D.M. Barnett and J. Lothe, Synthesis of the sextic and the integral formalism for dislocations, Green's functions, and surface waves in anisotropic elastic solids, *Physica Norveg.*, 7 (1973) 13–19.

D.M. Barnett, J.K. Lee, H.I. Aaronson and K.C. Russell, The strain energy of a coherent ellipsoidal precipitate, *Scrip. Metall.*, 8 (1974) 1447–1450.

D.M. Barnett, The singular nature of the self-stress field of a plane dislocation loop in an anisotropic elastic medium, *Phys. stat. sol.* (*a*), 38 (1976) 637–646.

J.L. Bassani, Yield characterization of metals with transversely isotropic plastic properties, *Int. J. Mech. Sci.*, 19 (1977) 651–660.

R.D. Bhargava and H.C. Radhakrishna, Elliptic inclusions in a stressed matrix, *Proc. Camb. Phil. Soc.*, 59 (1963) 821–832.

R.D. Bhargava and H.C. Radhakrishna, Elliptic inclusions in orthotropic medium, *J. Phys. Soc. Japan*, 19 (1964) 396–405.

B.A. Bilby, R. Bullough and E. Smith, Continuous distributions of dislocations: A new application of the methods of non-Riemannian geometry, *Proc. Roy. Soc.*, A231 (1955) 263–273.

B.A. Bilby and J.W. Christian, *The Mechanism of Phase Transformation in Metals*, The Institute of Metals, (1956) 121.

B.A. Bilby and E. Smith, Continuous distributions of dislocations III, *Proc. Roy. Soc.*, A236 (1956) 481–505.

B.A. Bilby, R. Bullough, L.R.T. Gardner and E. Smith, Continuous distributions of dislocations IV. Single glide and plane strain, *Proc. Roy. Soc.*, A244 (1958) 538–557.

B.A. Bilby and L.R.T. Gardner, Continuous distributions of dislocations V. Twisting under conditions of single glide, *Proc. Roy. Soc.*, A247 (1958) 92–108.

B.A. Bilby, L.R.T. Gardner and E. Smith, The relation between dislocation density and stress, *Act. Metall.*, 6 (1958) 29–33.

B.A. Bilby, Continouus distribution of dislocations, in *Progress in Solid Mechanics* 1, eds. I.N. Sneddon and R. Hill, North-Holland, (1960) 329–398.

B.A. Bilby, A.H. Cottrell and K.H. Swinden, The spread of plastic yield from a notch, *Proc. Roy. Soc.*, A272 (1963) 304–314.

B.A. Bilby, A.H. Cottrell, E. Smith and K.H. Swinden, Plastic yielding from sharp notch, *Proc. Roy. Soc.*, A279 (1964) 1–9.

B.A. Bilby and K.H. Swinden, Representation of plasticity of notches by linear dislocation arrays, *Proc. Roy. Soc.*, A285 (1965) 22–23.

B.A. Bilby, L.R.T. Gardner, A. Grinberg and M. Zorawski, Continuous distributions of dislocations VI. Non-metric connexions, *Proc. Roy. Soc.*, A292 (1966) 105–121.

B.A. Bilby and J.D. Eshelby, Microscopic and macroscopic fundamentals, in *An Advanced Treatise on Fracture*, 1, ed. H. Liebowitz, Academic Press, (1968) 99–182.

B.A. Bilby and P.T. Heald, Crack growth in notch fatigue, *Proc. Roy. Soc.*, A305 (1968) 429–439.

B.A. Bilby, J.D. Eshelby and A.K. Kundu, The changes of shape of a viscous ellipsoidal region embedded in a slowly deforming matrix having a different viscosity, *Tectonophysics*, 28 (1975) 265–274.

J.F.W. Bishop and R. Hill, A theory of the plastic distortion of a polycrystalline aggregate under combined stresses, *Phil. Mag.*, 42 (1951) 414–427.

J.F.W. Bishop and R. Hill, A theoretical derivation of the plastic properties of a polycrystalline face-centered metal, *Phil. Mag.*, 42 (1951) 1290–1307.

W. Boas and E. Schmid, Zur Berechnung physikalischer Konstanten quasi-isotroper Vielkristalle, *Helv. Phys. Acta*, 7 (1934) 628–632.

W. Boas, Zur Berechnung des Torsionsmoduls quasiisotroper Vielkristalle aus den Einkristallkonstanten, *Helv. Phys. Acta*, 8 (1935) 674–681.

S.R. Bodner, A hardness law for inelastic deformation, *Lett. Appl. Engng. Sci.*, 16 (1978) 221–230.

B.A. Boley and J.H. Weiner, *Theory of Thermal Stresses*, Wiley, (1960).

B.A. Boley, ed., *Thermoinelasticity*, International Union of Theoretical and Applied Mechanics, Springer-Verlag, (1970).

M. Born and K. Huang, *Dynamic Theory of Crystal Lattices*, Oxford, (1954).

S.C. Bose, On the torsion of a transversely isotropic circular cylinder having a small spherical elastic inclusion of an isotropic material, *Z. angew. Math. Mech.*, 45 (1965) 133–135.

S.K. Bose and A.K. Mal, Longitudinal shear waves in a fiber-reinforced composite, *Int. J. Solids Structures*, 9 (1973) 1075–1085.

J. Boussinesq, *Applications des Potentiels*, Gauthier-Villars, Paris, (1885).

K.B. Broberg, The propagation of a brittle crack, Arkiv för Fysik, Stockholm, 18 (1960) 159–192.

K.B. Broberg, Discussion of fracture from the energy point of view, in *Recent Progress in Applied Mechanics*, eds. B. Broberg, J. Hult and F. Niordson, Almquist and Wiksell, Stockholm, John Wiley and Sons, (1967) 125–151.

D. Broek, *Elementary Engineering Fracture Mechanics*, Noordhoff, (1974).

H. Bross, Zur Theorie Bewegter Versetzungen, *Phys. stat. sol.*, 5 (1964) 329–342.

H. Bross, Das Fundamentalintegral der Elastizitätstheorie für kubische Medien, *Z. angew. Math. Phys.*, 19 (1968) 434–446.

J.S. Bowle and J.K. Mackenzie, The crystallography of martensite transformations I, *Act. Metall.*, 2 (1954) 129–137.

L.M. Brown, A proof of Lothe's theorem, *Phil. Mag.*, 15 (1967) 363–370.

L.M. Brown and W.M. Stobbs, The work hardening of copper-silica, a model based on internal stress, with no plastic relaxation, *Phil. Mag.*, 23 (1971) 1185–1199.

L.M. Brown, Back-stresses, image stresses, and work-hardening, *Act. Metall.*, 21 (1973) 879–885.

L.M. Brown, R.H. Cook, R.K. Ham and G.R. Purdy, Elastic stabilization of arrays of precipitates, *Scrip. Metall.*, 7 (1973) 815–820.

L.M. Brown and D.R. Clarke, Work hardening due to internal stresses in composite materials, *Act. Metall.*, 23 (1975) 821–830.

D.A.G. Bruggeman, Berechnung der elastischen Moduln für die verschiedenen Texturen der regularen Metalle, *Z. Phys.*, 92 (1934) 561–588.

B. Budiansky and T.T. Wu, Theoretical prediction of plastic strains of polycrystals, *Proc. 4th U.S. Nat. Congr. Appl. Mech.*, (1962) 1175–1185.

B. Budiansky, On the elastic moduli of some heterogeneous materials, *J. Mech. Phys. Solids*, 13 (1965) 223–227.

B. Budiansky and J.R. Rice, Conservation laws and energy-release rates, *J. Appl. Mech.*, 40 (1973) 201–203.

B. Budiansky and R.J. O'Connell, Elastic moduli of a cracked solid, Int. *J. Solids Structures*, 12 (1976) 81–97.

B. Budiansky and J.W. Hutchinson, Analysis of closure in fatigue crack growth, *J. Appl. Mech.*, 45 (1978) 267–275.

B. Budiansky, J.W. Hutchinson, and S. Lutsky, Void growth and collapse in viscous solids, *Mechanics of Solids*, The Rodney Hill 60th Anniversary Volume, eds. H.G. Hopkins and M.J. Sewell, Pergamon Press (1980).

B. Budiansky and J.W. Hutchinson, *Self-similar shapes and asymptotic dilation-rates for voids in viscous solids*, Proc. XVth Int. Congr. Theor. Appl. Mech., Toronto (1980).

H.D. Bui, Évolution de la frontiére du domaine élastique des métaux avec l'écrouissage plastique et comportement élasto-plastique d'un agregat de cristaux cubiques, *Memorial Artilleri franc. Sci. Tech. Armement*, 1 (1970) 141–165.

H.D. Bui, Application des potentiels élastiques à l'étude des fissures planes de forme arbitraire en milieu tridimensionnel, *C.R. Acad. Sci.*, 280 (1975) 1157–1160.

H.D. Bui, An integral equations method for solving the problem of a plane crack of arbitrary shape, *J. Mech. Phys. Solids*, 25 (1977) 29–39.

R. Bullough and B.A. Bilby, Uniformly moving dislocations in anisotropic media, *Proc. Phys. Soc.*, B67, (1954) 615–624.

R. Bullough and B.A. Bilby, Continuous distributions of dislocations and the crystallography of martensitic transformation, *Proc. Phys. Soc.*, B69 (1956) 1276–1286.

J.M. Burgers, Some considerations on the fields of stress connected with dislocations in a regular crystal lattice I, II, *Proc. Kon. Nederl. Akad. Wetensch.*, 42 (1939) 293–324, (1939) 378–399.

W.G. Burgers, On the process of transition of the cubic-body-centered modification into the hexagonal-close-pucked modification of zirconium, *Physica*, 1 (1934) 561–586.

E. Byskov, The calculation of stress intensity factors using the finite element method with cracked element, *Int. J. Fracture Mech.*, 6 (1970) 159–167.

C. Callias and X. Markenscoff, The non-uniform motion of a supersonic dislocation, *Quart. Appl. Math.* 38 (1980) 323–330.

G.A. Campbell and R.M. Foster, *Fourier Integrals for Practical Applications*, D. Van Nostrand, (1948).

E. Cesáro, Sulle Formole del Volterra, fundamentali nella teoria delle distorsioni elastiche, Rendiconti dell'accademia della scienze fisiche et matematiche, Societa Real di Napoli, July and August, (1906) 311–321.

C.C. Chamis and G.P. Sendeckyj, Critique on theories predicting thermoelastic properties of fibrous composites, *J. Compos. Mater.*, 2 (1968) 332–358.

S.K. Chan, I.S. Tuba and W.K. Wilson, On the finite element method in linear fracture mechanics, *Engng. Fracture Mech.*, 2 (1970) 1–17.

P. Charsley and M.P.E. Desvaux, The behavior of copper–12% aluminium under simple reversed stresses, *Materials Sci. Engng.*, 4 (1969) 211–220.

C.H. Chen and S. Cheng, Mechanical properties of fiber reinforced composites, *J. Compos. Mater.*, 1 (1967) 30–41.

C.H. Chen, Fiber-reinforced composites under longitudinal shear loading, *J. Appl. Mech.*, 37 (1970) 198–201.

I.W. Chen and A.S. Argon, Grain boundary and interphase boundary sliding in power-law creep, *Act. Metall.*, 27 (1979) 749–754.

I.W. Chen and A.S. Argon, Steady state power-law creep in heterogeneous alloys with coarse microstructures, *Act. Metall.*, 27 (1979) 785–791.

W.T. Chen, On a spherical elastic inclusion in a transversely isotropic material under an axisymmetric torsion field, *J. Appl. Mech.*, 33 (1966) 944–945.

W.T. Chen, Some aspects of a flat elliptical crack under shear stress, *J. Math. Phys.*, 45 (1966) 213–223.

W.T. Chen, On an elliptic elastic inclusion in an anisotropic medium, *Q. J. Mech. Appl. Math.*, 20 (1967) 307–313.

W.T. Chen, Axisymmetric stress field around spheroidal inclusions and cavities in a transversely isotropic material, *J. Appl. Mech.*, 35 (1968) 770–773.

P.C. Cheng, *Elastic field of inclusion and crack in anisotropic media*, Ph.D. thesis, Northwestern University, (1977).

G.P. Cherepanov, On quasibrittle fracture, *J. Appl. Math. Mech.*, 32 (1968) 1050–1058.

Y.P. Chiu, On the stress field due to initial strains in cuboid surrounded by an infinite elastic space, *J. Appl. Mech.*, 44 (1977) 587–590.

Y.P. Chiu, On the stress field and surface deformation in a half space with a cuboidal zone in which initial strains are uniform, *J. Appl. Mech.*, 45 (1978) 302–306.

T.W. Chou, Elastic behavior of disclinations in nonhomogeneous media, *J. Appl. Phys.*, 42 (1971) 4931–4835.

T.W. Chou and T.L. Lu, Elastic behaviors of twist disclination loops near a free surface, *J. Appl. Phys.*, 43 (1972) 2562–2565.

T.W. Chou and T.L. Lu, Elastic behaviors of wedge disclination loops near a free surface, *J. Appl. Phys.*, 43 (1972) 2562–2565.

T.W. Chou and Y.C. Pan, Elastic energies of disclinations in hexagonal crystals, *J. Appl. Phys.*, 44 (1973) 63–65.

T.W. Chou and Y.C. Pan, Elastic energy of wedge disclination loops in hexagonal crystals, *J. Appl. Phys.*, 44 (1973) 3889–3891.

T.W. Chou and M. Ichikawa, A crystal defect theory approach to micro-biomechanics, *J. Biomech.*, 12 (1979) 321–331.

Y.T. Chou, Linear dislocation arrays in heterogeneous materials, *Act. Metall.*, 13 (1965) 779–783.

Y.T. Chou, Dislocation pile-ups against a locked dislocation of different Burgers vector, *J. Appl. Phys.*, 38, (1967) 2080–2085.

Y.T. Chou and J.C.M. Li, Theory of dislocation pile-ups, in *Mathematical Theory of Dislocations*, ed. T. Mura, ASME, (1969) 116–177.

J.W. Christian, Accommodation strains in martensite formation and the use of a dilatation parameter, *Act. Metall.*, 6 (1958) 377–379.

J.W. Christian, *The Theory of Transformations in Metals and Alloys*, Pergamon Press, (1965) 802–931.

J.W. Christian, *Thermodynamics and kinetics of martensite*, Proc. Int. Conf. Martensitic Trans. at Boston, in press, (1979).

T.J. Chuang, K.I. Kagawa, J.R. Rice and L.B. Sills, Non-equilibrium models for diffusive cavitation of grain interfaces, *Act. Metall.*, 27 (1979) 265–284.

P.E. Cladis, M. Kléman and P. Pieranski, Physique cristalline–Sur une nouvelle méthode de décoration de la phase mésomorphe du p. n-méthoxy-benzilidène p-butyaniline, *C.R. Acad. Sci., Paris*, 273 (1971) 275–277.

D.L. Clements, A crack between dissimilar anisotropic media, *Int. J. Engng. Sci.*, 9 (1971) 257–265.

L.F. Coffin, A study of the effects of cyclic thermal stresses on ductile metals, *Trans. ASME*, 76 (1954) 931–950.

W.D. Collins, Some axially summetric stress distributions in elastic solid containing penny-shaped cracks, *Proc. Roy. Soc.*, A266, (1962) 359–386.

R.J. Clifton and X. Markenscoff, Elastic precursor decay and radiation from non-uniformly moving dislocations, *J. Mech. Phys. Solids*, to 29 (1981) 227–251.

G. Colonnetti, Per una teoria generale delle coazioni elastiche, Atti R. Acad. Sci. Torino: *Cl. Sci. fis. mat. natur.*, 56 (1921) 188–198.

M. Comninou and J. Dundurs, Long-range interaction between a screw dislocation and a spherical inclusion, *J. Appl. Phys.*, 43 (1972) 2461–2463.

M. Comninou and J. Dundurs, The angular dislocation in a half space, *J. Elasticity*, 5 (1975) 203–216.

M. Comninou, The angular disclination, *J. Elasticity*, 7 (1977) 61–66.

M. Comninou, The interface crack, *J. Appl. Mech.*, 44 (1977) 631–636.

M. Comninou, A property of interface dislocations, *Phil. Mag.*, 36 (1977) 1281–1283.

M. Comninou, The interface crack in a shear field, *J. Appl. Mech.*, 45 (1978) 287–290.

M. Comninou and D. Schmueser, The interface crack in a combined tension-compression and shear field, *J. Appl. Mech.*, 46 (1979) 345–348.

A.H. Cottrell, Theory of dislocations, in *Progress in Metal Physics*, 1 Butterworth's Scientific Publ., (1949).

A.H. Cottrell, *Dislocations and Plastic Flow in Crystals*, Clarendon Press, (1953).

S.C. Cowin and M. Satake, eds., *Proc. U.S.-Japan Seminar on Continuum Mechanics and Statistical Approaches in the Mechanics of Granular Materials*, Gahujutsu Bunken Fukyu-kai, (1978).

J.W. Cragls, On the propagation of a crack in an elastic-brittle material, *J. Mech. Phys. Solids*, 8 (1960) 66–75.

T.A. Cruse, Numerical solutions in three dimensional elastostatics, *Int. J. Solids Structures*, 5 (1969) 1259–1274.

T.A. Cruse, Application of the boundary-integral equation method to three dimensional stress analysis, *Comput. and Struct.*, 3 (1973) 509–527.

E.S.P. Das, R. deWit, R.W. Armstrong and M.J. Marcinkowski, Kinks, jogs, and a regeneration mechanism for Volterra disclinations, *J. Appl. Phys.*, 44 (1973) 4804–4806.

E.S.P. Das, M.J. Marcinkowski and R.W. Armstrong, The movement of Volterra disclinations and the associated mechanical forces, *Phil. Mag.*, 27 (1973) 369–391.

S.C. Das, Stress concentration around a small spherical or spheroidal inclusion on the axis of a circular cylinder in torsion, *J. Appl. Mech.*, 21 (1954) 83–87.

S.K. Datta, Diffraction of plane elastic waves by ellipsoidal inclusions, *J. Acoust. Soc. Amer.*, 61 (1977) 1432–1438.

S.K. Datta, Scattering by a random distribution of inclusions and effective elastic properties, in *Continuum Models of Discrete Systems*, ed. J.W. Provan, Univ. of Waterloo Press, (1978) 111–127.

M.P.E. Desvaux and P. Charsley, Slip lines on pure copper deformed in tension and compression, *Materials Sci. Engng.*, 4 (1969) 221–230.

R. deWit, The continuum theory of stationary dislocations, in *Solid State Physics* 10, eds. F. Seitz and D. Turnbull, Academic Press, (1960) 249–292.

R. deWit, Linear theory of static disclinations, in *Fundamental Aspects of Dislocation Theory*, eds. J.A. Simmons, R. deWit and R. Bullough, National Bureau of Standards, (1969) 651–680.

R. deWit, Relation between dislocations and disclinations, *J. Appl. Phys.*, 42 (1971) 3304–3308.

R. deWit, The relation between continouus and discrete disclinations, *Arch. Mech. Stos.*, 24 (1972) 499–510.

R. deWit, Continuous distribution of disclination loops, *Phys. stat. sol.* (a), 18 (1973) 669–681.

R. deWit, Theory of disclinations: II. Continuous and discrete disclinations in anisotropic elasticity, J. Res. National Bureau of Standards, *Phys. and Chem.*, 77A (1973) 49–100.

J.P. Dixson and L.P. Pook, Stress intensity factors calculated generally by the finite element technique, *Nature*, 224 (1979) 166–167.

D.C. Drucker and W. Prager, Solid mechanics and plastic analysis or limit design, *Q. Appl. Math.*, 10 (1952) 157–165.

D.S. Dugdale, Yielding of steel sheets containing slits, *J. Mech. Phys. Solids*, 8 (1960) 100–104.

J. Dundurs and T. Mura, Interaction between an edge dislocation and a circular inclusion, *J. Mech. Phys. Solids*, 12 (1964) 177–189.

J. Dundurs and D.L. Guell, Center of dilation and thermal stresses in joined elastic half-spaces, *Proc. 2nd Southwestern Conf. Theoret. and Appl. Mech., Devel. in Theoret. and Appl. Mech.*, 2 (1965) 199–211.

J. Dundurs and M. Hetényi, Transmission of force between two semi-infinite solids, *J. Appl. Mech.*, 32 (1965) 671–674.

J. Dundurs and G.P. Sendeckyj, Edge dislocation inside a circular inclusion, *J. Mech. Phys. Solids*, 13 (1965) 141–147.

J. Dundurs, Effect of elastic constants on stress in a composite under plane deformation, *J. Compos. Mater.*, 1 (1967) 310–322.

J. Dundurs, On the interaction of a screw dislocation with inhomogeneities, in *Recent Advances in Engineering Science*, 2, ed. A.C. Eringen, Gordon and Breach, (1967) 223–233.

J. Dundurs, Edge-bonded dissimilar orthogonal elastic wedges under normal and shear loading, *J. Appl. Mech.*, 36 (1969) 650–652.

J. Dundurs and A.C. Gangadharan, Edge dislocation near an inclusion with a slipping interface, *J. Mech. Phys. Solids*, 17 (1969) 459–471.

J. Dundurs, Some properties of elastic stresses in a composite, in *Recent Advances in Engineering Science*, 5, ed. A.C. Eringen, Gordon and Breach, (1970) 203–216.

J. Dundurs and M.S. Lee, Stress concentrations due to edge dislocation pile-ups in two-phase materials, *Scrip. Metall.*, 4 (1970) 313–314.

J. Dundurs and N.J. Salamon, Circular prismatic dislocation loop in a two-phase material, *Phys. stat. sol.* (b), 50 (1972) 125–133.

J. Dundurs and M. Comninou, Some consequences of the inequality conditions in contact and crack problems, *J. Elasticity*, 9 (1979) 71–82.

F.W. Dyson, The potentials of ellipsoids of variable densities, *Q. J. Pure and Appl. Math.*, 25 (1891) 259–288.

Y.Y. Earmme and J.H. Weiner, Breakdown phenomena in high-speed dislocation, *J. Appl. Phys.*, 45 (1974) 603–609.

K.E. Easterling and H.M. Miekk-oja, The martensitic transformation of iron precipitates in a copper matrix, *Act. Metall.*, 15 (1967) 1133–1141.

K.E. Easterling and A.R. Thölén, The nucleation of martensite in steel, *Act. Metall.*, 24 (1976) 333–341.

M.L. Ebner and W.A. Backofen, Fatigue in single crystals of copper, *Trans. AIME*, 215 (1959) 510–520.

T. Edil, T. Mura and R.J. Krizek, A micromechanistic formulation for stress-strain response of clay, *Int. J. Engng. Sci.*, 13 (1975) 831–840.

R.H. Edward, Stress concentrations around spheroidal inclusions and cavities, *J. Appl. Mech.*, 18 (1951) 19–30.

M.A. Eisenberg, On the relation between continuum plasticity and dislocation theories, *Int. J. Engng. Sci.*, 8 (1970) 261–271.

N.G. Einspruch, E.J. Witterholt and R. Truell, Scattering of a plane transverse wave by a spherical obstacle in an elastic medium, *J. Appl. Phys.*, 31 (1960) 806–818.

H.A. Elliott, Three-dimensional stress distributions in hexagonal aeolotropic crystals, *Proc. Camb. Phil. Soc.*, 44 (1948) 522–533.

A.H. England, A note on cracks under longitudinal shear, *Mathematika*, 10 (1963) 107–113.

A.H. England and A.E. Green, Some two-dimensional punch and crack problems in classical elasticity, *Proc. Camb. Phil. Soc.*, 59 (1963) 489–500.

A.H. England, A crack between dissimilar media, *J. Appl. Mech.*, 32 (1965) 400–402.

F. Erdogan, On the stress distribution in plates with collinear cuts under arbitrary loads; *Proc. 4th U.S. Nat. Congr. Appl. Mech.*, (1962) 547–553.

F. Erdogan, Stress distribution in a nonhomogeneous elastic plane with cracks, *J. Appl. Mech.*, 30 (1963) 232–237.

F. Erdogan, Crack-propagation theories, in *Fracture*, 2, ed. H. Liebowitz, Academic Press, (1968) 497–590.

F. Erdogan and G. Gupta, On the numerical solution of singular integral equations, *Q. Appl. Math.*, 29 (1972) 525–539.

F. Erdogan, Mixed boundary-value problems in mechanics, in *Mechanics Today*, ed. S. Nemat-Nasser, 4 (1978) 1–86.

A.C. Eringen, Linear theory of micropolar elasticity, *J. Math. Mech.*, 15 (1966) 909–924.

J.D. Eshelby, Dislocations as a cause of mechanical damping in metals, *Proc. Roy. Soc.*, A197 (1949) 396–416.

J.D. Eshelby, Uniformly moving dislocations, *Proc. Phys. Soc.*, A62 (1949) 307–314.

J.D. Eshelby, The force on an elastic singularity, *Phil. Trans. Roy. Soc.*, A244 (1951) 87–112.

J.D. Eshelby, F.C. Frank and F.R.N. Nabarro, The equilibrium of linear arrays of dislocations, *Phil. Mag.*, 42 (1951) 351–364.

J.D. Eshelby, The equation of motion of a dislocation, *Phys. Rev.*, 90 (1953) 248–255.

J.D. Eshelby, W.T. Read and W. Shockley, Anisotropic elasticity with applications to dislocation theory, *Act. Metall.*, 1 (1953) 251–259.

J.D. Eshelby, Supersonic dislocations and dislocations in dispersive media, *Proc. Phys. Soc.*, B69 (1956) 1013–1019.

J.D. Eshelby, The continuum theory of lattice defects, in *Solid State Physics* 3, eds. F. Seitz and D. Turnbull, Academic Press, (1956) 79–144.

J.D. Eshelby, The determination of the elastic field of an ellipsoidal inclusion, and related problems, *Proc. Roy. Soc.*, A241 (1957) 376–396.

J.D. Eshelby, The elastic field outside an ellipsoidal inclusion, *Proc. Roy. Soc.*, A252 (1959) 561–569.

J.D. Eshelby, Elastic inclusions and inhomogeneities, in *Progress in Solid Mechanics* 2, eds. I.N. Sneddon and R. Hill, North-Holland, Amsterdam, (1961) 89–140.

J.D. Eshelby, The interaction of kinks and elastic waves, *Proc. Roy. Soc.*, A266 (1962) 222–246.

J.D. Eshelby, Energy relations and the energy-momentum tensor in continuum mechanics, in *Inelastic Behavior of Solids*, eds. M.F. Kanninen, W.F. Adler, A.R. Rosenfield and R.I. Jaffee, Battell Inst. Mater. Sci. Colloquia, McGraw-Hill, (1970) 77–115.

P.H. Eurin, J.M. Penisson and A. Bornett, Étude du reseau de germes dans l'alliage CoPt équiatomique par microscopie électronique et diffraction optique, *Act. Metall.*, 21 (1973) 559–570.

G. Faivre, Déformations de cohérence d'un précipite quadratique, *phys. stat. sol.*, 35 (1964) 249–259.

N.M. Ferrers, On the potentials of ellipsoids, ellipsoidal shells, elliptic laminae and elliptic rings of variable densities, *Q. J. Pure and Appl. Math.*, 14 (1877) 1–22.

J.E. Flinn, A continuum dislocation theory model for predicting the wall thickness changes of hollow draw tubing, *J. Basic Engng, ASME Trans.*, 92 (1970) 101–104.

A.J.E. Foreman, Dislocation energies in anisotropic crystals, *Act. Metall.*, 3 (1955) 322–330.

F.C. Frank, On the equation of motion of crystal dislocations, *Proc. Phys. Soc.*, A62 (1949) 131–134.

F.C. Frank, On the theory of liquid crystals, *Disc. Faraday Soc.*, 25 (1958) 19–28.

I. Fredholm, Sur les équations de l'équilibre d'un corps solide élastique, *Act. Math.*, 23 (1900) 1–42.

L.B. Freund, Crack propagation in an elastic solid subjected to general loading I. Constant rate of extension, *J. Mech. Phys. Solids*, 20 (1972) 129–140.

L.B. Freund, Crack propagation in an elastic solid subjected to general loading II. Nonuniform rate of extension, *J. Mech. Phys. Solids*, 20 (1972) 141–152.

L.B. Freund, Energy flux into the tip of an extending crack in an elastic solid, *J. Elasticity*, 2 (1972) 341–349.

L.B. Freund, Dynamic crack propagation, in *The Mechanics of Fracture*, ed. F. Erdogan, *ASME AMD*, 19 (1975) 105–134.

L.B. Freund, The analysis of elastodynamics crack Tip stress fields, in *Mechanics Today*, ed. S. Nemat-Nasser, 3 (1976) 55–91.

J. Friedel, *Dislocations*, Addison-Wesley, (1964).

J. Friedel and M. Kléman, Application of dislocation theory to liquid crystals, in *Fundamental Aspects of Dislocation Theory*, eds. J.A. Simmons, R. deWit and R. Bullough, National Bureau of Standards, 1 (1969) 607–636.

W.S. Fu and L.H. Keer, Coplanar circular cracks under shear loading, *Int. J. Engng. Sci.*, 7 (1969) 361–374.

W.S. Fu, Thermal stresses in an elastic solid weakened by coplanar circular cracks. *Int. J. Engng. Sci*, 11 (1973) 317–330.

Y.C. Fung, *Foundations of Solid Mechanics*, Prentice-Hall, (1965).

R. Furuhashi and T. Mura, On the equivalent inclusion method and impotent eigenstrains, *J. Elasticity*, 9 (1979) 263–270.

R. Furuhashi, N. Kinoshita and T. Mura, Periodic distributions of inclusions, *Int. J. Engng. Sci.*, 19 (1981) 231–236.

B. Galerkin, Contribution à la solution générale du problème de la théorie de l'élasticité dans le cas de trois dimensions, *C. R. Acad. Sci.*, 190 (1930) 1047–1048.

S.D. Gavazza and D.M. Barnett, The elastic interaction between a screw dislocation and a spherical inclusion, *Int. J. Engng. Sci.*, 12 (1974) 1025–1043.

S.D. Gavazza and D.M. Barnett, The self-force on a planar dislocation loop in an anisotropic linear-elastic medium, *J. Mech. Phys. Solids*, 24 (1976) 171–185.

R.W. Gerstner and J. Dundurs, Representation of stress concentration factors for a composite in plane strain, *J. Compos. Mater.*, 3 (1969) 108–115.

J.F. Geyer and S. Nemat-Nasser, Experimental investigation of thermally induced interacting cracks in brittle solids, *Int. J. Solids. Structures*, to appear (1981).

F. Ghahremani, Effect of grain boundary sliding on anelasticity of polycrystals, *Int. J. Solids and Structures*, 16 (1980) 825–845.

J.J. Gilman, Progress in the microdynamical theory of plasticity, *Proc. 5th U.S. Nat. Congr. Appl. Mech.*, (1966) 353–403.

A.A. Golebiewska-Lasota, Nonlocal theory of dislocations, in *Continuum Models of Discrete Systems*, ed. J.W. Provan, Univ. of Waterloo Press, (1978) 537–547.

A.A. Golebiewska-Lasota, Dislocations and gauge invariance, *Int. J. Engng. Sci.*, 17 (1979) 329–333.

I.V. Goncharyuk, On the elastic-plastic torsion of rods of complex profile, *Sov. Appl. Mech.*, 8 (1972) 755–759.

J.N. Goodier, Concentration of stress around spherical and cylindrical inclusions and flows, *ASME Trans.*, 55 (1933) 39–44.

J.N. Goodier, On the integration of the thermoelastic equations, *Phil. Mag.*, 23 (1937) 1017–1032.

J.N. Goodier and F.A. Field, Plastic energy dissipation in crack propagation, in *Fracture of Solids*, eds. D.C. Drucker and J.J. Gilman, Gordon and Breach, (1963) 103–118.

D. Gould, P.B. Hirsch and F.J. Humphreys, The Bauschinger effect, work hardening and recovery in dispersion hardened copper crystals, *Phil. Mag.*, 30 (1974) 1353–1377.

A. Granato and K. Lücke, Application of dislocation theory to internal friction phenomena at high frequencies, *J. Appl. Phys.*, 27 (1956) 789–804.

A.E. Green, Stress systems in isotropic and aeolotropic plates V, *Proc. Roy. Soc.*, A184 (1945) 231–252.

A.E. Green and W. Zerna, *Theoretical Elasticity*, Oxford Univ. Press, (1954).

A.A. Griffith, The phenomena of rupture and flow in solids, *Trans. Roy. Soc.*, A221 (1921) 163–179.

J.C. Grosskrentz, The mechanisms of metal fatigue, *phys. stat. sol. (b)*, 47 (1971) 359–396.

J.E. Gubernatis, E. Domany, J.A. Krumhansl and M. Huberman, The Born approximation in the theory of the scattering of elastic waves by flaws, *J. Appl. Phys.*, 48 (1977) 2812–2819.

J.E. Gubernatis, Long-wave approximations for the scattering of elastic waves from flaws with applications to ellipsoidal voids and inclusions, *J. Appl. Phys.*, 50 (1979) 4046–4058.

J.E. Gubernatis and E. Domany, Rayleigh scattering of elastic waves from cracks, *J. Appl. Phys.*, 50 (1979) 818–824.

J.E. Gubernatis, E. Domany and J.A. Krumhansl, Elastic wave scattering theory with application to nondestructive evaluation, in *Physical Principles of Nondestructive Testing*, ed. D.O. Thompson, Springer-Verlag, (1980).

D.L. Guell and J. Dundurs, Further results on center of dilation and residual stresses in joined elastic half-spaces, *Devel. of Theoret. and Appl. Mech.*, 3 (1967) 105–115.

H. Günther, *Zur Nichtlinearen Kontinuumstheorie bewegter Versetzungen*, Akademie-Verlag, Berlin, (1967).

H. Günther, Überschallbewegung von Eigenspannungsquellem in der Kontinuumstheorie, *Ann. Phys.*, 21 (1968) 93–105.

H. Günther, Zur Theorie der Überschall-Eigenspannungsquellen, *Ann. Phys.*, 24 (1969) 82–93.

H. Günther, Zur Kinematik von Disklinationen, *phys. stat. sol. (b)*, 49 (1972) 551–559.

I.S. Gradshteyn and I.M. Ryzhik, *Table of Integrals, Series and Products*, Academic Press, (1965).

G.T. Hahn and A.R. Rosenfield, Local yielding and extension of a crack under plane stress, *Act. Metall.*, 13 (1965) 293–306.

H.T. Hahn and W. Jaunzemis, A dislocation theory of plasticity, *Int. J. Engng. Sci.*, 11 (1973) 1065–1078.

J. Hardiman, Elliptic elastic inclusion in an infinite elastic plate, *Q. J. Mech. Appl. Math.*, 7 (1954) 226–230.

W.F. Harris, Topological restriction on the distribution of defects in surface crystals and possible biophysical application, in *Fundamental Aspects of Dislocation Theory*, eds. J.A. Simmons, R. deWit and R. Bullough, National Bureau of Standards, 1 (1969) 579–592.

W.F. Harris and L.E. Scriven, Function of dislocation in cell walls and membranes, *Nature*, 228 (1970) 827–829.

W.F. Harris and S.L. Thomas, Twist disclinations of rotation 2π radians and topological changes of membranous systems, *Phil. Mag.*, 32 (1975) 929–936.

E.W. Hart, Lattice resistance to dislocation motion at high velocity, *Phys. Rev.*, 98 (1955) 1775–1776.

E.W. Hart, Theory of dispersion hardening in metals, *Act. Metall.*, 20 (1972) 275–289.

Z. Hashin and S. Shtrikman, On some variational principles in anisotropic and nonhomogeneous elasticity, *J. Mech. Phys. Solids*, 10 (1962) 335–343.

Z. Hashin and S. Shtrikman, A variational approach to the elastic behaviour of multiphase materials, *J. Mech. Phys. Solids*, 11 (1963) 127–240.

Z. Hashin, Theory of mechanical behavior of heterogeneous media, *Appl. Mech. Rev.*, 17 (1964) 1–9.

Z. Hashin, Viscoelastic behavior of heterogeneous media, *J. Appl. Mech.*, 32 (1965) 630–636.

Z. Hashin and S. Shtrikman, A variational approach to the theory of the elastic behaviour of polycrystals, *J. Mech. Phys. Solids*, 10 (1962) 343–352.

K.S. Havner and A.H. Shalaby, A simple mathematical theory of finite distortional latent hardening in single crystals, *Proc. Roy. Soc. Lon.*, A358 (1977) 47–70.

K.S. Havner and A.H. Shalaby, Further investigation of a new hardening law in crystal plasticity, *J. Appl. Mech.*, 45 (1978) 500–506.

K.S. Havner, G.S. Baker and R.F. Vause, Theoretical latent hardening in crystals I. General equations for tension and compression with application to F.C.C. crystals in tension, *J. Mech. Phys. Solids*, 27 (1979) 33–50.

K. Hayashi and H. Abe, Stress intensity factors for a semi-elliptical crack in the surface of a semi-infinite solid, *Int. J. Fracture* 3 (1980) 275–285.

K. Hayashi and S. Nemat-Nasser, Energy release rate and crack kinking, *Int. J. Structures Solids*, (1981).

K. Hayashi and S. Nemat-Nasser, On branched, interface cracks, *J. Appl. Mech.*, (1981).

P.M. Hazzledine and P.B. Hirsch, A coplanar Orawan loops model for dispersion hardening, *Phil. Mag.*, 30 (1974) 1331–1351.

A.K. Head, Edge dislocations in inhomogeneous media, *Proc. Phys. Soc.*, B66 (1953) 793–801.

A.K. Head, Interaction of dislocations and boundaries, *Phil. Mag.*, 44 (1953) 92–94.

A.K. Head and N. Louat, The distribution of dislocations in linear arrays, *Austral. J. Phys.*, 8 (1955) 1–7.

A.K. Head and P.F. Thomson, On the method of Eshelby, Frank and Nabarro for calculating the equilibrium positions of dislocations, *Phil. Mag.*, 7 (1962) 439–449.

H.L. Heinisch and G. Sines, A comparison of crystal defect models II. Interactions of vacancies with 111 screw dislocations in body-centered cubic metals, *Phil. Mag.*, 34 (1976) 961–971.

A.V. Hershey, The elasticity of an isotropic aggregate of anisotropic cubic crystals, *J. Appl. Mech.*, 21 (1954) 236–241.

A.V. Hershey, The plasticity of an isotropic aggregate of anisotropic face-centered cubic crystals, *J. Appl. Mech.*, 21 (1954) 241–249.

R. Hill, The elastic behavior of a crystalline aggregate, *Proc. Phys. Soc.*, A65 (1952) 349–354.

R. Hill, *The Mathematical Theory of Plasticity*, Oxford, (1960).

R. Hill, Discontinuity relations in mechanics of solids, in *Progress in Solid Mechanics* 2, eds. I.N. Sneddon and R. Hill, North-Holland, Amsterdam, (1961) 245–276.

R. Hill, Elastic properties of reinforced solids: Some theoretical principles, *J. Mech. Phys. Solids*, 11, (1963) 347–372.

R. Hill, Continuum micro-mechanics of elastoplastic polycrystals, *J. Mech. Phys. Solids*, 13 (1965) 89–101.

R. Hill and J.R. Rice, Constitutive analysis of elastic-plastic crystals at arbitrary strain, *J. Mech. Phys. Solids*, 20 (1972) 401–413.

J.E. Hilliard, Spinodal decomposition, in *Phase Transformations*, ed. H.I. Aaronson, American Society for Metals, (1970) 497–560.

P.B. Hirsch, A. Howie and M.J. Whelan, A kinematical theory of diffraction contrast of electron transmission microscope images of dislocations and other defects, *Phil. Trans. Roy. Soc.*, A252 (1960) 499–529.

J.P. Hirth and J. Lothe, *Theory of Dislocations*, McGraw-Hill, (1968).

E.W. Hobson, *Theory of Spherical and Ellipsoidal Harmonics*, Cambridge, (1931).

A. Hoenig, The behavior of a flat elliptical crack in an anisotropic elastic body, *Int. J. Solids Structures*, 14 (1978) 925–934.

E.F. Hollander, The basic equations of the dynamics of the continuous distribution of dislocations I. General theory II. Interpretation of general theory III. Special problems, *Czech. J. Phys.*, B10 (1960) 409–418, 479–487, 551–560.

M.R. Horne, The behaviour of an assembly of rotund, rigid, cohesionless particles II, *Proc. Roy. Soc.*, A310 (1965) 79–97.

I.C. Howard and P. Brierley, On the finite deformation of an inhomogeneity in a viscous liquid, *Int. J. Engng. Sci.*, 14 (1976) 1151–1159.

W. Huang and T. Mura, Elastic fields and energies of a circular edge disclination and a straight screw disclination, *J. Appl. Phys.*, 41 (1970) 5175–5179.

W. Huang and T. Mura, Elastic energy of an elliptical-edge disclination, *J. Appl. Phys.*, 43 (1972) 239–241.

A. Huber and E. Schmid, Bestimmung der elastischen Eigenschaften quasi-isotroper Vielkristalle durch Mittelung, *Helv. Phys. Acta*, 7 (1934) 620–627.

D. Hull, *Introduction to Dislocations*, Pergamon Press, (1965).

J.A.H. Hult, Fatigue crack propagation in torsion, *J. Mech. Phys. Solids*, 6 (1957) 47–52.

J.A.H. Hult and F.A. McClintock, Elastic-plastic stress and strain distributions around sharp notches under repeated shear, *Proc. 9th Int. Congr. Appl. Mech.*, 8 (1957) 51–64.

J.W. Hutchinson, Plastic stress-strain relations of F.C.C. polycrystalline metals hardening according to Taylor's rule, *J. Mech. Phys. Solids*, 12 (1964) 11–24.

J.W. Hutchinson, Plastic deformation of B.C.C. polycrystals, *J. Mech. Phys. Solids*, 12 (1964) 25–33.

J.W. Hutchinson, Singular behaviour at the end of a tensile crack in a hardening material, *J. Mech. Phys. Solids*, 16 (1968) 13–31.

J.W. Hutchinson, Plastic stress and strain fields at a crack tip, *J. Mech. Phys. Solids*, 16 (1968) 337–347.

J.W. Hutchinson, Elastic-plastic behavior of polycrystalline metals and composites, *Proc. Roy. Soc.*, A319 (1970) 247–272.

M. Ichikawa and T.W. Chou, Analysis of deformation and contraction in cylindrical crystals due to dispirations or dislocation, *J. Theor. Biol.*, 78 (1979) 129–142.

V.L. Indenbom, Internal stress in crystals, in *Theory of Crystal Defects*, Academia Prague, (1966) 257–274.

V.L. Indenbom and S.S. Orlov, Construction of Green's function in terms of Green's function of lower dimension, *J. Appl. Math. Mech.*, 32 (1968) 414–420.

C.E. Inglis, Stress in a plate due to the presence of cracks and sharp corners, *Trans. Inst. Nav. Arch.*, 44 (1913) 219–230.

G.R. Irwin, Fracture dynamics, in *Fracturing of Metals*, American Society for Metals, (1948) 147–166.

G.R. Irwin, Relation of stress near a crack to the crack extension force, *Proc. 9th Int. Congr. Appl. Mech.*, 8 (1957) 245–251.

G.R. Irwin, Analysis of stresses and strains near the end of a crack transversing a plate, *J. Appl. Mech.*, 24 (1957) 361–364.

G.R. Irwin, Fracture, *Handbuch der Physik*, 6, Springer-Verlag, (1958) 551–590.

M. Isida, Effect of width and length on stress intensity factors of internally cracked plates under various boundary conditions, *Int. J. Fracture Mech.*, 7 (1971) 301–316.

Y. Izumi and M.E. Fine, Role of plastic work in fatigue crack propagation in metals, *Engng. Fracture Mech.*, 11 (1979) 791–804.

K.H. Jack, The occurrence of the crystal structure of α''-iron nitride; a new type of interstitial alloy formed during the tempering of nitrogen martensite, *Proc. Roy. Soc. Lon.*, A208 (1951) 216–224.

M.A. Jaswon and R.D. Bhargava, Two-dimensional elastic inclusion problems, *Proc. Camb. Phil. Soc.*, 57 (1961) 669–680.

T. Jimma and T. Masuda, An analysis of elastic-plastic waves by the continuum theory of dislocations, in *IUTAM Symposium on High Velocity Deformation of Solids*, eds. K. Kawata and J. Shioiri, Springer, (1977) 289–294.

T. Jimma and K. Kawai, The observation of the strain field in form rolling process and the analysis of the residual stress in its field, Bull. Prec. Mach. Electron., *Tokyo Inst. Tech.*, 44 (1979) 39–47.

W.C. Johnson, Y.Y. Earmme and J.K. Lee, Approximation of the strain field associated with an inhomogeneous precipitate I. Theory, *J. Appl. Mech.*, 47 (1980) 775–780.

W.C. Johnson, Y.Y. Earmme and J.K. Lee, Approximation of the strain field associated with an inhomogeneous precipitate II. The cuboidal inhomogeneity, *J. Appl. Mech.*, 47 (1980) 781–788.

K. Kagannadham and M.J. Marcinkowski, Loss of coherency across an interface, *J. Appl. Phys.*, 48 (1977) 3788–3797.

A. Kamei and T. Yokobori, Some results on stress intensity factors of the cracks and/or slip bands system, Rep. Res. Inst. Strength and Fract. of Mater., *Tohoku Univ.*, 10 (1974) 29–93.

M.F. Kanninen, An estimate of the limiting speed of a propagating ductile crack, *J. Mech. Phys. Solids*, 16 (1968) 215–228.

M.F. Kanninen, A.K. Mukherjee, A.R. Rosenfield and G.T. Hahn, The speed of ductile crack propagation and the dynamics of flow in metals, in *Mechanical Behavior of Materials Under Dynamic Loads*, ed. U.S. Lindholm, Springer-Verlag, (1968) 96–133.

M.F. Kanninen, A critical appraisal of solution techniques in dynamic fracture mechanics, in *Numerical Methods in Fracture Mechanics*, eds. A.R. Luxmoore and D.R.J. Owen, Univ. of Swansea, (1978) 612–633.

B.L. Karihaloo, Dislocation mobility and the concept of flow stress, *Proc. Roy. Soc.*, A344 (1975) 375–385.

B.L. Karihaloo, Fracture characteristics of solids containing doubly-periodic arrays of cracks, *Proc. Roy. Soc.*, A360 (1978) 373–387.

B.L. Karihaloo, Yielding on inclined planes at the tip of a crack, *Arch. Mech. Stos.*, 31 (1979) 649–656.

B.L. Karihaloo, Fracture of solids containing arrays of cracks, *Engng. Fracture Mech.*, 12 (1979) 49–77.

M.K. Kassir and G.C. Sih, Three-dimensional stress distribution around an elliptical crack under arbitrary loadings, *J. Appl. Mech.*, 33 (1966) 601–611.

M.K. Kassir and G.C. Sih, Geometric discontinuities in elastostatics, *J. Math. Mech.*, 16 (1967) 927–948.

M.K. Kassir and G.C. Sih, Griffith's theory of brittle fracture in three dimensions, *Int. J. Engng. Sci.*, 5 (1967) 899–918.

M.K. Kassir and G.C. Sih, Three-dimensional stress around elliptical cracks in transversely isotropic solids, *Engng. Fracture Mech.*, 1 (1968) 327–345.

M.K. Kassir and G.C. Sih, *Three-Dimensional Crack Problems*, *Mechanics of Fracture*, 2, Noordhoff, (1975).

L.M. Keer and T. Mura, Stationary crack and continuous distribution of dislocations, *Proc. 1st Int. Conf. Fracture*, 1 (1966) 99–115.
L.M. Keer, Mixed boundary-value problems for an elastic half-space, *Proc. Camb. Phil. Soc.*, 63 (1967) 1379–1386.
L.M. Keer, *A note on dislocation pile-up in two phase materials*, Private communication, (1968).
L.M. Keer, S.H. Chen and M. Comninou, The interface penny-shaped crack reconsidered, *Int. J. Engng. Sci.*, 16 (1978) 675–772.
L.M. Keer, S. Nemat-Nasser, and A. Oranratnachai, Unstable growth of thermally induced interacting cracks in brittle solids: Further results, *Int. J. Solids Structures*, 15 (1979) 111–126.
R.E. Keith and J.J. Gilman, Progress report on dislocation behavior in lithium fluoride crystals during cyclic loading, *ASTM Special Tech. Publ.*, No. 237, (1959) 3–20.
J.M. Keller, Unpublished (1957), see Kröner's Book, p. 164 (1958).
Lord Kelvin, Note on the integration of the equations of equilibrium of an elastic solid, in *Mathematical and Physical Papers*, 1, Cambridge Univ. Press, (1882) 97–98.
A.P. Kfouri and J.R. Rice, Elastic/plastic separation energy rate for crack advance in finite growth steps, in *Fracture* 1977, 1, ed. D.M.R. Taplin, Univ. of Waterloo Press, (1977) 43–59.
A.P. Kfouri, Crack separation energy-rates for the DBCS model under biaxial modes of loading, *J. Mech. Phys. Solids*, 27 (1979) 135–150.
A.G. Khachaturyan, Some questions concerning the theory of phase transformations in solids, *Sov. Phys. Solid State*, 8 (1967) 2163–2168.
A.G. Khachaturyan, Elastic strains during decomposition of homogeneous solid solutions – Periodic distribution of decomposition products, *phys. stat. sol.*, 35 (1969) 119–132.
A.G. Khachaturyan, A theory of spatially periodic phase distribution (modulated structures) produced in the decomposition of solid solutions, *Sov. Phys. JETP*, 31 (1970) 98–106.
A.G. Khachaturyan and V.M. Airapetyan, Spatially periodic distributions of new phase inclusions caused by elastic distortions, *phys. stat. sol.* (*a*), 26 (1974) 61–69.
N. Kikuchi, Elastic anisotropy and its temperature variation of single crystals of 18–12 type stainless steel, *J. Japan Inst. Metals*, 35 (1971) 518–522.
T.J. Kim and V.G. Ukadgaonker, Plane stress analysis of two rigid circular inclusions, *J. AIAA*, 9 (1971) 2294–2296.
N. Kinoshita and T. Mura, On the boundary value problem of elasticity, *Res. Rep. Faculty of Engng.*, Meiji Univ., No. 8 (1956) 4.29, 1–7.

N. Kinoshita and T. Mura, Elastic fields of inclusions in anisotropic media, *phys. stat. sol. (a)*, 5 (1971) 759–768.

K.R. Kinsman, J.W. Sprys and R.J. Asaro, Structure of martensite in very small iron-rich precipitates, *Act. Metall.*, 23 (1975) 1431–1442.

J. Kiusalaas and T. Mura, On the elastic field around an edge dislocation with application to dislocation vibration, *Phil. Mag.*, 9 (1964) 1–7.

J. Kiusalaas and T. Mura, On the motion of a screw dislocation, in *Recent Advances in Engineering Science*, 1, ed. A.C. Eringen, Gordon and Breach, (1964) 543–564.

G. Kneer, Über die Berechnung der Elastizitätsmoduln vielkristalliner Aggregate mit Textur, *phys. stat. sol.*, 9 (1965) 825–838.

J.F. Knott and A.H. Cottrell, Notch brittleness in mild steel, *J. Iron Steel Inst.*, 201 (1941) 249–260.

J.K. Knowles and E. Sternberg, On a class of conservation laws in linearized and finite elastostatics, *Arch. Ration, Mech. Anal.*, 44 (1972) 187–211.

S. Kobayashi and S.M. Ohr, In situ studies of crack propagation in BCC metals, *37th Ann. Proc. Electron Microscopy Soc. Amer.*, ed. G.W. Bailey, (1979) 424–425.

J.S. Koehler, On the dislocation theory of plastic deformation, *Phys. Rev., II*, 60 (1941) 397–410.

J.S. Koehler, The influence of dislocations and impurities on the damping and the elastic constants of metal single crystals, in *Imperfections in Nearly Perfect Crystals*, eds. W. Shockley, J.H. Hollomon, R. Maurer and F. Seitz, Wiley, (1952) 197–216.

R.C. Koeller and R. Raj, Diffusional relaxation of stress concentration at second phase particles, *Act. Metall.*, 26 (1978) 1551–1558.

W. Kohn, J.A. Krumhansl and E.H. Lee, Variational methods for dispersion relations and elastic properties of composite materials, *J. Appl. Mech.*, 39 (1972) 327–336.

W.T. Koiter, Rectangular tensile sheet with symmetric edge cracks, *J. Appl. Mech.*, 32 (1965) 327–327.

K. Kondo, A proposal of a new theory concerning the yielding of materials based on Riemannian geometry, *J. Japan Soc. Appl. Mech.*, 2 (1949) 123–128, 146–151.

K. Kondo, Non-holomonic geometry of plasticity and yielding, in *Memoirs of the Unifying Study of the Basic Problems in Engineering Sciences by Means of Geometry* 1, ed. K. Kondo, Gakujutsu Bunken Fukyu-Kai, Tokyo, (1955) 453–562.

K. Kondo, Two general diakoptical processes included in the non-

Riemannian treatments of plastic phenomena, *Matrix and Tensor Quart.*, 11 (1960) 1–10.

K. Kondo, Non-Riemannian and Finslerian approaches to the theory of yielding, *Int. J. Engng. Sci.*, 1, (1963) 71–88.

K. Kondo, On the analytical and physical foundations of the theory of dislocations and yielding by the differential geometry of continua, *Int. J. Engng. Sci.*, 2, (1964) 219–251.

T. Kontorova and Y.I. Frenkel, Theory of plastic deformation and twinning, *Zh. eksp. téor. Fiz.*, 8 (1938) 89–95.

J. Korringa, Theory of elastic constants of heterogeneous media, *J. Math. Phys.*, 14 (1973) 509–513.

J. Korringa and D.D. Thompson, Comment on the self-consistent imbedding approximation in the theory of elasticity of porous media, *J. Geophys. Res.*, 82 (1977) 933–934.

A.M. Kosevich, The deformation field in an isotropic elastic medium containing moving dislocations, *Sov. Phys. JETP*, 15 (1962) 108–115.

A.M. Kosevich, The equation of motion of a dislocation, *Sov. Phys. JETP*, 16 (1963) 455–462.

A.M. Kosevich and V.C. Natsik, Elastic field of continuously distributed moving dislocation loops, *Sov. Phys. Solid State*, 6 (1964) 181–186.

A.M. Kosevich, Dynamical theory of dislocations, *Sov. Phys. USPEKHI*, 7 (1965) 837–854.

M.F. Koshinen, Elastic-plastic deformation of a single grooved flat plate under longitudinal shear, *Trans. ASME*, Series D, 85 (1963) 585–594.

E. Kossecka, Defects as surface distributions of double forces, *Arch. Mech. Stos.*, 21 (1969) 167–190.

E. Kossecka and R. deWit, Disclination kinematics, *Arch. Mech.*, 29 (1977) 633–651.

E. Kossecka and R. deWit, Disclination dynamics, *Arch. Mech.*, 29 (1977) 749–767.

A.D. Kovalenko, *Thermoelasticity*, trans. D.B. Macvean and J.B. Alblas, Wolters-Noordhoff, (1969).

E. Kröner, Das Fundamentalintegral der anisotropen elastischen Differentialgleichungen, *Z. Phys.*, 136 (1953) 402–410.

E. Kröner, Über die Berechnung der Verzerrungsenergie bei Keimbildung in Kristallen, *Act. Metall.*, 2 (1954) 302–309.

E. Kröner, Die Spannungsfunktionen der dreidimensionalen anisotropen Elastizitätstheorie, *Z. Phys.*, 141 (1955) 386–398.

E. Kröner, Der fundamentale Zusammenhang zwischen Versetzungsdichte und Spannungs funktionen, *Z. Phys.*, 142 (1955) 463–475.

E. Kröner, Die Versetzung als elementare Eigenspannungsquelle, *Z. Naturforschung*, 2a (1956) 969–985.

E. Kröner and G. Rieder, Kontinuumstheorie der Versetzungen, *Z. Phys.*, 145 (1956) 424–429.

E. Kröner, Berechnung der elastischen Konstanten des Vielkristalls aus den Konstanten des Einkristalls, *Z. Phys.*, 151 (1958) 504–518.

E. Kröner, *Kontinuumstheorie der Versetzungen und Eigenspannungen*, Springer-Verlag, (1958).

E. Kröner, Allgemeine Kontinuumstheorie der Versetzungen und Eigenspannungen, *Arch. Rat. Mech. Anal.*, 4 (1960) 273–334.

E. Kröner, Zur plastischen Verformung des Vielkristalls, *Act. Metall.*, 9 (1961) 155–161.

E. Kröner, Zum statischen Grundgesetz der Versetzungstheorie, *Annal. Phys.*, 11 (1963) 13–21.

E. Kröner, Dislocation: a new concept in the continuum theory of plasticity, *J. Math. Phys.*, 42 (1963) 27–37.

E. Kröner, ed., *Mechanics of Generalized Continua*, Springer-Verlag, (1968).

E. Kröner and K. H. Anthony, Dislocations and disclinations in material structures: The basic topological concepts, in *Annual Review of Materials Science* 5, eds. R.A. Huggins, R.H. Bube and R.W. Roberts, (1975) 43–72.

E. Kröner, Bounds for effective elastic moduli of disordered materials, *J. Mech. Phys. Solids*, 25 (1977) 137–155.

F. Kroupa, Circular edge dislocation loop, *Czech. J. Phys.*, B10 (1960) 284–293.

F. Kroupa, Continuous distribution of dislocation loops, *Czech. J. Phys.*, B12 (1962) 191–201.

J.G. Kuang and T. Mura, Dislocation pile-up in half space, *J. Appl. Phys.*, 40 (1969) 5017-5021.

I.A. Kunin, Non-local theory of elasticity, in *Theory of Elasticity with Microstructures*, Nauka, Moscow, (1975).

H.H. Kuo and T. Mura, Elastic field and strain energy of a circular wedge disclination, *J. Appl. Phys.*, 43 (1972) 1454–1457.

H.H. Kuo and T. Mura, Circular disclinations and interface effects, *J. Appl. Phys.*, 43 (1972) 3936–3942.

H.H. Kuo and T. Mura, Circular twist disclination in viscoelastic materials, *J. Appl. Phys.*, 44 (1973) 3586–3588.

H.H. Kuo, T. Mura and J. Dundurs, Moving circular twist disclination loop in homogeneous and two-phase materials, *Int. J. Engng. Sci.*, 11 (1973) 193–201.

V.C. Kupradze, *Potential Methods in Theory of Elasticity*, Israel Program Sci. Trans., Jerusalem, (1965).
M. Kuriyama, Theory of X-ray diffraction by a distorted crystal, *J. Phys. Soc. Japan*, 23 (1967) 1369–1379.
A.H. Lachenbruch, Depth and spacing of tension cracks, *J. Geophys. Res.*, 66 (1961) 4273–4292.
J.C. Lacht and J.O. Watson, Effective numerical treatment of boundary integral equations; a formulation for three-dimensional elastostatics, *Int. J. Numer. Meth. Engng.*, 10 (1976) 991–1005.
C. Laird, The fatigue limit of metals, *Mat. Sci. Eng.*, 22 (1976) 231–236.
C. Laird and D.J. Duquette, Mechanisms of fatigue crack nucleation, *Corrosion Fatigue*, eds. O.F. Devereaux, A.J. McEvily and R.W. Staehle, NACE 2, (1972) 88–117.
J.H. Lambermont, On the unification of the mathematical theory of plasticity and dislocation theory by using thermodynamics I. Basic theory, *Int. J. Engng. Sci.*, 12 (1974) 937–959.
M.G. Lamé, Leçons sur la théorie mathématique de l'elasticité des corps solides, Gauthier-Villars, Paris, (1866).
H.L. Langhaar, *Dimensional Analysis and Theory of Models*, Wiley, (1951).
R.W. Lardner, A dislocation model for fatigue crack growth in metals, *Phil. Mag.*, 17 (1968) 71–82.
R.W. Lardner, Dislocation dynamics and the theory of the plasticity of single crystals, *J. Appl. Math. Phys. (ZAMP)*, 20 (1969) 514–529.
R.W. Lardner, *Mathematical Theory of Dislocations and Fracture*, University of Toronto Press (1974).
F. László, Tessellated stresses, *J. Iron Steel Inst.*, 164 (1950) 5–26.
T. Laub and J.D. Eshelby, The velocity of a wave along a dislocation, *Phil. Mag.*, 11 (1966) 1285–1293.
N. Laws, The determination of stress and strain concentrations at an ellipsoidal inclusion in an anisotropic material, *J. Elasticity*, 7 (1977) 91–97.
N. Laws and R. McLaughlin, Self-consistent estimates for the viscoelastic creep compliances of composite materials, *Proc. Roy. Soc.*, A359 (1978) 251–273.
N. Laws and R. McLaughlin, The effect of fibre length on the overall moduli of composite materials, *J. Mech. Phys. Solids*, 27 (1979) 1–13.
J.D. Lee and H. Liebowitz, Considerations of crack growth and plasticity in finite element analysis, Comp. Structures, 8 (1978) 403–410.
J.K. Lee and W.C. Johnson, Elastic strain energy and interactions of

thin square plates which have undergone a simple shear, *Scrip. Metall.*, 11 (1977) 477–484.
J.K. Lee, D.M. Barnett and H.I. Aaronson, The elastic strain energy of coherent ellipsoidal precipitates in anisotropic crystalline solids, *Metall. Trans.*, 8A (1977) 963–970.
J.K. Lee and W.C. Johnson, Re-examination of the elastic strain energy of an incoherent ellipsoidal precipitate, *Act. Metall.*, 26 (1978) 541–545.
J.K. Lee and W.C. Johnson, Calculation of the elastic strain field of a cuboidal precipitate in an anisotropic matrix, *phys. stat. sol. (a)*, 46 (1978) 267–272.
M.S. Lee, *Uniform moving dislocations*, Ph.D. Dissertation, Northwestern University, Evanston, (1972).
M.S. Lee and J. Dundurs, Edge dislocation in a surface layer, *Int. J. Engng. Sci.*, 11 (1973) 87–94.
N.G. Lee, Elastic Green's function for an infinite half-space of a hexagonal continuum with its basal plane as surface, *Int. J. Engng. Sci.*, 17 (1979) 681–689.
G. Leibfried and H.D. Dietze, Zur theorie der Schraubenversetzung, *Z. Phys.*, 126 (1949) 790–808.
G. Leibfried, Uber den Einfluss thermisch angeregter Schallwellen auf die plastische Deformation, *Z. Phys.*, 127 (1950) 344–356.
G. Leibfried, Verteilung von Versetzungen in statischen Gleichgewicht, *Z. Phys.*, 130 (1951) 214–226.
G. Leibfried, Versetzungen in anisotropen Material, *Z. Phys.*, 135 (1953) 23–43.
L. Lejček, The Green function of the theory of elasticity in an anisotropic hexagonal medium, *Czech. J. Phys.*, B19, (1969) 799–803.
S.G. Lekhnitskii, *Anisotropic Plates*, Gordon and Breach, (1968).
A. Leon, *Störungen der Spannungsverteilung*, Vienna, (1908).
M. Lévy, Mémoire sur les équations générales des mouvements intérieurs des corps solids ductiles au delà des limites où l'élasticité pourrait les ramener a leur premier état, *C. R. Acad. Sci.*, 70 (1870) 1323–1325.
J.C.M. Li, Pileup of dissociated dislocations and the strength-grain size relationships, *Phil. Mag.*, 19 (1969) 189–198.
J.C.M. Li and J.J. Gilman, Disclination loops in polymers, *J. Appl. Phys.*, 41 (1970) 4248–4256.
K.H.C. Lie and J.S. Koehler, The elastic stress field produced by a point force in a cubic crystal, *Adv. Phys.*, 17 (1968) 421–478.
I.M. Lifshitz and L.N. Rozenzweig, On the construction of the Green's

tensor for the basic equation of the theory of elasticity of an anisotropic infinite medium, *Zh. Eksp. Teor. Fiz.*, 17 (1947) 783–791.
M.J. Lighthill, *Introduction to Fourier Analysis and Generalized Functions*, Cambridge, (1964).
S.C. Lin and T. Mura, Elastic fields of inclusions in anisotropic media (II), *phys. stat. sol. (a)*, 15 (1973) 281–285.
S.C. Lin and T. Mura, Long-range elastic interaction between a dislocation and an ellipsoidal inclusion in cubic crystals, *J. Appl. Phys.*, 44 (1973) 1508–1514.
S.C. Lin, T. Mura, T. Mori and M. Shibata, The work-hardening behavior of anisotropic media by non-deforming particles or fibres, *Act. Metall.*, 21 (1973) 505–516.
T.H. Lin, Analysis of elastic and plastic strains of a face-centered cubic crystal, *J. Mech. Phys. Solids*, 5 (1957) 143–149.
T.H. Lin and T.K. Tung, The stress field produced by localized plastic slip at a free surface. *J. Appl. Mech.*, 29 (1962) 523–534.
T.H. Lin, Slip and stress fields of a polycrystalline aggregate at different stages of loading, *J. Mech. Phys. Solids*, 12 (1964) 391–408.
T.H. Lin and M. Ito, Theoretical plastic stress-strain relationship of a polycrystal and the comparisons with the von Mises and the Tresca plasticity theories, *Int. J. Engng. Sci.*, 4 (1966) 543–561.
T.H. Lin and Y.M. Ito, Mechanics of fatigue crack nucleation mechanism, *J. Mech. Phys. Sol.*, 17 (1969) 511–523.
T.H. Lin, Physical theory of plasticity, *Adv. Appl. Mech.*, 11 (1971) 255–311.
C.B. Ling and K.L. Yang, On symmetrical strain in solids of revolution in spherical coordinates, *J. Appl. Mech.*, 18 (1951) 367–370.
C.B. Ling, Torsion of a cylinder having a spherical cavity, *Q. Appl. Math.*, 10 (1952) 149–156.
R.D. List, A two-dimensional circular inclusion problem, *Proc. Camb. Phil. Soc.*, 65 (1969) 823–830.
G.C.T. Lin and J.C.M. Li, Strain energies of disclination loops, *J. Appl. Phys.*, 42 (1971) 3313–3315.
J. Lothe, Aspects of the theories of dislocation mobility and internal friction, *Phys. Rev.*, 117 (1960) 704–708.
J. Lothe, Theory of dislocation mobility in pure slip, *J. Appl. Phys.*, 33 (1962) 2116–2125.
J. Lothe, Dislocation bends in anisotropic media, *Phil. Mag.*, 15 (1967) 353–362.
A.E.H. Love, *A Treatise on the Mathematical Theory of Elasticity*, Cambridge Univ. Press, (1927).

M. Lowengrub and I.N. Sneddon, The effect of internal pressure on a penny-shaped crack at the interface of two bonded dissimilar elastic half-spaces, *Int. J. Engng. Sci.*, 12 (1974) 387–396.

P. Lukás, M. Klesnil, and J. Polák, High cycle fatigue life of metals, *Materials Sci. Engng.*, 15 (1974) 239–245.

A.T. Lurie, The state of stress around an ellipsoidal cavity, *Dokl. Akad. Nauk*, 87 (1952) 709–710.

W.D. MacMillan, *The Theory of the Potential*, McGraw-Hill, 78 (1930).

A.K. Mak, L.M. Keer, S.H. Chen and J.L. Lewis, *A no slip interface crack*, *J. Appl. Mech.*, 47 (1980) 347–350.

A.K. Mal and L. Knopoff, Elastic wave velocities in two-component systems, *J. Inst. Math. Appl.*, 3 (1967) 376–387.

K. Malén, On the analysis of the force on a moving dislocation, *phys. stat. sol.*, 37 (1970) 267–274.

K. Malén, Numerical analysis of properties of dislocations in anisotropic media, *phys. stat. sol.*, 38 (1970) 259–269.

K. Malén and J. Lothe, Explicit expressions for dislocation derivatives, *phys. stat. sol.*, 39 (1970) 287–296.

K. Malén, A unified six-dimensional treatment of elastic Green's functions and dislocations, *phys. stat. sol. (b)*, 44 (1971) 661–672.

J. Mandel, Sur les lignes de glissement et la calcul de déplacements dans la déformation plastique, *C. R. Acad. Sci.*, 225 (1947) 1272–1273.

J. Mandel, Contribution théorique á l'étude de l'écrouissage et des lois de l'écoulement plastique, *Proc. 11th Int. Congr. Appl. Mech.* Munich, Springer-Verlag, (1966) 502–509.

G. Mandl and R.F. Luque, Fully developed plastic shear flow of granular materials, *Geotechnique*, 20 (1970) 277–307.

E. Mann, R. v. Jan and A. Seeger, Das elastische Verschiebungsfeld von Einzel- und Doppelkräften in anisotropen Medien, *phys. stat. sol.*, 1 (1961) 17–29.

S.S. Manson, Behavior of materials under conditions of thermal stress, *NACA TN* 2933 (1954).

A.A. Maradudin, Screw dislocations and discrete elastic theory, *J. Phys. Chem. Solids*, 9 (1958) 1–20.

X. Markenscoff, The transient motion of a nonuniformly moving dislocation, *J. Elasticity*, 10 (1980) 193–201.

X. Markenscoff and R.J. Clifton, The non-uniform moving edge dislocation, *J. Mech. Phys. Solids*, 29 (1981) 253–262.

D.E. Martin, An energy criterion for low-cycle fatigue, Trans. ASME, *J. Basic Engng.*, 83 (1961) 565–571.

M.D. Martynenko, Some three-dimensional problems on the equilibrium of an elastic body weakened by a crack, *Sov. Appl. Mech.*, 6 (1970) 1107–1111.

K. Marukawa, Characteristic tensor of a crystal disclination, *J. Phys. Soc. Japan*, 37 (1974) 1476.

W.P. Mason, Phonon viscosity and its effect on acoustic wave attenuation and dislocation motion, *J. Acoust. Soc. Am.*, 32 (1960) 458–472.

E.N. Mastrojannis, L.M. Keer and T. Mura, Stress intensity factor for a plane crack under normal pressure, *Int. J. Fracture*, 15 (1979) 247–258.

R.A. Masumura and M.E. Glicksman, The elastic energy of interacting edge dislocations in an infinite array, *J. Appl. Mech.*, 44 (1977) 595–598.

L.N. McCartney and B. Gale, A generalized theory of fatigue crack propagation, *Proc. Roy. Soc. Lon.*, A322 (1971) 223–241.

F.A. McClintock, Ductile fracture instability in shear, *J. Appl. Mech.*, 25 (1958) 582–588.

F.A. McClintock and G.R. Irwin, Plasticity aspects of fracture mechanics, in Symposium on *Fracture Toughness Testing and Its Applications*, ASTM Special Tech. Publ. No. 381 (1965) 84–113.

F.A. McClintock, On the mechanics of fracture from inclusions, in *Ductility*, American Society for Metals, (1968) 255–277.

J.J. McCoy, On the dynamic response of disordered composites, *J. Appl. Mech.*, 40 (1973) 511–517.

A.J. McEvily and R.C. Boettner, On fatigue crack propagation in F.C.C. metals, *Act. Metall.*, 11 (1963) 725–743.

M.M. Mehrabadi and S.C. Cowin, Initial planar deformation of dilatant granular materials, *J. Mech. Phys. Solids*, 26 (1978) 269–284.

J.H. Michell, On the direct determination of stress in an elastic solid, with application to the theory of plates, *Proc. Lon. Math. Soc.*, 31 (1899) 100–124.

J.H. Michell, The stress in an aeolotropic elastic solid with an infinite plane boundary, *Proc. Lon. Math. Soc.*, 32 (1900) 247–258.

S.G. Mikhlin, *Multidimensional Singular Integrals and Equations*, Pergamon Press, 219 (1965).

L.M. Milne-Thomson, *Anatiplane Elastic Systems*, Springer-Verlag, (1962).

S. Minagawa, On the force exerted upon continuously distributed dislocations in a Cosserat continuum, *phys. stat. sol.*, 39 (1970) 217–222.

S. Minagawa, A contribution to the theory of electromagnetic analogy for the field of moving dislocations, *phys. stat. sol. (b)*, 47 (1971) 197–202.

S. Minagawa, Elastic fields of dislocations and disclinations in an isotropic micropolar continuum, *Lett. Appl. Engng. Sci.*, 5 (1977) 85–94.

S. Minagawa, Stress and couple-stress fields produced by Frank disclinations in an isotropic elastic micropolar continuum, *Int. J. Engng. Sci.*, 15 (1977) 447–453.

R.D. Mindlin, Note on the Galerkin and Papkovich stress functions, *Bull. Amer. Math. Soc.*, 42 (1936) 373–376.

R.D. Mindlin and D.H. Cheng, Thermoelastic stress in the semi-infinite solid, *J. Appl. Phys.*, 21 (1950) 931–933.

R.D. Mindlin and H.J. Cooper, Thermoelastic stress around a cylindrical inclusion of elliptic cross section, *J. Appl. Mech.*, 17 (1950) 265–268.

R.D. Mindlin, Force at a point in the interior of a semi-infinite solid, in *Midwestern Conf. Solid Mech.*, (1953) 56–59.

R.D. Mindlin, Mechanics of granular media, *Proc. 2nd U.S. Nat. Congr. Appl. Mech.*, (1954) 13–20.

R.D. Mindlin, Micro-structure in linear elasticity, *Arch. Rat. Mech. Anal.*, 16 (1964) 51–78.

R.D. Mindlin, Private communication, June (1975).

L. Mirandy and B. Paul, Stresses at the surface of a flat three-dimensional ellipsoidal cavity, *J. Engng. Mater. Tech.*, 98 (1976) 159–163.

R. von Mises, Mechanik der plastischen Formänderung von Kristallen, *Zeit. angew. Math. Mech.*, 8 (1928) 161–185.

H. Miyamoto, Stress concentration around a heterogeneous insertion of ellipsoid in an infinite body I., *J. Japan Soc. Appl. Mech.*, 3 (1950) 15–19.

H. Miyamoto, Stress concentration around a heterogeneous insertion of spheroid in an elastic body III, (Torsion), *Proc. 1st Japan Congr. Appl. Mech.*, (1951) 43–46.

H. Miyamoto, Stress concentration around a spherical cavity in an infinite elastic body, *Trans. Japan Soc. Mech. Engng.*, 19 (1953) 60–66.

H. Miyamoto, On the problem of the theory of elasticity for a region containing more than two spherical cavities, *Bull. JSME*, 1 (1958) 103–108.

H. Miyamoto, *Three-Dimensional Elasticity*, Shōka Bō, Tokyo (1967).

H. Miyamoto and T. Miyoshi, Analysis of stress intensity factor for surface-flawed tension plate, Proc. IUTAM symposium on *High Speed Computing of Elastic Structures*, Univ. of Liege, (1971) 137–155.

H. Miyamoto and K. Kageyama, Extension of J-integral to the general elasto-plastic problem and suggestion of a new method for its evaluation, Proc. 1st Int. Conf. on *Numerical Meth. In Fracture Mech.*, eds. A.R. Luxmoore and D.R.J. Owen, Swansea, (1978) 322–335.

H. Miyata and S. Kusumoto, A method of evaluation of three-dimensional stress intensity factor using the finite element method, *Proc. Joint JSME-ASME Western Conf. on Appl. Mech.*, (1975) 383–390.

T. Mogami, A statistical approach to the mechanics of granular materials, *Soils and Foundations*, 5 (1965) 26–36.

F.C. Moon and Y.H. Pao, The influence of the curvature of spherical waves on dynamic stress concentration, *J. Appl. Mech.*, 34 (1967) 373–379.

F.C. Moon and Y.H. Pao, Interactions of point defects and elastic inclusions, *J. Appl. Phys.*, 38 (1967) 595–601.

T. Mori and K. Tanaka, Average stress in matrix and average energy of materials with misfitting inclusions, *Act. Metall.*, 21 (1973) 571–574.

T. Mori and K. Narita, Decrease of back stress and softening of work hardened copper-silica crystals by recovery, *Act. Metall.*, 23 (1975) 85–91.

T. Mori, P.C. Cheng and T. Mura, Interaction among interstitial solute atoms in iron and tetragonal ordering, Proc. 1st JIM International Symposium on *New Aspects of Martensitic Transformation*, Japan Inst. of Metals, (1976) 281–286.

T. Mori and T. Mura, Slip morphology in dispersion hardened materials, *Mater. Sci. Engng.*, 26 (1976) 89–93.

T. Mori and H. Tokushige, Effect of particle size on low temperature softening of work hardened Cu-SiO_2 crystals, *Act. Metall.*, 25 (1977) 635–641.

T. Mori, P.C. Cheng, M. Kato and T. Mura, Elastic strain energies of precipitates and periodically distributed inclusions in anisotropic media, *Act. Metall.*, 26 (1978) 1435–1441.

T. Mori and T. Mura, Calculation of back stress decrease caused by climb of Orowan loops in a dispersion hardened alloy, *Act. Metall.*, 26 (1978) 1199–1204.

T. Mori and T. Osawa, Softening of work hardened Al-Si crystals and low temperature creep, *Phil. Mag.*, 40 (1979) 445–457.

T. Mori, M. Okabe and T. Mura, Diffusional relaxation around a second phase particle, *act. Metall.*, 28 (1980) 319–325.

S. Moriguti, Fundamental theory of dislocation in an elastic body, *Appl. Math. Mech.* (Ōyō Sūgaku Rikigaku), 1 (1947) 29–36.

P.R. Morris, Elastic constants of polycrystals, *Int. J. Engng. Sci.*, 8 (1970) 49–61.

P.M. Morse and H. Feshbach, *Methods of Theoretical Physics*, McGraw-Hill, (1958).

Z.A. Moschovidis, *Two ellipsoidal inhomogeneities and related problems treated by the equivalent inclusion method*, Ph.D. Dissertation, Northwestern University, Evanston, (1975).

Z.A. Moschovidis and T. Mura, Two-ellipsoidal inhomogeneities by the equivalent inclusion method, *J. Appl. Mech.*, 42 (1975) 847–852.

V.I. Mossakovskii and M.T. Rykba, Generalization of Griffith-Sneddon criterion on the case of inhomogeneous body, *Prikl. Mat. Mekh.*, 28 (1964) 1061–1069.

N.F. Mott, A theory of the origin of fatigue cracks, *Act. Metall.*, 6 (1958) 195–197.

H. Mughrabi, Microscopic mechanisms of metal fatigue, in *Strength of Metals and Alloys*, eds. P. Haasen, V. Gerold and G. Kostorz, Pergamon Press, 3 (1980) 1615–1638.

R. Muki and E. Sternberg, On the stress analysis of overlapping bonded elastic sheets, *Int. J. Solids Structures*, 4 (1968) 75–94.

T. Mura, Continuous distribution of moving dislocations, *Phil. Mag.*, 8 (1963) 843–857.

T. Mura, On dynamic problems of continuous distribution of dislocations, *Int. J. Engng. Sci.*, 1 (1963) 371–381.

T. Mura, Periodic distribution of dislocations, *Proc. Roy. Soc.*, A280 (1964) 528–544.

T. Mura, Continuous distribution of dislocations and the mathematical theory of plasticity, *phys. stat. sol.*, 10 (1965) 447–453.

T. Mura, Continuous distribution of dislocations and the mathematical theory of plasticity (II), *phys. stat. sol.*, 11 (1965) 683–688.

T. Mura, The continuum theory of dislocations, in *Advances in Materials Research* 3, ed. H. Herman, Interscience Publ., (1968) 1–107.

T. Mura, Continuum theory of dislocations and plasticity, in Proc. IUTAM Symposium on *Mechanics of Generalized Continua*, ed. E. Kröner, Springer, (1968) 269–278.

T. Mura, Method of continuously distributed dislocations, in *Mathematical Theory of Dislocations*, ed. T. Mura, ASME, (1969) 25–48.

T. Mura, ed., *Mathematical Theory of Dislocations*, American Society of Mechanical Engineering, (1969).

T. Mura, Individual dislocations and continuum mechanics, in *Inelastic Behavior of Solids*, eds. M.F. Kanninen, W.F. Adler, A.R.

Rosenfield and R.I. Jaffee, Battel Inst. Mater. Sci. Colloq., McGraw-Hill, (1970) 211–230.

T. Mura and N. Kinoshita, Green's functions for anisotropic elasticity, *phys. stat. sol. (b)*, 47 (1971) 607–618.

T. Mura, Displacement and plastic distortion fields produced by dislocations in anisotropic media, *J. Appl. Mech.*, 38 (1971) 865–868.

T. Mura, Dynamic response of dislocations in solids, in *Dynamic Response of Structures*, eds. G. Herrmann and N. Perrone, Pergamon, (1972) 345–370.

T. Mura, Semi-microscopic plastic distortion and disclinations, *Arch. Mech. Stos.*, 24 (1972) 449–456.

T. Mura, A variational method for micromechanics of composite materials, in Proc. Int. Conf. on *Mechanical Behavior of Materials*, 5, Society of Materials Science, Japan (1972) 12–18.

T. Mura and C.T. Lin, Theory of fatigue crack growth for work hardening materials, *Int. J. Fracture*, 10 (1974) 284–287.

T. Mura and S.C. Lin, Thin inclusions and cracks in anisotropic media, *J. Appl. Mech.*, 41 (1974) 209–214.

T. Mura, A. Novakovic and M. Meshii, A mathematical model of cyclic creep acceleration, *Mater. Sci. Engng.*, 17 (1975) 221–225.

T. Mura and T. Mori, Elastic fields produced by dislocation in anisotropic media, *Phil. Mag.*, 33 (1976) 1021–1027.

T. Mura, T. Mori and M. Kato, The elastic field caused by a general ellipsoidal inclusion and the application to martensite formation, *J. Mech. Phys. Solids*, 24 (1976) 305–318.

T. Mura and P.C. Cheng, The elastic field outside an ellipsoidal inclusion, *J. Appl. Mech.*, 44 (1977) 591–594.

T. Mura, Eigenstrains in lattice theory, in *Continuum Models of Discrete Systems*, eds. J.W. Provan and H.H.E. Leipholz, Univ. of Waterloo Press, (1978) 503–519.

T. Mura and P.C. Cheng, An equivalent inclusion method for a three-dimensional lens-shaped crack in anisotropic media, *Fracture* 1977, 3, ed. D.M.R. Taplin, Univ. of Waterloo Press (1978) 191–196.

T. Mura and N. Kinoshita, The polynomial eigenstrain problem for an anisotropic ellipsoidal inclusion, *phys. stat. sol. (a)*, 48 (1978) 447–450.

T. Mura, Recent results in micromechanics, in IUTAM Symposium on *High Velocity Deformation of Solids*, ed. K. Kawata, Springer, (1979) 295–304.

T. Mura and C. Vilmann, Fatigue crack propagation related to a dislocation distribution, in *Proc. Sympos. on Defects and Fracture*, ed. G.C. Sih, Noordhoff (1981).

Y. Murakami, A simple procedure for the accurate determination of stress intensity factors by finite element method, *Engng. Fracture Mech.*, 8 (1976) 643–655.

Y. Murakami, A simple procedure for the accurate determination of stress intensity factors by finite element method, Proc. 1st Int. Conf. on *Numerical Methods in Fracture Mechanics*, eds. A.R. Luxmoore and D.R.J. Owen, Univ. of Swansea, (1978) 322–335.

Y. Murakami, Application of the body force method to the calculation of stress intensity factors for a crack in the arbitrarily shaped plate, *Engng. Fracture Mech.*, 10 (1978) 499–513.

Z. Mróz, Deformation and flow of granular materials, *Theoretical and Applied Mechanics*, Preprints IUTAM Toronto, eds. F.P.J. Rimrott and B. Tabarrok, North-Holland, (1980) 119–132.

N.I. Muskhelishvili, *Singular Integral Equations*, Noordhoff, (1953).

N.I. Muskhelishvili, *Some Basic Problems of the Mathematical Theory of Elasticity*, Noordhoff, (1953).

N.O. Mykelstad, Two problems of thermal stress in the infinite solid, *ASME Trans.*, 64 (1942) A136–143.

F.R.N. Nabarro, The strains produced by precipitation in alloys, *Proc. Roy. Soc.*, A175 (1940) 519–538.

F.R.N. Nabarro, The interaction of screw dislocations and sound waves, *Proc. Roy. Soc.*, A209 (1951) 278–290.

F.R.N. Nabarro, *Theory of Crystal Dislocations*, Oxford, (1967).

F.R.N. Nabarro, The disclination structure of insect muscle, in *Physics of Strength and Plasticity*, ed. A.S. Argon, M.I.T. Press, (1969) 97–110.

F.R.N. Nabarro, Disclinations in surface, in *Fundamental Aspects of Dislocation Theory*, eds. J.A. Simmons, R. deWit and R. Bullough, National Bureau of Standards, 1 (1969) 593–606.

F.R.N. Nabarro, ed., *Dislocations in Solids*, I, II, III, IV, V, North-Holland, (1979).

Y. Nakada, W.C. Leslie and T.P. Chray, Stress-orienting of $Fe_{16}N_2$ precipitates in an Fe–N alloy, *Trans. Amer. Soc. Metals*, 60 (1967) 223–227.

S. Nemat-Nasser and S. Minagawa, Harmonic waves in layered composites: Comparison among several schemes, *J. Appl. Mech.*, 42 (1975) 699–704.

S. Nemat-Nasser, Stability of a system of interacting cracks, Letter in *Appl. and Eng. Sci.*, 16 (1978) 277–285.

S. Nemat-Nasser and A. Shokooh, On finite plastic flows of compressible materials with internal friction, *Int. J. Solids Structures*, 16 (1980) 495–514.

S. Nemat-Nasser, L.M. Keer and K.S. Parihar, Unstable growth of thermally induced interacting cracks in brittle solids, *Int. J. Solids Structures*, 14 (1978) 409–430.

S. Nemat-Nasser and C.O. Horgan, Variational methods for eigenvalue problems with discontinuous coefficients, *Mechanics Today*, ed. S. Nemat-Nasser, 5 (1980) 365–376.

P. Neményi, Selbstspannungen elastischer Gebilde, Z. angew. Math. Mech., 11 (1931) 59–70.

H. Neuber, Ein neuer Ansatz zur Lösung raumlicher Probleme der Elastizitätstheorie, *Z. angew. Math. Mech.*, 14 (1934) 203–212.

H. Neuber, *Kerbspannungslehre*, Springer, (1937).

W. Niesel, *Über elastische ellipsoidische Einschlüsse in homogenen Medien*, Inaugural Dissertation, Technische Hochschule, Karlsruhe, (1953).

T. Ninomiya and S. Ishioka, Dislocation vibration: Effective mass and line tension, *J. Phys. Soc. Japan*, 23 (1967) 361–372.

H. Nisitani and Y. Murakami, Interaction of elasto-plastic cracks subjected to a uniform tensile stress in an infinite or semi-infinite plate, Proc. Int. Conf. *Mech. Behavior of Mater.*, 1, Soc. Mater. Sci., Japan (1972) 346–356.

H. Nisitani and Y. Murakami, Stress intensity factors of an elliptical crack or a semi-elliptical crack subject to tension, *Int. J. Fracture*, 10 (1974) 353–368.

W. Nowacki, *Thermoelasticity*, Addison-Wesley, (1962).

A.S. Nowick and B.S. Berry, *Anelastic Relaxation in Crystalline Solids*, Academic Press, (1975) 235.

R.J. Nuismer and J.D. Achenbach, Dynamic induced fracture, *J. Mech. Phys. Solids*, 20 (1972) 203–222.

J.F. Nye, Some geometrical relations in dislocated crystals, *Act. Metall.*, 1 (1953) 153–162.

R.J. O'Connell and B. Budiansky, Seismic velocities in dry and saturated cracked solids, *J. Geophys. Res.*, 79 (1974) 5412–5426.

M. Oda, Initial fabrics and their relations to mechanical properties of granular materials, *Soils and Foundations*, 12 (1972) 17–36.

M. Ohnami, K. Shiozawa and A. Kamitani, X-ray study on the influence of pressure soaking on the structural change of plastically deformed polycrystalline metal, in *Proc. 15th Japan Congr. Mater. Res.* (1972) 50–55.

S.M. Ohr and J. Narayan, Electron microscope observation of shear cracks in stainless single crystals, *Phil. Mag.*, 41 (1980) 81–89.

M. Okabe and T. Mori, Recovery of work hardening and low temperature creep in a dispersion strengthened Cu alloy with rod shaped particles, *Act. Metall.*, 27 (1979) 1373–1380.

M. Okabe, T. Mochizuki and T. Mori, Low temperature recovery creep of a Cu–SiO_2 alloy, *Phil. Mag.*, 41 (1980) 615–618.

M. Okabe, T. Mori and T. Mura, Internal friction caused by diffusion around a second phase particle. Al-Si alloy, *Phil. Mag.*, 44 (1981) 1–12.

E. Orowan, Causes and effects of internal stresses, in *Internal Stresses and Fatigue in Metals*, eds. G.M. Rassweiler and W.L. Grube, Elsevier Publ. (1959) 59–80.

E. Orowan, *Zur Kristallplastizität*, I, II, III, Z. Phys., 89 (1934) 605–659.

E. Orowan, Problems of plastic gliding, *Proc. Phys. Soc.*, 52 (1940) 8–22.

E. Orowan, Notch brittleness and the strength of metals, *Trans. Inst. Engnrs. and Shipbuilders in Scotland*, 89 (1945) 165–204.

E. Orowan, Discussion in *Symposium on Internal Stress in Metals and Alloys*, Institute of Metals, London, (1948).

E. Orowan, Fracture and Strength of solids, in *Reports on Progress in Physics*, 12, ed. Physical Society, London, (1949) 185–232.

E. Orowan, Fundamentals of brittle behavior in metals, in MIT Symposium on *Fatigue and Fracture of Metals*, Wiley, (1950).

Orsay Liquid Crystal Group, Existence and magnetic instability of double disclination lines in cholesteric liquid crystals, *Phys. Lett.*, 28A (1969) 687–688.

N. Oshima, Asymmetrical stress tensor and its application to a granular medium, *Proc. 3rd Japan Nat. Congr. Appl. Mech.* (1953) 77–83.

D.R.J. Owen and T. Mura, Periodic dislocation distributions in a half-space, *J. Appl. Phys.*, 38 (1967) 1999–2009.

D.R.J. Owen and T. Mura, Dislocation configurations in cylindrical coordinates, *J. Appl. Phys.* 38 (1967) 2818–2825.

D.R.J. Owen, The application of dislocation theory to the determination of stress-strain relationships of composite materials, *Int. J. Non-Linear Mech.*, 6 (1971) 167–175.

D.R.J. Owen, Solutions to arbitrary oriented periodic dislocation and eigenstrain distributions in a half-space, *Int. J. Solids Structures*, 7 (1971) 1343–1361.

K. Palaniswamy and W.G. Knauss, On the problem of crack extension in brittle solids under general loading, in *Mechanics Today*, ed. S. Nemat-Nasser, Pergamon Press (1978) 87–148.

Y.C. Pan and T.W. Chou, Point force solution for an infinite transversely isotropic solid, *J. Appl. Mech.*, 43 (1976) 608–612.

Y.C. Pan and T.W. Chou, Green's function solutions for semi-infinite transversely isotropic materials, *Int. J. Engng. Sci.*, 17 (1979) 545–551.

Y.C. Pan and T.W. Chou, Green's functions for two-phase transversely isotropic materials, *J. Appl. Mech.*, 46 (1979) 551–556.

V. V. Panasyuk, *Limiting Equilibrium of Brittle Solids with Fractures*, Naukova Dumka, (1968).

Y.H. Pao and C.C. Mow, *Diffraction of Elastic Waves and Dynamics Stress Concentrations*, Crane Russak, (1971).

P.F. Papkovich, Solution générale des équations différentielles fondamentales d'élasticité, exprimée par trois fonctions harmoniques, *C.R. Acad. Sci.*, 195 (1932) 513–515.

P.F. Papkovich, Expressions générales des composantes des tensions, ne renfermant comme fonctions arbitraires que des fonctions harmoniques, *C.R. Acad. Sci.*, 195 (1932) 754–756.

P.C. Paris, The fracture mechanics approach to fatigue, in *Fatigue – An Interdisciplinary Approach*, Syracuse Univ. Press, (1964).

P.C. Paris and G.C. Sih, Stress analysis of cracks, in *Fracture Toughness Testing and Its Applications*, ASTM Special Publ. No. 381 (1964) 30–81.

H. Parkus, *Thermoelasticity*, Blaisdell, (1968).

M.O. Peach and J.S. Koehler, The force exerted on dislocations and the stress field produced by them, *Phys. Rev.*, 80 (1950) 436–439.

B. Pegel, Strahlungslose Eigenschwingungen von Versetzungen, *phys. stat. sol.*, 14 (1966) K165–167.

R.E. Peierls, The size of a dislocation, *Proc. Phys. Soc.*, 52 (1949) 34–37.

R.E. Peterson, *Stress Concentration Design Factors*, John Wiley and Sons (1962).

T.H.H. Pian, P. Tong and C.H. Luk, Elastic crack analysis by a finite element hybrid method, *Proc. 3rd Conf. Matrix Meth. Struct. Mech.*, (1971) 690–711.

M. Polanyi, Über eine Art Gitterstörung, die einen Kristall plastisch machen könnte, *Z. Phys.*, 89 (1932) 660–664.

W. Prager and P.G. Hodge, *Theory of Perfectly Plastic Solids*, John Wiley & Sons, (1951).

L. Prandtl, Spannungsverteilung in plastischen Korperen, *Proc. 1st Int. Cong. Appl. Mech. Delft*, eds. C.B. Biezeno and J.M. Burgers, (1924) 43–54.

R. Raj and M.F. Ashby, On grain boundary sliding and diffusional creep, *Metall. Trans.*, 2 (1971) 1113–1127.

R. Raj and M.F. Ashby, Grain boundary sliding and the effect of particles on its rate, *Metall. Trans.*, 3 (1972) 1937–1942.

R. Raj and M.F. Ashby, Intergranular fracture at elevated temperature, *Act. Metall.*, 23, (1975) 653–666.

R. Raj, Transient behavior of diffusion induced creep and creep rupture, *Metall. Trans.*, 6 (1975) 1499–1509.

R. Raj, Crack Initiation in grain boundaries under conditions of steady-state and cyclic creep, ASME Trans., *J. Engng. Mater. Tech.*, 98 (1976) 132–139.

A.K. Rao, I.S. Raju and A.V. Krishna Murty, A powerful hybrid method in finite element analysis, *Int. J. Numer. Meth. Engng.*, 3 (1971) 389–403.

W.T. Read, *Dislocations in Crystals*, McGraw-Hill, (1953).

W.T. Read, Dislocation theory of plastic bending, *Act. Metall.*, 5 (1957) 83–88.

R.E. Reed-Hill, *Physical Metallurgy Principles*, Van Nostrand Reinhold, (1964).

H. Reissner, Eigenspannungen und Eigenspannungsquellen, *Z. angew. Math. Mech.*, 11 (1931) 1–8.

A. Reuss, Berechnung der Fliessgrenze von Mischkristallen auf Grund der Plastizitätsbedingung für Einkristalle, *Z. angew. Math. Mech.*, 9 (1929) 49–58.

A. Reuss, Berückstichtigung der elastischen Formänderung in der Plastizitatshtheorie, *Z. angew. Math. Mech.*, 10 (1930) 266–274.

J.R. Rice and G.C. Sih, Plane problems of cracks in dissimilar media, *J. Appl. Mech.*, 32 (1965) 418–423.

J.R. Rice, Contained plastic deformation near cracks and notches under longitudinal shear, *Int. J. Fracture Mech.*, 2 (1966) 416–447.

J.R. Rice, An examination of the fracture mechanics energy balance from the point of view of continuum mechanics, *Proc. 1st Int. Conf. on Fracture at Sendai*, 1, eds. T. Yokobori, T. Kawasaki and J.L. Swedlow, (1966) 309–340.

J.R. Rice, Plastic yielding at a crack tip, *Proc. 1st Int. Conf. on Fracture at Sendai*, 1, eds. T. Yokobori, T. Kawasaki and J. L. Swedlow, (1966) 283–308.

J.R. Rice, Stress due to a sharp notch in a work-hardening elastic-plastic material loaded by longitudinal shear, *J. Appl. Mech.*, 34 (1967) 287–298.

J.R. Rice, Mathematical analysis in the mechanics of fracture, *Fracture*, 2, ed. H. Liebowitz (1968) 191–311.

J.R. Rice, Mechanics of crack tip deformation and extension by fatigue, in *Fatigue Crack Propagation*, ASTM Special Tech. Publ. No. 415, (1968) 247–309.

J.R. Rice, A path independent integral and the approximate analysis of strain concentration by notches and cracks, *J. Appl. Mech.*, 35 (1968) 379–386.

J.R. Rice and G.F. Rosengren, Plane strain deformation near a crack tip in a power-law hardening material, *J. Mech.. Phys. Solids*, 16 (1968) 1–12.

J.R. Rice and R. Thomson, Ductile versus brittle behavior of crystals, *Phil. Mag.*, 29 (1974) 73–97.

J.R. Rice and E.P. Sorensen, Continuity crack-tip deformation and fracture for plane-strain crack growth in elastic-plastic solids, *J. Mech. Phys. Solids*, 26 (1976) 163–186.

J.R. Rice and J.W. Rudnicki, Earthquake precursory effects due to pore fluid stabilization of a weakening fault zone, *J. Geophys. Res.*, 84 (1979) 2177–2193.

J.R. Rice, J.W. Rudnicki and D.A. Simons, Deformation of spherical cavities and inclusions in fluid-infiltrated elastic materials, *Int. J. Solids Structures*, 14 (1978) 289–303.

H. Riedel, Plastic yielding on inclined slip-planes at a crack tip, *J. Mech. Phys. Solids*, 24 (1976) 277–289.

C. Robinson, J.C. Ward and R.B. Beevers, Liquid crystalline structure in polypeptide solutions, Part 2, *Disc. Faraday Soc.*, 25 (1958) 29–42.

K. Robinson, Elastic energy of an ellipsoidal inclusion in an infinite solid, *J. Appl. Phys.*, 22 (1951) 1045–1054.

L. Rongved, Force interior to one of two joined semi-infinite solids, *Proc. 2nd Midwestern Conf. Solid Mech.*, (1955) 1–13.

E.J. Routh, Theorems on the attraction of ellipsoids for certain laws of force other than the inverse space, *Phil. Trans. Roy. Soc.*, 186-II (1895) 897–950.

J.W. Rudnicki, The inception of faulting in a rock mass with a weakened zone, *J. Geophys. Res.*, 82 (1977) 844–854.

V.L. Rvachev, Method of R-functions in boundary-value problems, *Sov. Appl. Mech.*, 11 (1975) 345–354.

G. Saada, Elastic field of planar periodic dislocation networks in anisotropic materials, *Phil. Mag.* 34 (1976) 639–641.

G. Saada, The elastic field of periodic dislocation network, *Phil. Mag.* 33 (1976) 825–841.

R.A. Sack, Extension of Griffith's theory of rupture to three dimensions, *Proc. Roy. Soc.*, 58 (1946) 729–736.

M.A. Sadowsky and E. Sternberg, Stress concentration around an ellipsoidal cavity in an infinite body under arbitrary plane stress perpendicular to the axis of revolution of cavity, *ASME Trans.*, 69 (1947) A191–201.

M.A. Sadowsky and E. Sternberg, Stress concentration around a triaxial ellipsoidal cavity, *ASME Trans.*, 71 (1949) A149–157.

K.Saito, R.O. Bozkurt and T. Mura, Dislocation stresses in a thin film due to the periodic distributions of dislocations, *J. Appl. Phys.*, 43 (1972) 182–188.

N.J. Salamon and J. Dundurs, Elastic fields of a dislocation loop in a two-phase material, *J. Elasticity*, 1 (1971) 153–164.

K. Salma and G.A. Alers, Elastic moduli variations at the martensitic transformation of Fe-Ni alloys, *J. Appl. Phys.*, 39 (1968) 4857–4859.

R. Sankaran and C. Laird, Interfacial structure of platelike precipitates, *Phil. Mag.*, 29 (1974) 179–215.

R. Sankaran and C. Laird, Deformation field of a misfitting inclusion, *J. Mech. Phys. Solids*, 24 (1976) 251–262.

S.L. Sass, T. Mura and J.B. Cohen, Diffraction contrast from non-spherical distortions – in particular a cuboidal inclusion, *Phil. Mag.*, 16 (1967) 679–690.

M. Satake, Constitution of mechanics of granular materials, *Proc. 26th Japan Nat. Cong. Appl. Mech.*, (1978) 257–266.

A. Sato, T. Mori and Y. Tanaka, Stress distribution around an inclusion in an Fe-N single crystal, *Act. Metall.*, 27 (1979) 131–136.

G.N. Savin, *Stress Concentration Around Holes*, Pergamon Press (1961).

R.O. Scattergood and D.J. Bacon, The Orowan mechanism in an isotropic crystal, *Phil. Mag.*, 31 (1975) 179–198.

H. Schaefer, The basic affine connection in a Cosserat continuum, in *Mechanics of Generalized Continua*, ed. E. Kröner, Springer-Verlag, (1968).

H. Schaefer, Eine Feldtheorie der Versetzungen in Cosserat-Kontinuum, *Z. angew. Math. Phys.*, 20 (1969) 891–899.

A. Seeger, Theorie der Gitterfehlstellen, *Handbuch der Physik*, VII, Part 1: *Crystal Physics* 1, Springer-Verlag, (1955).

H. Sekine and T. Mura, A dislocation dipole in an anisotropic medium, *Phil. Mag.*, 40 (1979) 183–191.

H. Sekine and T. Mura, The elastic field around an elliptical crack in an anisotropic medium under an applied stress of polynomial forms, *Int. J. Engng. Sci.*, 17 (1979) 641–649.

A.P.S. Selvadurai, The load-deflexion characteristics of a deep rigid anchor in an elastic medium, *Geotechnique*, 26 (1976) 603–612.

A.P.S. Selvadurai, The response of a deep rigid anchor due to undrained

elastic deformation of the surrounding solid medium, *Int. J. Numer. Anal. Meth. Geomech.*, 2 (1978) 189–197.
A.P.S. Selvadurai, The time-dependent response of a deep rigid anchor in a viscoelastic medium, *Int. J. Rock Mech. Min. Sci. Geomech. Abstr.*, 15 (1978) 11–19.
G.P. Sendeckyj, Ellipsoidal inhomogeneity problem, Ph.D. Dissertation, Northwestern University, Evanston, (1967).
G.P. Sendeckyj, Longitudinal shear deformation of composites – effective other modulus, *J. Compos. Mater.*, 4 (1970) 500–512.
G.P. Sendeckyj, Screw dislocations near circular inclusions, *phys. stat. sol. (a)*, 3 (1970) 529–535.
G.P. Sendeckyj, Elastic inclusion problems in plane elastostatics, *Int. J. Solids Structures*, 6 (1970) 1535–1543.
G.P. Sendeckyj, Longitudinal shear deformation of composites II. Stress distribution, *J. Compos. Mater.*, 5 (1971) 82–93.
K. Seo and T. Mura, The elastic field in a half space due to ellipsoidal inclusions with uniform dilatational eigenstrains, *J. Appl. Mech.*, 46 (1979) 568–572.
R.C. Shah and A.S. Kobayashi, Stress intensity factors for an elliptical crack approaching the surface of a semi-infinite solid, *Int. J. Fracture*, 9 (1973) 133–146.
A.H. Shalaby and K.S. Havner, A general kinematical analysis of double slip, *J. Mech. Phys. Solids*, 26 (1978) 79–92.
G.S. Shapiro, Axisymmetric deformations around an ellipsoidal cavity, *Dokl. Akad. Nauk.*, 58 (1947) 1309–1312.
D.I. Sherman, On the problem of plane strain in non-homogeneous media, in *Non-Homogeneity in Elasticity and Plasticity*, ed. W. Olszak, Pergamon, (1959) 3–20.
D.K. Shetty, T. Mura and M. Meshii, Analysis of creep deformation under cyclic loading conditions, *Mater. Sci. Engng.*, 20 (1975) 261–266.
M. Shibata and K. Ono, The strain energy of a spheriodal inclusion and its application to b.c.c.-h.c.p. martensitic transformation, *Act. Metall.*, 23 (1975) 587–597.
M. Shibata and K. Ono, On the minimization of strain energy in martensitic transformation of titanium, *Act. Metall.*, 25 (1977) 35–42.
R.T. Shield, Notes on problems in hexagonal aerolotropic materials, *Proc. Camb. Phil. Soc.*, 47 (1951) 401–409.
R.T. Shield, Mixed boundary value problems in soil mechanics, *Q. Appl. Math.*, 11 (1953) 61–75.

R.T. Shield, Stress and velocity fields in soil mechanics, *J. Math. Phys.*, 33 (1954) 144–156.

T. Shioya and J. Shioiri, Elastic-plastic analysis of the yield process in mild steel, *J. Mech. Phys. Solids*, 24 (1976) 187–204.

K. Shiozawa and M. Ohnami, Study on flow stress of metal by means of the geometrical aspect of the continuously dislocated continuum, in *Mechanical Behavior of Materials*, Society of Materials Science, Japan, (1974) 93–104.

G.C. Sih and H. Liebowitz, Mathematical theory of brittle fracture, in *An Advanced Treatise on Fracture*, 2, ed. H. Liebowitz, Academic Press, (1968) 67–190.

G. Simmons and H. Wang, *Single Crystal Elastic Constants and Calculate Aggregate Properties*, M.I.T. Press, (1971).

J.A. Simmons, R. deWit and R. Bullough, eds., *Fundamental Aspects of Dislocation Theory*, I, II, NBS Special Publ. No. 317, (1970).

G. Sines and R. Kikucki, The elastic interaction between point defects and clusters of point defects in metals, *Act. Metall.*, 6 (1958) 500–508.

J.C. Slater and N.H. Frank, *Introduction to Theoretical Physics*, McGraw-Hill, 69 (1933).

C.S. Smith, Metallographic studies of metals after explosive shock, *Trans. Met. Soc. AIME*, 218 (1958) 574–589.

E. Smith, Screw dislocation arrays in heterogeneous materials, *Act. Metall.*, 15 (1967) 249–252.

E. Smith, The by-passing of an obstacle by a group of coplanar dislocations, *Phil. Mag.*, 22 (1970) 1161–1173.

E. Smith, Note on the growth of a penny-shaped crack in a general uniform applied stress, *Int. J. Fracture Mech.*, 7 (1971) 339–342.

J. Smith and C.J. Bechtoldt, Multiple twin structures in electro-deposited silver dendrites, *Electrochem. Sci.*, 115 (1968) 371–374.

I.N. Sneddon, The distribution of stress in the neighborhood of a crack in an elastic solid, *Proc. Roy. Soc.*, A187 (1946) 229–260.

I.N. Sneddon, *Crack Problems in the Mathematical Theory of Elasticity*, North Carolina State College, Raleigh, (1961).

I.N. Sneddon, *Mixed Boundary Value Problems in Potential Theory*, North-Holland, (1966).

I.N. Sneddon and M. Lowengrub, *Crack Problems in the Classical Theory of Elasticity*, Wiley, (1969).

K. Sobezyk, Elastic wave propagation in a discrete random medium, *Acta Mech.*, 25 (1976) 13–28.

I.S. Sokolnikoff, *Mathematical Theory of Elasticity*, McGraw-Hill, (1956).

I.S. Sokolnikoff, *Tensor Analysis*, Wiley, (1964).
V.V. Sokolovsky, Plane equilibrium of plastic wedge, *PMM*, 14 (1950) 391–404.
C. Somigliana, Sulla teoria delle distorsioni elastiche, *Rend. R. Acc. Lincei*, Ser. 5, 23 (1914) 463–472.
R.V. Southwell and H.J. Gough, On the concentration of stress in the neighborhood of a small spherical flaw; and on the propagation of fatigue fractures in "statically isotropic" materials, *Phil. Mag.*, 1 (1926) 71–97.
A.J.M. Spencer, A theory of the kinematics of ideal soils under plane strain conditions, *J. Mech. Phys. Solids*, 12 (1964) 337–351.
J.W. Steeds, *Introduction to Anisotropic Elasticity Theory of Dislocations*, Clarendon Press, (1973).
J.A. Steketee, Some geophysical applications of the elasticity theory of dislocations, *Can. J. Phys.*, 36 (1958) 1168–1198.
E. Sternberg, R.A. Eubanks and M.A. Sadowsky, On the axisymmetric problem of elasticity theory for a region bounded by two concentric spheres, *Proc. 1st U.S. Nat. Cong. Appl. Mech.* (1952) 209–215.
E. Sternberg and M.A. Sadowsky, On the axisymmetric problem of the theory of elasticity for an infinite region containing two spherical cavities, *J. Appl. Mech.*, 19 (1952) 19–27.
E. Sternberg, Three-dimensional stress concentrations in the theory of elasticity, *Appl. Mech. Rev.*, 11 (1958) 1–4.
G.C. Stokes, Dynamic theory of diffraction, *Trans. Camb. Phil. Soc.*, 9 (1849).
A.N. Stroh, The formation of cracks as a result of plastic flow, *Proc. Roy. Soc.*, A223, (1954) 404–414.
A.N. Stroh, The formation of cracks in plastic flow II., *Proc. Roy. Soc.*, A232, (1955) 548–560.
A.N. Stroh, Dislocations and cracks in anisotropic elasticity, *Phil. Mag.*, 3 (1958) 625–646.
A.N. Stroh, Steady state problem in anisotropic elasticity, *J. Math. Phys.*, 41 (1962) 77–103.
M. Suezawa, A note on the thickness of transformation twin, *Trans. Japan Inst. Metals*, 19 (1978) 459–465.
Y. Sumi, S. Nemat-Nasser, and L.M. Keer, A new combined analytical and finite-element solution method for stability analysis of the growth of interacting tension cracks in brittle solids, *Int. J. Engng. Sci.*, 18 (1980) 211–224.
C.T. Sun, J.D. Achenbach and G. Herrmann, Continuum theory of a laminated medium, *J. Appl. Mech.*, 35 (1968) 467–475.

H. Suzuki, *Introduction to Dislocation Theory, Agne Publ., Tokyo* (1967).

V.A. Sveklo, Concentrated force in a transversely-isotropic half-space and in a composite space, *J. Appl. Math. Mech.*, 33 (1969) 517–523.

J.L. Synge, *The Hypercircle in Mathematical Physics*, Cambridge Univ. Press, (1957).

H. Tada, *The Stress Analysis of Cracks Handbook*, Del. Research Co., Hellertown, PA, (1973).

S. Taira, K. Tanaka and M. Hoshina, Grain size effect on crack nucleation and growth in long-life fatigue of low-carbon steel, *ASTM Special Tech. Publ.*, No. 675 (1979) 135–173.

T. Takeuchi, Stress distribution around a crack tip in non-hardening materials under plane strain, *J. Phys. Soc. Japan*, 47 (1979) 1216–1223.

Y. Takeuti, *Analysis of Thermal Stresses*, Nitshin Publ., Tokyo, (1971).

O. Tamate, The effect of a circular inclusion on the stresses around a line crack in a sheet under tension, *Int. J. Fracture Mech.*, 4 (1968) 257–266.

O. Tamate, Behavior of a screw dislocation near a partially bonded bimetallic interface, *Int. J. Fracture Mech.*, 4 (1968) 357–364.

O. Tamate and T. Kurihara, Behavior of an edge dislocation near a partially bonded bimetallic interface, *Int. J. Fracture Mech.*, 6 (1970) 341–351.

O. Tamate and H. Sekine, Elastic interaction of a screw dislocation with an interface micro crack in bimetallic orthotropic media, *Tech. Rep. Tohoku Univ.*, 37 (1972) 69–85.

O. Tamate, Elastic interaction of a screw dislocation with a circular cylindrical inclusion in a semi-infinite body, *Tech. Rep. Tohoku Univ.*, 38 (1973) 259–268.

I. Tamura, H. Yoshimura, M. Ibaraki and M. Tagaya, Hardness and structure of ausformed Fe-Ni and Fe-Ni-C alloys, *Trans. Japan Inst. Metals*, 5 (1964) 47–52.

K. Tanaka and T. Mori, The hardening of crystals by non-deforming particles and fibres, *Act. Metall.*, 18 (1970) 931–941.

K. Tanaka and T. Mori, Note on volume integrals of the elastic field around an ellipsoidal inclusion, *J. Elasticity*, 2 (1972) 199–200.

K. Tanaka and T. Mura, *Fatigue crack initiation at inclusions, Metall. Trans.* (1981) in press.

K. Tanaka, K. Wakashima and T. Mori, Plastic deformation anisotropy and work-hardening of composite materials, *J. Mech. Phys. Solids*, 21 (1973) 207–214.

Y. Tanaka, A. Sato and T. Mori, Stress assisted nucleation of α'' precipates in Fe-N single crystals, *Act. Metall.*, 26 (1978) 529–540.
K. Tanaka and T. Mura, *A dislocation model for fatigue crack initiation*, *J. Appl. Mech.*, 48 (1981) 97–103.
M. Taya and E.D. Seidel, Void growth in a viscous metal, *Int. J. Engng. Sci.*, 19 (1981) 1083–1094.
G.I. Taylor, The mechanism of plastic deformation of crystals, *Proc. Roy. Soc.*, A145 (1934) 362–415.
G.I. Taylor, Analysis of plastic strain in a cubic crystal, in *S. Timoshenko 60th Anniversary Volume*, Macmillan, (1938) 218–224.
G.I. Taylor, Plastic strain in metals, *J. Inst. Metals*, 62 (1938) 307–324.
C. Teodosiu, A dynamic theory of dislocations and its applications to the theory of the elastic-plastic continuum, in *Fundamental Aspects of Dislocation Theory*, eds. J.A Simmons, R. deWit and R. Bullough, National Bureau of Standards Special Publ. No. 317, 2 (1970) 837–876.
L.J. Teutonico, Dynamic behavior of dislocations in anisotropic media, *Phys. Rev.*, 124 (1961) 1039–1045.
L.J. Teutonico, Moving edge dislocations in cubic and hexagonal materials, *Phys. Rev.*, 125 (1962) 1530–1533.
L.J. Teutonico, Uniformly moving dislocations of arbitrary orientation in anisotropic media, *Phys. Rev.*, 127 (1962) 413–418.
L.J. Teutonico, Dislocations in the primary slip systems of face-centered cubic crystals, *Act. Metall.*, 11 (1963) 391-397.
L.J. Teutonico, Dynamic behavior of dislocations in the primary slip systems of body-centered cubic crystals, *J. Appl. Phys.*, 34 (1963) 950–955.
P.S. Theocaris and E.E. Gdoutos, The modified Dugdale–Barenblatt model adapted to various fracture configurations in metals, *Int. J. Fracture*, 10 (1974) 549–563.
P.S. Theocaris and N.I. Ioakimidis, The inclusion problem in plane elasticity, *Q.J. Mech. Appl. Math.*, 30 (1977) 437–448.
A.W. Thompson and W.A. Backofen, The effect of grain size on fatigue, *Act. Metall.*, 19 (1971) 597–606.
S. Timoshenko and J.N. Goodier, *Theory of Elasticity*, McGraw-Hill, (1951).
H. Tokushige, *The dispersion hardening*, M.S. Thesis, Tokyo Institute of Technology, (1976).
B. Tomkins, Fatigue crack propagation – An analysis, *Phil. Mag.*, 13 (1968) 1041–1066.
M. Toya, Interfacial de-bonding caused by a screw dislocation pile-up at

a circular cylindrical rigid inclusion, *J. Mech. Phys. Solids*, 24 (1976) 1–18.

M. Toyoshima and S. Nagakura, Modulated structure of iron-carbon martensite, in AIP Conf. Proc. *Modulated Structures* – 1979, eds. J.M. Cowley, J.B. Cohen, M.B. Salamon and B.J. Wuensh, American Inst. Phys., (1979) 263–265.

M. Toyoshima, Ph.D. Dissertation, Tokyo Institute of Technology, (1980).

H. Träuble and U. Essmann, Fehler im Flussliniengitter von Supraleitern zweiter Art, *phys. stat. sol.*, 25 (1968) 373–392.

M.O. Tucker, Cracked dislocations at boundaries in two-material solids, *Phil. Mag.*, 28 (1973) 343–362.

V.K. Varadan and V.V. Varadan, Frequency dependence of elastic (SH-) wave velocity and attenuation in anisotropic two phase media, *Wave Motion*, 1 (1979) 53–63.

V. Varatharajulu and Y.H. Pao, Scattering matrix for elastic waves I. Theory, *J. Acoust. Soc. Amer.*, 60 (1976) 556–566.

C. Vilmann and T. Mura, Fracture related to a dislocation distribution, *J. Appl. Mech.*, 46 (1979) 817–820.

W.M. Visscher, in Proc. ARPA/AFML Review of Progress in Quantitative NDE, ed. D.O. Thompson, Air Force Mater. Lab. Tech. Rep. AFML-TR-205, 424 (1978).

V. Vitek, Yielding from a crack with finite root-radius loaded in uniform tension, *J. Mech. Phys. Solids*, 24 (1976) 263–275.

V. Vitek, Yielding on inclined planes at the tip of a crack loaded in uniform tension, *J. Mech. Phys. Solids*, 24 (1976) 263–275.

W. Voigt, Über die Beziehung zwischen den beiden Elastizitätskonstanten isotroper Körper, *Wied. Ann.*, 38 (1889) 573–587.

W. Voigt, *Lehrbuch der Kristallphysik*, B.G. Teubner, Leipzig und Berlin, (1910).

V. Volterra, Sur l'équilibre des carps élastiques multiplement connexes, *Annal. Sci. de l'École Norm. Super.*, Paris, 24 (1907) 401–517.

L.J. Walpole, On bounds for the overall elastic moduli of inhomogeneous systems – I, *J. Mech. Phys. Solids*, 14 (1966) 151–162.

L.J. Walpole, On bounds for the overall elastic moduli of inhomogeneous systems – II, *J. Mech. Phys. Solids*, 14 (1966) 289–301.

L.J. Walpole, The elastic field of an inclusion in an anisotropic medium, *Proc. Roy. Soc.*, A300 (1967) 270–289.

L.J. Walpole, On the overall elastic moduli of composite materials, *J. Mech. Phys. Solids*, 17 (1969) 235–251.

L.J. Walpole, Strengthening effects in elastic solids, *J. Mech. Phys. Solids*, 18 (1970) 343–358.

J.B. Walsh, Attenuation in partially melted materials, *J. Geophys. Res.*, 73 (1968) 2209–2216.

J.B. Walsh, New Analysis of attenuation in partially melted rock, *J. Geophys. Res.*, 74 (1969) 4333–4337.

P.C. Waterman, New formulation of acoustic scattering, *J. Acoust. Soc. Amer.*, 45 (1969) 1417–1429.

P.C. Waterman, Symmetry, unitarity, and geometry in electromagnetic scattering, *Phys. Rev.*, D3 (1971) 825–839.

P.C. Waterman, Matrix theory of elastic wave scattering, *J. Acoust. Soc. Amer.*, 60 (1976) 567–580.

G.N. Watson, *A Treatise on the Theory of Bessel Function*, Cambridge Univ. Press, (1952).

V.B. Watwood, The finite element method for prediction of crack behavior, *Nucl. Engng. Design*, 11 (1969) 323–332.

C.M. Wayman, *Introduction to the Crystallography of Martensitic Transformations*, Macmillan, (1964).

C.M. Wayman, The crystallography of martensitic transformations in alloys of iron, in *Advances in Materials Research*, 3, ed. H. Herman, Instersci. Publ., (1968) 147–304.

J. Weaver, Three-dimensional crack analysis, *Int. J. Solids Structures*, 13 (1977) 321–330.

M.S. Wechsler, D.S. Lieberman and T.A. Read, On the theory of the formation of martensite, *AIME Trans.*, 197 (1953) 1503–1515.

R. Weeks, J. Dundurs and M. Stipps, Exact analysis of an edge dislocation near a surface layer, *Int. J. Engng. Sci.*, 6 (1968) 365–372.

J. Weertman, Possiblity of the existence of attractive forces between dislocations of like sign, *Phys. Rev.*, 119 (1960) 1871–1872.

J. Weertman, High velocity dislocations, in *Response of Metals to High Velocity Deformation*, eds. P.G. Shewmon and V.F. Zackay, Intersci. Publ., (1961) 205–246.

J. Weertman, Fast moving edge dislocations in the (110) plane in anisotropic body-centered cubic crystals, *Phil. Mag.*, 7 (1962) 617–631.

J. Weertman, Fast moving edge dislocations on the (111) plane in anisotropic face-centered cubic crystals, *J. Appl. Phys.*, 33 (1962) 1631–1635.

J. Weertman, Dislocations moving uniformly on the interface between isotropic media of different elastic properties, *J. Mech. Phys. Solids*, 11 (1963) 197–204.

J. Weertman and J.R. Weertman, *Elementary Dislocation Theory*, Macmillan, (1964).

J. Weertman, Coalescence energy of two moving dislocations of like sign, *J. Appl. Phys.*, 37 (1966) 4925–4927.

J. Weertman, Rate of growth of fatigue cracks as calculated from the theory of infinitesimal dislocations distributed on a plane, *Proc. 1st Int. Conf. Fracture*, 1 (1966) 153–164.

J. Weertman, Uniformly moving transonic and supersonic dislocations, *J. Appl. Phys.*, 38 (1967) 5293–5301.

J. Weertman, Dislocations in uniform motion on slip or climb planes having periodic force laws, in *Mathematical Theory of Dislocations*, ed. T. Mura, ASME, (1969).

J. Weertman, Theory of fatigue crack growth based on a BCS crack theory with work hardening, *Int. J. Fracture*, 9 (1973) 125–131.

J. Weertman, A unified view for Griffith-Irwin-Orowan cracks, *Act. Metall.*, 26 (1978) 1731–1738.

J.H. Weiner, Thermoelastic dissipation due to high-speed dislocations, *J. Appl. Phys.*, 29 (1958) 1305–1307.

J.H. Weiner, Thermally activated motion of dislocations, in *Mathematical Theory of Dislocations*, ed. T. Mura, ASME (1969) 203–209.

G. Weingarten, Sulle superfici di discontinuitá nella teoria dell'elasticita dei carpi solidi, *Rend. R. Acc. Lincei*, Ser. 5, 10 (1901) 57–60.

G.J. Weng, Self-consistent determination of time-dependent behavior of metals, *J. Appl. Mech.*, 48 (1981) 41–46.

G.J. Weng and A. Phillips, On the kinematics of continuous distribution of dislocations in plasticity, *Int. J. Engng. Sci.*, 14 (1976) 65–73.

G.J. Weng and A. Phillips, An investigation of yield surfaces based on dislocation mechanics – I, *Int. J. Engng. Sci.*, 15 (1977) 45–59.

G.J. Weng and A. Phillips, An investigation of yield surfaces based on dislocation mechanics – II, *Int. J. Engng. Sci.*, 15 (1977) 61–70.

G.J. Weng and A. Phillips, The stress fields of continuous distribution of dislocations and of their movement in a polycrystalline aggregate, *Int. J. Solids Structures*, 14 (1978) 535–544.

R.W. Werne and J.M. Kelly, A dislocation theory of isotropic polycrystalline plasticity, *Int. J. Engng. Sci.*, 16 (1978) 951–965.

H.M. Westergaard, Bearing pressures and cracks, *J. Appl. Mech.*, 6 (1939) 49–53.

P. Wheeler and T. Mura, Dynamic equivalence of composite material and eigenstrain problems, *J. Appl. Mech.*, 40 (1973) 498–502.

E.T. Whittaker and G.N. Watson, *Modern Analysis*, Cambridge, (1962).

L.A. Wigglesworth, Stress distribution in a notched plate, *Mathematika*, 4 (1957) 76–96.
C.E. Williams, Helical disclination lines in semectics A, *Phil. Mag.*, 32 (1975) 313–321.
J.C. Williams, Critical review, kinetic and phase transformations, in *Titanium, Science and Technology*, 3, eds. R.I. Jaffee and H.M. Burte, Plenum, (1973) 1433–1494.
M.L. Williams, On the stress distribution at the base of a stationary crack, *J. Appl. Mech.*, 24 (1957) 104–114.
M.L. Williams, The stresses around a fault or crack in dissimilar media, *Bull. Seism. Soc. Amer.*, 49 (1959) 199–204.
J.R. Willis, Anisotropic elastic inclusion problems, *Q. J. Mech. Appl. Math.*, 17 (1964) 157–174.
J.R. Willis, Dislocations and inclusions, *J. Mech. Phys. Solids*, 13 (1965) 377–395.
J. R. Willis, The elastic interaction energy of dislocation loops in anisotropic media, *Q. J. Mech. Appl. Math.*, 18 (1965) 419–433.
J.R. Willis, Second-order effects of dislocations in anisotropic crystals, *Int. J. Engng. Sci.*, 5 (1967) 171–190.
J.R. Willis, The stress field around an elliptical crack in an anisotropic medium, *Int. J. Engng. Sci.*, 6 (1968) 253–263.
J.R. Willis and R. Bullough, The interaction of finite gas bubbles in a solid, *J. Nucl. Mater.*, 32 (1969) 76–87.
J.R. Willis, Stress fields produced by dislocations in anisotropic media, *Phil. Mag.*, 21 (1970) 931–949.
J.R. Willis, The penny-shaped crack on an interface, *Q.J. Mech. Appl. Math.*, 25 (1972) 367–385.
J.R. Willis, M.R. Hayns and R. Bullough, The dislocation void interaction, *Proc. Roy. Soc.*, A329 (1972) 121–126.
J.R. Willis, The interaction of gas bubbles in an anisotropic elastic solid, *J. Mech. Phys. Solids*, 23 (1975) 129–138.
J.R. Willis and J.R. Acton, The overall elastic moduli of a dilute suspension of spheres, *Q. J. Mech. Appl. Math.*, 29 (1976) 163–177.
J.R. Willis, Bounds and self-consistent estimates for the overall properties of anisotropic composites, *J. Mech. Phys. Solids*, 25 (1977) 185–202.
M.P. Wnuk, Accelerating crack in a viscoelastic solid subject to subcritical stress intensity, Proc. Int. Conf. on *Dynamic Crack Propagation*, ed. G. Sih, Lehigh Univ., (1972) 273–280.
M.P. Wnuk, Quasi-static extension of a tensile crack contained in a visco-elastic-plastic solid, *J. Appl. Mech.*, 41 (1974) 243–248.

M.P. Wnuk, Occurrence of catastrophic fracture in fully yielded components; stability analysis, *Int. J. Fracture*, 15 (1979) 553–581.

T.T. Wu, The effect of inclusion shape on the elastic moduli of a two-phase material, *Int. J. Solids Structures*, 2, (1966) 1–8.

N. Yamamoto, K. Yagi and G. Honjo, Electron microscopic studies of ferroelectric and ferroelastic $Gd_2(MoO_4)_3$, *phys. stat. sol.* (*a*), 42 (1977) 257–265.

Y. Yamamoto, Finite element approach with the aid of analytical solutions, in *Recent Advances in Matrix Method of Structural Analysis and Design*, eds. R.H. Gallagher, Y. Yamada and J.T. Oden, Univ. of Alabama Press (1971) 85–103.

Y. Yamamoto and Y. Sumi, Stress intensity factors for three-dimensional cracks, Int. J. Fracture, 14 (1978) 17–38.

Y. Yamamoto and Y. Sumi, Finite element analysis for stress intensity factors, in *Computing Methods in Applied Science and Engineering*, 1977, *Lecture Notes in Mathematics*, 704, eds. R. Glowinski and J.L. Lions, Springer, (1979) 154–168.

H.C. Yang and Y.T. Chou, Generalized plane problems of elastic inclusions in anisotropic solids, *J. Appl. Mech.*, 98 (1976) 424–430.

H.C. Yang and Y.T. Chou, Antiplane strain problems of an elliptic inclusion in an anisotropic medium, *J. Appl. Mech.*, 44 (1977) 437–441.

C.F. Ying and R.Truell, Scattering of a plane longitudinal wave by a spherical obstacle in an isotropically elastic solid, *J. Appl. Phys.*, 27 (1956) 1086–1097.

E.H. Yoffé, The moving Griffith crack, *Phil. Mag.*, 42 (1951) 739–750.

E.H. Yoffé, The angular dislocation, *Phil. Mag.*, 5 (1960) 161–175.

E.H. Yoffé, A dislocation at a free surface, *Phil. Mag.*, 6 (1961) 1147–1155.

T. Yokobori and M. Ichikawa, The energy principle as applied to the elastic-plastic crack model for fracture criterion, in *Rep. Res. Inst. Strength and Fracture of Mater.*, 2, Tohoku Univ., (1966) 21–28.

T. Yokobori, Criteria for nearly brittle fracture, *Int. J. Fracture Mech.*, 4 (1968) 179–205.

T. Yokobori, A. Kamei and M. Ichikawa, A criterion for unstable elastic-plastic fracture based on energy balance considerations, *Rep. Res. Inst. Strength and Fracture of Mater.*, 4, Tohoku Univ., (1968) 1–10.

M.H. Yoo, Elastic interaction of small defects and defect clusters in hexagonal crystals, *phys. stat. sol.* (*b*) 61 (1974) 411–418.

M.H. Yoo and S.M. Ohr, Elastic interaction of a point defect with a prismatic dislocation loop in hexagonal crystals, *J. Appl. Phys.* 43 (1972) 4477–4482.

I.W. Yu and G.P. Sendeckyj, Multiple circular inclusion problems in plane elastostatics, *J. Appl. Mech.*, 41 (1974) 215–221.

J. Zarka, Generalization de la theorie du potential plastique multiple en viscoplasticite, *J. Mech. Phys. Solids*, 20 (1972) 179–195.

J. Zarka, J.J. Engel and G. Inglebert, On a simplified inelastic analysis of structures, *Nucl. Eng. Design*, 57 (1980) 333–368.

J. Zarka, Étude du comportement des monocristaux métalliques – Application á la traction du monocristal C.F.C., *J. Mécanique*, 12 (1973) 275–318.

C. Zener, Kinetics of the decomposition of austenite, *Trans. AIME*, 167 (1946) 550–590.

C. Zener, *Elasticity and Anelasticity of Metals*, Univ. of Chicago Press, (1948).

O.C. Zienkiewicz, *The Finite Element Method in Engineering Science*, McGraw-Hill, (1971).

H. Zorski, Theory of discrete defects, *Arch. Mech. Stos.*, 18 (1966) 301–372.

Author index

Author index

Author index

Subject index

GPSR Compliance

The European Union's (EU) General Product Safety Regulation (GPSR) is a set of rules that requires consumer products to be safe and our obligations to ensure this.

If you have any concerns about our products, you can contact us on ProductSafety@springernature.com

In case Publisher is established outside the EU, the EU authorized representative is:

Springer Nature Customer Service Center GmbH
Europaplatz 3
69115 Heidelberg, Germany

Batch number: 10607220

Printed by Printforce, the Netherlands